T0140354

Smart Innovation, Systems and Technologies

Volume 57

About this Series

The Smart Innovation, Systems and Technologies book series encompasses the topics of knowledge, intelligence, innovation and sustainability. The aim of the series is to make available a platform for the publication of books on all aspects of single and multi-disciplinary research on these themes in order to make the latest results available in a readily-accessible form. Volumes on interdisciplinary research combining two or more of these areas is particularly sought.

The series covers systems and paradigms that employ knowledge and intelligence in a broad sense. Its scope is systems having embedded knowledge and intelligence, which may be applied to the solution of world problems in industry, the environment and the community. It also focusses on the knowledge-transfer methodologies and innovation strategies employed to make this happen effectively. The combination of intelligent systems tools and a broad range of applications introduces a need for a synergy of disciplines from science, technology, business and the humanities. The series will include conference proceedings, edited collections, monographs, handbooks, reference books, and other relevant types of book in areas of science and technology where smart systems and technologies can offer innovative solutions.

High quality content is an essential feature for all book proposals accepted for the series. It is expected that editors of all accepted volumes will ensure that contributions are subjected to an appropriate level of reviewing process and adhere to KES quality principles.

More information about this series at http://www.springer.com/series/8767

Ireneusz Czarnowski · Alfonso Mateos Caballero
Robert J. Howlett · Lakhmi C. Jain
Editors

Intelligent Decision Technologies 2016

Proceedings of the 8th KES International Conference on Intelligent Decision Technologies (KES-IDT 2016) – Part II

Springer

Editors
Ireneusz Czarnowski
Gdynia Maritime University
Gdynia
Poland

Alfonso Mateos Caballero
Artificial Intelligence Department
Universidad Politécnica de Madrid
Madrid
Spain

Robert J. Howlett
KES International
Shoreham-by-sea
UK

Lakhmi C. Jain
University of Canberra
Canberra
Australia

and

Bournemouth University
Poole
UK

and

KES International
Shoreham-by-sea
UK

ISSN 2190-3018 ISSN 2190-3026 (electronic)
Smart Innovation, Systems and Technologies
ISBN 978-3-319-81926-6 ISBN 978-3-319-39627-9 (eBook)
DOI 10.1007/978-3-319-39627-9

Softcover reprint of the hardcover 1st edition 2016

Printed on acid-free paper

This Springer imprint is published by Springer Nature
The registered company is Springer International Publishing AG Switzerland

Preface

This volume contains the proceedings (Part II) of the 8th International KES Conference on Intelligent Decision Technologies (KES-IDT 2016) held in Puerto de la Cruz, Tenerife, Spain, in June 15–17, 2016.

The conference was organized by KES International. The KES-IDT Conference series is a subseries of the KES Conference series.

KES-IDT is a well-established international annual conference, an interdisciplinary conference in nature. It provides excellent opportunities for the presentation of interesting new research results and discussion about them, leading to knowledge transfer and generation of new ideas.

This edition, KES-IDT 2016, attracted a substantial number of researchers and practitioners from all over the world, who submitted their papers for the main track and 12 special sessions. All papers have been reviewed by 2–3 members of the International Program Committee and International Reviewer Board. Following a rigorous review process, only the highest quality submissions were accepted for inclusion in the conference. The 80 best papers have been selected for oral presentation and publication in the two volumes of the KES-IDT 2016 proceedings.

We are very satisfied with the quality of the program and would like to thank the authors for choosing KES-IDT as the forum for the presentation of their work. Also, we gratefully acknowledge the hard work of the KES-IDT international program committee members and of the additional reviewers for taking the time to review the submitted papers and selecting the best among them for the presentation at the conference and inclusion in its proceedings.

We hope and intend that KES-IDT 2016 significantly contributes to the fulfillment of the academic excellence and leads to even greater successes of KES-IDT events in the future.

June 2016

Ireneusz Czarnowski
Alfonso Mateos Caballero
Robert J. Howlett
Lakhmi C. Jain

KES-IDT 2016 Conference Organization

Honorary Chairs

Lakhmi C. Jain, University of Canberra, Australia and Bournemouth University, UK
Gloria Wren-Phillips, Loyola University, USA
Junzo Watada, Waseda University, Japan

General Chair

Ireneusz Czarnowski, Gdynia Maritime University, Poland

Executive Chair

Robert J. Howlett, KES International and Bournemouth University, UK

Program Chair

Alfonso Mateos Caballero, Universidad Politécnica de Madrid, Spain

Publicity Chair

Izabela Wierzbowska, Gdynia Maritime University, Poland

Special Sessions

Specialized Decision Techniques for Data Mining, Transportation and Project Management
Piotr Jędrzejowicz, Gdynia Maritime University, Poland
Ireneusz Czarnowski, Gdynia Maritime University, Poland

Intelligent Decision Technologies for Water Resources Management
Wojciech Froelich, University of Silesia, Sosnowiec, Poland
Ewa Magiera, University of Silesia, Sosnowiec, Poland

Intelligent Methods for Eye Movement Data Processing and Analysis
Katarzyna Harezlak, Silesian University of Technology, Poland
Paweł Kasprowski, Silesian University of Technology, Poland

Intelligent Data Analysis and Applications
Urszula Stańczyk, Silesian University of Technology, Gliwice, Poland
Beata Zielosko, University of Silesia, Katowice, Poland

Intelligent Decision Making for Uncertain Unstructured Big Data
Bharat Singh, Big Data Analyst, Hildesheim, Germany
Neel Mani, ADAPT Centre for Digital Content Technology, Dublin City University, Dublin, Ireland
Pandian Vasant, PETRONAS University of Technology, Malaysia
Junzo Watada, Waseda University, Japan

Decision Making Theory for Economics
Eizo Kinoshita, Meijo University, Japan

New Advances of Soft Computing in Industrial and Management Engineering
Shing Chiang Tan, Multimedia University, Malaysia
Chee Peng Lim, Deakin University, Australia
Junzo Watada, Waseda University, Japan

Interdisciplinary Approaches in Business Intelligence Research and Practice
Ivan Luković, University of Novi Sad, Serbia

Recent Advances in Fuzzy Systems
Jose L. Salmeron, University Pablo de Olavide, Spain
Wojciech Froelich, University of Silesia, Sosnowiec, Poland
Elpiniki Papageorgiou, Technological Education Institute (TEI) of Central Greece, Lamia, Greece

Pattern Recognition for Decision Making Systems
Paolo Crippa, Università Politecnica delle Marche
Claudio Turchetti, Università Politecnica delle Marche

Pattern Recognition in Audio and Speech Processing
Claudio Turchetti, Università Politecnica delle Marche
Paolo Crippa, Università Politecnica delle Marche

Reasoning-Based Intelligent Systems
Kazumi Nakamatsu, University of Hyogo, Japan
Jair M. Abe, Paulista University, Brazil

International Program Committee

Mohamed Arezki Mellal, M'Hamed Bougara University, Boumerdes, Algeria
Dariusz Barbucha, Gdynia Maritime University, Poland
Alina Barbulescu, Ovidius University of Constanta, Romania
Mokhtar Beldjehem, University of Ottawa, Ontario
Monica Bianchini, Department of Information Engineering and Mathematics, Italy
Gloria Bordogna, CNR—National Research Council, Italy
Janos Botzheim, Tokyo Metropolitan University, Japan
Wei Cao, School of Economics, HeFei University of Technology, China
Michele Ceccarelli, Qatar Computing Research Institute, HBKU
Shing Chiang Tan, Multimedia University, Malaysia
Marco Cococcioni, University of Pisa, Italy
Paolo Crippa, Università Politecnica delle Marche, Ancona, Italy
Alfredo Cuzzocrea, University of Trieste, Italy
Ireneusz Czarnowski, Gdynia Maritime University, Poland
Ana de Almeida, ISCTE Instituto Universitário de Lisboa, Portugal
Dawn E. Holmes, University of California, USA
Margarita Favorskaya, Siberian State Aerospace University, Russian Federation, Russia
Antonio Fernández-Caballero, Universidad de Castilla-La Mancha, Spain
Wojciech Froelich, University of Silesia, Sosnowiec, Poland
Marcos G. Quiles, Federal University of São Paulo (UNIFESP), Brazil
Mauro Gaggero, National Research Council of Italy
Daniela Godoy, Unicen University, Argentina
Raffaele Gravina, DIMES, University of Calabria, Italy
Christos Grecos, Sohar University, Oman
Katarzyna Harezlak, Silesian University of Technology, Gliwice, Poland
Ioannis Hatzilygeroudis, University of Patras, Department of Computer Engineering and Informatics, Greece
Katsuhiro Honda, Osaka Prefecture University, Japan
Daocheng Hong, Fudan University, China
Tzung-Pei Hong, National University of Kaohsiung, Taiwan
Yuh-Jong Hu, National Chengchi University, Taipei, Taiwan
Yuji Iwahori, Chubu University, Japan

International Referee Board

Dragan Ivanović, Shohel Sayeed
Tomasz Jach, Ralf Seepold
Nikita Jain, Roman Simiński
Przemyslaw Juszczuk, Krzysztof Simiński
Tomasz Kajdanowicz, Aleksander Skakovski
Jarosław Koźlak, Rafał Skinderowicz
Robert Koprowski, Fred Stefan
Aleksandar Kovačević, Choo Jun Tan
Jan Kozak, Kay Meng Tay
Michał Kozielski, Magdalena Tkacz
Agnieszka Lijewska, Katarzyna Trynda
Way Soong Lim, Marco Vannucci
Chee Lim, Robert Wójcicki
Pei-Chun Lin, Berlin Wu
Agnieszka Lisowska, Tomasz Xięski
Vannucci Marco, Zhenyuan Xu
Joanna Marnik, Raimondas Zemblys

Contents

Recent Advances in Fuzzy Systems

Intelligent Methods for Eye Movement Data Processing and Analysis

Intelligent Decision Technologies for Water Resources Management

Intelligent Data Analysis and Applications

Comparison of Four Methods of Combining Classifiers on the Basis of Dispersed Medical Data

Małgorzata Przybyła-Kasperek

Abstract The main aim of the article is to compare the results obtained using four different methods of combining classifiers in a dispersed decision-making system. In the article the following fusion methods are used: the majority vote, the weighted majority vote, the Borda count method and the highest rank method. Two of these methods are used if the individual classifier generates a class label and two are used in the case when the individual classifier produces ranking of classes instead of unique class choice. All of these methods were tested in a situation when we have access to data from medical field and this data are in a dispersed form. The use of dispersed medical data is very important because it is common situation that medical data from one domain are collected in many different medical centers. It would be good to be able to use all this accumulated knowledge at the same time.

Keywords Decision-making system · Global decision · Fusion method · Majority vote · Weighted majority vote · Borda count · Highest rank

1 Introduction

The use of an ensemble of classifiers is a very popular method of decision-making. In this paper, this approach is used in the context of the application of dispersed knowledge. A dispersed knowledge is a knowledge that is stored in separate, independently gathered knowledge bases. The use of dispersed knowledge is extremely important in medical field, because many medical units separately collect knowledge in the same field. In a combining of base classifiers very important issue is method of fusion of outputs generated by base classifiers. In this study, four different fusion method are used in a dispersed decision-making system. The concept of a dispersed decision-making system is being considered by the author for several years.

M. Przybyła-Kasperek (✉)
Institute of Computer Science, University of Silesia, Będzińska 39,
41-200 Sosnowiec, Poland
e-mail: malgorzata.przybyla-kasperek@us.edu.pl

I. Czarnowski et al. (eds.), *Intelligent Decision Technologies 2016*,
Smart Innovation, Systems and Technologies 57,
DOI 10.1007/978-3-319-39627-9_1

In the first stage of studies the considerations were directed to a system with a static structure [14, 20]. In recent papers a system with a dynamic structure has been proposed [15, 16]. In this paper an approach proposed in the article [16] is used. The aim of the paper is to investigate the use of four selected fusion methods in the system with dynamically generated clusters.

The issue of combining classifiers is a very important aspect in the literature [6, 9–11]. This concept has various terms: combination of multiple classifiers [17], classifier fusion [3] or classifier ensembles [5, 18]. But always, the aim of the issue is to improve the quality of classification by combining the results of prediction of base classifiers. There are two basic approaches to this topic: classifier selection and classifier fusion. In the classifier selection approach each classifier is an expert in some local area. If the object from the classifier area appears, the classifier is responsible for assign the class label to the object [1, 8]. In the classifier fusion approach it is assume that all classifiers are trained over the whole feature space, and they are competitive rather than complementary [13]. In this paper, the classifier fusion approach is considered.

2 A Brief Overview of Decision-Making System Using Dispersed Knowledge

A dispersed decision-making system, which is used in the paper, was proposed by the author in the article [16]. During the construction of this system's structure a negotiation stage is used. The main assumptions of the system are very briefly described below. We do not give a detailed definition because it is not the subject of this paper. A detailed discussion can be found in the paper [16].

We assume that the knowledge is available in a dispersed form, which means in a form of several decision tables. The set of local knowledge bases that contain data from one domain is pre-specified. The only condition which must be satisfied by the local knowledge bases is to have common decision attributes. We assume that each local decision table $D_{ag} = (U_{ag}, A_{ag}, d_{ag})$ is managed by one agent, which is called a resource agent ag. We want to designate homogeneous groups of resource agents. The agents who agree on the classification for a test object into the decision classes will be combined in the group. It is realized in two steps. At first initial coalitions are created. Then the negotiation stage is implemented. These two steps are based on the test object classification carried out by the resource agents. For more details, please refer to the paper [16]. After completion of the second stage of the process of clustering we get the final form of clusters. For each cluster that contains at least two resource agents, a superordinate agent is defined, which is called a synthesis agent, as_j, where j is the number of cluster. The synthesis agent, as_j, has access to knowledge that is the result of the process of inference carried out by the resource agents that belong to its subordinate group. As is a finite set of synthesis agents. A formal definition of a dispersed decision-making system is as follows.

Definition 1 By a dispersed decision-making system with dynamically generated clusters we mean $WSD_{Ag}^{dyn} = \langle Ag, \{D_{ag} : ag \in Ag\}, \{As_x : x \text{is a classified object}\}, \{\delta_x : x \text{is a classified object}\}\rangle$ where Ag is a finite set of resource agents; $\{D_{ag} : ag \in Ag\}$ is a set of decision tables of resource agents; As_x is a finite set of synthesis agents defined for clusters dynamically generated for the test object x, $\delta_x : As_x \rightarrow 2^{Ag}$ is a injective function that each synthesis agent assigns a cluster generated due to classification of the object x.

On the basis of the knowledge of agents from one cluster, local decisions are taken.

An important problem that occurs is to eliminate inconsistencies in the knowledge stored in different knowledge bases. In previous papers the approximated method of the aggregation of decision tables have been used to eliminate inconsistencies in the knowledge [14–16]. In this paper, we also use this method. In the method for every cluster, a kind of combined information is determined. This combined information is in the form of aggregated decision table. Object of this table are constructed by combining relevant object from decision tables of the resource agents that belong to one cluster. Based on the aggregated decision tables global decisions are taken using the fusion method.

3 Fusion Methods

In this article, we use four different fusion methods: the majority vote, the weighted majority vote, the Borda count method and the highest rank method. These methods are discussed in the papers [7, 10, 11]. The majority vote and the weighted majority vote method are used if the individual classifiers produce unique class choices. While the Borda count and the highest rank method are used if the individual classifiers generate rankings of classes. Therefore at first, on the basis of each aggregated decision table a ranking of classes or a decision class is generated (depending on the method, that is used).

In the first step, a c-dimensional vector of values $[\mu_{j,1}(x), \ldots, \mu_{j,c}(x)]$ is generated for each jth cluster, where c is the number of all of the decision classes. The value $\mu_{j,i}(x)$ determines the level of certainty with which the decision v_i is taken by agents for a given test object x belonging to the cluster j. This vector will be defined on the basis of relevant objects. From each aggregated decision table and from each decision class, the smallest set containing at least m_2 objects for which the values of conditional attributes bear the greatest similarity to the test object is chosen. The value of the parameter m_2 is selected experimentally. The value $\mu_{j,i}(x)$ is equal to the average value of the similarity of the test object to the relevant objects form jth aggregated decision table, belonging to the decision class v_i. On the basis of the vector of values defined above, for each cluster a vector of the rank is specified. The vector of rank is defined as follows: rank 1 is assigned to the values of the decision attribute that are taken with the maximum level of certainty. Rank 2 is assigned to the next most certain decisions, etc. Proceeding in this way for each jth cluster, the vector of rank $[r_{j,1}(x), \ldots, r_{j,c}(x)]$ will be defined. In this way, for each cluster the ranking of classes is generated. The class choices of cluster are obtained by choosing the decision with the maximum support for the hypothesis that object x comes from this class. For each jth synthesis agent a set is selected $\{v_i : \mu_{j,i}(x) = \max_{k\in\{1,\ldots,c\}} \mu_{j,k}(x)\}$. The result of prediction is defined as a set because it can happen that more than one decision received maximum support. In this case, method of resolving ties is not used.

Majority vote
In the majority vote method, the test object is assigned to this class, which was indicated by the highest number of base models. The set of global decisions generated using the method of majority vote is defined as follows: $\hat{d}_{WSD^{dyn}_{Ag}}(x) = \arg\max_{i\in\{1,\ldots,c\}} \sum_{j=1}^{L} I(v_i \in \hat{d}_j(x))$, where $\hat{d}_{WSD^{dyn}_{Ag}}(x)$ denotes the set of global decisions taken by the dispersed decision-making system WSD^{dyn}_{Ag} for the test object x; $I(v_i \in \hat{d}_j(x)) = 1$, when $v_i \in \hat{d}_j(x)$ and $I(v_i \in \hat{d}_j(x)) = 0$, when $v_i \notin \hat{d}_j(x)$.

Weighted majority vote
In the weighted majority vote method greater weights are assigned to the classifiers that make more accurate decisions. The prediction result is determined according to the formula: $\hat{d}_{WSD^{dyn}_{Ag}}(x) = \arg\max_{i\in\{1,\ldots,c\}} \sum_{j=1}^{L} \omega_j I(v_i \in \hat{d}_j(x))$, where ω_j is a weight for jth synthesis agent. The problem of selecting appropriate values of weights was considered in the paper [2]. It was proven that if the base classifiers are independent then the optimal values of weights are proportional to $\omega_j \propto \log\left(\frac{1-e_{as_j}}{e_{as_j}}\right)$, where e_{as_j} is a individual error rate of the jth base classifier, in the considered case the jth synthesis agent. Due to the computational complexity the error rate of synthesis agent is estimated based on the error rates of resource agents that belong to its subordinate cluster. At first for each resource agent the error rate e_{ag_i} was determined based on the training set. This process is carried out only once. When the set of synthesis agents As_x is designated for a test object x the error rate of synthesis agent $as_j \in As_x$ is determined as follows $e_{as_j} = \frac{1}{card\{\delta_x(as_j)\}} \sum_{ag_i\in\delta_x(as_j)} e_{ag_i}$. These calculations are performed each time for a new, dynamically generated, system's structure.

The Borda count method
The Borda count method consists in the designation for each decision class the sum of the number of classes ranked below it by each cluster. Thus for each decision class v_i the value is determined $\sum_{j\text{th cluster}}(card\{V^d\} - r_{j,i}(x))$. The decision classes sorted according to their Borda count gives the final ranking. For the purposes of dispersed decision-making system it is assumed that the set of decisions taken by the system is the set of classes which have the maximum value of the Borda count. The Borda count method requires no training and all classifiers are treated equally.

The highest rank method

The highest rank method is simple, has low computational complexity and requires no training. The method has the following steps: for a given test object and for each cluster the ranking of classes is determined. Then, for each decision class the minimum (highest) of these ranks is assigned. Thus for each decision class v_i the value $\min_{j\text{th cluster}} r_{j,i}(x)$ is calculated. The classes are then sorted by these scores to derive a combined ranking for that test object. The set of decisions taken by the dispersed system is the set of classes which have the highest rank. A big disadvantage of this method is that the combined ranking may have many ties. The number of classes sharing the same ranks depends on the number of classifiers used. During the experiments with the dispersed decision-making system it was noted that there are some weak clusters, which assigns the highest rank to several decision classes at the same time. This significantly increases the number of ties, and made uninteresting results. Therefore, for the highest rank method only the individual classifiers were taken into consideration which made unambiguous decisions. That is the highest rank was assigned to one decision class.

4 Experiments

The aim of the experiments is to examine the quality of the classification made on the basis of dispersed medical data by the decision-making system using four different fusion methods. The majority vote method, the weighted majority vote method, the Borda count method and the highest rank method were considered.

4.1 Datasets

For the experiments the following data, which are in the UCI repository (archive.ics.uci.edu/ml/), were used: Lymphography data set, Primary Tumor data set. Both sets of data was obtained from the University Medical Centre, Institute of Oncology, Ljubljana, Yugoslavia (M. Zwitter and M. Soklic provided this data). Lymphography is a medical imaging technique in which a radiocontrast agent is injected, and then an X-ray picture is taken to visualize structures of the lymphatic system. In the Primary Tumor data set, on the basis of values of attributes such as histologic-type, supraclavicular etc. a decision is taken where (of 22 organs) the cancer cells are located. In order to determine the efficiency of inference each data set was divided into two disjoint subsets: a training set and a test set. A numerical summary of the data sets is as follows: Lymphography: # The training set—104; # The test set—44; # Conditional—18; # Decision—4; Class Distribution: #Class1—2, #Class2—81, #Class3—61, #Class4—4; Primary Tumor: # The training set—237; # The test set—102; # Conditional—17; # Decision—22; Class Distribution (given in order): 84, 20, 9, 14, 39, 1, 14, 6, 0, 2, 28, 16, 7, 24, 2, 1, 10, 29, 6, 2, 1, 24.

We want to investigate the effectiveness of system, that uses dispersed knowledge. We assume that in a few medical centers different knowledge bases from one domain are collected independently. We want to use all of these bases simultaneously in the process of inference. This approach not only allows the use of all available knowledge, but also should improve the efficiency of inference. In order to consider the discussed situation it is necessary to provide the knowledge stored in the form of a set of decision tables. Therefore, the training set was divided into a set of decision tables. Divisions with a different number of decision tables were considered. For each of the data sets used, the decision-making system with five different versions (with 3, 5, 7, 9 and 11 resource agents) were considered. For these systems, we use the following designations: WSD^{dyn}_{Ag1}—3 resource agents; WSD^{dyn}_{Ag2}—5 resource agents; WSD^{dyn}_{Ag3}—7 resource agents; WSD^{dyn}_{Ag4}—9 resource agents; WSD^{dyn}_{Ag5}—11 resource agents.

All used conflict analysis methods have one disadvantage. The final decision may have ties. In order to analyze these properties the appropriate classification measures were applied, which are adapted to this situation. The measures of determining the quality of the classification are: *estimator of classification error* e in which an object is considered to be properly classified if the decision class used for the object belonged to the set of global decisions generated by the system; *estimator of classification ambiguity error* e_{ONE} in which object is considered to be properly classified if only one, correct value of the decision was generated to this object; *the average size of the global decisions sets* $\overline{d}_{WSD^{dyn}_{Ag}}$ generated for a test set. Note that the definition of estimator of classification error that is used in the paper is different from that known from the literature. The definition was modified to be suitable for the case when the classifier produces a set of decisions instead of one decision. This measure, combined with the estimator of classification ambiguity error and the average size of the global decisions sets, gives full information needed to analyze the obtained results. Other measures known from the literature e.g. precision or recall would not be appropriate in this situation, since for each test object only one decision is correct, and the mentioned measures are used in situations where the set of relevant decisions occur not just one relevant decision.

In the description of the results of experiments for clarity some designations for algorithms and parameters have been adopted: m_1—parameter which determines the number of relevant objects that are selected from each decision class of the decision table and are then used in the process of cluster generation; p—parameter which occurs in the definition of friendship, conflict and neutrality relations; $A(m)$—the approximated method of the aggregation of decision tables; $C(m_2)$—the method of conflict analysis (the majority vote method, the weighted majority vote method, the Borda count method and the highest rank method), with parameter which determines the number of relevant objects that are used to generate decision of one cluster.

4.2 Results

At the beginning of experiments the process of parameters optimization was carried out. A series of tests for different parameter values were performed: $m_1 \in \{1, 4, 7, 10, 13\}$, $m_2, m_3 \in \{1, \ldots, 10\}$ and $p \in \{0.05, 0.1, 0.15, 0.2\}$. From all of the obtained results, one was selected that guaranteed a minimum value of estimator of classification error (e), while maintaining the smallest possible value of the average size of the global decisions sets ($\overline{d}_{WSD^{dyn}_{Ag}}$). In tables presented below the best results, obtained for optimal values of the parameters, are given. In the tables the following information is given: the name of dispersed decision-making system (System); the selected, optimal parameter values (Parameters); the algorithm's symbol (Algorithm); the three measures discussed earlier e, e_{ONE}, $\overline{d}_{WSD^{dyn}_{Ag}}$; the time t needed to analyse a test set expressed in minutes. The tables show the results for four different fusion methods, the best results in terms of the measures e and $\overline{d}_{WSD^{dyn}_{Ag}}$ are bolded.

The results of the experiments with the Lymphography data set are presented in Table 1. The Lymphography data set has 4 decision classes and because of that the results with the average number of global decisions sets substantially greater than 1.5 are uninteresting. As can be seen, for the Lymphography data set, in the case of the weighted majority vote and the Borda count method the average size of the global decisions sets is very close to 1. For the majority vote and the highest rank method we receive larger sets of decisions, which means that these methods are more ambiguous. The average frequency that decision occurs in the generated data set was also checked. These values are as follows #Class1—0.042, #Class2—0.639, #Class3—0.559, #Class4—0.003. Based on the results it can be concluded that the best methods are the majority vote and the highest rank method. This probably stems from the fact that these methods produce larger sets of global decisions. Among the methods that generate unambiguous decisions the best method is the Borda count method.

The results of the experiments with the Primary Tumor data set are presented in Table 2. The Primary Tumor data set has 22 decision classes and because of that even results with the average number of global decisions sets less than 4 are interesting. As can be seen, for the Primary Tumor data set all considered methods generate ambiguous decisions, which means that the average size of the global decisions sets is significantly greater than 1. The largest sets of global decisions are generated by the highest rank method. It was noted that the number of ties depends on the number of classifiers used—the more resource agents are present, the more ties occur. The average frequency that decision occurs in the generated data set was also checked. These values are as follows (given in order) 0.907, 0.002, 0.096, 0.216, 0.570, 0, 0.293, 0.044, 0, 0, 0.309, 0.219, 0.082, 0.214, 0.029, 0, 0.066, 0.260, 0, 0, 0.002, 0.218. Based on the results it can be concluded that the best methods are the majority vote and the weighted majority vote. The worst is the Borda count method. This probably stems from the fact that this method produce the smallest sets of global decisions.

Table 1 Summary of experiments results with the Lymphography data set

System	Parameters	Algorithm	e	e_{ONE}	$\overline{d}_{WSD_{Ag}}$	t
Majority vote						
WSD_{Ag1}	$m_1 = 10, p = 0.05$	$A(8)C(7)$	0.045	0.409	1.386	0.04
WSD_{Ag2}	$m_1 = 1, p = 0.05$	$A(2)C(8)$	**0.091**	0.273	**1.182**	0.10
WSD_{Ag3}	$m_1 = 1, p = 0.05$	$A(2)C(3)$	**0.091**	0.318	**1.273**	0.08
WSD_{Ag4}	$m_1 = 1, p = 0.05$	$A(8)C(3)$	**0.091**	0.523	**1.523**	0.21
WSD_{Ag5}	$m_1 = 7, p = 0.05$	$A(1)C(1)$	0.182	0.545	1.386	2.36
Weighted majority vote						
WSD_{Ag1}	$m_1 = 12, p = 0.05$	$A(2)C(5)$	0.136	0.136	1	0.04
WSD_{Ag2}	$m_1 = 1, p = 0.05$	$A(4)C(4)$	0.136	0.136	1	0.07
WSD_{Ag3}	$m_1 = 11, p = 0.05$	$A(1)C(4)$	0.159	0.295	1.136	0.10
WSD_{Ag4}	$m_1 = 1, p = 0.05$	$A(1)C(1)$	0.182	0.341	1.159	0.25
WSD_{Ag5}	$m_1 = 3, p = 0.05$	$A(1)C(1)$	0.227	0.500	1.273	2.41
Borda count						
WSD_{Ag1}	$m_1 = 10, p = 0.05$	$A(5)C(7)$	0.068	0.341	1.273	0.06
WSD_{Ag2}	$m_1 = 1, p = 0.05$	$A(2)C(8)$	0.091	0.295	1.205	0.07
WSD_{Ag3}	$m_1 = 1, p = 0.05$	$A(4)C(4)$	0.091	0.409	1.318	1.21
WSD_{Ag4}	$m_1 = 4, p = 0.05$	$A(1)C(2)$	0.136	0.341	1.205	0.18
WSD_{Ag5}	$m_1 = 13, p = 0.05$	$A(9)C(9)$	0.182	0.295	1.114	3.02
Highest rank method						
WSD_{Ag1}	$m_1 = 13, p = 0.05$	$A(8)C(7)$	**0.045**	0.386	**1.364**	0.06
WSD_{Ag2}	$m_1 = 1, p = 0.05$	$A(5)C(5)$	0.091	0.455	1.364	0.07
WSD_{Ag3}	$m_1 = 1, p = 0.05$	$A(2)C(3)$	0.091	0.523	1.523	1.21
WSD_{Ag4}	$m_1 = 4, p = 0.05$	$A(3)C(1)$	0.114	0.432	1.409	0.18
WSD_{Ag5}	$m_1 = 13, p = 0.05$	$A(1)C(1)$	**0.159**	0.523	**1.386**	3.02

Based on the results of the experiments for both data it can be concluded that the Borda count method generates the smallest sets of global decisions and the highest rank method generates the largest sets of global decisions. In most cases, the majority vote method produces very good results. It is very surprising to see how well the simple method—majority vote compete with the more sophisticated ones. This is probably the reason that this method is so popular.

The papers [4, 12] also shows the experiments with the Lymphography and the Primary Tumor data set. Data in the non-dispersible form were examined. Table 3 presents the results given in these papers. Presented, in this paper results cannot be compared uniquely with the results shown in Table 3, because the decision-making system described in the paper generates a set of decisions, while Table 3 shows the results of the algorithms that generate one decision. It should be noted that for the Lymphography data set the average size of the global decisions sets is small, since it is close to the value 1. In the case of the Primary Tumor data set the average size of

Table 2 Summary of experiments results with the Primary Tumor data set

System	Parameters	Algorithm	e	e_{ONE}	$\overline{d}_{WSD_{Ag}}$	t
Majority vote						
WSD_{Ag1}	$m_1 = 1, p = 0.1$	$A(2)C(1)$	0.402	0.843	2.696	0.15
WSD_{Ag2}	$m_1 = 4, p = 0.1$	$A(1)C(3)$	**0.324**	0.843	**3.353**	0.07
WSD_{Ag3}	$m_1 = 13, p = 0.15$	$A(1)C(1)$	0.343	0.922	3.853	0.26
WSD_{Ag4}	$m_1 = 4, p = 0.05$	$A(3)C(1)$	0.353	0.882	3.686	4.14
WSD_{Ag5}	$m_1 = 4, p = 0.2$	$A(3)C(1)$	**0.304**	0.931	**4.343**	2.18
Weighted majority vote						
WSD_{Ag1}	$m_1 = 2, p = 0.1$	$A(2)C(1)$	**0.392**	0.824	**2.539**	0.13
WSD_{Ag2}	$m_1 = 4, p = 0.1$	$A(1)C(4)$	0.333	0.804	3.039	0.15
WSD_{Ag3}	$m_1 = 13, p = 0.15$	$A(1)C(1)$	**0.343**	0.922	**3.814**	0.22
WSD_{Ag4}	$m_1 = 1, p = 0.05$	$A(1)C(3)$	0.353	0.892	3.696	0.23
WSD_{Ag5}	$m_1 = 1, p = 0.05$	$A(2)C(1)$	0.314	0.922	4.294	3.55
Borda count						
WSD_{Ag1}	$m_1 = 1, p = 0.1$	$A(1)C(1)$	0.422	0.833	2.549	0.06
WSD_{Ag2}	$m_1 = 13, p = 0.05$	$A(2)C(1)$	0.343	0.882	3.196	0.07
WSD_{Ag3}	$m_1 = 13, p = 0.15$	$A(1)C(1)$	0.343	0.912	3.833	1.21
WSD_{Ag4}	$m_1 = 1, p = 0.05$	$A(1)C(3)$	0.353	0.882	3.686	0.18
WSD_{Ag5}	$m_1 = 4, p = 0.2$	$A(1)C(2)$	0.314	0.931	4.275	3.02
Highest rank method						
WSD_{Ag1}	$m_1 = 1, p = 0.2$	$A(2)C(1)$	0.412	0.833	2.539	0.06
WSD_{Ag2}	$m_1 = 10, p = 0.2$	$A(3)C(1)$	0.333	0.902	4.157	0.17
WSD_{Ag3}	$m_1 = 1, p = 0.1$	$A(1)C(2)$	0.353	0.912	4.098	0.13
WSD_{Ag4}	$m_1 = 1, p = 0.1$	$A(1)C(2)$	**0.333**	0.892	**3.902**	0.20
WSD_{Ag5}	$m_1 = 7, p = 0.1$	$A(1)C(3)$	0.284	0.980	5.284	6.04

Table 3 Results of experiments from other papers

Lymphography				Primary Tumor			
Algorithm	Error	Algorithm	Error	Algorithm	Error	Algorithm	Error
Bayes	0.17	AQ15	0.18	Bayes	0.61	AQ15	0.59
AQR	0.24	Human experts	0.15	AQR	0.65	Human experts	0.58
CN2	0.22	Random choice	0.75	CN2	0.63	Random choice	0.95

the global decisions sets is between 3 and 4, note that there are 22 decision classes. As can be seen the use of dispersed decision-making system with the considered fusion methods results in significant improvement in the efficiency of inference. Probably also the approval of ambiguity decision has affected for improvement. However, in medicine it is important information when we can say, with great certainty, that only 4 decision classes from 22 classes are probable. Because maybe it will be possible to carry out additional tests towards these 4 probable decisions. Moreover, very important advantage of the proposed decision-making system is the possibility of using dispersed knowledge, which are collected in different medical centers.

5 Conclusions

In this article, four different fusion methods were used in the dispersed decision-making system: the majority vote, the weighted majority vote, the Borda count method and the highest rank method. In the experiments, which are presented, dispersed medical data have been used: Lymphography data set and Primary Tumor data set. The use of the dispersed data in medical field is particularly important. Often in many medical institutions knowledge, from the same domain, is accumulated independently. The use of all available knowledge increase the correctness of decisions taken. The general conclusions, that were reached based on the results of experiments are as follows. If it is important to get an unequivocal decision it is best to use the Borda count method. If we want to receive the best possible quality of inference and we allow greater ambiguity the best method is the majority vote method.

References

1. Alpaydin, E., Jordan, M.I.: Local linear perceptrons for classification. IEEE Trans. Neural Netw. **7**(3), 788–794 (1996)
2. Barabash, Y.L.: Collective Statistical Decisions in Recognition. Radio i Sviaz, Moscow (1983)
3. Cho, S.-B., Kim, J.H.: Combining multiple neural networks by fuzzy integral for robust classification. IEEE Trans. Syst. Man Cybern. **25**(2), 380–384 (1995)
4. Clark, P., Niblett, T.: Induction in noisy domains. In: Bratko I., Lavrac N. (eds.) Progress in Machine Learning, pp. 11–30 (1987)
5. Drucker, H., Cortes, C., Jackel, L.D., LeCun, Y., Vapnik, V.: Boosting and other ensemble methods. Neural Comput. **6**(6), 1289–1301 (1994)
6. Gatnar, E.: Multiple-Model Approach to Classification and Regression. PWN, Warsaw (2008). (in Polish)
7. Ho, T.K., Hull, J.J., Srihari, S.N.: Decision combination in multiple classifier systems. IEEE Trans. Pattern Anal. Mach. Intell. **16**(1), 66–75 (1994)
8. Jacobs, R.A., Jordan, M.I., Nowlan, S.J., Hinton, G.E.: Adaptive mixtures of local experts. Neural Comput. **3**(1), 79–87 (1991)
9. Kittler, J., Hatef, M., Duin, R.P.W., Matas, J.: On combining classifiers. IEEE Trans. Pattern Anal. Mach. Intell. **20**(3), 226–239 (1998)
10. Kuncheva, L., Bezdek, J.C., Duin, R.P.W.: Decision templates for multiple classifier fusion: an experimental comparison. Pattern Recognit. **34**(2), 299–314 (2001)
11. Kuncheva, L.: Combining pattern classifiers methods and algorithms. John Wiley & Sons (2004)
12. Michalski, R., Mozetic, I. Hong, J., Lavrac, N.: The multi-purpose incremental learning system AQ15 and its testing applications to three medical domains. In: Proceedings of the 5th National Conference on Artificial Intelligence, pp. 1041–1045 (1986)
13. Ng, K.-C., Abramson, B.: Probabilistic multi-knowledge-base systems. Appl. Intell. **4**(2), 219–236 (1994)

14. Przybyła-Kasperek, M., Wakulicz-Deja, A.: Application of reduction of the set of conditional attributes in the process of global decision-making. Fund. Inform. **122**(4), 327–355 (2013)
15. Przybyła-Kasperek, M., Wakulicz-Deja, A.: Global decision-making system with dynamically generated clusters. Inf. Sci. **270**, 172–191 (2014)
16. Przybyła-Kasperek, M., Wakulicz-Deja, A.: A dispersed decision-making system—the use of negotiations during the dynamic generation of a systems structure. Inf. Sci. **288**, 194–219 (2014)
17. Rogova, G.L.: Combining the results of several neural network classifiers. Neural Netw. **7**(5), 777–781 (1994)
18. Shoemaker, L., Banfield, R.E., Hall, L.O., Bowyer, K.W., Kegelmeyer, W.P.: Using classifier ensembles to label spatially disjoint data. Inf. Fusion **9**(1), 120–133 (2008)
19. Ślęzak, D., Wróblewski, J., Szczuka, M.: Neural network architecture for synthesis of the probabilistic rule based classifiers. In: ENTCS 82, Elsevier (2003)
20. Wakulicz-Deja, A., Przybyła-Kasperek, M.: Application of the method of editing and condensing in the process of global decision-making. Fund. Inform. **106**(1), 93–117 (2011)

Estimation of Coefficient of Static Friction of Surface by Analyzing Photo Images

Hitoshi Tamura and Yasushi Kambayashi

Abstract We propose a method to estimate the coefficient of static friction of floor surfaces by analyzing photo image of the floor tiles. The image features that we use to estimate the coefficient are micro-shape features and micro-depth features. We extract the difference between the flash images and the non-flash images of floor tiles. We have composed an equation by applying multiple linear regression analysis that sets the image features as explanatory variables and the measurements of the tile images as objective values. As the result, we have obtained an estimate equation that coefficient of determination R^2 is 0.97 and we observed the two-sided 95 % confidence interval ±0.053. We can say that the equation is good enough for practical use.

Keywords Coefficient of static friction · Texture analysis · Image measurements · Shape-pass filter · Micro shape feature · Micro depth feature

1 Introduction

Coefficient of static friction of a floor is an important factor for controlling a robot. It is difficult to measure the friction coefficient without contacting the floor. Watanabe et al. proposed a control method for the grasping device that doesn't use the friction coefficient [1]. If we can estimate the coefficient of the floor in front of a robot only though image sensors, we can increase the stability of controlling robots. In this paper, we propose a method to estimate coefficient of static friction of a floor

---

H. Tamura (✉)
Department of Innovative Systems Engineering,
Nippon Institute of Technology, Miyashiro, Japan
e-mail: tamura@nit.ac.jp

Y. Kambayashi
Department of Computer and Information Engineering,
Nippon Institute of Technology, Miyashiro, Japan
e-mail: yasushi@nit.ac.jp

I. Czarnowski et al. (eds.), *Intelligent Decision Technologies 2016*,
Smart Innovation, Systems and Technologies 57,
DOI 10.1007/978-3-319-39627-9_2

tiles by analyzing two photo images of the floor tiles. We are not aware of many researches for measuring frictional properties by only through visual information. Even though Kim et al. proposed a method for classifying terrains and for predicting friction coefficient on terrains by applying visual information, their method doesn't measure frictional properties directly [2].

Authors have treated a photo image of the floor tile as a random texture image and have applied general-purpose texture analysis filters, which are called shape-pass filters, to the texture [3, 4]. The shape-pass filters are nonlinear filter banks that extract micro-shape features from the random texture image.

We set an assumption that image features have a correlation with frictional properties. Image features we are using are the micro-shape features and the micro-depth features on a surface. We have composed an equation to estimate the coefficient of static friction of the floor tiles by applying multiple linear regression analysis that set micro-shape features and reflection features as explanatory variables.

The structure of the balance of this paper is follows. In Sect. 2, we explain both the micro-shape features and the micro-depth features. In Sect. 3, we compose the estimate equation from the sample images and friction data. We then confirm our equation through discriminant analysis and estimation of the coefficient of friction in Sect. 4. In Sect. 5, we examine the frictions according to the directions. Finally we conclude our discussion in Sect. 6.

2 Features in Images

2.1 Micro Shape Features

Research scientists have proposed various approaches for texture analyses [5–9]. In our method, we interpret a texture as a collection of tiny elementary shapes and classify the texture images by analyzing which portions in the texture image contain certain elementary shapes. We extract these shapes by applying nonlinear procedures that use only local pixels in the given texture image. When we scan the entire image, we use one local domain as a window and apply the nonlinear procedures to each local domain as the filtered area. Each local domain corresponds to a point spread area for a linear image filtering.

There is no widely accepted standard that defines what shapes are elementary for characterizing textures. The authors propose the five shapes, namely black-line, black-pepper, black-roof, black-snake, and cliff as the elementary micro-shape features as shown in Fig. 1. Both the line and the snake consist of one thin black area but the latter is curved. The pepper consists of an isolated small black area. The roof is characterized as containing a black area fanning from the center of a specified small area in some angle narrower than 90°. The last one represents about fifty-fifty partition of black and white area. Since a negative photograph of a general

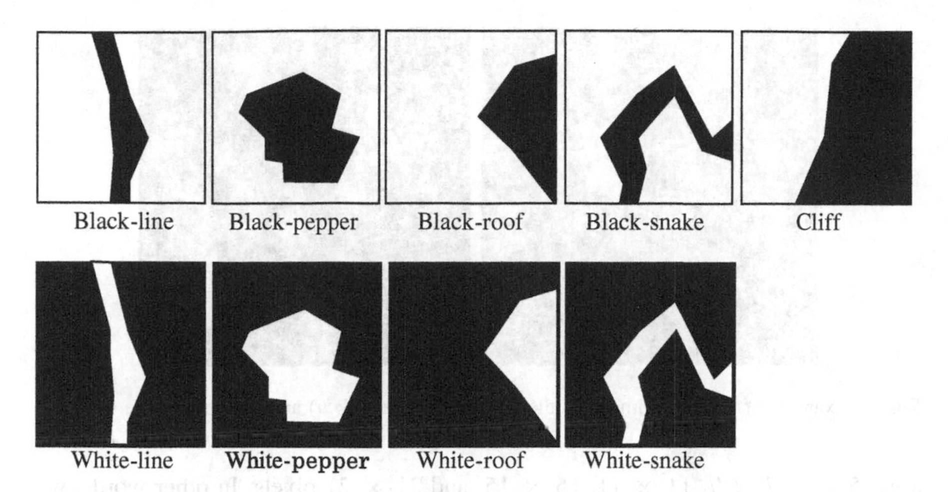

Fig. 1 Micro-shape features

texture from positive one can be recognized as different features, we also introduce four shapes in which black and white areas are reversed as shown in Fig. 1. Thus we set the black roof, the black line, the black snake, the black pepper, the white roof, the white line, the white snake, the white pepper, and the cliff are the elementary shapes.

We do not employ the traditional simple pattern matching to extract the micro shapes. Instead, we nonlinearly extract the micro shapes by applying well-defined procedures. We define one extract procedure for each micro-shape. In the procedures, we classify the pixels in the filter area into two values, and then we determine the shape of the black pixels that exists at the center of the local domain. For example, the procedure for the black pepper detects the existence of a small black isolated area.

The filtering procedure determines the value at the center of the local domain as the output value of the area. The output value is not simple black-and-white value obtained from the filtered area but the average of brightness of the original image of the area. The filtering procedure scans the entire image and produces the output value of each position of the image. Figure 2 shows an example. The left figure is the original input image, and the right figure is the output image obtained by applying the black-roof filter procedure to the input image. The white dots in the output image are not just white but gray scaled.

Output images of the filter contain features of shapes. The representing value of the output images is RMS (root mean square value) of the output image rather than a simple average value. Only brightness information on each pixel is used here because the procedure pays attention about the shape features. The filtering procedure does not use any information on colors, such as Chroma.

The size of the local domain affects to determine the shape feature of the given filter area. Therefore, we extract each micro shape feature by five sizes of the filter

Fig. 2 Example of an output image of the Black-roof filter (*right*) and an input image (*left*)

area: 5×5, 7×7, 11×11, 15×15, and 21×21 pixels. In other words, we apply the procedures to forty-five features, i.e. nine shapes $\times$ five sizes, and stream into the regression analysis to obtain the estimation equation.

2.2 Micro-Depth Features

In addition to the micro shape features, we use the micro-depth features for estimation. The micro depth features are characterized by existence micro unevenness on the surface. The features are obtained by observing the difference between the flash images and the non-flash images. The flash image is an image that is taken with flash, and the non-flash image is an image that is taken without flash. When the examined area is uneven, the flash produces some shadows. Thus we can measure the micro-depth by examining the difference between the flash image and the non-flash image as shown in Fig. 3.

We prepare the two images; the flash image and the non-flash image for each of forty-five micro shape features that are obtained from one floor tile. We also obtain forty-five micro depth features by taking the difference between the flash images and the non-flash images. Therefore we obtain in total 135 (= 45 + 45 + 45) features for one floor tile. When obtaining images of a floor tile, using the flash affects the determination of the micro shape features greatly, especially when there is a small ruggedness on the surface of the floor tile. It is clear that the ruggedness on the surface influences frictional properties. The reason why we extract the micro shape features from not only non-flash images but also flash images is to detect this micro ruggedness on the surface by taking the difference between the micro shape features in the flash and the non-flash images. A part of these features are selected for estimation of coefficient of static friction of the floor tile surface by applying the multiple linear regression analysis to the sample images.

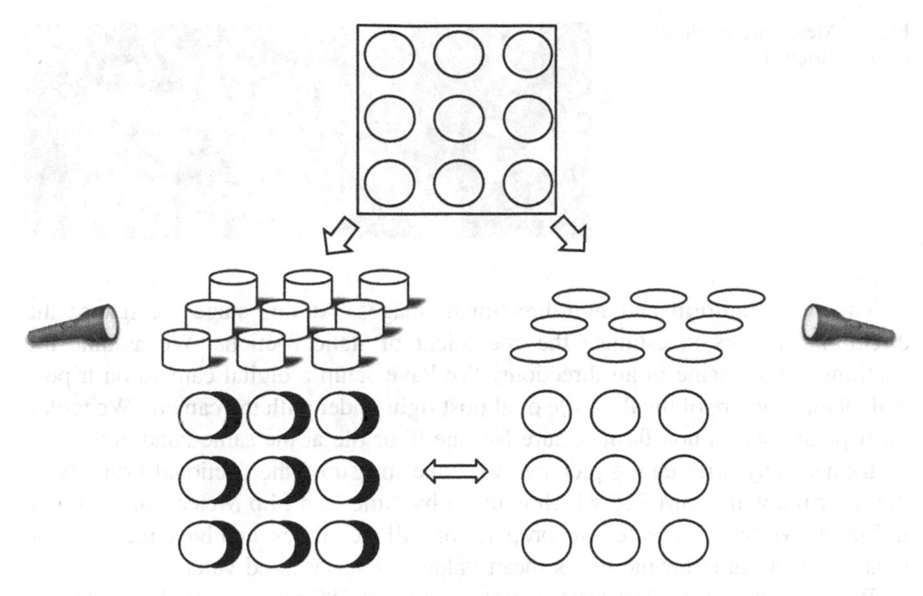

Fig. 3 Micro depth feature can be extracted by examining the shadow produced by unevenness

3 Composing Estimate Equation

3.1 Sample Images

We have selected twelve samples of the floor tiles we have found on our campus. They are several kinds of linoleums, stone materials, wood surfaces, fiber carpets and concrete. Figure 4 shows them. We use those twelve samples for the analysis and experiments. We have taken forty pairs (flash and non-flash) of images for each sample tile; therefore we prepare 480 pairs of images in total.

Fig. 4 Twelve sample images

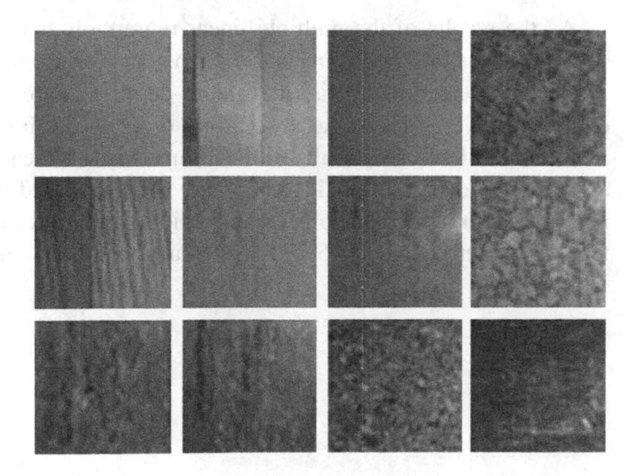

Fig. 5 Measuring coefficient of static friction

These are uniform and non-directional images. At this stage we ignore the directional factors to estimate the coefficient of static frictions. We assume the frictions are the same in all directions. We have setup a digital camera on tripod, and taken a picture of the floor tile of almost right under with the camera. We took a flash picture and a non-flash picture for one floor tile at the same condition.

Immediately after taking pictures, we have measured the frictional property of the floor tile with ASM 725 which is made by American Slip Meter that as shown in Fig. 5. We have measured the property of a tile ten times, and have recorded the measurement value for the tile is mean value of ten observed values.

We have employed 240 pairs of images among 480 pair of sample images for composing the estimate equation, and we have used the other half of the images for evaluation of the equation. We have composed the equation by applying the multiple linear regression analysis that sets both the micro-shape features and the micro depth features of the tile image as explanatory variables and the measurement of the tile image as objective value. We have performed the analysis by using the forward selection method that set 2.0 to F-in value and that set 2.0 to F-out value. Selecting those values in the restricted range *D*, we have composed the following polynomial equation, where *a* is the partial regression coefficient, *x* is the explanatory variable.

$$\mathrm{f} = \sum_{i \in D} a_i x_i \qquad (1)$$

As the results of the multiple linear regression analysis, we have obtained the partial regression coefficient and explanatory variables. The results are shows in Table 1, where (depth) is the micro-depth feature. In Table 1, we set the explanatory variables in the descending order of the F values.

Figure 6 shows the results of the comparison. We can observe that the multiple correlation coefficient R is 0.90, and the two-sided 95 % confidence interval is ±0.053. The difference of 0.05 in frictional properties makes the maximum static friction power change by 5 %. This accuracy is good enough for practical use.

Table 1 Result of the multiple linear regression analysis

Explanatory var.	Partial regression coefficient	F value
Black-line 5 × 5 (depth)	−0.043	86.43
Black-line 5 × 5	0.043	82.33
Cliff 11 × 11	−0.010	67.14
White-line 15 × 5	0.064	52.58
Black-line 21 × 21	−0.010	48.14
Black-line 7 × 7	−0.041	46.43
White-line 5 × 5	−0.149	46.23
White-snake 21 × 21 (depth)	0.018	45.24
White-roof 21 × 21	0.014	38.76
Black-line 15 × 15	0.021	38.22
Black-snake 7 × 7	0.036	28.98
White-roof 15 × 15	0.013	28.83
Black-roof 15 × 15	0.011	25.75
White-line 5 × 5 (depth)	−0.075	25.36
White-roof 21 × 21 (depth)	0.007	23.25
Black-pepper 5 × 5	−0.064	22.68
Black-pepper 7 × 7	0.037	19.66
…	…	…
Constant	0.704	—

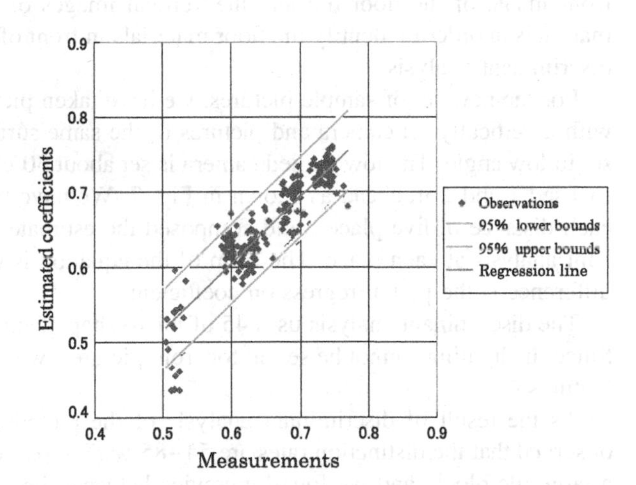

Fig. 6 Scatter diagram plotted the estimation values and the observation values the coefficients

4 Discriminant Analysis of the Floor Surface

As described in the previous section, we have confirmed that the estimation of the coefficient of friction using the images of the floor that are taken in vertical angle is quite precise. In order to apply this method to robotics, however, we need to extend

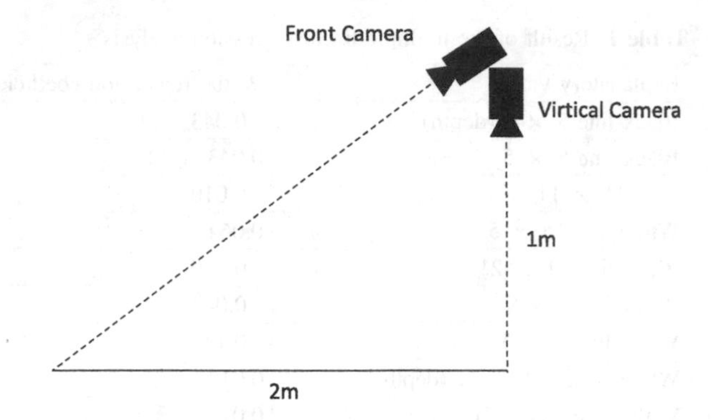

Fig. 7 Setting of the front camera and the vertical camera

the method so that it can use the images of the floor surface obtained by a camera in front of the robot.

The method we examined in the previous section depends on the uniform-density of the micro shapes in the picture of surface taken in vertical angle. The images taken by a camera in front of the robot cannot be uniform-density, because the camera captures distant shapes smaller and interprets them denser than the near shapes.

Since we can make accurate estimates with the images taken by the vertical camera, if we can determine the materials of the floor in front of the robot, we can estimate the friction of the floor in front of the robot. Therefore, we can compare the front image of the floor tile and the vertical images of the twelve kinds of floor materials in order to identify the floor materials in front of the robot by applying the discriminant analysis.

For the new set of sample pictures, we have taken pictures of the floor surfaces with a vertically set camera and pictures of the same surfaces with another camera set in low angle. The low angled camera is set about 30 cm high from the floor and focused 1 and 2 m ahead as shown in Fig. 7. We have taken fifty-four pictures in each distance of five places, and composed the estimate equation by the discriminant analysis at each place. The form of the equation is the same as (1). The only difference is the partial regression coefficients.

The discriminant analysis uses 45 of micro shape features (9 shapes $\times$ 5 sizes). Since the lighting cannot be set up for front pictures, we have not used micro depth features.

As the result of discriminant analysis of the pictures of five places, we have observed that the distinction rates are 54–85 %. The particularly inaccurate place is a large tile block, and we found a crevice between tiles influenced aggravation of accuracy greatly. The results from the other places, however, show that the accuracy is practically good enough.

5 Estimation of the Directional Coefficient of Friction

It is well known that the frictions are directionally different from each other even with the same floor materials. Therefore it is desirable that we can estimate the maxi-mum coefficient friction and minimum coefficient friction and their directions.

We have obtained the Fourier power-spectrum pictures from the floor surface pictures, as shown in Fig. 8. Then we have computed the average brightness according to direction, and obtained the maximum direction of each floor materials. The obtained maximum direction is called the lengthwise direction and it is known that the length-wise direction produces the maximum coefficient friction.

Forty-five of micro shape features and forty-five of micro depth features, ninety features in total, are used for computing the estimated frictions. Here, the micro depth features are obtained from the difference of the pictures flashed along the direction of the maximum coefficient of friction and along the direction of the minimum coefficient of friction.

The number of the floor surfaces we have used for the experiment is twenty-three. Six pictures were taken at each place, 138 pictures in total.

We have conducted multiple linear regression analysis over the set of maximum measured values of frictions and minimum measured values of frictions as the objective variables.

The results of the regression analysis by using the estimate equation for maxi-mum coefficient and the estimate equation for minimum coefficient are shown in Figs. 9 and 10, respectively.

The coefficient of determination shows 0.86 with the estimate equation for maximum coefficient, and 0.88 with the estimate equation for minimum coefficient.

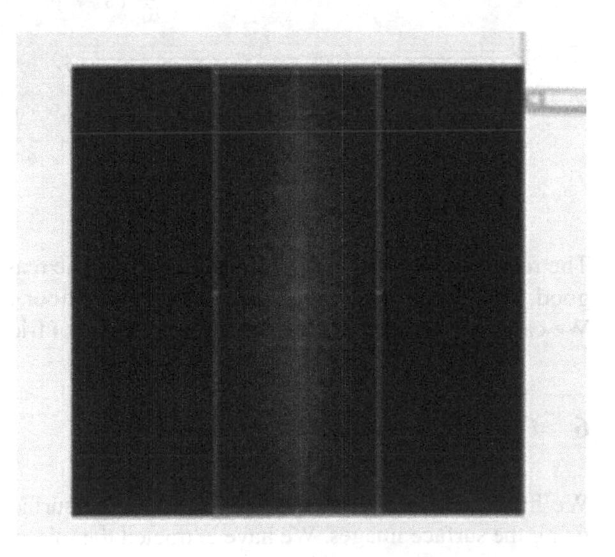

Fig. 8 Fourier power spectrum

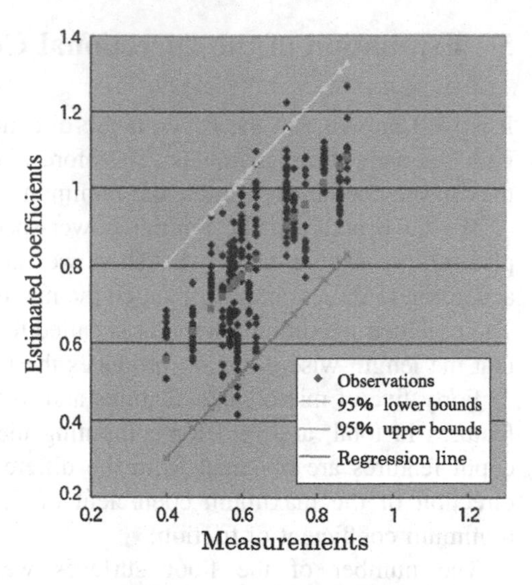

Fig. 9 Regression line for maximum coefficient of static friction

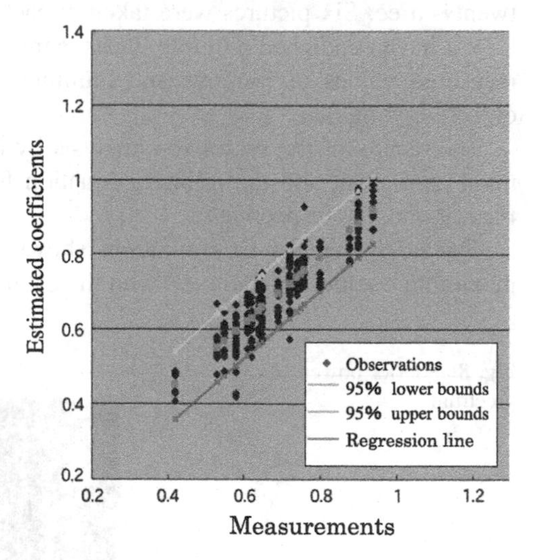

Fig. 10 Regression line for minimum coefficient of static friction

The results are not as good as shown in Fig. 6. The reason why the results are not so good may be that we did not use the smoother floor materials in the experiments. We can say, however, that at least the coefficient of friction has directional property.

6 Conclusion and Discussion

We have proposed a technique to estimate the surface coefficient of friction only using the surface images. We have extracted the micro shape features and the micro depth features from a uniform texture of the surface image and estimate the surface coefficient of friction for each surface image. The proposed method consists of the following three functions.

1. The function that estimates the coefficient of frictions of the surface right below the camera with sufficient accuracy.
2. The function that identifies the floor material observed from the low angled camera.
3. The function that estimate the maximum and the minimum coefficients of friction with the pictures that have the directional property.

We have prepared forty-five micro shape features for each floor tile. Then, for each floor tile, we obtained two images; the flash image and the non-flash image. Also we have obtained forty-five micro depth features by taking the difference between the flash image and the non-flash image. In total, we obtained and used 135 (= 45 + 45 + 45) features in the floor tile for applying the multiple regression analysis.

We have observed the following results. Multiple correlation coefficient R is 0.90, and the two-sided 95 % confidence interval was ±0.053 for the first function. Although the second function makes misjudgments for some materials, it can distinguish the floor materials with practically enough accuracy. The third function shows that the maximum and the minimum coefficients of frictions can be estimated from pictures with directions.

For the computational complexity, we use the least squares regression with N training examples and C features, and $N > C$. We can say that the computational complexity of matrix multiplications is $O(C^2N)$, and it is not significant in our setting: $C = 135$ and $N = 240$. Thus the estimation of the frictions can be performed by a notebook computer on a small mobile robot.

We are planning to integrate this algorithm into the control software for the electric wheelchair we are developing [10]. Successful inclusion of the algorithm should prove the usability of our algorithm in real-time setting.

Acknowledgments This work was supported by Japan Society for Promotion of Science (JSPS), with the basic research program (C) (No. 25420416 and 26350456), Grant-in-Aid for Scientific Research.

References

1. Watanabe, N., Obinata, G.: Grip force control based on the degree of slippage using optical tactile sensor. In: International Symposium on Micro-Nano Mechatronics and Human Science, pp. 466–471. IEEE Press, New York (2007)
2. Kim, J., Kim, D., Lee, J., Lee, J., Joo, H., Kweon, I.: Non-contact terrain classification for autonomous mobile robot. In: IEEE International Conference on Robotics and Biomimetics, pp. 824–829. IEEE Press, New York (2009)
3. Tamura, H., Atoda, O.: A nonlinear filter bank for shape-based texture analysis. In: Third Asia-Pacific Conference on Control and Measurement, pp. 58–62 (1998)

4. Tamura, H., Atoda, O.: A bank of "shape-pass" nonlinear filters extracting pictorial features of natural textures. In: Transactions of the Institute of Electronics, Information and Communication Engineers, vol. J82-D-II, no. 12, pp. 2260–2270 (1999) (in Japanese)
5. Haralick, R.M.: Statistical and structural approaches to textures. IEEE Proc. **67**(5), 786–804 (1979)
6. Salari, E., Ling, Z.: Texture segmentation using hierarchical wavelet decomposition. Pattern Recogn. **28**(12), 1819–1824 (1995)
7. Wang, L., He, D.C.: Texture classification using texture spectrum. Pattern Recogn. **23**(8), 905–910 (1990)
8. Ganesan, L., Bhattacharyya, P.: A Statistical design of experiments approach for texture description. Pattern Recogn. **28**(1), 99–105 (1995)
9. Ojala, T., Pietikainen, M., Harwood, D.: A comparative study of texture measures with classification based on feature distributions. Pattern Recogn. **29**(1), 51–59 (1994)
10. Tamura, H., Kambayashi, Y.: Design of intuitive interfaces for electric wheelchairs to prevent accidents. In: Fifteenth International Conference on HCI International, vol. 6, LNCS 8009, pp. 592–601. Springer, Heidelberg (2013)

Decision Case Management for Digital Enterprise Architectures with the Internet of Things

Alfred Zimmermann, Rainer Schmidt, Dierk Jugel, Kurt Sandkuhl, Christian Schweda, Michael Möhring and Justus Bogner

Abstract The Internet of Things (IoT), Enterprise Social Networks, Adaptive Case Management, Mobility systems, Analytics for Big Data, and Cloud services environments are emerging to support smart connected products and services and the digital transformation. Biological metaphors of living and adaptable ecosystems with service-oriented enterprise architectures provide the foundation for self-optimizing and resilient run-time environments for intelligent business services and related distributed information systems. We are investigating mechanisms for flexible adaptation and evolution for the next digital enterprise architecture systems in the context of the digital transformation. Our aim is to support flexibility and agile transformation for both business and related enterprise systems through adaptation and dynamical evolution of digital enterprise architectures. The present research paper investigates mechanisms for decision case management in the context of multi-perspective explorations of enterprise services and Internet of Things architectures by extending original enterprise architecture reference models with state of art elements for architectural engineering for the digitization and architectural decision support.

Keywords Decision support · Decision case management · Internet of Things · Adaptive enterprise architecture · Digital transformation

A. Zimmermann (✉) · D. Jugel · C. Schweda · J. Bogner
Herman Hollerith Center, Reutlingen University, Reutlingen, Germany
e-mail: alfred.zimmermann@reutlingen-university.de

R. Schmidt · M. Möhring
Munich University of Applied Sciences, Munich, Germany
e-mail: Rainer.Schmidt@hm.edu

M. Möhring
e-mail: michael.burch@visus.uni-stuttgart.de

D. Jugel · K. Sandkuhl
University of Rostock, Rostock, Germany

J. Bogner
Hewlett Packard Enterprise, Böblingen, Germany

I. Czarnowski et al. (eds.), *Intelligent Decision Technologies 2016*,
Smart Innovation, Systems and Technologies 57,
DOI 10.1007/978-3-319-39627-9_3

1 Introduction

Information, data and knowledge are fundamental concepts of our everyday activities. Social networks, smart portable devices, and intelligent cars, represent only a few instances of a pervasive, information-driven vision [25] for the next wave of the digital economy and the digital transformation. Digitization is the collaboration of human beings and autonomous objects beyond their local context using digital technologies. Digitization further increases the importance of information, data and knowledge as fundamental concepts of our everyday activities. By exchanging information human beings and intelligent objects are able to make decisions in a broader context and with higher quality.

Smart connected products and services expand physical components from their traditional core by adding information and connectivity services using the Internet. Internet of Things covers often-small intelligent physical components, which are connected over the Internet. Smart products and services amplify the basic value and capabilities and offer exponentially expanding opportunities [2]. Smart connected products combine three fundamental elements: physical components, smart components, and connectivity components. A challenging example of digital transformation for smart products results from the capabilities of the Internet of Things (IoT) [33]. Major trends for the digital transformation of digitized products and services are investigated in [10]. The Internet of Things enables a large number of physical devices to connect each other to perform wireless data communication and interaction using the Internet as a global communication environment.

The technological and business architectural impact of digitization has multiple aspects, which directly affect adaptable digital enterprise architectures and their related systems. Enterprise Architecture Management [19, 38] for Services Computing is the approach of choice to organize, build and utilize distributed capabilities for Digital Transformation [1, 25]. They provide flexibility and agility in business and IT systems. The development of such applications integrates the Internet of Things, Web and REST Services, Cloud Computing and Big Data management, among other frameworks and methods, like software architecture [3] and architectural semantic support.

Our current research paper focuses on the following research questions:

RQ1: What is the architectural decision context from Internet of Things architecture for the digital transformation of products and services?
RQ2: How digital enterprise architecture management should be holistically tailored to include Internet of Things architectures as the decisional context for architectural analytics and optimization efforts?
RQ3: How can collaborative decision support mechanisms be specifically designed by introducing decision case management models for digital enterprise architecture?

The following Sect. 2 sets the fundamental architectural context for Digital Transformation with the Internet of Things approach. Section 3 describes our research platform for digital enterprise architecture, which was extended by concepts from adaptive case management, architectural adaptation mechanisms and a specific model integration method. Section 4 presents our collaborative architectural engineering and transformation approach and links it with specific decisional and prediction mechanisms. Finally, we summarize in Sect. 5 our research findings and limitations, our ongoing work in academic and practical environments and our future research plans.

2 Architecting the Internet of Things

The Internet of Things maps and integrates real world objects into the virtual world, and extends the interaction with mobility systems, collaboration support systems, and systems and services for big data and cloud environments. Sensors, actuators, devices as well as humans and software agents interact and communicate data to implement specific tasks or more sophisticated business or technical processes. Therefore, smart products as well as their production are supported by the Internet of Things and can help enterprises to create more customer-oriented products. Furthermore, the Internet of Things is an important influence factor of the potential use of Industry 4.0 [28].

The Internet of Things (IoT) fundamentally revolutionizes today's digital strategies with disruptive business operating models [26], and holistic governance models for business and IT [34], in context of current fast changing markets [33]. With the huge diversity of Internet of Things technologies and products organizations have to leverage and extend previous enterprise architecture efforts to enable business value by integrating the Internet of Things into their classic business and computational environments.

The Internet of Things is the result of a convergence of visions [2, 10] like, a Things-oriented vision, an Internet-oriented vision, and a Semantic-oriented vision. The Internet of Things supports many connected physical devices over the Internet as a global communication platform. A cloud centric vision for architectural thinking of a ubiquitous sensing environment is provided by [10]. The typical configuration of the Internet of Things includes besides many communicating devices a cloud-based server architecture, which is required to interact and perform remote data management and calculations. A main question of current and further research is, how the Internet of Things architecture fits in a context of a services-based enterprise-computing environment? A service-oriented integration approach for the Internet of Things was elaborated in [29]. A layered Reference Architecture for the Internet of Things is proposed in [36] and (Fig. 1). Layers can be instantiated by suitable technologies for the Internet of Things.

The Internet of Things maps and integrates real world objects into the virtual world, and extends the interaction with mobility systems, collaboration support systems, and systems and services for big data and cloud environments. Sensors, actuators, devices as well as humans and software agents interact and communicate

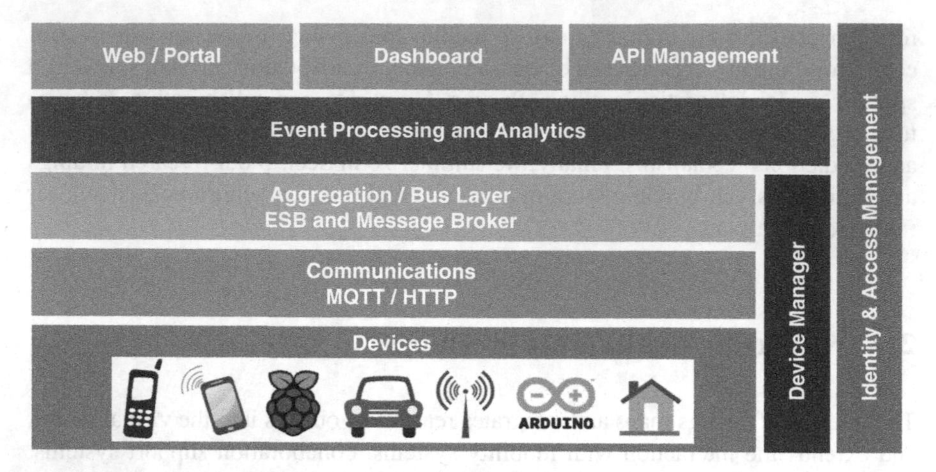

Fig. 1 Internet of Things reference architecture [36]

data to implement specific tasks or more sophisticated business or technical processes.

The Internet of Things architecture has to support a set of generic as well as some specific requirements. Generic requirements result from the inherent connection of a magnitude of devices via the Internet, often having to cross firewalls and other obstacles. Having to consider so many and a dynamic growing number of devices we need an architecture for scalability. Because these devices should be active in a 24 × 7 timeframe we need a high-availability approach [9], with deployment and auto-switching across cooperating datacenters in case of disasters and high scalable processing demands. Additionally, an Internet of Thing architecture has to support automatic managed updates and remotely managed devices. Often connected devices collect and analyze personal or security relevant data. Therefore, it is mandatory to support identity management, access control and security management on different levels: from the connected devices through the holistic controlled environment.

An inspiring approach for the development for the Internet of Things environments is presented in [24]. This research has a close link to our work about leveraging the integration of the Internet of Things into a decision framework for digital enterprise architectures. The main contribution considers a role-specific development methodology, and a development framework for the Internet of Things. The development framework contains a set of modeling languages for a vocabulary language to describe domain-specific features of an IoT application, an architecture language for describing application-specific functionality, and a deployment language for deployment features. Associated with this language set are suitable automation techniques for code generation, and linking to reduce the effort for developing and operating device-specific code. The metamodel for Internet of Things applications defines elements of an Internet of Things architectural reference model like, IoT resources of type: sensor, actuator, storage, and user interface.

3 Digital Enterprise Architecture

Enterprise Architecture Management (EAM) [19, 4] defines today with frameworks, standards [22, 23], tools and practical expertise a quite large set of different views and perspectives. We argue in this paper that a new refocused digital enterprise architecture approach should support digitization of products and services, and should be both holistic [37] and easily adaptable [38] to support IoT [39] and the digital transformation with new business models and technologies like social software, big data, services and cloud computing, mobility platforms and systems, security systems, and semantics support.

In this paper we extend our service-oriented enterprise architecture reference model for the context of managed adaptive cases and decisions [30], which are supported by case services of a collaborative case framework. Additionally, we have extended our architectural metamodel integration approach [39] to support enterprise architectures for digital transformations and the integration of Internet of Things.

ESARC—Enterprise Services Architecture Reference Cube [38] (Fig. 2) is an architectural reference model for an extended view on evolved digital enterprise architectures. ESARC is more specific than existing architectural standards of EAM —Enterprise Architecture Management [22, 23] and extends these architectural standards for digital enterprise architectures with services and cloud computing. ESARC provides a holistic classification model with eight integral architectural domains. These architectural domains cover specific architectural viewpoint descriptions [7] in accordance to the orthogonal dimensions of both architectural layers and architectural aspects [37, 38]. ESARC abstracts from a concrete business scenario or technologies, but it is applicable for concrete architectural instantiations to support digital transformations. The Open Group Architecture Framework [22]

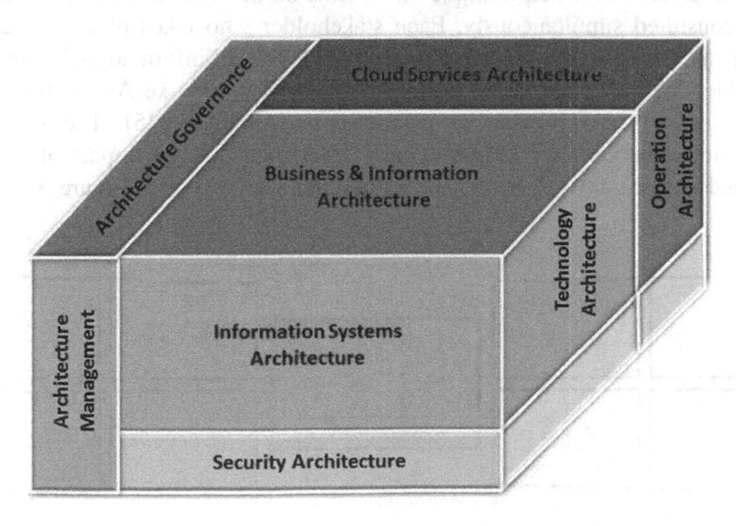

Fig. 2 Enterprise services architecture reference cube [37, 38]

provides the basic blueprint and structure for our extended service-oriented enterprise architecture domains.

Our research in progress main question asks, how we can dynamically federate these EA-IoT-Mini-Descriptions to a global high scalable EA model and information base by promoting a mixed automatic and collaborative decision process [15–17, 27]. For the automatic part we currently extend model federation and transformation approaches from [8] and [32] by introducing sematic-supported architectural representations, e.g. by using partial and federated ontologies and ontology-supported model transformations as well as associated mapping rules—as universal enterprise architectural knowledge representation, which are combined with special inference mechanisms.

Adaptation drives the survival [31] of digital enterprise architectures [38], platforms and application ecosystems. Volatile technologies and markets typically drive the evolution of ecosystems. The alignment of Architecture-Governance [34, 26] shapes resiliency, scalability and composability of components and services for distributed information systems.

4 Decision Case Management

A Decision Support System (DSS) is a system "to help improve the effectiveness of managerial decision making in semi-structured tasks" [18, p. 255]. In particular, knowledge intensive management activities, like Enterprise Architecture Management (EAM), can benefit from a DSS to improve architectural decision-making. We are exploring in our current research, how an enterprise architecture cockpit [15–17] can be leveraged and extended to a DSS for EAM. A cockpit presents a facility or device via which multiple viewpoints on the system under consideration can be consulted simultaneously. Each stakeholder who takes place in a cockpit meeting can utilize a viewpoint that displays the relevant information. Thereby, the stakeholders can leverage views that fit the particular role like Application Architect, Business Process Owner or Infrastructure Architect [35]. The viewpoints applied simultaneously are linked to each other such that the impact of a change performed in one view can be visualized in other views as well. Figure 3 gives the

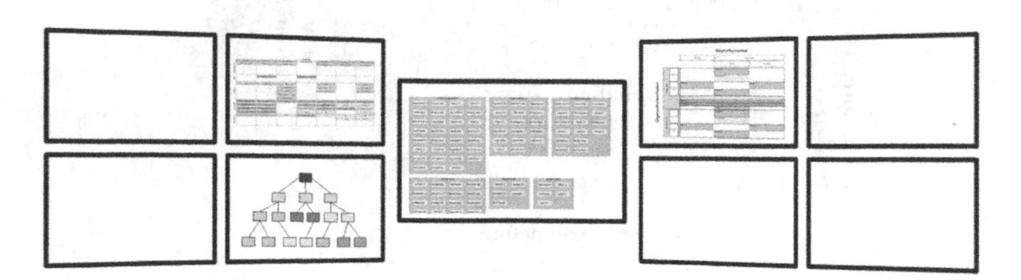

Fig. 3 Example: enterprise architecture cockpit [16, 17]

idea of multi-perspectives of a collaborative enterprise architecture cockpit and social-based processes [27].

Jugel et al. [16] present a collaborative approach for decision-making for EA management. They identify decision making in such complex environment as a knowledge-intensive process strongly depends on the participating stakeholders. Therefore, the collaborative approach presented is built based on the methods and techniques of adaptive case management (ACM), as defined in [30].

The Case Management Modeling Notation (CMMN) [20] is a notation for ACM that describes mandatory and optional tasks (DiscretionaryItem), and thereby supports flexible processes. In line with Jugel et al. [17], we utilize CMMN to describe a collaborative decision-making case for EAM, cf. Fig. 4.

The *Issue* is the starting point of a collaborative decision-making case. This issue describes the problem space of the decision-making activity, which aligns with the perspective of Mayring [13]. We further assume that goals and success criterions, as required by Johnson et al. [14], have already been defined as part of strategic management activities. The issue is the reason why the EA has to be analyzed and decided upon. Based on this issue, involved stakeholders choose architectural viewpoints [7] that they need to analyze the issue.

The *decision-making step* [21] is the central activity of the decision-making case, as presented in Fig. 4. This step can involve different optional activities in which different kinds of quantitative and qualitative analysis techniques [5, 6] are applied:

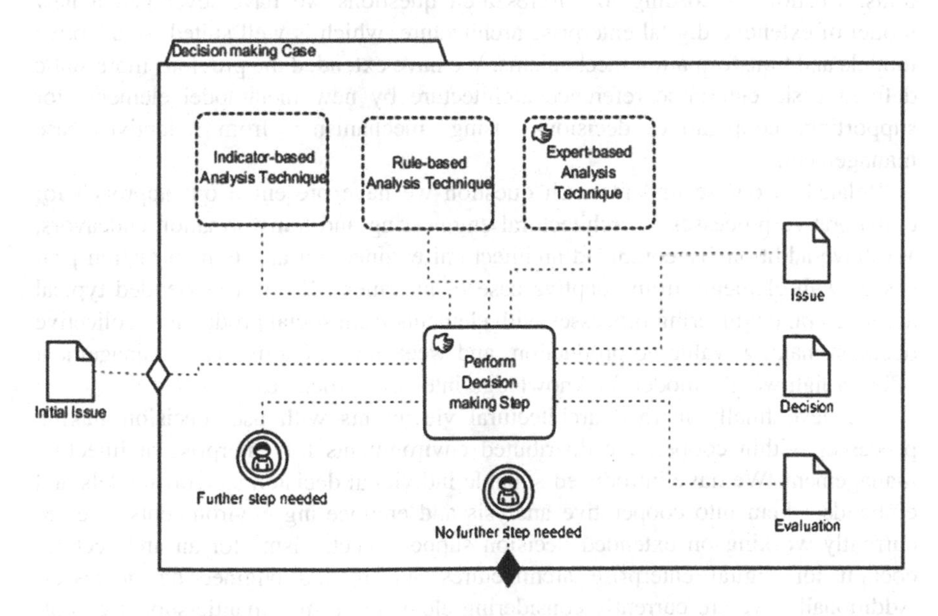

Fig. 4 CMMN model of a collaborative decision making case [17]

- *Expert-based analysis* techniques are dependent on expert knowledge and tacit information of the involved stakeholders. Jugel et al. [16] identify these techniques with interactive functions like "graphical highlighting and filtering".
- *Rule-based analysis* techniques [11] correspond to algorithms that are used to identify patterns in the EA.
- *Indicator-based analysis* techniques [12] are formal methods that compute indicators from properties of the EA.

The *decision-making step* is based on case data consisting of an EA model and additional insights elicited in previous steps. Consequently, the insights gained during each step contribute to the *case file* (CaseFile) of the decision-making case. Derived values, like the values of KPIs are thereby not considered additional information, but only a different way of representing and aggregating existing information. Stakeholder decisions represent new information, which is added to the case file.

5 Conclusion

We have identified in this paper the need for an integral understanding and support of collaborative decisions in the process of architectural adaptation and enterprise transformation. According to our research questions we have leveraged a new model of extended digital enterprise architecture, which is well suited for adaptive models and transformation mechanisms. We have extended the previous more static defined basic enterprise reference architecture by new metamodel elements for supporting cooperative decisions using mechanisms from adaptive case management.

Related to our second research question we have presented our approach for collaborative processes in architectural engineering and transformation endeavors. We have additionally combined architectural engineering and transformation processes with elements from adaptive case management. We have extended typical architectural engineering processes with elements from social production, collective decision-making, value co-production, and week ties. Adaptive case management offers a lightweight model for knowledge-intensive processes.

We have finally merged architectural viewpoints with user decision-making processes within cooperative distributed environments for enterprise architecture management. We have introduced suitable individual decision support models and embedded them into cooperative analysis and engineering environments. We are currently working on extended decision support mechanisms for an architectural cockpit for digital enterprise architectures and related engineering processes. Additionally, we are currently considering elements from semantic-supported collaborative systems.

We have contributed to the current IS literature by introducing this new perspective for decision support in the context of digital enterprise architectures with

IoT. EA managers can benefit from new knowledge about adaptable enterprise architectures and can use it for decision support and can therefore reduce operational risks. Some limitations (e.g. use and adoption in different sectors, or the IoT integration technologies) must be considered. There is a need to integrate more analytics based decisions support and context-data driven architectural decision-making. By considering the context of service-oriented enterprise architecture, we have set the foundation for integrating metamodels and related ontologies for orthogonal architecture domains of our integrated Enterprise Architecture Management approach for the Internet of Things. Our results can help practical users to understand the integration of EAM with the Internet of Things and to support architectural decisions. Limitations can be found e.g. in the field of practical multi-level evaluation of our approach as well as domain-specific adoptions.

Future work will extend both mechanisms for adaptation and flexible integration of digital enterprise architectures as well as will extend decisional processes by rationales and explanations. Future work will also include conceptual work to federate EA-IoT-Mini-Descriptions to a global EA model and enterprise architecture repository by promoting a semi-automatic and collaborative decision process. We are currently extending our architectural model federation and transformation approaches with basic research for ontology-based model transformations and elements from related work. We are researching about semantic-supported architectural representations, as universal enterprise architectural knowledge representations, which are combined with special inference mechanisms. Additional improvement opportunities will focus on methods for visualization of architecture artifacts and control information to be operable in a multi-perspective architecture management cockpit.

References

1. Aier, S., Buckl, S., Gleichauf, B., Matthes, F., Winter, R.: Towards a more integrated EA planning: linking transformation planning with evolutionary change. In: Proceedings of EMISA 2011, Hamburg, Germany, pp. 23–36 (2011)
2. Atzori, L., Iera, A., Morabito, G.: The internet of things: a survey. J. Comput. Netw. **54**, 2787–2805 (2010)
3. Bass, C., Clements, P., Kazman, R.: Software Architecture in Practice. Addison Wesley (2013)
4. Bente, S., Bombosch, U., Langade, S.: Collaborative Enterprise Architecture. Morgan Kaufmann (2012)
5. Buckl, S., Matthes, F., Schweda, C.M.: Classifying enterprise architecture analysis approaches. In: The 2nd IFIP WG5.8 Workshop on Enterprise Interoperability (IWEI'2009), Valencia, Spain, pp. 66–79 (2009)
6. Buckl, S., Gehlert, A., Matthes, F., Schulz, C., Schweda, C.M.: Modeling the supply and demand of architectural information on enterprise level. 15th IEEE International EDOC Conference 2011, Helsinki, Finland, pp. 44–51 (2011)

7. Emery, D., Hilliard, R.: Every architecture description needs a framework: expressing architecture frameworks using ISO/IEC 42010. In: IEEE/IFIP WICSA/ECSA, pp. 31–39 (2009)
8. Farwick, M., Pasquazzo, W., Breu, R., Schweda, C.M., Voges, K., Hanschke, I.: A meta-model for automated enterprise architecture model maintenance. EDOC **2012**, 1–10 (2012)
9. Ganz, F., Li, R., Barunaghi, P., Harai, H.: A resource mobility scheme for service-continuity in the Internet of Things. GreenCom **2012**, 261–264 (2012)
10. Gubbi, J., Buyya, R., Marusic, S., Palaniswami, M.: Internet of Things (IoT): a vision, architectural elements, and future directions. Future Gener. Comput. Syst. **29**(7), 1645–1660 (2013)
11. Hanschke, I.: Strategisches Management der IT-Landschaft: Ein praktischer Leitfaden für das Enterprise Architecture Management, 3rd edn. Hanser Verlag, München (2013)
12. Matthes, F.: EAM KPI Catalog v.1.0. Technical report, Technical University Munich, Germany (2011)
13. Mayring, P.: Qualitative Inhaltsanalyse, 11th edn. Beltz (2010)
14. Johnson, P., Ekstedt, M.: Enterprise Architecture—Models and Analyses for Information Systems Decision Making. Studentliteratur (2007)
15. Jugel, D., Schweda, C.M.: Interactive functions of a cockpit for enterprise architecture planning. In: International Enterprise Distributed Object Computing Conference Workshops and Demonstrations (EDOCW), Ulm, Germany, 2014, pp. 33–40. IEEE (2014)
16. Jugel, D., Schweda, C.M., Zimmermann, A.: Modeling decisions for collaborative enterprise architecture engineering. In: 10th Workshop Trends in Enterprise Architecture Research (TEAR), held on CAISE 2015, pp. 351–362. Stockholm, Sweden, Springer (2015)
17. Jugel, D., Kehrer, S., Schweda, C.M., Zimmermann, A.: A decision-making case for collaborative enterprise architecture engineering. In: Cunningham, D., Hofstedt, P., Meer, K., Schmitt, I. (eds.) Informatik 2015, Lecture Notes in Informatics (LNI) (2015)
18. Keen, P.G.W.: Decision support systems: the next decade. In: Decision Support Systems, vol. 3(3), pp. 253–265. Elsevier (1987)
19. Lankhorst, M.: Enterprise Architecture at Work: Modelling, Communication and Analysis. Springer (2013)
20. Object Management Group: Case Management Modeling Notation 1.0. OMG (2014)
21. Object Management Group: Decision Model and Notation 1.0—Beta 1. OMG (2014)
22. Open Group: TOGAF Version 9.1. Van Haren Publishing (2011)
23. Open Group: ArchiMate 2.0 Specification. Van Haren Publishing (2012)
24. Patel, P., Cassou, D.: Enabling high-level application development for the Internet of Things. submitted to J. Syst. Softw. (2015). arXiv:1501.05080
25. Porter, M.E., Heppelmann, J.E.: How smart connected products are transforming competition. Harvard Bus. Rev. 1–23 (2014)
26. Ross, J.W., Weill, P., Robertson, D.: Enterprise Architecture as Strategy—Creating a Foundation for Business Execution. Harvard Business School Press (2006)
27. Schmidt, R., Zimmermann, A., Möhring, M., Jugel, D., Bär, F., Schweda, C.M.: Social-software-based support for enterprise architecture management processes. In: Business Process Management Workshops, pp. 452–462. Springer (2014)
28. Schmidt, R., Möhring, M., Härting, R.-C., Reichstein, C., Neumaier, P., Jozinovic, P.: Industry 4.0—potentials for creating smart products: empirical research results. In: 18th Conference on Business Information Systems, Poznan 2015, Lecture Notes in Business Information Processing. Springer (2015)
29. Spiess, P., Karnouskos, S., Guinard, D., Savio, D., Baecker, O., Sá des Souza, L.M., Trifa, V.: SOA-based integration of the Internet of Things in enterprise services. In: ICWS 2009, pp. 968–975 (2009)
30. Swenson, K.D.: Mastering the Unpredictable: How Adaptive Case Management Will Revolutionize the Way that Knowledge Workers Get Things Done. Meghan-Kiffer Press (2010)

31. Tiwana, A.: Platform Ecosystems: Aligning Architecture, Governance, and Strategy. Morgan Kaufmann (2013)
32. Trojer, T., et al.: Living modeling of IT architectures: challenges and solutions. In: Software, Services, and Systems 2015, pp. 458–474 (2015)
33. Walker, M.J.: Leveraging enterprise architecture to enable business value with IoT innovations today. In: Gartner Research. http://www.gartner.com/analyst/49943 (2014)
34. Weill, P., Ross, J.W.: IT Governance: How Top Performers Manage It Decision Rights for Superior Results. Harvard Business School Press (2004)
35. Wißotzki, M., Köpp, C., Stelzer, P.: Rollenkonzepte im enterprise architecture management. In: Zimmermann, A., Rossmann, A. (eds.) DEC 15, 25–26 June 2015, Böblingen, Germany. Lecture Notes in Informatics, vol. P-244, pp. 127–138 (2015)
36. WSO2 White Paper: Reference Architecture for the Internet of Things. Version 0.8.0. http://wso2.com (2015)
37. Zimmermann, A., Buckow, H., Groß, H.-J., Nandico, F.O., Piller, G., Prott, K.: Capability diagnostics of enterprise service architectures using a dedicated software architecture reference model. In: IEEE International Conference on Services Computing (SCC 2011), Washington DC, USA, pp. 592–599 (2011)
38. Zimmermann, A., Gonen, B., Schmidt, R., El-Sheikh, E., Bagui, S., Wilde, N.: Adaptable enterprise architectures for software evolution of SmartLife ecosystems. In: Proceedings of the 18th IEEE International Enterprise Distributed Object Computing Conference Workshops (EDOCW), Ulm/Germany, pp. 316–323 (2014)
39. Zimmermann, A., Schmidt, R., Sandkuhl, K., Wißotzki, M., Jugel, D., Möhring, M.: Digital enterprise architecture—transformation for the Internet of Things. In: Kolb, J., Weber, B., Hall, S., Mayer, W., Ghose, A.K., Grossmann, G. (eds.) EDOC 2015 with SoEA4EE, 21–25 Sept 2015, Adelaide, Australia, IEEE Proceedings, pp. 130–138 (2015)

Exploiting Emoticons to Generate Emotional Dictionaries from Facebook Pages

Hanen Ameur, Salma Jamoussi and Abdelmajid Ben Hamadou

Abstract During the first events of the Tunisian revolution, the social network, Facebook, played a key role in Tunisia and everywhere in the world. It became the first political tool that allows the Tunisian people to share trending news in actual time. Facebook provides the opportunity for users to comment on the news by expressing their sentiments. In this paper, we focus on emotion analysis of Tunisian Facebook pages. To do this, we first collect comments from the Facebook pages in order to analyze sentiments written in Tunisian dialect. Then, we propose a new method for emotional dictionaries construction. In fact, we distinguish nine emotional classes: surprised, satisfied, happy, gleeful, romantic, disappointed, sad, angry and disgusted. At this step, we focus on the use of emotion symbols as indicators of sentiment polarity. Finally, we present the experimental results of our method. Our system achieves effective and consistent results.

Keywords Sentiment analysis · Emotion analysis · Emotional dictionaries · Tunisian dialect · Emotion symbols · Political lexicon

1 Introduction

Since the beginning of political upheavals and the triggering of the Tunisian revolution, social networks, especially Facebook, play a leading role in the political life in Tunisia. In fact, Facebook allows exchanging news in actual time, everywhere in the

H. Ameur (✉) · S. Jamoussi · A. Ben Hamadou
Multimedia InfoRmation Systems and Advanced Computing Laboratory,
MIRACL-Sfax University, Sfax-Tunisia Technopole of Sfax,
Av.Tunis Km 10 B.P. 242, 3021 Sfax, Tunisia
e-mail: ameurhanen@gmail.com

S. Jamoussi
e-mail: salma.jamoussi@isimsf.rnu.tn

A. Ben Hamadou
e-mail: abdelmajid.benhamadou@isimsf.rnu.tn

I. Czarnowski et al. (eds.), *Intelligent Decision Technologies 2016*,
Smart Innovation, Systems and Technologies 57,
DOI 10.1007/978-3-319-39627-9_4

world. Today, we are witnessing the appearance of many popular Facebook pages that have political aspect. These pages allow the users to ask important questions about news and express their opinions and sentiments freely. Hence, the large quantity of exchanged political texts encouraged us to use sentiment analysis techniques to collect and treat the users' sentiments about the discussed political subjects. It seems pressing to develop tools for analyzing and investigating the conflicts of ideas and the variability of sentiments of Tunisian people. In addition, it is very interesting to know the lexicon of words used by commentators in some Facebook pages. For this reason, we propose, in this paper, to use Facebook comments as a source of textual data for the construction of emotional dictionaries (lexicon). Therefore, we first thought of collecting automatically Facebook comments to realize a fine sentiment classification task "emotion analysis".

The main difficulty of handling corpus collected from Facebook pages, is the wide variety of comments. Indeed, the comments present a great diversity, whatsoever, on the level of the writing style or their size. Furthermore, Facebook offers to users, another writing style which is closer to SMS language than to the language used by journalists and professionals. This new style is characterized by the presence of emoticons, elongated specific words and by a simplified syntax with misspellings and unpronounced characters. In addition, in the Tunisian pages, users write their comments using a mixture of three languages (French, standard Arabic and Tunisian dialect) and they are able to easily switch between them. Tunisian dialect is generally written by using Latin alphabets and numeric digits. All this diversity presents many challenges in sentiment analysis when it deals with Tunisian comments. The sentiment dictionaries "lexicon" creation is a very important and difficult task to achieve. In fact, most of the sentiments analysis methods are based on sentiment lexicons to classify subjective comments. However, from our best knowledge, there is no existing sentiment lexicon for the Tunisian language. Furthermore, we cant use English external resources (like WordNet affect[1]) because it requires a translation step which can affect the actual meaning of words. In this paper, we propose a new automatic method based on the emotion symbols for the construction of emotional dictionaries. Due to the richness of sentiments in the emotion symbols, we consider nine emotional classes (*surprised*, *satisfied*, *happy*, *gleeful*, *romantic*, *disappointed*, *sad*, *angry* and *disgusted*) based on expert judgments.

This paper is organized as follows. We review the related work in the next section. We then present our corpus acquired from the Tunisian Facebook pages and the step of its preprocessing in the Sect. 3. Next, in Sect. 4 we elaborate on the principle of our proposed method for automatic construction of emotional dictionaries in Tunisian dialect. Finally, we report the experimental results and conclude the paper with future works.

[1]http://wndomains.fbk.eu/wnaffect.html.

2 Related Work

There are many research studies based on using social networks to analyze sentiments and on the construction of sentiment dictionaries lexicon of words. To our best knowledge, there hasn't been any study conducted on Tunisian words and how to classify them according to the sentiments which they express in social networks, especially Facebook. In the literature, many works try to address the problem of sentiment dictionaries construction by extracting at least two vocabularies groups. One expresses the positive sentiments and the other expresses the negative sentiments. For this purpose, there are four techniques: the manual method, the dictionaries-based method, the corpora-based method and the method combining the last two ones.

Some sentiment lexicon has been constructed manually by experts [15]. As these lexicons often contain thousands of words, their manual creation is therefore very difficult, expensive and time-intensive. Other researchers (e.g. [9]) have proposed to construct the sentiment lexicon based on external linguistic resources that handle semantic relations (synonymy and antonymy), such as WordNet, SentiWordNet and ANEW. The idea is to classify words based on other words whose semantic orientation is known (called seed), by applying bootstrapping algorithms [14]. Kamps and Marx [9] have proposed a semantic distance to measure the shortest distance between the examined word and the seed words whose their valences are known. To deal with the lack of information about semantic relations between words, other researchers have proposed based on the information present in a corpus (annotated or not). Douglas and Christopher [6] have concentrated on the coordinating conjunctions present between the words such as: and, but, either-or, or, etc. In other words, if for example, two words are separated by the conjunction "and", they necessarily have a similar polarity. They consist of counting the number of times which the examined word appears beside the words already classified "seeds". Kim and Hovy [10] and Ameur and Jamoussi [3] have proposed a hybrid method combining techniques based on dictionaries and those based on corpus to construct a sentiment lexicon. Kim and Hovy [10] constructed a lexicon containing a large number of words (verbs and adjectives) carrying sentiments from three defined sets of words. Thereafter, they merged these three sets using an averaging method. Ameur and Jamoussi [3] used emoticons to differentiate between the positive sentiments (indicated by positive emoticons) and the negative sentiments (indicated by negative emoticons).

Instead of classifying the text into three classes only (i.e. positive, negative and neutral), other studies extended the sentiment analysis methods in order to treat and analyze emotions [12]. Mihalcea and Liu [11] have classified blog posts into two particular emotion classes (happiness and unhappiness). These blog posts are self-annotated by the blog writers with happy and sad mood labels.

Some researchers have analyzed the emotions of text focused on the six basic emotions identified by Ekman [8] (positive, negative, fear, joy, surprise, hate, disgust). Alena et al. [2] has used a rule-based method for determining Ekman's basic emotions in the sentences in blog posts. Balabantaray et al. [4] took into account all these basic emotions, with the addition of the neutral class in order to analyze the subjectivity of the text. Duyu et al. [7] have proposed a method which aims to classify emotion bearer in tweets as (happy, sad, angry or surprise) using pseudo-labelled data with emoticons. Solakidis et al. [13] applied the emotion classification task on multilingual data, focusing on documents written in Greek. They identified the polarity of the text (neutral, negative or positive); and the emotion expressed through the positive sentiments (joy and love) and the negative sentiments (anger and sadness).

Recently, there has been some works on Arabic sentiment analysis mainly concerning about the construction sentiment lexicons (e.g. [1]). Abdul-Mageed and Diab [1] presented SANA, a subjectivity and sentiment lexicon for Arabic. The lexicon combines pre-existing lexicons and involves automatic machine translation, manual annotations and gloss matching across several resources such as THARWA [5].

In this paper, we propose a new method allowed to distinguish nine emotional classes (emotional states), using Tunisian Facebook data. This method is based on the presence of emoticons in the corpus and without using external linguistic resources.

3 Tunisian Corpus Collection

In order to construct our Facebook corpus using the Tunisian dialect, we employed the APIs provided by Facebook.[2] We extracted the textual information from very active political Tunisian pages in the period [1-Jan-2010, 31-Dec-2013]. We used 13 political pages among the most popular in Tunisia. From these Facebook pages, we obtained 60,000 political comments and about 780 K words.

We present our collected corpus as a set of multilingual comments organized in a well-structured XML file to facilitate their handling. In The Tunisian Facebook pages, most users comment using free language as colloquial dialect. Furthermore, the Tunisian dialect is characterized by the presence of a mixture of languages such as French, Standard Arabic, Tunisian dialect which is an Arabic text written in Latin characters and numbers, etc.

Before performing processing on Facebook comments, we need a pretreatment and a shaping step to homogenize them. This step of corpus pre-processing aims to select relevant and the most significant words. Thereby, it allows us to facilitate the construction of our emotional dictionaries. In this step, we performed *character normalization* by replacing specific unpronounced characters with a space and removing accents, stars "*" and others. In order to avoid the presence of segments of words evoking no interest (such as hyperlinks and @target_user), we performed a *filtering* step. This step keeps only the words that reflect the semantic and sentimental content of the comments.

[2] https://developers.facebook.com/docs/reference/apis/.

In this paper, we proposed to construct a lexicon dictionary for each used language on the Tunisian Facebook. In order to distinguish between these languages, we use the language identification tool proposed by Cybozy Labs; hosted on Google Developers.[3] This tool allows us to identify the French and Standard Arabic comments. When the identified language is different from these two languages, we consider that the comment is written using the Tunisian dialect. Concerning the emotion symbols present in the dictionaries, we considered them as multilingual symbols. Therefore, they can also be added in the Tunisian dictionaries. Then, we performed a *lemmatization* step to encompass the words having the same primary entity "lemma". This step was applied only on the French text. However, it wasn't possible to apply lemmatization on the Arabic text because it was full of grammatical errors. For Tunisian dialect, the morphological analyser has not been available yet. To further remove all words deemed unnecessary and keep the interesting words, we eliminated the stop words. To do this, we prepared our own stop-list file containing the grammatical words and the linking words of the three languages (French, standard Arabic, Tunisian dialect). Then, we proceeded to apply a normalization technique for all lengthened words and emoticons, such as "::))))" which is replaced by ":)".

4 Emotional Dictionaries Construction

The goal of the automatic construction of sentiment dictionaries is to list the lexicon words in dictionaries by distinguishing between positive and negative.

Actually, the human sentiments are not limited to positive and negative expressions, but they contain several emotional states. In our case, we consider that there are ***four*** states of positive sentiments: satisfied, happy, gleeful, romantic; ***four*** other states of negative sentiments: disappointed, sad, angry and disgusted, and ***one*** surprised state (non-neutral and not positive or negative). In this paper, we generate a dictionary for each emotional state. In other words, we propose to train ***9 dictionaries***. However, the determination of these dictionaries is a very delicate and complicated task. Moreover, the distinction between emotional states is not well understood with used textual data.

Facebook users use, in their comments, emotion symbols (emoticons ":), :(, etc.", acronyms "lol, mdr, etc." and exclamation words "pf, hh, etc.") to emphasize on their sentiments. In fact, these emotion symbols provide an important contextual value to determine the general sentiment of the text. We take advantage of the integrity of these symbols to distinguish between 9 emotional states that we concluded from the used emoticons (see Fig. 1).

In order to create dictionaries pertaining to these emotional states, we elaborate two stages: the first one is the initial dictionaries construction and the second one is the enrichment of these dictionaries.

[3] https://code.google.com/p/language-detection/.

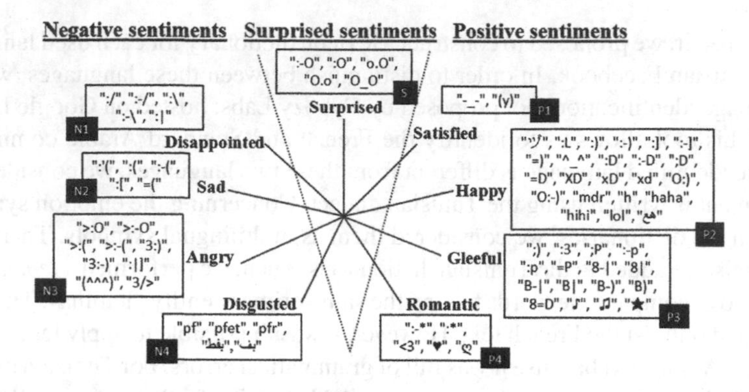

Fig. 1 The emotion symbols used on Facebook to express the nine emotions

4.1 Initial Construction of Emotional Dictionaries

We proposed a method to extract an initial version of the emotional dictionaries by using the comments that contain emotion symbols (4132 comments). This method is based essentially on the presence of emotion symbols in the comments. These emotion symbols reflect the sentiment expressed by the words that precede them in the comments. Indeed, a word has the same polarity as the first emotion symbol appears after it in a comment. In the case that the comment contains one emotion symbol, all words will be attached to this emotion symbol.

Each lexicon word can be attached to several dictionaries according to their appearance with the emotion symbols. Hence, we assigned to each word a valence value. This valence consists to divide the frequency *j* of the word by the sum of the frequencies of all words present in the dictionary *j*. The word valences are calculated using (1):

$$valence(w)_j = \frac{frequency(w)_j}{\sum_{i=0}^{n} frequency(w_i)_j} \times 1000 \tag{1}$$

where: *n* is the number of words in the dictionary *j*. $j \in$ (surprised, satisfied, happy, gleeful, romantic, disappointed, sad, angry, disgusted). $frequency(w)_j$ denotes the cooccurrence of the word *w* in comments with emoticons expressing the sentiment *j*.

The negation words (e.g. not, no, never, non, ne, etc.) play a specific and important role in the sentimental orientation of words. For this reason, we have taken advantage the presence of negation particles in our comments, in order to reverse the polarity of all words directly preceded by one of these particles (e.g. Im not happy :)). Thereby, when we calculate the cooccurrence frequency of word and emotion symbols, we test if the word is preceded by a negation particle, we decrement its cooccurrence frequency by 1. Otherwise, we increment it by 1.

To determine the emotion brought by every lexicon word, we simply compare its valences in each dictionary. After the initial dictionaries construction step, we obtained *15,576 words* in the sentiment dictionaries. It is clear that in the initial construction step, we kept all the words, even the weakly appeared ones. In fact, these words are likely to be encountered in other comments. For this reason, we proceed to an enrichment step, the objective of the next section.

4.2 Emotional Dictionaries Enrichment

The dictionaries enrichment step aims to use the rest of the comments that do not contain emotion symbols (*50,917 comments*), for two objectives: (i) Settle and adjust the valences of words present in the dictionaries and (ii) Extend the initial dictionaries by adding new words appeared in other comments.

In order to achieve these two objectives, we must firstly determine the dictionary to enrich from a comment. To do this, we calculate the 9 valences of each comment C using the Eq. 2:

$$valence(C)_j = \frac{\sum_{i=0}^{p} frequency(m_i)_j}{\sum_{l=0}^{q} frequency(m_l)_j} \times 1000 \quad (2)$$

where: p is the number of known comment's words. q is the number of words in the dictionary j. $j \in$ (surprised, satisfied, happy, gleeful, romantic, disappointed, sad, angry, disgusted). However, sometimes we cannot calculate these valences, when the words of the comment are all unknown (do not exist in our dictionaries).

Then, we compare the values of these valences to identify the comment polarity. Thus, we enrich the dictionary having the same polarity of the comment by all words in the treated comment.

The principle of the enrichment step is to browse the words of the comment. If the word is inserted in the dictionary to enrich, we increase its frequency in this dictionary by the percentage of polarity of the comment, and then recalculate its valences according to formula (1). Whereas, if the word is new, we add it to the studied dictionary by initializing its frequency by the percentage of polarity of the comment and we then compute its valence (using (1)). As the negation particles occur in *5910 comments* that did not have any emotion symbols, we handle the presence of negation words. In fact, in the case where the treated word is preceded by a negation word, we use the inverse of the percentage ($-percentage(C)_j$). The percentage of polarity of the comment is calculated using (3).

$$percentage(C)_j = \frac{valence(C)_j}{\sum_{k=0}^{8} valence(C)_k} \quad (3)$$

where, $j \in$ (surprised, satisfied, happy, gleeful, romantic, disappointed, sad, angry, disgusted).

At this stage, we obtained nine dictionaries that cover the majority of the words of our corpus. In fact, the number of words in the enriched dictionaries equals to *131,937*. This shows that the enrichment step is allowed to widen the initial dictionaries.

5 Results and Interpretations

By applying our method of emotional dictionaries construction on our corpus, we have obtained encouraging results. Figure 2 shows for our studied corpus, the number of words attached to emotional dictionaries in each language. We notice that the number of Arabic words expressing disappointment and disgust is very important compared with the other languages. In fact, Tunisians use long Arabic texts in order to emphasize on their sentiments when they are disappointed or when they want to oppose something. However, they prefer to write short texts in Tunisian dialect for expressing fun and amusement.

In order to evaluate our method of emotional dictionaries construction, we used a test corpus containing *755 words* manually labelled by **three experts**. We applied the external evaluation techniques (recall, Accuracy and F-score) to measure the adequacy of the classification of words by our system and that made by the experts. In the Table 1, we presented the obtained results without taking into account the negation and with the negation handling at the level of generation of initial and enriched dictionaries. From these results, we obviously notice the interest of handling negation particles included in our comments and the usefulness of enrichment step. In fact, the best results are obtained with the enrichment method by taking account of the negation. The negation handling has an amplified effect when considered during the enrichment step. This allowed achieving well-adjusted valences and polarities of words. We obtain a F-score of **81.01** % compared to reference dictionaries "textbf-Expert 2".

To analyze the performance of emotional dictionaries generated by our method, we propose to test the validity of the top-*n* words classified in each emotion. Indeed, we take the *n* words having the highest valance values in each emotional class and we annotate them manually depending on expert judgments. Thus, we compare them to the reference words in order to measure the success rate (see (4)) and the error rate (see (5)). In our experimentation, we try several *n* values [$n = 10, 20$ *and* 30] (see Fig. 3).

Dictionaries	Surprised	Satisfied	Happy	Gleeful	Romantic	Disappointed	Sad	Angry	Disgusted
Tunisian	12403	356	7417	14866	2793	2129	12818	23665	229
Arabic	200	29518	7041	386	19456	28058	708	102	29986
French	2749	50	736	1013	355	180	884	1371	38

Fig. 2 The number of words present in sentiment dictionaries in 3 languages: Tunisian dialect, Standard Arabic and French

Table 1 The precision, recall and F-score obtained with the method for initial construction and enrichment dictionaries

Expert	External evaluation measures	Sentiment dictionaries			
		Without negation		With negation	
		Initial (%)	Enriched (%)	Initial (%)	Enriched (%)
Expert 1	Precision	47.96	55.67	49.42	79.72
	Recall	21.71	60.72	21.80	78.89
	F-score	20.41	56.13	20.53	**80.36**
Expert 2	Precision	48.11	57.68	49.56	80.65
	Recall	21.76	62.48	21.85	79.45
	F-score	20.62	57.97	20.74	**81.01**
Expert 3	Precision	48.01	57.07	49.70	79.18
	Recall	21.83	62.92	21.92	78.10
	F-score	20.39	57.30	20.52	79.65

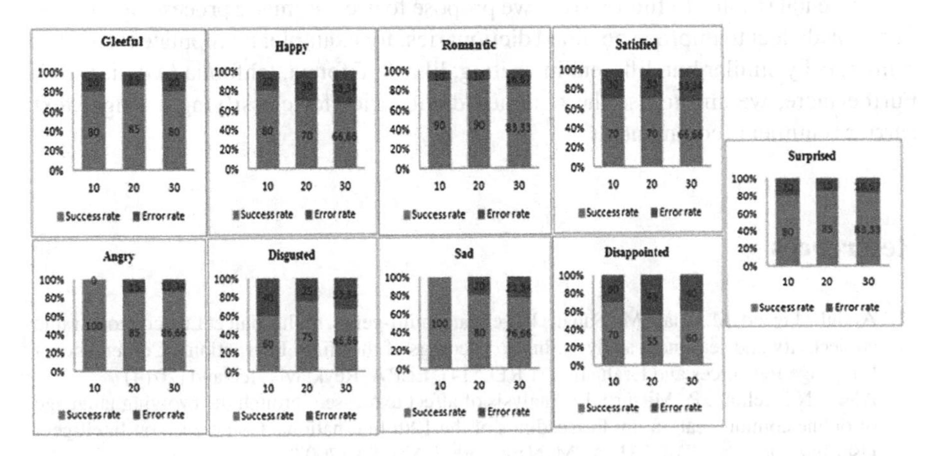

Fig. 3 The success rate and the error rate obtained by our classification method of the top-n words in the 9 emotional dictionaries (n = 10, 20 and 30)

$$Success\_rate = \frac{nbre_{Correct}}{n} \times 100 \tag{4}$$

where: $nbre_{Correct}$ is the number of words correctly classified. n is the chosen number for the test (the total number of words).

$$Error\_rate = 100 - Success\_rate \tag{5}$$

The goal of this evaluation is to verify the correctness of our method of calculating the valences of words to assign them to dictionaries. We note from Fig. 3 that, for every emotional class and whatsoever the value of n, the success rate remains high. This shows that the n words which have the highest valance values correctly express the emotion of the attached dictionary.

6 Conclusion

In this paper, we presented a new method to construct dynamically emotional dictionaries. Our method is essentially based on the emotion symbols which can be used to express sentiments when commenting on social networks. Our corpus of comments was collected from the Tunisian Facebook pages. Thus, we proposed to create initial dictionaries from the comments having emotion symbols. Moreover, using the rest of the comments, we proposed to enrich these dictionaries. Finally, we discussed the experimental results. In future work, we propose to use automatic processing tools of Tunisian dialect to improve obtained dictionaries, for example: manipulate the words semantically similar but different in writing, like: "ta7founa, tahfouna (wonderful)". Furthermore, we aim to use the obtained dictionaries for classifying a longer text carrier sentiment (comment).

References

1. Abdul-Mageed, M., Diab, M.: Sana: a large scale multi-genre, multi-dialect Lexicon for Arabic subjectivity and sentiment analysis. In: Proceedings of the Ninth International Conference on Language Resources and Evaluation (LREC'14). ELRA, Reykjavik, Iceland (2014)
2. Alena, N., Helmut, P., Mitsuru, I.: Analysis of affect expressed through the evolving language of online communication. In: Proceedings of the 12th International Conference on Intelligent User Interfaces, pp. 278–281. ACM, New York, NY, USA (2007)
3. Ameur, H., Jamoussi, S.: Dynamic construction of dictionaries for sentiment classification. In: 13th IEEE International Conference on Data Mining Workshops. ICDM Workshops, pp. 896–903. TX, USA (2013)
4. Balabantaray, R.C., Mohammad, M., Sharma, N.: Article: Multi-class twitter emotion classification: a new approach. Int. J. Appl. Inf. Syst. **4**(1), 48–53 (2012)
5. Diab, M., Albadrashiny, M., Aminian, M., Attia, M., Elfardy, H., Habash, N., Hawwari, A., Salloum, W., Dasigi, P., Eskander, R.: Tharwa: A large scale dialectal Arabic—standard Arabic—English Lexicon. In: Proceedings of the Ninth International Conference on Language Resources and Evaluation (LREC'14). ELRA, Reykjavik, Iceland (2014)
6. Douglas, R.R., Christopher, Z.: Corpus-based dictionaries for sentiment analysis of specialized vocabularies. In: New Directions in Analyzing Text as DataWorkshop (2013)
7. Duyu, T., Bing, Q., Ting, L., Zhenghua, L.: Learning sentence representation for emotion classification on microblogs. In: Natural Language Processing and Chinese Computing—Second CCF Conference, pp. 212–223. Chongqing, China (2013)
8. Ekman, P.: An argument for basic emotions. Cogn. Emot. **6**, 169–200 (1992)
9. Kamps, J., Marx, M.: Words with attitude. In: 1st International WordNet Conference, pp. 332–341. Mysore, India (2002)

10. Kim, S.M., Hovy, E.: Determining the sentiment of opinions. In: Proceedings of the 20th International Conference on Computational Linguistics. ACL, Stroudsburg, PA, USA (2004)
11. Mihalcea, R., Liu, H.: A corpus-based approach to finding happiness. In: Proceedings of the AAAI Spring Symposium on Computational Approaches to Weblogs (2006)
12. Mohammad, S.M.: Sentiment analysis: detecting valence, emotions, and other affectual states from text. In: Meiselman, H. (ed.) Emotion Measurement. Elsevier (2016)
13. Solakidis, G., Vavliakis, K., Mitkas, P.: Multilingual sentiment analysis using emoticons and keywords. In: 2014 IEEE/WIC/ACM International Joint Conferences on Web Intelligence (WI) and Intelligent Agent Technologies (IAT), pp. 102–109. Warsaw, Poland (2014)
14. Taboada, M., Anthony, C., Voll, K.: Methods for creating semantic orientation dictionaries. In: Conference on Language Resources and Evaluation (LREC), pp. 427–432 (2006)
15. Wilson, T., Wiebe, J., Hoffmann, P.: Recognizing contextual polarity in phrase-level sentiment analysis. In: Proceedings of the Conference on Human Language Technology and Empirical Methods in Natural Language Processing, pp. 347–354. ACL, Stroudsburg, PA, USA (2005)

Multiple Ontology-Based Indexing of Multimedia Documents on the World Wide Web

Mohammed Maree, Mohammed Belkhatir, Fariza Fauzi, Aseel B. Kmail, Ahmad Ewais and Muath Sabha

Abstract In order to cope with the growing need to search multimedia documents with precision on the Web, we propose a multimedia conceptual indexing framework incorporating semantic relations between annotation words. To do this, we utilize our DOM Tree-based Webpage segmentation algorithm to automatically extract surrounding textual information of the multimedia documents in Webpages. Next, we employ knowledge represented in multiple ontologies to discover the latent semantic dimensions of the surrounding textual information. As a consequence, indexes (represented as semantic networks) are constructed where nodes of each network capture words that exist in the ontologies and edges represent the semantic relations that hold between those words. To address the semantic heterogeneity problem between the produced networks, we employ a multi-level merging algorithm that combines heterogeneous networks into a more coherent network. Additionally, we utilize concept-relatedness measures to address the issue of unrecognized entities by the ontologies. We evaluate the techniques of the

M. Maree (✉) · M. Sabha
Faculty of Engineering and Information Technology, Multimedia Technology Department, The Arab American University, Jenin, Palestine
e-mail: mohammad.maree@aauj.edu

M. Sabha
e-mail: muath.sabha@aauj.edu

M. Belkhatir
University of Lyon, Campus de La Doua, France
e-mail: mohammed.belkhatir@univ-lyon1.fr

F. Fauzi
Department of Engineering Science, University of Auckland, Auckland, New Zealand
e-mail: w.paizi@auckland.ac.nz

A.B. Kmail · A. Ewais
Faculty of Engineering and Information Technology, Computer Science Department, The Arab American University, Jenin, Palestine
e-mail: aseel.kmail@aauj.edu

A. Ewais
e-mail: ahmad.ewais@aauj.edu

I. Czarnowski et al. (eds.), *Intelligent Decision Technologies 2016*,
Smart Innovation, Systems and Technologies 57,
DOI 10.1007/978-3-319-39627-9_5

proposed framework using three different multimedia dataset types. Experimental results indicate that the proposed techniques are effective and precise.

Keywords Multimedia indexing • Webpage segmentation • Ontology

1 Introduction

With the prevalence of digital imaging, video and audio devices and the increasing usage of the Internet, more and more multimedia documents are now available on the Web. The development of precision-based semantic indexing and search systems over these multimedia documents remains a pressing issue. Among the existing solutions, one approach is to index multimedia documents based on their low-level features such as color, texture, and shape. For querying, the user has to submit a query document to describe his information needs. In this case, documents that are similar with respect to their low-level features are retrieved without considering the semantic similarity of their visual content. In addition, since a large collection of multimedia documents is available on the Web, automatically extracting low-level features from the content of these documents demands high computational costs. Another approach is to index multimedia documents based on their surrounding textual information, a.k.a., surrounding contextual information such as titles and textual content of Webpages. In this approach, keywords are automatically extracted from the textual content that surrounds multimedia documents, and the retrieval process is carried out by matching keywords in the user's query to their correspondences from the contextual information. This approach has proved to better facilitate user interaction and overcome the limitations of Content-based Multimedia Retrieval models [1]. However, it ignores the latent semantic dimensions of both the user's query and the surrounding contextual information of multimedia documents. To address this issue, ontologies (machine readable resources that capture knowledge about different domains) have been exploited for indexing purposes. Examples of such ontologies are WordNet, OpenCyc and YAGO. Previous research works have utilized these ontologies individually. Hence, the extracted concepts and their corresponding semantic relations from the surrounding contextual information are limited to those that are defined in the used ontology. Clearly, the precision of multimedia indexing systems is highly dependent on the domain coverage of the used ontology. Starting from this position, we propose using multiple ontologies to ensure having wider domain coverage and thus providing a richer source of semantic information. Based on each of the exploited ontologies, we derive concepts and semantic relations from the surrounding contextual information of the multimedia documents and organize them into semantic networks. Subsequently, all resulting semantic networks are merged using ontology merging techniques into a single coherent network. As such, the produced merged networks represent the initial semantic indexes for multimedia documents. Albeit the potential increment in domain coverage, we acknowledge that it is possible to still be faced with missing concepts or semantic relations, even when using multiple ontologies. For this reason, we utilize statistical techniques based on the Normalized Retrieval Distance (NRD) function to tackle this issue. Accordingly, we summarize the contributions of our work as follows:

- Employing knowledge represented in multiple ontologies to derive the latent semantic dimensions of the multimedia documents on the WWW.
- Constricting semantic-based indexes using ontology merging techniques.

The rest of this paper is organized as follows: Sect. 2 provides an overview of the related work. The overall architecture of the proposed conceptual indexing framework is presented in Sect. 3. Section 4 shows how the semantic indexes can be constructed and further enriched by combining semantic and statistical modules of the proposed framework. The results of our experiments are presented in Sect. 5. Finally, we discuss the conclusions and the future work in Sect. 6.

2 Related Work

Several systems have been proposed to index and retrieve multimedia documents on the WWW. In the systems proposed by the authors of [2, 3], the retrieval process is based on the combination of low-level features extracted from the content of multimedia documents and keywords extracted from the surrounding textual content. However, both modules are integrated within a loosely-coupled indexing and retrieval architecture, i.e. there is no clear mapping mechanism between the extracted features and the surrounding textual information of the multimedia documents. Other systems such as [4, 5] apply Relevance Feedback (RF) techniques to improve the retrieval accuracy. However, RF engages the user in a tedious task involving several interaction loops. More recent approaches propose to exploit ontologies to improve the retrieval performance by taking into account the semantic dimension of the document contextual information. The authors of [6] propose a prototype system for retrieving images by combining both Content-Based Image Retrieval (CBIR) techniques and WordNet ontology. In this system, both query formulation and ranking of the retrieved results are performed according to the conceptual hierarchy of WordNet. The system proposed in [7] uses a manually constructed ontology as the core component for semantic image annotation and retrieval. The used ontological model provides the terminology and concepts for characterizing the image metadata. The system exploits the semantic relations defined between the concepts of the ontology to recommend other results that may be of interest to the user. Wang et al. [8] exploit a multi-modality ontology for image retrieval on the Web. The process of building the multi-modality ontology consists of two main phases. In the first phase, a high-level textual ontology is automatically constructed based on the Wikipedia encyclopedia. In the second phase, visual words or concepts are built based on the low-level features of the images. Clearly, these approaches depend solely on the domain coverage of the used ontology. This means that the effectiveness of such approaches is largely based on the number of recognized entities (concept, and relations) by the used ontologies. As such, they will fail in discovering the latent semantic dimensions of the images' contextual information because the domain coverage of their backbone components (ontologies) is limited.

3 General Overview of the Proposed System's Architecture

As depicted in Fig. 1, we propose a conceptual indexing framework to index multimedia documents on the Web by employing knowledge represented by multiple ontologies. In this section, we present a high-level overview of the

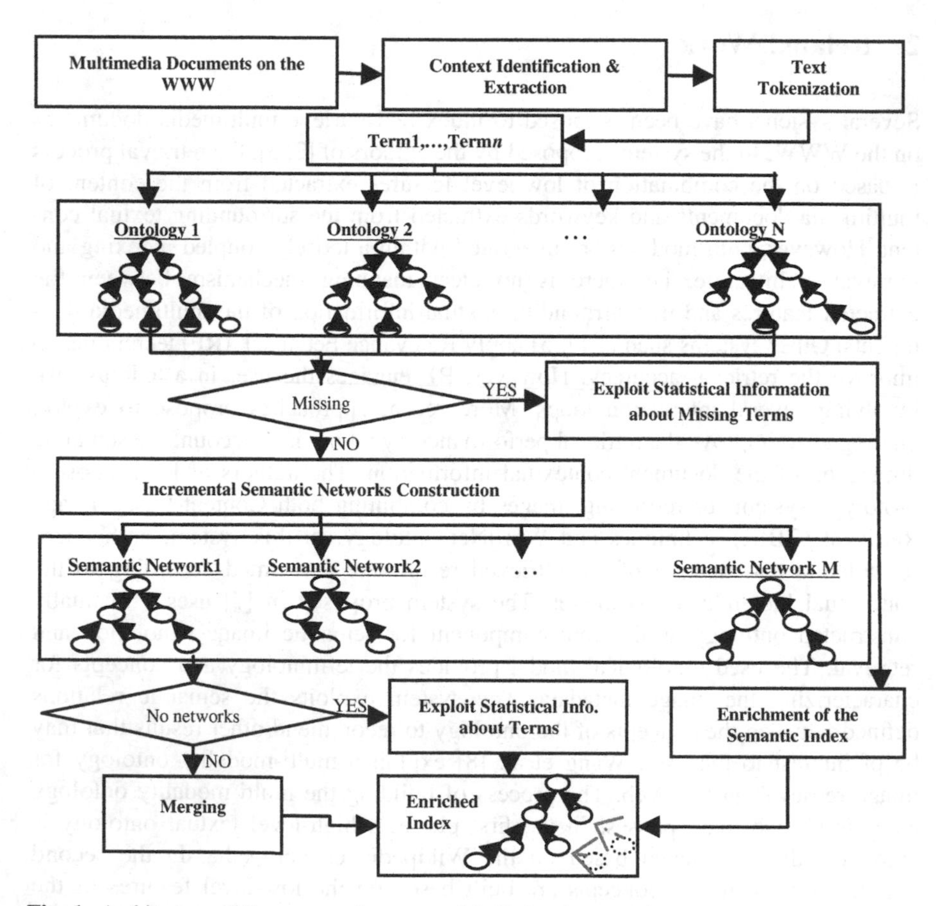

Fig. 1 Architecture of the proposed conceptual indexing framework

architecture of the proposed framework. The framework comprises the following modules:

- Context identification and extraction—followed by a Natural Language Pre-processing (NLP) module.
- Incremental semantic networks construction.
- Merging the initially proposed semantic indexes.
- Unrecognized entities handler.

As shown in Fig. 1, we start by identifying and extracting surrounding contextual information of the multimedia documents on the Web. This step is carried out through utilizing our previously proposed Webpage segmentation algorithm [9]. Next, we refine the extracted information by employing several NLP steps such as stop words removal, n-gram text tokenization, and part-of-speech (POS) tagging. After this step, and based on the processed content, semantic networks are incrementally constructed based on the used ontologies. These networks represent the initially suggested semantic indexes for multimedia documents. To build these networks, each n-gram token is submitted to each of the ontologies to find whether it is defined in it or not. Recognized tokens by the ontologies are represented by nodes in the semantic networks. An edge is inserted between two nodes if there is a semantic relation between them according to their definition in the ontologies. In this context, each ontology may produce zero, one, or more networks. Since we are using multiple ontologies, there may be conceptual and terminological difference between the produced networks. Therefore, we utilize the merging techniques proposed in our previous work [10] to combine them into a single coherent network. The merged semantic networks represent a semantic index that is built upon a cooperative decision made by the exploited ontologies. On the other hand, the unrecognized entities handler is employed to tackle cases of missing entities in all of the exploited ontologies. To do this, we utilize the NRD function [11] to measure the semantic relatedness between the missing entities and other entities that are represented by the nodes of the merged semantic network. Based on the returned result, we enrich the merged semantic index by attaching missing concepts that have strong semantic relatedness measures. In the next section, we provide detailed description of the methods and techniques that we employ in our proposed framework.

4 Our Approach to Conceptual Multimedia Indexing

To automatically extract multimedia documents and their relevant surrounding contextual information from the Web, we use the Document Object Model (DOM) Tree-based Webpage segmentation algorithm proposed in [9]. As depicted in Fig. 2, the segmentation process is based on the DOM Tree structure of

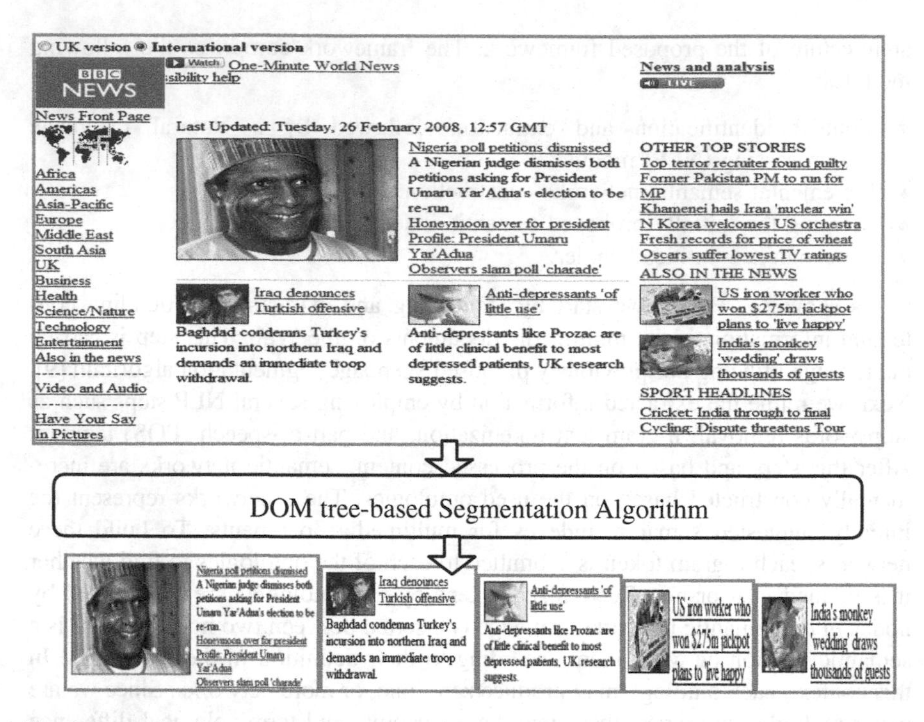

Fig. 2 Example of the output of the segmentation algorithm

Webpages. Using this algorithm, multimedia documents on the Web are classified into three categories: Listed, Semi-listed, and Unlisted documents.

Regarding the traversal of the DOM trees, each tree structure is examined to discover different DOM Tree patterns for each class of Web multimedia document. In this context, Unlisted documents are standalone or random documents that appear anywhere on a page, for example, profile photos in personal homepages, company logos, advertisements etc. The corresponding DOM Tree for such multimedia documents and their surrounding context is consistently a node with its surrounding text as text node siblings, with a root HTML tag representing the boundary of this document segment. Listed documents are two or more multimedia documents that are systematically ordered within the webpage. Examples are list of product images, news videos, etc. The associated DOM Trees for such document segments are characteristically the document node with its surrounding text nodes that are a sub-tree under a root HTML tag defining the segment boundary. Other siblings under this root HTML tag share similar sub-tree structure. Semi-listed multimedia documents are visually similar to listed documents. The difference is characterized by their DOM tree. Their DOM tree is similar to a DOM Tree of an unlisted documents in the sense the node with its surrounding text nodes are under a root HTML tag that represents the segment boundary but along with those nodes, there are other document nodes with their own surrounding texts nodes as well on the same level. For every multimedia document extracted, the segmentation method

Table 1 Context identification and extraction—sample segment

Segment	Multimedia document
Java (Indonesian: Jawa) is an island of Indonesia and the site of its capital city, Jakarta. Once the centre of powerful Hindu-Buddhist kingdoms, Islamic sultanates, and the core of the colonial Dutch East Indies, Java now plays a dominant role in the economic and political life of Indonesia	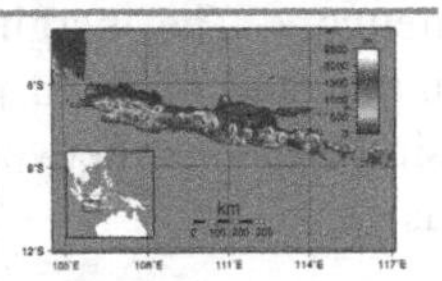

only searches the surrounding region making it more efficient and scalable for large websites containing huge amount of multimedia documents. As we pointed out in Sect. 3, the produced results of the context identification and extraction module are further refined using several NLP steps such as: stop words removal, text tokenization, and part-of-speech tagging. The following example illustrates the details of these NLP steps when applied on an extracted segment from Wikipedia Webpage about the concept (java).

To process the segment in Table 1, the stop words removal function first removes stop words based on a pre-defined list that includes stop words such as: a, the, an,…etc. Then, the text tokenization algorithm tokenizes the input text into n-gram tokens of lengths from 1 to 4. Examples of such tokens are: *Java*, *Java Indonesian*, *Java Indonesian Jawa*, and *Java Indonesian Jawa island*. Clearly, we can find that not all of the produced tokens have a significant contribution to the meaning of the extracted segment. To address this issue, we refer to the used ontologies (in this example, we use WordNet [12] and YAGO [13]) wherein the algorithm attempts to produce only meaningful n-gram tokens (also referred to as the set of *correct tokens*). For instance, the token "Dutch East Indies" is declared as a meaningful token when submitted to WordNet. Next, by utilizing the POS tagger, meaningful tokens are then classified into the grammatical category that they belong to. Following to the NLP-based processing, the semantic networks construction algorithm takes the set of correct tokens as input and starts producing semantic networks by automatically identifying the semantic relations that may hold between the tokens based on their definitions in the used ontologies. Such networks are then suggested as the initial indexes for the multimedia documents. In the context of our work, we refer to such networks as the Initial Semantic Indexes. Formally, we define a semantic index as:

Definition 1: Initial Semantic Index: is a triad $\zeta := \langle T, R, A \rangle$ where:

- T is the set of terms encoded in the nodes of ζ. These terms are extracted from the surrounding contextual information of the multimedia documents and exist at least in one of the exploited ontologies.
- R represents the relations that link elements of T. These are derived from the ontologies and used in ζ.
- A is the set of axioms defined over (T and R).

As the resulting semantic networks are constructed based on multiple ontologies, the probability that these networks are semantically heterogeneous is high.

Therefore, to overcome this issue, we utilize the merging techniques proposed in our previous work, namely in [10] to merge these networks into a single coherent network. The merging algorithm takes pairs of semantic networks (ζ_1 and ζ_2) as input, and finds all possible mapping elements between their nodes, producing a single merged network ζ_{merged} as output. We formally define the mapping function as:

Definition 2: Mapping Element: Given two semantic networks ζ_1 and ζ_2, compute the N1 × N2 mapping elements $\langle Id_{ij}, n_{1i}, n_{2j}, R \rangle$ where:

- Id_{ij} is a unique identifier of the mapping element
- n_{1i} is a node in the first semantic network ($n_{1i} \in \zeta_1$)
- n_{2j} is a node in the second semantic network ($n_{2i} \in \zeta_1$)
- R is a semantic relation such as Synonymy (≡), Disjointness (⊥), Hyponymy, etc. that holds between the nodes n_{1i} and n_{2j}. These relations are derived from the exploited ontologies.
- i, j (i = 1, …, N1), (j = 1, …, N2)

Despite the fact that using multiple ontologies guarantees covering a broader set of entities in different domains, it is important to highlight that, in some cases; we have found entities that are not recognized by any of the used ontologies. To address such cases, we utilize concept-relatedness measures (in the context of our work, we use the *unrecognized entities handler*) to find whether we can suggest those entities for potential enrichment of the merged semantic network. To do this, we utilize the concept-relatedness based enrichment algorithm—formally defined as:

Definition 3: Concept-relatedness based Enrichment: This algorithm takes a given set of terms $T = \{t_1, t_2, t_3, \ldots, t_n\}$ that are not defined in ζ_{merged} and produces for each t ∈ in ζ_{merged} a set of S(t) ⊆ T as output, where:

- S(t) is the set of suggested enrichment candidates for t.

The set S(t) can be obtained using the Normalized Retrieval Distance (NRD) function proposed in [11] and based on a threshold value v using Eq. 1.

$$S(\mathrm{t}, v) := \{w \in W | \, \mathrm{NRD}(\mathrm{t}, w) \leq v\} \quad (1)$$

Based on the provided suggestions from the previous step, we attach the missing entity(s) to the merged semantic network and update its hierarchy.

5 Experimental Results

This section describes the experiments carried out to evaluate the effectiveness of the techniques of the proposed multimedia indexing framework. In these experiments, we exploited two publically available ontologies, namely WordNet [12] and YAGO [13]. We would like to point out that we implement all solutions in Java and experiments are performed on a PC with dual-core CPU (3000 GHz) and (4 GB RAM). The operating system is OpenSuse 11.1.

5.1 Experiments Using One Against Multiple Ontologies

In order to validate our proposal of employing knowledge represented by multiple ontologies to derive the semantic aspects of the surrounding contextual information of multimedia documents on the Web, we have manually selected a collection of 300 Webpages that contain different types of multimedia documents (100 per multimedia type). We would like to point out that the selected Webpages were obtained randomly but under the condition that their content should include one of the three multimedia document types (image, video, or audio). From a larger collection of Webpages, we have retained only those (300 Webpages) who have a proper structure that the DOM tree algorithm was able to process correctly. Then, we employed the methods of the proposed framework to automatically index the multimedia documents according to their surrounding contextual information using one against using multiple ontologies. We evaluated the precision of the proposed system (using the conventional Precision/Recall indicators reported in [14]) by comparing our ground truth (i.e. manually-assigned relevance scores) to their corresponding scores that were automatically produced by the system. Tables 2, 3, and 4 are provided to demonstrate the effectiveness of the employed techniques when dealing with Image, Audio, and Video datasets respectively. In the current version of the system's prototype we are using six queries per multimedia type.

As reported in the above tables, for the three different types of multimedia documents, the precision of the indexing framework has improved when combining multiple ontologies. It is important to point out that employing YAGO only has led to better (i.e. more precise) results when compared to using WordNet only. This is due to the fact that the coverage of WordNet is very limited compared to YAGO ontology.

Table 2 Precision of the proposed indexing framework when using a single against multiple ontologies—images dataset

Ouery_Id	Single-WordNet		Single-YAGO		Multiple-YAGO and WordNet	
	P (%)	R (%)	P (%)	R (%)	P (%)	R (%)
1	25	50	33	50	40	***100***
2	33	50	33	50	40	***100***
3	***0***	***0***	***66***	***100***	***66***	***100***
4	***66***	***50***	***66***	50	***66***	***100***
5	33	50	***100***	***100***	***66***	***100***
6	33	50	25	50	40	***100***

Table 3 Precision of the proposed indexing framework when using a single against multiple ontologies—audio dataset

Ouery_Id	Single-WordNet		Single-YAGO		Multiple-YAGO and WordNet	
	P (%)	R (%)	P (%)	R (%)	P (%)	R (%)
1	25	50	33	50	40	***100***
2	***0***	***0***	***100***	***100***	***100***	***100***
3	***100***	50	***100***	50	***100***	***100***
4	66	50	66	50	***80***	***100***
5	50	33	66	66	***75***	***100***
6	50	33	50	33	33	33

Table 4 Precision of the proposed indexing framework when using a single against multiple ontologies—video dataset

Ouery_Id	Single-WordNet		Single-YAGO		Multiple-YAGO and WordNet	
	P (%)	R (%)	P (%)	R (%)	P (%)	R (%)
1	***0***	***0***	***100***	***100***	***100***	***100***
2	66	50	66	50	80	***100***
3	***0***	***0***	***100***	***100***	***100***	***100***
4	***66***	***100***	***0***	***0***	66	***100***
5	50	33	66	66	75	***100***
6	66	50	80	100	80	***100***

It is also important to highlight that for some results the precision and recall values where very similar. This is because the used ontologies—despite of their domain coverage—still suffer from semantic knowledge incompleteness problems. To address this issue, we plan to extend the current experiments in the future work by incorporating the missing background knowledge handle to validate its significance in improving the produced results by the proposed framework.

6 Conclusions and Future Work

In this research work, we have introduced a fully-automated framework for indexing multimedia documents on the WWW through exploiting multiple ontologies. In order to derive the latent semantic dimensions of the surrounding information of multimedia document, we have employed two publically available ontologies, namely WordNet and YAGO. Additionally, we have further expanded the proposed semantic indexes of multimedia documents by utilizing concept-relatedness enrichment techniques. We have empirically evaluated the effectiveness of the proposed framework using three different multimedia datasets. The produced results demonstrated that using multiple ontologies increased the precision of the system by a remarkable percentage. Nevertheless, the conducted experiments have also shown that, in some cases, using a single ontology produced comparable results to using more than one ontology. This was due to the fact that the precision of the system was highly dependent on the domain depth and coverage level of each ontology. Also, we would like to highlight that our aim of the currently proposed techniques was to explore the impact of using multiple semantic resources on the precision/effectiveness of the system. However, we believe that the time complexity is another important factor that we need to take into consideration. We will consider this important factor and incorporate complexity experiments in the updated version of the prototype.

In the future work, we plan to extend the current approach by incorporating additional ontologies. We will investigate the impact of using domain-specific against generic ontologies on the effectiveness of the proposed framework. Furthermore, we will expand the current datasets by increasing the number of multimedia documents as well as queries in each dataset. This will also open the door for exploring how the performance (efficiency) of the indexing framework will be affected when dealing with large-scale real-world multimedia datasets. An equally important issue that we plan to investigate is the incorporation of HTML5-based Webpages in the experiments. We will in investigate the impact of incorporating such Webpages in our future experiments.

References

1. Amato, F., et al.: Content-based multimedia retrieval. In: Colace, F., et al. (eds.) Data Management in Pervasive Systems, pp. 291–310. Springer International Publishing (2015)
2. Wattanarachothai, W., Patanukhom, K.: Key frame extraction for text based video retrieval using Maximally Stable Extremal Regions. In: Industrial Networks and Intelligent Systems (INISCom), vol. 2, no. 4, pp. 29–37, Mar 2015
3. Yang, H., Meinel, C.: Content based lecture video retrieval using speech and video text information. IEEE Trans. Learn. Technol. **7**(2), 142–154 (2014)
4. Gao, Y., Wang, M., Zha, Z.J., Shen, J.L.: Visual-textual joint relevance learning for tag-based social image search. IEEE Trans. Image Process. **22**(1), 363–376 (2013)
5. Zhang, Y., Yang, X., Mei, T.: Image search reranking with query-dependent click-based relevance feedback. IEEE Trans. Image Process. **23**(10), 4448–4459 (2014)
6. Popescu, A., Moëllic, P., Millet, C.: SemRetriev—an ontology driven image retrieval system. In: CIVR, Amsterdam, The Netherlands (2007)
7. Manzoor, U., Ejaz, N., Akhtar, N.: Ontology based image retrieval. In: Proceedings of the International Conference for Internet Technology and Secured Transactions, pp. 288–293 (2012)
8. Wang, H., Chia, L., Gao, S.: Wikipedia-assisted concept thesaurus for better web media understanding. In: MIR10. Pennsylvania, USA, pp. 349–358 (2010)

9. Fauzi, F., Hong, J., Belkhatir, M., Hong, D.: Webpage segmentation for extracting images and their surrounding contextual information. In: ACM Multimedia'09, Beijing, China, pp. 649–652 (2009)
10. Maree, M., Belkhatir, M.: A Coupled statistical/semantic framework for merging heterogeneous domain-specific ontologies. In: 22nd International Conference on Tools with Artificial Intelligence (ICTAI'10), Arras, France, pp. 159–166 (2010)
11. Maree, M., Belkhatir, M.: Addressing semantic heterogeneity through multiple knowledge base assisted merging of domain-specific ontologies. Knowl.-Based Syst. **73**, 199–211 (2015)
12. Miller, G.A.: WordNet: A lexical database for English. Commun. ACM 409–409 (1995)
13. Fabian, M.S., Gjergji, K., Gerhard, W.: YAGO: a core of semantic knowledge unifying WordNet and wikipedia. In: Proceedings of the 16th International World Wide Web Conference, WWW, pp. 697–706 (2007)
14. Suchanek, M.F., Sozio, M., Weikum, G.: SOFIE: a self-organizing framework for information extraction. In: WWW09, pp. 631–640 (2009)

Feature Selection Methods Based on Decision Rule and Tree Models

Wiesław Paja

Abstract Feature selection methods, as a preprocessing step to machine learning, is effective in reducing dimensionality, removing irrelevant data, increasing learning accuracy, and improving result comprehensibility. However, the recent increase of dimensionality of data poses a severe challenge to many existing feature selection methods with respect to efficiency and effectiveness. In this work, a novel concepts of relevant feature selection based on information gathered from decision rule and decision tree models were introduced. A new measures *DRQualityImp* and *DTLevelImp* were additionally defined. The first one is based on feature presence frequency and rule quality, while the second is based on feature presence on different levels inside decision tree. The efficiency and effectiveness of that method is demonstrated through the exemplary use of five real-world datasets. Promising initial results of classification efficiency could be gained together with substantial reduction of problem dimensionality.

Keywords Feature selection · Feature ranking · Decision rules · Dimensionality reduction · Relevance and irrelevance

1 Introduction

In the era of the acquisition of vast amounts of data, different domain information databases, efficient analysis and retrieval of regularity has become an extremely important task. The issue of classification and object recognition is applied in many fields of human activity. Data mining is fraught with many aspects which hinder it like a very large number of observations, too many attributes, the insignificance of the part of variables for the classification process, mutual interdependence of conditional variables, the simultaneous presence of variables with different types, the

---

W. Paja (✉)
Faculty of Mathematics and Natural Sciences, Department of Computer Science,
University of Rzeszów, 1 Prof. S. Pigonia Street, 35-310 Rzeszów, Poland
e-mail: wpaja@ur.edu.pl

I. Czarnowski et al. (eds.), *Intelligent Decision Technologies 2016*,
Smart Innovation, Systems and Technologies 57,
DOI 10.1007/978-3-319-39627-9_6

presence of undefined values of variables, the presence of erroneous values of the variables, uneven distribution of categories for the target variable. Thus, the development of efficient methods for significant feature selection is valid.

This kind of methods are frequently used as a preprocessing steps to machine learning experiments. It could be defined as a process of choosing a subset of original features so that the feature space is optimally reduced according to a certain evaluation criterion. Feature selection has been a fruitful field of research and development since 1970s and proven to be effective in removing irrelevant features, increasing efficiency in learning tasks, improving learning performance like predictive accuracy, and enhancing comprehensibility of learned results [1].

The feature selection methods are typically presented in three classes based on how they combine the selection algorithm and the model building: filter, wrapper and embedded FS methods. Filter methods select features regardless of the model. They are based only on general features like the correlation with the variable to predict. These methods select only the most interesting variables. Then selected subset will be part of a classification model. Such methods are effective in computation time and robust to overfitting [2]. But, some redundant, but relevant, features are not recognized. In turn, wrapper methods evaluate subsets of features which allows to detect the possible interactions between variables [1, 3, 4]. However, the increasing overfitting risk when the number of observations is insufficient could be possible. Additionally, the significant computation time when the number of variables is large highly increase. The third type called embedded methods devotes to reduce the classification of learning. These methods try to combine the advantages of both previous methods. Thus, the learning algorithm takes advantage of its own variable selection algorithm. So, it needs to know initially what a good selection is, which limits their exploitation [5].

Kohavi and John [1] observed that there are several definitions of relevance that may be contradictory and misleading. They proposed that two degrees of relevance (*strong* and *weak*) are required to encompass all notions that are usually associated with this term. In their approach the relevance is defined in the absolute terms, with the help of ideal Bayes classifier. In this context a feature X is *strongly relevant* when removal of X alone from the data always results in deterioration of the prediction accuracy of the ideal Bayes classifier. In turn, feature X is *weakly relevant* if it is not strongly relevant and there exists a subset of features S, such that the performance of ideal Bayes classifier on S is worse than the performance on $S \cup \{X\}$. A feature is *irrelevant* if it is neither strongly or weakly relevant.

Nilsson and co-workers [6] introduced the formal definition of two different feature selection problems *Minimal Optimal Feature Selection (MOSF)* and *All Relevant Feature Selection (ARFS)*. *MOSF* means identification of minimal set of features to obtain optimum quality classification. In turn, *ARFS* devotes to find all the variables that may, under certain conditions, improve the classification. There are two important differences between these problems. The first one is detection of attributes with low importance (*ARFS*) [7], which may be completely obscured by other, more important attributes from the point of view of the classifier (*MOFS*). The second difference is to find the boundary between the variables poorly, but realistically related to the decision and those for whom such a relation is created as a result of random fluctuations. The formal definition of the problem of all relevant feature selection (*ARFS*) as a distinct problem from the classical minimal optimal feature selection (*MOFS*), was proposed as recently as 2007 [6].

Here, two different wrapper methods of feature importance calculation are presented. The first one apply frequency of each feature occurrence inside rules, and the second method comply decision tree structure during calculation. Similar methods for selecting and evaluating most important rule features based on the rule accuracy, frequency of the elementary condition in the discovered rule set, and its influence for the quality of the whole set of generated rules was extensively analyzed in [8]. Another approach [9] devotes to method for evaluating the importance of gene ontology terms which compose multi-attribute rules. The obtained ranking is used to generate a new set of rules that provide additional information about the biological function of genes.

Additionally, in this research, to distinguish between relevant and irrelevant features the *contrast variable* concept [7] were applied. It is a variable that does not carry information on the decision variable by design that is added to the system in order to discern relevant and irrelevant variables. Here, it is obtained from the real variables by random permutation of values between objects. The use of contrast variables was for the first time proposed by Stoppiglia and co-workers [10] and then by Tuv and co-workers [11].

2 Methods and Algorithms

During experiments the following general procedure was applied:

1. *Step: Selection of dataset and features for investigation (10-folds)*
    - *Addition of contrast features to original data*
    - *Application of a set of ranking measures to calculate importance for each feature*
    - *Calculation of DRQualityImp (or DTLevelImp) importance parameter for each feature*
    - *Definition (selection) of the most important feature subset*
2. *Step: Application of different machine learning algorithms for classification of unseen objects (the same 10-folds like in Step 1)*
    - *Using all original features*
    - *Using only selected, important features*
3. *Step: Comparison of gathered results using evaluation measures*

In the first step, dataset and feature for investigation were defined. Different ranking measures were applied to estimate importance of each feature. In order to check specificity of the feature selection, the dataset was extended by adding contrast variables. It means that each original variable was duplicated and its values were randomly permuted between all objects. Hence a set of non-informative by design shadow variables was added to original variables. The variables that were selected as important significantly than random, were examined further, using different test. To define level of feature importance six well-known ranking measures were applied: *ReliefF, Information Gain, Gain Ratio, Gini Index, SVM weight* and *RandomForest*. Additionally, new measures based on decision rules and decision trees, called *DRQualityImp* and *DTLevelImp*, were introduced. The first one is based on the frequency of presence of the different feature in rule model that is generated from original dataset and also takes into consideration the quality of rules in which this feature occurs. Thus, the *DRQualityImp* of the *ith* attribute could be presented in the Eq. 1.

$$DRQualityImp_{A_i} = \sum_{j=1}^{n} Q_{R_j}\{A_i\} \tag{1}$$

where n is a number of rules inside the model, Q_{R_j} defines classification quality of the rule R_j and A_i describe the presence of the *ith* attribute, usually *1* (feature occurred) or *0* (feature didnt occur). In turn, quality of rule is defined in the Eq. 2.

$$Q_{R_j} = \frac{E_{corr}}{E_{corr} + E_{incorr}} \tag{2}$$

where E_{corr} depicts the number of correctly matched learning objects by the *jth* rule and E_{incorr} depicts the number of incorrectly matched learning objects by this rule.

Moreover, second new measure, based on decision trees, called *DTLevelImp* were defined. It is based on the presence of different feature in the decision tree nodes generated from original dataset and also takes into consideration the product of weight W_j assigned to a given level j of the tree and the number of cases *Inst(node)* classified in a given *node* at this level in which feature A_i occurs. Thus, the *DTLevelImp* of the *ith* attribute could be presented in the Eq. 3.

$$DTLevelImp_{A_i} = \sum_{j=1}^{l} \sum_{node=1}^{x} W_j * Inst(node) * \{A_i\} \tag{3}$$

where l is the number of levels inside the model, x is the number of nodes inside given level and A_i describe the presence of the *ith* attribute, usually *1* (feature occurred) or *0* (feature didnt occur). In turn, weight W of level j is defined in the Eq. 4.

$$W_j = \begin{cases} 1 & j = 1, j \in N \\ \frac{W_{j-1}}{2} & 1 < j \leq l \end{cases} \tag{4}$$

During the second step the test probing the importance of variables was performed by analyzing the influence of variables used for model building on the prediction quality.

Table 1 Summary characteristic of benchmark datasets

Dataset	# Instances	# Features	# Classes
Breast cancer	286	9	2
Heart disease	303	13	2
Lung cancer	32	56	3
Primary tumor	339	17	21
Skin cancer	548	13	4

Six different machine learning algorithms were applied to build different predictors for the original set of features and for selected features: *Classification Tree (CT), Random Forest (RF), CN2 decision rules algorithm (CN2), Naive Bayes (NB), k Nearest Neighbors (kNN)* and *Support Vector Machine (SVM)*. During this step the 10-fold cross validation paradigm were also applied using the same folds as it was in the first step. Nine known evaluation measures were applied in each predictor: *Classification Accuracy (CA), Sensitivity, Specificity, Area Under ROC curve (AUC), Information Score (IS), F1 score (F1), Precision, Brier measure* and *Matthew Coefficient Correlation (MCC)* [12]. Finally, two of them were summarized in Table 4.

3 Investigated Datasets

Initial investigations were focused on applying developed algorithms on several real-world datasets. Five datasets have been used during experiments. Four of them are gathered from UCI ML repository, and the fifth set have been developed earlier by the author [13]. The summary of datasets is presented in Table 1. These datasets have diverse number of objects, features and their types and also classes.

4 Results and Conclusions

To illustrate proposed methodology only results for Breast cancer datasets will be presented in details. The first step of the experiment revealed three features, that were recommended as important by all, or nearly all, ranking measures. In Table 2, we can observe that *deg-malig, node-caps,* and *irradiat* features create stable and core set of features which have the highest rank values using most of eight measures of importance, particularly using *DRQualityImp* measure, introduced in this investigation. In the same table, comparison with importance of contrast values (*"contrast"* index) is also presented. The most important contrast feature is *irradiat (contrast)* for which *DRQualityImp* measure is equal to *4.59*. In this way, it is also treated as a threshold that separates the core, relevant set of attributes from other less infor-

Table 2 Ranking of features using eight different measures

Feature	ReliefF	Inf. gain	Gain ratio	Gini	SVM weight	RF	DRQuality Imp	DTLevel Imp
deg-malig	−0.02	**0.08**	**0.05**	**0.02**	0.07	**2.03**	**8.06**	**235.46**
node-caps	**0.03**	**0.06**	**0.08**	**0.02**	0.06	**1.98**	**7.94**	24.94
irradiat	**0.01**	0.03	**0.03**	0.01	0.02	**0.69**	**5.64**	3.78
irradiat (contrast)	−0.05	0	0	0	0.05	**0.12**	**4.59**	5.21
inv-nodes	**0.03**	**0.07**	**0.05**	**0.02**	0.06	0.07	4.52	17.62
breast	−0.08	0	0	0	0.02	**0.33**	3.66	3.44
menopause (contrast)	**−0.01**	0.01	0.01	0	0.07	−0.01	3.44	9.82
menopause	−0.06	0	0	0	0.03	0	3.21	8.85
node-caps (contrast)	−0.02	0	0	0	0.03	0.07	2.78	21.01
inv-nodes (contrast)	−0.05	0.02	**0.01**	0	**0.17**	−0.02	2.39	12.42
breast-quad (contrast)	−0.12	0.01	0	0	0.06	−0.02	2.10	13.43
deg-malig (contrast)	−0.07	0	0	0	0.01	0	1.89	3.71
age (contrast)	−0.11	0.02	0.01	0	0.14	0.1	1.85	21.08
breast (contrast)	−0.06	0	0	0	0.03	0.1	1.71	3.45
breast-quad	−0.11	0.01	0.01	0	0.13	0.1	1.48	**45.50**
tumor-size	−0.13	**0.06**	**0.02**	**0.01**	0.1	0.01	1.32	**58.38**
tumor-size (contrast)	−0.16	**0.03**	0.01	**0.01**	0.11	−0.01	0.88	**29.6**
age	−0.1	0.01	0.01	0	0.05	0.06	0	7.36

mative attributes. Most of the measures (except *SVM weight*) used in this approach show that selected set of features has higher values of these parameters than gathered threshold value (underlined values). These values are denoted in bold style in Table 2. Hereby, we can observe that different measures give different threshold.

It should be stressed that using the *DTLevelImp* parameter the selected set of features is different: *deg-malig, breast-quad* and *tumor-size*. This selected set of features achieve results of classification similar to original one (see Table 4). Thus, some redundant information could be recognized.

The second step of experiment devoted to evaluation of prediction quality of utilized machine learning algorithms described in Sect. 2. During this step six different algorithms were applied using 10-fold cross validation method. Average results for the *Breast cancer* dataset are collected in Table 3. Three types of results is presented: achieved using original dataset, achieved using the cuted set of features by application of *DRQualityImp* and *DTLevelImp* indicator.

This procedure was applied to two specified sets:

- the original dataset containing all descriptive features,
- the dataset containing only selected features according to their importance calculated in the first step.

Finally, all average results for *Breast cancer* dataset are collected in Table 3. Based on these results, it could stressed that set of selected features which contains only *3* from *9* attributes has similar (even better) prediction quality (*CA* and *AUC*)

Table 3 Average results of classification quality for the *Breast cancer* dataset

Model	CA	Sens	Spec	AUC	IS	F1	Prec	Brier	MCC
On original data									
CT	0.68	0.57	0.57	0.57	0.00	0.59	0.58	0.49	0.14
CN2	0.74	0.61	0.61	0.71	0.06	0.61	0.72	0.37	0.30
SVM	0.75	0.61	0.61	0.68	0.04	0.65	0.76	0.37	
RF	0.76	0.60	0.60	0.69	0.03	0.67	0.78	0.37	
kNN	0.73	0.63	0.63	0.65	0.15	0.63	0.68	0.46	0.31
NB	0.74	0.67	0.67	0.69	0.12	0.67	0.69	0.43	0.36
On data selected using *DRQualityImp*									
CT	0.74	0.61	0.61	0.69	0.08	0.65	0.69	0.37	0.29
CN2	0.75	0.64	0.64	0.70	0.08	0.67	0.74	0.36	
SVM	0.76	0.62	0.62	0.66	0.06	0.65	0.78	0.38	
RF	0.76	0.62	0.62	0.71	0.06	0.65	0.78	0.37	
kNN	0.70	0.62	0.62	0.61	0.00	0.62	0.67	0.43	0.28
NB	0.75	0.66	0.66	0.72	0.11	0.66	0.73	0.37	0.38
On data selected using *DTLevelImp*									
CT	0.73	0.63	0.63	0.63	0.02	0.63	0.67	0.41	0.30
CN2	0.68	0.53	0.53	0.66	−0.01	0.63	0.61	0.38	
SVM	0.70	0.51	0.51	0.68	−0.04	0.80	0.69	0.39	
RF	0.71	0.57	0.57	0.69	0.00	0.70	0.61	0.38	
kNN	0.72	0.63	0.63	0.66	0.07	0.62	0.68	0.44	0.29
NB	0.72	0.63	0.63	0.66	0.03	0.66	0.64	0.40	0.26

Table 4 Summary results of feature selection and classification

Dataset	Measure	Original set	DRQualityImpset	DTLevelImp set
Breast cancer	#Features	**9**	**3** (33.3 %)	**3** (33.3 %)
	CA	0.73 ± 0.03	0.74 ± 0.02	0.71 ± 0.02
	AUC	0.67 ± 0.05	0.68 ± 0.04	0.66 ± 0.04
Heart disease	#Features	**13**	**8** (61.5 %)	**6** (46.2 %)
	CA	0.80 ± 0.03	0.79 ± 0.03	0.80 ± 0.02
	AUC	0.87 ± 0.05	0.86 ± 0.04	0.88 ± 0.04
Lung cancer	#Features	**56**	**3** (5.4 %)	**4** (7.1 %)
	CA	0.52 ± 0.09	0.52 ± 0.09	0.53 ± 0.07
	AUC	0.70 ± 0.05	0.70 ± 0.05	0.73 ± 0.05
Skin cancer	#Features	**13**	**8** (61.5 %)	**9** (69.2 %)
	CA	0.82 ± 0.02	0.79 ± 0.01	0.79 ± 0.02
	AUC	0.96 ± 0.02	0.95 ± 0.01	0.95 ± 0.01
Primary tumor	#Features	**17**	**13** (76.5 %)	**12** (70.6 %)
	CA	0.42 ± 0.04	0.42 ± 0.03	0.42 ± 0.04
	AUC	0.83 ± 0.04	0.83 ± 0.04	0.83 ± 0.04

as it was observed with all original attributes. Furthermore, all other measures in Table 3 also increased a little. With the exception of *Brier* score, which decreased, but the lower the *Brier* score is for a set of predictions, the better the predictions are calibrated [14].

Similar results were obtained for other investigated datasets (see Table 4). All number of features in selected sets are significantly less than in original one. It is average about *45 %* of original features selected. Using these selected sets promising initial results of classification efficiency could be gained together with substantial reduction of problem dimensionality.

Acknowledgments This work was supported by the Center for Innovation and Transfer of Natural Sciences and Engineering Knowledge at the University of Rzeszw.

References

1. Kohavi, R., John, G.H.: Wrappers for feature subset selection. Artif. Intell. **97**, 273–324 (1997)
2. Bermingham, M.L., Pong-Wong, R., Spiliopoulou, A., Hayward, C., Rudan, I., Campbell, H., Wright, A.F., Wilson, J.F., Agakov, F., Navarro, P., Haley, C.S.: Application of high-dimensional feature selection: evaluation for genomic prediction in man. Sci. Rep. **5**, (2015)
3. Phuong, T.M., Lin, Z., Altman, R.B.: Choosing SNPs using feature selection. In: Proceedings of 2005 IEEE Computational Systems Bioinformatics Conference, CSB 2005, pp. 301–309 (2005)
4. Paja, W., Wrzesien, M., Niemiec, R., Rudnicki, W.R.: Application of all-relevant feature selection for the failure analysis of parameter-induced simulation crashes in climate models. Geosci. Model Dev. **9**, 1065–1072 (2016)
5. Zhu, Z., Ong, Y.S., Dash, M.: Wrapper-filter feature selection algorithm using a memetic framework. IEEE Trans. Syst. Man, Cybern. Part B Cybern. **37**, 70–76 (2007)
6. Nilsson, R., Peña, J.M., Björkegren, J., Tegnér, J.: Detecting multivariate differentially expressed genes. BMC Bioinf. **8**, 150 (2007)
7. Rudnicki, W.R., Wrzesień, M., Paja, W.: All Relevant feature selection methods and applications. In: Stańczyk, U., Lakhmi, C.J. (eds.) Feature Selection for Data and Pattern Recognition, pp. 11–28. Springer-Verlag, Berlin Heidelberg, Berlin (2015)
8. Greco, S., Słowinski, R., Stefanowski, J.: Evaluating importance of conditions in the set of discovered rules. In: RSFDGrC'07: Proceedings of the 11th International Conference on Rough Sets, Fuzzy Sets, Data Mining and Granular Computing, Toronto, Ontario, Canada, pp. 314–321 (2007)
9. Sikora, M., Gruca, A.: Quality improvement of rules based gene groups descriptions using information about GO terms importance occurring in premises of determined rules. Int. J. Appl. Math. Comput. Sci. **20**(3), 555–570 (2010)
10. Stoppiglia, H., Dreyfus, G., Dubois, R., Oussar, Y.: Ranking a random feature for variable and feature selection. J. Mach. Learn. Res. **3**, 1399–1414 (2003)
11. Tuv, E., Borisov, A., Torkkola, K.: Feature selection using ensemble based ranking against artificial contrasts. In: International Symposium on Neural Networks, pp. 2181–2186 (2006)
12. Fawcett, T.: An introduction to ROC analysis. Pattern Recogn. Lett. **27**, 861–874 (2006)
13. Hippe, Z.S., Bajcar, S., Blajdo, P., Grzymala-Busse, J.P., Grzymala-Busse, J.W., Knap, M., Paja, W., Wrzesien, M.: Diagnosing skin melanoma: current versus future directions. TASK Q. **7**, 289–293 (2003)
14. Hernández-Orallo, J., Flach, P., Ferri, C.: A unified view of performance metrics: translating threshold choice into expected classification loss. J. Mach. Learn. Res. **13**, 2813–2869 (2012)

Automatic Categorization of Email into Folders by Ant Colony Decision Tree and Social Networks

Urszula Boryczka, Barbara Probierz and Jan Kozak

Abstract This paper presents a new approach to an automatic categorization of email messages into mailbox folders. The aim of this paper is to create an algorithm that would allow one to improve the classification of emails into folders by using solutions that have been applied in Ant Colony Decision Tree (ACDT). Additionally, elements of Social Network Analysis (SNA) were included in this algorithm. The new algorithm that is proposed here was tested on the publicly available Enron E-mail data set and all experiments were conducted on uncleaned data. For the purpose of comparing the results, additional tests were carried out by using selected classifiers which were generally available. The obtained results confirm that the proposed approach allows one to improve the accuracy with which new emails are assigned to particular folders based on an analysis of previous correspondence, even when uncleaned data sets are used.

Keywords Ant colony optimization · Social network analysis · Enron E-mail

1 Introduction

The history of email messages began almost half a century ago, i.e. when Louis Pouzin, Glenda Schroeder and Pat Crisman sent an email from one user to another in 1965. Unfortunately, this email service only made it possible to leave such a message for the other users of the same computer, whereas an email address had not been invented yet.

U. Boryczka · B. Probierz (✉) · J. Kozak
Institute of Computer Science, University of Silesia, Będzińska 39, 41–200
Sosnowiec, Poland
e-mail: barbara.probierz@us.edu.pl

U. Boryczka
e-mail: urszula.boryczka@us.edu.pl

J. Kozak
e-mail: jan.kozak@us.edu.pl

I. Czarnowski et al. (eds.), *Intelligent Decision Technologies 2016*,
Smart Innovation, Systems and Technologies 57,
DOI 10.1007/978-3-319-39627-9_7

It was not until 1971 that the American engineer and computer programmer Raymond S. Tomlinson came up with an idea that allowed one to send an email message from one computer to another. In order to separate the user name from the computer name, Raymond S. Tomlinson picked the @ symbol, which was only used occasionally at that time. On this basis, members of the Internet Engineering Task Force agreed on the standard syntax for email communication in 1973, i.e. "username@hostname", which is still used to this day.

The presented paper deals with research that is focused on creating decision tables in accordance with the authors' idea and testing classical algorithms by using these tables. The aim of this paper is to create an algorithm that would allow one to improve the classification of emails into folders (so. E-mail Foldering Problem) by using solutions that have been applied in Ant Colony Decision Tree (ACDT). Additionally, elements of Social Network Analysis (SNA) were included in this algorithm. A comparison of the results that have been obtained allows one to even more precisely determine the usefulness of the Ant Colony Optimization algorithm that is proposed here. It is one of artificial inteligence methods used in data mining [9]. The authors of the presented paper intend to show that emails can be classified into folders with a satisfactory accuracy, even if uncleaned data sets are included.

This article is organized as follows. Section 1 comprises an introduction to the subject of this article. In Sect. 2, characteristic of social networks is presented. Section 3 describes Ant Colony Decision Tree algorithm. Section 4 focuses on the presented, new version of the ACO approach based on ACDT algorithm and Social Network Analysis. Additionally, it shows the visualization of a network of contacts with objects. Section 5 presents the experimental study that has been conducted to evaluate the performance of the proposed algorithm, taking into consideration Enron e-mail dataset. Finally, we conclude with general remarks on this work and a few directions for future research are pointed out.

2 Characteristic of Social Networks

Social Networks Analysis (SNA) plays an extremely important role in studies of data sets containing email messages. Most of all, SNA provides a specific perspective on an analysis because it does not focus on individual units or macrostructures but studies the connections between particular units or groups.

SNA is a branch of sociology which deals with the quantitative assessment of the individuals role in a group or community by analyzing the network of connections between individuals. The first studies of social networks were conducted in 1923 by Jacob L. Moreno, who is regarded as one of the founders of social network analysis. Morenos 1934 book that is titled "Who Shall Survive?" presents the first graphical representations of social networks as well as definitions of key terms that are used in an analysis of social networks and sociometric networks [8].

A social network is usually represented as a graph. According to the mathematical definition, a graph is an ordered pair

$$G = (V, E), \tag{1}$$

where:
V denotes a finite set of a graphs vertices,
E denotes a finite set of all two-element subsets of set V that are called edges, which link particular vertices such that:

$$E \subseteq \left\{\{u, v\} \, : \, u, v \in V, u \neq v\right\}. \tag{2}$$

Vertices represent objects in a graph whereas edges represent the relations between these objects. Depending on whether this relation is symmetrical, a graph which is used to describe a network can be directed or undirected.

The degree of a vertex (indegree and outdegree) denotes the number of head endpoints or tail endpoints adjacent to a given node. Degree centrality is useful in determining which nodes are critical as far as the dissemination of information or the influence exerted on immediate neighbors is concerned. Centrality is often a measure of these nodes popularity or influence.

Social network analysis has a wide range of applications. It is primarily used in large organizations and companies as a tool for supporting strategic human resource management or knowledge management in an organization. SNA supports a companys innovativeness and an analysis of business processes as well as training needs. Additionally, it is used in marketing research for creating a map of a social network of customers. However, social network analysis primarily allows managers to familiarize themselves with the informal structure of an organization and the flow of information within a company.

Many studies that were carried out as part of SNA were aimed at finding correlation between a networks social structure and efficiency [5]. At the beginning, social network analysis was conducted based on questionnaires that were filled out by hand by the participants [4]. However, research carried out by using email messages has become popular over time [1]. Some of the studies found that research teams were more creative if they had more social capital [6]. Social networks are also associated with discovering communication networks. The database which was used in the experiments that are presented in this article can be used to analyze this problem. G. C. Wilson and W. Banzhaf, among others, discussed such an approach, which they described in their article [10].

3 Ant Colony Decision Tree

The Ant Colony Decision Tree algorithm (ACDT) is one of the most popular Ant Colony Optimization algorithms that are used in data mining. This algorithm combines the idea of Ant Colony Optimization algorithms with the idea of the CART algorithm and, as tests have shown, it produces very good quality classifiers for many standard problems related to data mining [2]. The ACDT algorithm is based on using Ant Colony Optimization algorithms in the process of optimizing the construction of decision trees. The execution of the algorithm involves choosing a test for each node based on two factors. The maximum value that is consistent with the splitting criteria used in the CART algorithm is one of these factors, and the additional information that is recorded in the form of the pheromone trail is the other factor [2, 7].

In the Ant Colony System a virtual ant decides on the next step based on a modified transition rule while being at a particular stage of problem—solving and at a specific point in time. For this purpose, it generates a random number q, $0 \leq q \leq 1$. If $q \leq q_0$ (q_0— the parameter of the algorithm that has been determined) then "the best" available decision option is chosen (exploitation); otherwise, the ant makes a random decision (exploration) by taking into account probabilities that are calculated in accordance with Eq. (3) [7] (Table 1).

$$j = \begin{cases} \arg\max_{r \in J_i^k} \{[\tau_{ir}(t)]^{\alpha} \cdot [\eta_{ir}]^{\beta}\}, & \text{if } q \leq q_0 \text{ (exploitation)} \\ S, & \text{otherwise (exploration),} \end{cases} \tag{3}$$

where:

τ_{ir}—value of the reward, i.e. the degree of usefulness of the decision option that is being considered (pheromone),

η_{ir}—value of the quality of a transition from state i to state r which was estimated heuristically,

α i β—parameters describing the importance of values $\tau_{ir}(t)$ i η_{ir},

S—the next step (decision) which was randomly selected by using the probabilities:

$$p_{ij}^k(t) = \begin{cases} \frac{\tau_{ij}(t) \cdot [\eta_{ij}]^{\beta}}{\sum_{r \in J_i^k} \tau_{ir}(t) \cdot [\eta_{ir}]^{\beta}}, & \text{if } j \in J_i^k \\ 0, & \text{otherwise,} \end{cases}$$

where:

J_i^k denotes the set of decisions that ant k can make while being in state i.

The value of the heuristic function is determined based on the splitting criteria used in the CART algorithm, i.e. in accordance with the following Eq. (4).

$$\arg\max_{a_j \leq a_j^R, j=1,\ldots,M} \left(\frac{P_l P_r}{4} \left[\sum_{k=1}^{K} |p(k|m_l) - p(k|m_r)| \right]^2 \right), \tag{4}$$

where:

$p(k|m_l)$—probability of the occurrence of decision class k in node m_l (in the left subtree),

$p(k|m_r)$—probability of the occurrence of decision class k in node m_r (in the right subtree),

Table 1 Parameters in data sets

Dataset	N. of objects	N. of class	Number of attributes						Parameters of social networks	
			from	word1	word2	word3	cc	Length	N. of edges	Frequency of information flow
germany-c	1013	19	207	382	419	340	2	835	25	208
haedicke-m	112	16	64	70	85	67	2	110	30	238
mann-k	1616	24	254	394	490	433	2	1248	18	123
rogers-b	1395	14	289	445	521	430	2	1101	13	26
scott-s	641	10	135	350	219	166	2	578	28	443
shackleton-s	1001	53	158	330	384	357	2	836	27	1656
shapiro-r	1970	75	325	720	856	754	2	1566	25	1050
steffes-j	625	23	157	242	341	300	2	555	19	401
symes-k	770	12	119	324	346	287	2	685	16	387
taylor-m	656	21	173	255	288	245	1	580	32	1256

P_l—probability of object transition to node m_l (in the left subtree),
P_r—probability of object transition to node m_r (in the right subtree),
K—decision classes.
The pheromone trail is updated by increasing pheromone levels on the edges connecting each tree node with its parent node (excepting the root):

$$\Delta\tau_{m,m_L}(t+1) = (1-\gamma)\cdot\tau_{m,m_L}(t) + Q(T), \tag{5}$$

where:
$Q(T)$ determines the evaluation function of the decision tree,
γ is a parameter representing the evaporation rate, which is equal to 0.1.

4 Proposed Algorithm

The proposed method entails using a modified version of the ACDT algorithm (which is described in Sect. 3) and transforming a data set of emails into a decision table. For such a data set the proposed algorithm was prepared; it contains elements of communication network analysis which entails analyzing the list of recipients.

The decision table that has been prepared consists of the following attributes:

- from—the sender;
- word1—the first word which is used in the subject of an email (with the exception of basic words and copulas); additionally, words which belong to the set of decision classes are supported;
- word2—the second word which is established similarly to word1;
- word3—the third word which is established similarly to word1 and word2;
- cc—the Boolean value which indicates whether the person who has received an email was added as a recipient of a copy of an email (if not then it means that the person was the addressee of an email);
- length—number of characters of the mail (with white spaces);
- category—a decision class, i.e. a folder, to which an email message is assigned.

Conditional attributes were selected to define the most important information about each message. They consist of the information from the sender field, the first three words from the email subject, information as a Boolean value, conditional attributes check also the length of the message and whether the person who received the message was added to a courtesy copy (CC). If not, it implies that was the recipient (To). In addition, from the email subject was omitted basic phrases and copula verbs, and there was additionally supported words, which belonged to a set of decision classes.

Application of the ACDT algorithm (which is based on a modification of this algorithm at the present stage) entails exploring the communication network between people if an email was sent to a group of persons, i.e. $cc = true$. The list of all recipients is analyzed, which has an influence on which decision class (email folder) a classifier will choose.

This decision is also influenced by the preferences of the group of users who contact one another; therefore, if the users contacted one another with the same frequency then the emails they received were classified in the same way. Because of that the network of interactions has been analyzed using well-know methodology (from the SNA field).

All of the mailboxes that have been selected contain uncleaned data, which is why in the sets there might be folders without any messages or folders that were automatically created by email programs. Uncleaned data sets may also contain unnecessary email messages which were sent many times as well as emails that have not been assigned to any folder. The number of decision classes depends on the case, which is analyzed.

These data sets are very large - they are composed of a large number of decision classes and have attributes with many values, mainly with continuous values. Therefore, modified Ant Colony Decision Tree algorithm was used to analyze this data set because they perform very well as far as such problems are concerned [2]. The way in which such algorithm work is presented based on the example of Algorithm 1.

Algorithm 1: Pseudo code of the proposed algorithm

1 dataset = prepare_dataset_from_email(person);
2 ph = initialization_pheromone_trail(); // $\tau_{m,m_L}(t=0) = \frac{\log_2(C)}{\sum_{att=1}^{|A|} |a_{att}|}$
3 best_constr._classifier = *NULL*;
4 **for** i=1 **to** number_of_iterations **do**
5   best_classifier = *NULL*;
6   **for** j=1 **to** number_of_ants **do**
7   new_classifier = build_prototype_classifier_EMAIL(ph, dataset);
8   new_classifier = check_contacts_SNA(new_classifier, dataset);
9   assessment_of_the_quality_classifier(new_classifier);
10   **if** new_classifier **is_higher_quality_than** best_classifier **then**
11   best_classifier = new_classifier;
12   **endIf**
13   **endFor**
14   update_pheromone_trail(best_classifier, ph);
15   **if** best_classifier **is_higher_quality_than** best_constr._classifier **then**
16   best_constr._classifier = best_classifier;
17   **endIf**
18 **endFor**
19 result = best_constr._classifier;

The proposed analysis of the network of contacts between individual employees is used to determine the leaders in terms of the spread of information or to influence the persons, who are in the immediate vicinity. Figure 1 shows a visualization of the network of contacts with objects selected from the dataset of 150 objects. The dataset was analyzed in terms of the frequency of sending e-mails. Additionally, Table 1 shows parameters for each dataset.

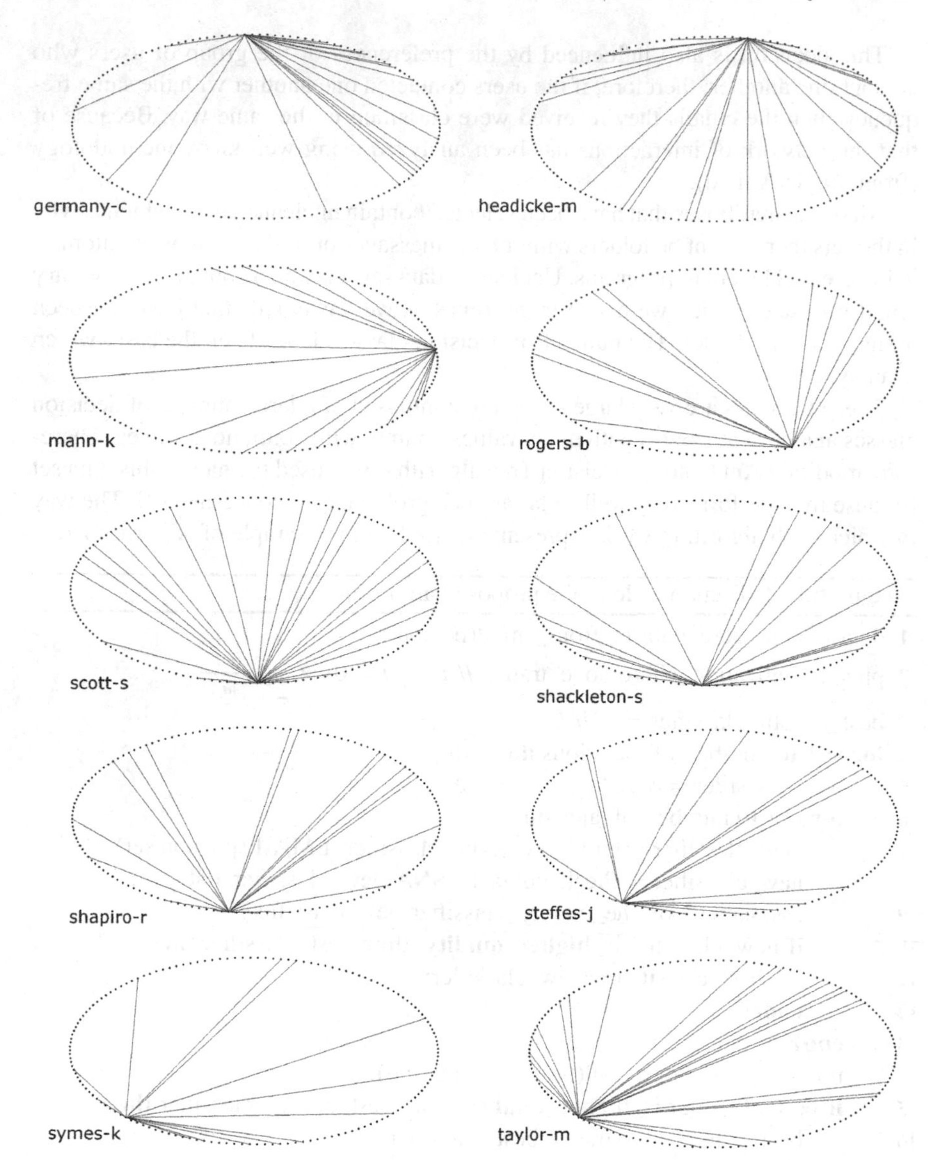

Fig. 1 Social networks for Enron dataset

5 Experiments

The proposed algorithm was implemented in C++. All computations were carried out on a computer with an Intel Core i5 2.5 GHz processor, 2.9 GB RAM, running on the Debian GNU/Linux operating system.

Table 2 Comparison of all approaches in terms of classification accuracy on uncleaned data sets

Dataset	Simple cart	Naive Bayes	Conjunctive rule	PART	Hyper pipes	LBR	Proposed algorithm
germany-c	0.386	0.626	0.415	0.576	0.549	0.626	**0.783**
haedicke-m	0.297	0.622	0.297	0.297	0.595	0.622	**0.623**
mann-k	0.245	0.708	0.178	–	0.684	0.712	**0.767**
rogers-b	0.510	0.772	0.443	0.746	0.738	0.774	**0.911**
scott-s	0.662	0.826	0.615	0.723	0.864	0.836	**0.936**
shackleton-s	0.565	0.667	0.291	0.682	0.628	0.673	**0.709**
shapiro-r	0.091	0.421	0.125	–	0.392	0.428	**0.605**
steffes-j	0.649	0.755	0.548	0.639	0.769	0.755	**0.841**
symes-k	0.324	0.789	0.457	0.723	0.781	0.785	**0.930**
taylor-m	0.367	0.757	0.399	0.321	0.757	0.761	**0.862**

The experiments were repeated 30 times for each data set with the same standard parameter settings which were related to Ant Colony Optimization algorithms. Given the size of the data set, the number of generations of the Ant Colony Optimization algorithm was initially restricted to 30 for a population of 5 ants. The run-time of the proposed algorithm ranged, depending on the data set, between 7 and 400 s for one run of the algorithm. This is, however, a time during which a classifier is created whereas classification itself is carried out very quickly.

In order to check the proposed algorithms adaptability, experiments were carried out so as to make it possible to compare the obtained results with the results that were produced by other classifiers. The algorithms that had been chosen constructed classifiers by using the same data sets as the SNA Ant Colony Optimization Algorithm that was proposed in [3]. Additionally, a larger number of data sets (email messages) were used to test the algorithms.

The experiments that are described in this section were conducted in order to check if the proposed method for classifying email messages into folders works correctly and also to verify whether the proposed method of creating decision tables can be used for any classifiers, even if uncleaned data sets are included. Out of all mailboxes obtained from the Enron E-mail data set a total of ten mailboxes were selected. The mailbox sizes ranged from 10 to 42MB, which could indicate that they contained a small number of emails and folders.

The proposed algorithm, which is described in Sect. 4, and selected algorithms that had been implemented in the Weka system (Waikato Environment for Knowledge Analysis) [11] were chosen for the purpose of carrying out the tests. All tests were conducted on uncleaned data sets. The obtained results are presented in Table 2 and Fig. 2.

The proposed algorithm each time generated better results for all data sets that had been created based on ten users. As for three data sets (mann-k, scott-s and steffes-j), the accuracy with which a folder is assigned to an email improved by 5–7 % points in

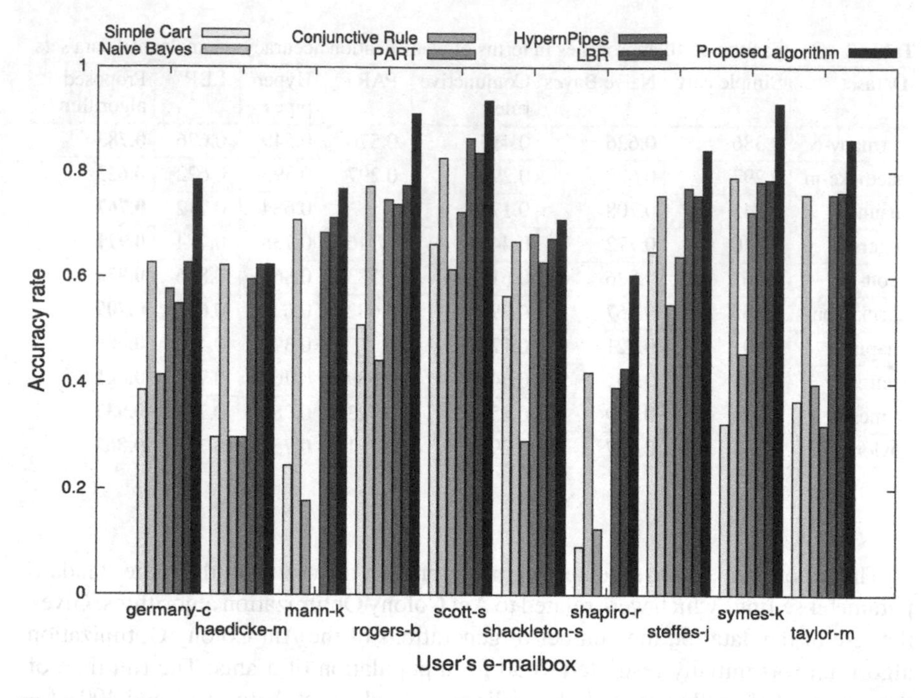

Fig. 2 The correctness of the proposed categorization method

relation to the best of the other methods that were compared. For three other data sets (rogers-b, symes-k and taylor-m) there was a large, 10–14 % points improvement, whereas for sets germany-c and shapiro-r there was a very large improvement, i.e. of 16–18 % points.

For the two remaining sets (haedicke-m and shackleton-s), all of the algorithms achieved the same level of classification accuracy. The results concerning classification accuracy achieved by using the PART algorithm are not provided for sets mann-k and shapiro-r because the run-time of this algorithm was too long.

Other elements that are related to analyzing algorithms also need to be compared, i.e. those which could not be compared at this stage. Nonetheless, the classification stage itself is very similar for all the methods; therefore, potential differences may only result from the complex structure of the classifiers.

6 Conclusions

Based on the experiments that were carried out, it was confirmed that the accuracy of classification, i.e. the correctness of an automatic categorization of email messages, was considerably improved when Ant Colony Optimization algorithms and Social Network Analysis were used. The aim of this article has been achieved.

The proposed approach led to a significant improvement in the classification of emails into folders. Creating a map of contacts in the form of social networks suggests that method will not only reduce the time spent on reading and replying to e-mails received, but above all is crucial to the process flow of information between employees of the company.

It has been noticed that the proposed method of creating decision tables makes it possible to use classical classifiers to categorize email messages. However, the proposed algorithm produces even better results due to its adaptability and the use of SNA elements.

In the future the authors of this article intend to adapt the social network mechanism for this purpose to a larger extent and to improve the process of creating decision tables. In future stages of the research, the incorporation of elements of text mining in an analysis of email message content and the direct coupling of these elements with the pheromone trail of the proposed algorithm should produce positive effects.

References

1. Aral, S., Van Alstyne, M.: Network structure & information advantage. In: Proceedings of the Academy of Management Conference, vol. 3, Philadelphia, PA. Citeseer (2007)
2. Boryczka, U., Kozak, J.: Ant Colony Decision Trees—a new method for constructing decision trees based on Ant Colony Optimization. In: Computational Collective Intelligence. Technologies and Applications, LNCS, vol. 6421, pp. 373–382. Springer (2010)
3. Boryczka, U., Probierz, B., Kozak, J.: An ant colony optimization algorithm for an automatic categorization of emails. Computational Collective Intelligence. Technologies and Applications, LNCS, vol. 8733, pp. 583–592. Springer, Berlin (2014)
4. Cummings, J.N., Cross, R.: Structural properties of work groups and their consequences for performance. Soc. Netw. **25**(3), 197–210 (2003)
5. Gloor, P.A.: Swarm creativity: competitive advantage through collaborative innovation networks. Oxford University Press (2005)
6. Gloor, P.A., Grippa, F., Putzke, J., Lassenius, C., Fuehres, H., Fischbach, K., Schoder, D.: Measuring social capital in creative teams through sociometric sensors. Int. J. Organ. Des. Eng. **2**(4), 380–401 (2012)
7. Kozak, J., Boryczka, U.: Enhancing the effectiveness of ant colony decision tree algorithms by co-learning. Appl. Soft Comput. **30**, 166–178 (2015)
8. Moreno, J.L.: Who shall survive? Foundations of Sociometry, Group Psychotherapy and Sociodrama. Beacon House (1953)
9. Tkacz, M.: Artificial neural networks in incomplete data sets processing. In: Intelligent Information Processing and Web Mining, pp. 577–583. Springer (2005)
10. Wilson, G., Banzhaf, W.: Discovery of email communication networks from the enron corpus with a genetic algorithm using social network analysis. In: IEEE Congress on Evolutionary Computation, 2009. CEC'09, pp. 3256–3263. IEEE (2009)
11. Witten, I.H., Frank, E., Hall, M.A.: Data Mining: Practical Machine Learning Tools and Techniques, 3rd edn. Morgan Kaufmann Publishers Inc. (2011)

Using Dissimilarity Matrix for Eye Movement Biometrics with a Jumping Point Experiment

Pawel Kasprowski and Katarzyna Harezlak

Abstract The paper presents studies on the application of the dissimilarity matrix-based method to the eye movement analysis. This method was utilized in the biometric identification task. To assess its efficiency four different datasets based on similar scenario ('jumping point' type) yet using different eye trackers, recording frequencies and time intervals have been used. It allowed to build the common platform for the research and to draw some interesting comparisons. The dissimilarity matrix, which has never been used for identifying people on the basis of their eye movements, was constructed with usage of different distance measures. Additionally, there were different signal transforms and metrics checked and their performance on various datasets was compared. It is worth mentioning that the paper presents the algorithm that was used during the BioEye 2015 competition and ranked as one of the top three methods.

Keywords Eye movement biometrics · Dissimilarity matrix · Fusion · Dynamic time warping

1 Introduction

Eye movement biometrics has been investigated for over 10 years, however there are still no commercial applications utilizing this modality. The main problem is lack of established and well understood methods that can be used to distinguish eye movement characteristics of different people.

Some effort has already been made to solve this problem. There are eye movement datasets available to download, and there are biometric contest organized like EMVIC 2012 [6], EMVIC 2014 [5] or BioEye 2015 [13]. But the methods used by contest participants are not always published and therefore are sometimes not reproducible. Moreover, because multiple submissions are possible during such a

P. Kasprowski (✉) · K. Harezlak
Silesian University of Technology, ul. Akademicka 16, 44-100 Gliwice, Poland
e-mail: pawel.kasprowski@polsl.pl; kasprowski@polsl.pl

I. Czarnowski et al. (eds.), *Intelligent Decision Technologies 2016*,
Smart Innovation, Systems and Technologies 57,
DOI 10.1007/978-3-319-39627-9_8

competition and typically some training data is available in advance, the methods are optimized for the competition's dataset and suffer from poor generalization on other datasets. Such a generalization requires the application of elaborated methods for datasets collected using different setups and for different users, which may help to find out a solution serving well for many eye movements collections.

This was the one of the motivating factors to apply the new feature extraction method—which was developed by authors for the BioEye 2015 competition and ranked as one of the top three methods—to various eye movement datasets. This method, based on dissimilarity matrix [2], has not yet been used for eye movement biometrics. We checked its performance using various signals, transforms and divisions of samples.

To overcome the lack of generalization, the usefulness of the solutions applied was tested on four different datasets recorded using three different eye trackers. It enabled us to draw some meaningful conclusions about efficiency of various approaches combinations and experiment's scenarios.

2 Eye Movement Biometrics Using a 'Jumping Point' Stimulus

All datasets which were used, were recorded using a jumping point—one of the most popular stimuli. During such an experiment a subject is instructed to follow with eyes a point displayed on a screen. The point's position changes periodically—that is why it is called a 'jumping point'. The advantage of such a stimulus is that eye movements are more or less predictable and comparable between trials. On the other hand such a stimulus forces a specific behavior so it measures more physiological patterns of a person than behavioral ones.

The first usage of such a kind of stimulus was reported in [7]. There were cepstral coefficients used as features for a classifier. In the work [8] the idea was extended with usage of Principal Component Analysis (PCA) to reduce the number of attributes. Another notable work was [10]. Authors extracted saccades and used training samples to create an Oculomotor Plant Mathematical Models (OPMM) [12]. The idea was extended in [11] where nine oculomotor plant characteristics (OPC) were empirically chosen. The OPC biometrics calculated for different subjects were compared using a voting version of Student t-test and the Hotelling T-square test. The results were fused using logical AND or OR techniques.

In 2012 there was the first Eye Movement Verification and Identification Competition (EMVIC) organized and it resulted in several publications [6]. There were four datasets presented—all created using a 'jumping point' stimulus. According to [6] the winner of the competition divided samples into parts and calculated 2D histograms of speed and direction. The second place holder extracted velocity and acceleration and compared their distributions using the multivariate Walf-Wolfowitz test [14].

In 2013 Holland and Komogortsev [3] compared results for different stimuli and devices using the same set of 14 features (named CEM features). The results were calculated for every feature and the fusion of all features.

Finally, in 2015 there was the BioEye competition announced with four datasets [13]. Two of them were based on jumping point stimulus and were used in the presented research.

3 Feature Extraction and Classification

Before any feature extraction method was applied to eye movement signal, each sample from the dataset was divided into events. An event was a part of a sample for which stimulus point's position was in the same place. Every event was described by a starting position—location of the point just before the event—and the ending position—location of the point during the event. A direction of an event was defined as a direction of a vector from its start position to end one.

3.1 Signals Extraction

On the basis of the raw eye positions, the first, second and third derivatives were extracted for every event independently. There were velocity (v), acceleration (a) and jerk (j) calculated as an absolute value and for both horizontal and vertical directions. It resulted in 9 signals for every event (Table 1).

Table 1 Set of signals extracted from eye movement

Signal	Formula	Description
vx, vy	$V_x = \frac{\partial x}{\partial t}, V_y = \frac{\partial y}{\partial t}$	The first derivative of x and y (i.e. vertical and horizontal velocities)
vxy	$V = \sqrt{V_x^2 + V_y^2}$	The first derivative for absolute velocity
ax, ay	$V'_x = \frac{\partial V_x}{\partial t}, V'_y = \frac{\partial V_y}{\partial t}$	The second derivative of x and y (i.e. vertical and horizontal accelerations)
axy	$V' = \sqrt{V'^2_x + V'^2_y}$	The derivative of vxy
jx, jy	$V''_x = \frac{\partial V'_x}{\partial t}, V''_y = \frac{\partial V'_y}{\partial t}$	The third derivative of x and y (jerk)
jxy	$V'' = \sqrt{V''^2_x + V''^2_y}$	The derivative of axy

where x, y—the raw coordinates

3.2 Signal Transformation

The next phase was the calculation of different transforms from each of the nine signals separately. There were four transformations used: Fourier transform (**F**) [8], Cepstrum transform (**C**) [7] Daub Wavelet transform (**W**) and signal normalization to 0–1 (**N**). Together with not transformed signal (**S**) it gave 5 different transforms and altogether, there were $9 \times 5 = 45$ different signals extracted.

3.3 Features Preparation

Signals obtained in the previous pre-processing phase were subsequently used to build feature sets with usage of the dissimilarity matrix-based method [2]. Similar method has already been used for behavioral biometrics [16], but using it for eye movement signal is our original contribution. While preparing feature sets, at first the eye movement dataset was divided by half into training and testing events. Because each of datasets used to evaluate the proposed method was built from samples collected during two sessions, there were always two samples for each subject. Events from the first user's sample were treated as training, while events from the second sample as testing ones.

Then, for every training event, its distances to all other training events were calculated. These distances formed a feature set, consisting of N features where N is the number of training events. This feature set was used as an input to a classification model building algorithm. The same procedure was used for testing events—at first distances to all training events were calculated and formed a set of attributes and then this set was used to classify the given event.

Given two signals S_a and S_b the distance between them may be calculated using different measures. In this research three of them were taken into account—Dynamic Time Warping (DTW) [1], Euclidean distance (EUC) and Earth Mover's Distance (EMD) [15]. As a result, a separate feature set for every combination of every signal and distance measure was prepared ($45 \times 3 = 135$ feature sets).

3.4 Classification

As it was stated above, every dataset used consisted of two sessions for each subject. During the classification, events from the first session of the subject were always used as training data and the events from the second session of the same subject as testing data. The K Nearest Neighbors algorithm with value of K equal to 1 was used as a classifier and every testing event was classified separately.

The final result for each testing sample was determined using a simple score fusion. The classification model, for every event e, returned a probability $p(e, c)$ that this event belongs to class c. For 1NN classifier this value was equal to 1 for one class and 0 for all other classes. Score for class c in every sample s was calculated as: $score(s, c) = \sum_{e=1}^{E} p(e, c)$, where E means all events belonging to the sample s. The final label for the sample was calculated as: $label(s) = argmax_c(score(s, c))$.

4 Datasets

The studies discussed in the paper were conducted using four datasets called JAZZ, VOG, RAN30 and RAN1Y. As it was mentioned above all of them were based on a jumping point stimulus, however the time of presentation and a number of point's positions displayed differed for given sets. Other differences regarded a type of an eye tracker used to record eye movements, numbers of users taking part in experiments and a time interval between sessions of an experiment. The detailed information for each set is provided below.

VOG dataset—VOG dataset was obtained with usage of the self-developed VOG head-mounted eye tracker with a single CMOS camera with USB 2.0 interface (Logitech QuickCam Express) possessing 352 × 288 sensor and lens with IR-Pass filter. The camera was mounted on the arm attached to head and was pointing at the right eye. The system generated 20–25 measurements of a center of a pupil per second. The dataset consisted of recordings collected for 26 participants during two sessions separated by three weeks interval. One recording of an eye movement referred to 30 points displayed on a screen, each for 3 seconds. There were 52 recordings in this dataset each including 1400–1500 samples.

JAZZ dataset—The second dataset was obtained using head mounted Jazz-Novo eye tracker (product by Ober-consulting) that records eye positions with frequency 1000 Hz. It uses direct Infra-Red Oculography (IROG) and utilizes pairs of IR emitters and sensors. The optoelectronic transducers are located between the eyes. This set included 48 recordings from two sessions related to 24 participants. A setup for a between session interval, a number and time of stimuli displayed was the same like for the VOG dataset. Each recording consisted of between 99000 to 100000 samples.

RAN30 and RAN1Y datasets—Both RAN30 and RAN1Y datasets were part of the BioEye competition and were recorded using an EyeLink eye-tracker working at 1000 Hz. The raw eye movement signals were subsampled to 250 Hz with the usage of an anti-aliasing filter.

RAN30 dataset was built on the basis of recordings of 153 subjects and was collected during two sessions organized one by one in 30 min (all together 306 recordings). During each session user's task was to follow with eyes 100 points, each of which was shown for one second, which gave 25000 samples for one recording.

RAN1Y dataset consisted of recordings of 37 subjects. The only difference between RAN30 and RAN1Y experiments was the interval between sessions—it was one year in the latter.

5 Comparison of Results

Results obtained from the classification process were studied in terms of an influence of pre-processing phases on a final accuracy of a classification. For this analysis purpose ANOVA test was used to check an existence of significant differences among groups of the above described feature creating methods. In case, when such a difference was found, Tukey's HSD test was applied to determine, which groups exactly differ from each other. Comparing these outcomes with an accuracy of a classification results allowed us to point out method yielding the best results.

In the first step of the analysis the transform type applied to each kind of signals was taken into account. In all four analyzed datasets results of the ANOVA test rejected the null hypothesis, that all groups had identical means.

Deeper studies of differences with usage of Tukey HSD test and classification results revealed that for all sets Wavelet and Cepstrum transforms gave significantly worse results than three other types—normalization (N), Fourier (F) and original signal (S). The latter group (S, N, F) provided better accuracy, however these results turned to be not significant between each other, with one exception, the VOG set.

Subsequently, our attention was paid to measures used for calculating dissimilarity matrix. As it was mentioned above, there were three different distance measures: Dynamic Time Warping, Euclidean distance and Earth Mover Distance (denoted during tests by D, E, M respectively) taken into account. The comparison of these methods using the ANOVA test in conjunction with studies of the classification accuracy revealed that for all four datasets DTW provided the best classification results.

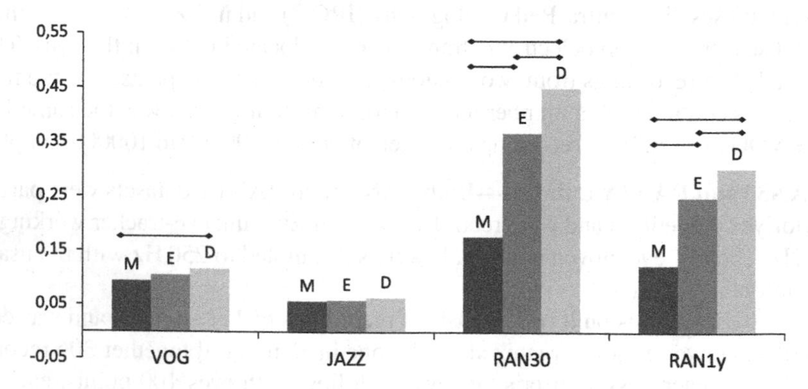

Fig. 1 Mean results of classification for all four sets and every distance (D—DTW distance, E—Euclidean distance, M—EMD distance). Significant differences were marked with *horizontal lines*

The statistical significance was confirmed for RAN30 and RAN1Y, and in regard to EMD method in VOG data set. Only in case of JAZZ set statistically significant differences were not found. All these discussed results are collected in Fig. 1.

It can be seen that for video based eye trackers the results are correlated with frequency sampling as the results for RAN30 and RAN1Y datasets are significantly better than for VOG dataset recorded with much lower frequency. On the other hand, the results of the same method for Jazz-Novo eye tracker—working differently—are worse, despite of its very high recording frequency (1000 Hz). It shows that the method presented in this paper is not sufficient for such kind of data.

6 Fusion of Feature Sets and Final Results

To check the real strength of that method to perform users' identification it was decided to combine feature sets results in a score level voting fusion. For every feature set at first a score for each sample s and class c was determined and then a score for fusion was calculated as: $score_{fus}(s, c) = \sum_{k=1}^{K} score_k(s, c)$ where K is the number of feature sets taken into account. The final label for a sample s was determined based on equation $label_{fus}(s) = argmax_c(score_{fus}(s, c))$.

The aim of this analysis was to find a combination of features, which provides the best classification results for all datasets. Because Wavelet and Cepstrum transforms gave in most cases results significantly worse than the other transforms, and Earth Mover Distance was the worst among distances—feature sets prepared using these pre-processing methods were omitted in the subsequent analysis. Additionally, because Dynamic Time Warping gave the best results for every dataset, it was decided to use feature sets based on this metric in every analyzed combination. As the result, only three transforms (S, N and F) and two distance measures (DTW and EUC) were taken into account. There were different combinations of feature sets checked with a number of feature sets ranging from 27 to 162. The results were also compared with the combination of all feature sets (see Table 2).

The obtained results were examined in terms of an existence of Pearson correlation between datasets. These studies confirmed it for RAN30 and RAN1Y results (0.42) and for RAN1Y and VOG (0.43). The correlation between VOG and RAN30 results is lower but still visible (0.37). What is interesting, the results for JAZZ dataset are negatively correlated with VOG (−0.49). The main reason of this fact is that, contrary to VOG, JAZZ dataset gave quite good results for Fourier based transform.

The best combination of feature sets was *SFN and D*—three transforms: (S) not transformed signal, (F) Fourier, (N) Normalization and: (D) DTW distance measure—with 46.55 % accuracy on average for all datasets. However, it is visible that differences in classification accuracy among datasets are significant and the results are reasonable only for RAN30 and RAN1Y datasets.

Additionally, false rejection and false acceptance rates for R number of tested recordings and different acceptance thresholds th were calculated using Eqs. (1) and (2).

Table 2 Accuracies and Equal Error Rates obtained for different combinations of transforms and distance functions

Transf./distance	Accuracy				EER			
	RAN30 (%)	RAN1Y (%)	VOG (%)	JAZZ (%)	RAN30 (%)	RAN1Y (%)	VOG (%)	JAZZ (%)
SFNCW/DEM	81.1	56.8	34.6	12.5	6.5	18.1	30.8	41.7
S/D	81.1	51.4	**38.5**	8.3	8.1	24.3	35.1	39.6
N/D	78.4	54.1	15.4	8.3	9.4	17.8	**30.8**	45.4
F/D	70.3	51.4	15.4	**16.7**	9.3	18.9	38.5	42.2
SN/D	83.8	54.1	34.6	8.3	6.4	19.8	33.2	38.7
SF/D	78.4	56.8	30.8	12.5	8.1	21.6	38.5	41.7
FN/D	78.4	59.5	23.1	12.5	8.1	16.8	36.4	41.9
SFN/D	81.1	**62.2**	34.6	8.3	6.1	19.6	35.6	42.8
S/DE	**89.2**	54.1	15.4	8.3	8.1	20.9	31.5	41.7
N/DE	73.0	54.1	19.2	12.5	9.8	**16.2**	33.5	41.7
F/DE	75.7	48.6	15.4	12.5	9.5	20.1	40.0	**37.5**
SN/DE	83.8	59.5	26.9	8.3	5.8	18.5	31.2	39.8
SF/DE	83.8	59.5	34.6	8.3	8.1	18.9	35.1	39.6
FN/DE	83.8	54.1	26.9	8.3	6.2	17.2	35.7	45.2
SFN/DE	**89.2**	59.5	26.9	8.3	**5.4**	18.2	33.4	40.7

S—not transformed signal, F—Fourier, N—Normalization, C—Cepstrum, W—Wavelet, D—DTW distance, E—Euclidean distance, M—EMD distance

$$FRR(th) = \frac{R - \sum_{s=1}^{R} a_{s,c(s)}}{R} \tag{1}$$

$$FAR(th) = \frac{\sum_{s=1}^{R} \sum_{j=1, j \neq c(s)}^{C} a_{s,j}}{(C-1)R} \tag{2}$$

where $c(s)$ denotes the class identifier the sample s belongs to and $a_{s,j}$ is given by Eq. (3):

$$a_{s,j}(th) = \begin{cases} 1 & score_{fus}(s,j) > th \\ 0 & otherwise \end{cases} \tag{3}$$

By changing the acceptance threshold, the Equal Error Rate (EER)—error value for the threshold for which FAR and FRR are equal—was calculated for each set. The results are presented in Table 2. The best combination is the same as for accuracy only for RAN30 dataset, there are differences in all other datasets. While accuracy measure deals only with one—the best—result, EER calculation takes into account all results so it may be treated as a better description of model's performance. Results obtained for RAN30 and RAN1Y datasets are acceptable as for eye movement biometric—in fact EER equal to 5.4 % is one of the best results published so far. However, the same method used for VOG and JAZZ datasets achieved significantly higher error rates.

7 Discussion

The primary aim of the experiments presented in this paper was to examine whether usage of exactly the same method for various datasets of eye movements would ensure the similar classification efficiency in all considered cases. It occurred that the results obtained for each of the four datasets used differed substantially in accuracy, however there were some common patterns visible, when comparing performance of different transforms and distance measures. The detailed discussion of these outcomes is provided below.

Analyzing results concerning distance measures it turns out that Dynamic Time Warping method proved to be the best choice for every dataset while Earth Mover's Distance function was the worst one. Additionally the usage of Wavelet and Cepstrum transforms did not offer any improvement to the results in any dataset.

The differences in the accuracy and EER for the same method and different datasets show that every new method to be applied for eye movement based identification—despite of achieving good results for some available dataset—should always be checked against other data collections before any general conclusions about its performance may be presented.

Another issue to explore was to check, which properties of datasets influenced the results. All datasets used were built using a very similar scenario (a jumping point stimulus). Therefore, it was possible to compare results directly. Two of the datasets (RAN30 and RAN1Y) were collected with usage of the same equipment and the correlation of results for these two datasets is visible. The results are also correlated with the VOG dataset, which was created with a similar technique using infrared camera (however with much lower frequency). Interestingly, it turned out that the results for the last of the datasets (JAZZ) are completely different and the correlation is even negative. As the latter dataset was gathered using a device utilizing a completely different technique (IROG) it may be supposed that a type of a device utilized to record data has a significant influence on classification results.

The other interesting conclusion may be the finding, that a pool of participants did not influence results significantly. There were different pools used for RAN30 and RAN1Y datasets and results were similar while almost the same pool of participants was used for both VOG and JAZZ datasets—and the results were different in this case.

Comparing RAN30 and RAN1Y datasets it is visible that, despite of similar distribution of the results, the results for RAN30 dataset are significantly better. The only reason for this may be a different time interval between sessions, which was 30 min for RAN30 and 1 year for RAN1Y dataset. It shows that short term repeatability of eye movement caused by current attitude or mood of a person may significantly (and artificially) improve classification results. It is in line with conclusions derived in [4, 9].

8 Summary

The studies conducted in this research were inspired by the awareness that the assessment of methods used for eye movement data processing and analysis should be done by their comparison with other studies conducted in the same field.

Having different collections of data it is possible to explore an influence of some data pre-processing methods on the final classification result, which was presented in the paper. The research confirmed the existence of both some differences and some patterns when various methods and results obtained for them are taken into account. It allows to suppose that continuing such a type of studies will enable to reach some general conclusions in field of eye movement data biometrics. Because the results presented in this work are far from perfect it indicates that there is a lot of work to be done to lower error rates.

Acknowledgments The authors would like to thank organizers of BioEye 2015 competition for publishing eye movement datasets that were used in this research. We also acknowledge the support of Silesian University of Technology grant BK/263/RAu2/2016.

References

1. Berndt, D.J., Clifford, J.: Using dynamic time warping to find patterns in time series. In: KDD Workshop, vol. 10, pp. 359–370. Seattle, WA (1994)
2. Duin, R.P., Pekalska, E.: The dissimilarity space: Bridging structural and statistical pattern recognition. Pattern Recogn. Lett. **33**(7), 826–832 (2012)
3. Holland, C.D., Komogortsev, O.V.: Complex eye movement pattern biometrics: Analyzing fixations and saccades. In: 2013 International Conference on Biometrics (ICB), pp. 1–8. IEEE (2013)
4. Kasprowski, P.: The impact of temporal proximity between samples on eye movement biometric identification. In: Computer Information Systems and Industrial Management, pp. 77–87. Springer (2013)
5. Kasprowski, P., Harezlak, K.: The second eye movements verification and identification competition. In: 2014 IEEE International Joint Conference on Biometrics (IJCB), pp. 1–6. IEEE (2014)
6. Kasprowski, P., Komogortsev, O.V., Karpov, A.: First eye movement verification and identification competition at btas 2012. In: 2012 IEEE Fifth International Conference on Biometrics: Theory, Applications and Systems (BTAS), pp. 195–202. IEEE (2012)
7. Kasprowski, P., Ober, J.: Eye movements in biometrics. In: Biometric Authentication, pp. 248–258. Springer (2004)
8. Kasprowski, P., Ober, J.: Enhancing eye-movement-based biometric identification method by using voting classifiers. In: Defense and Security, International Society for Optics and Photonics, pp. 314–323 (2005)

9. Kasprowski, P., Rigas, I.: The influence of dataset quality on the results of behavioral biometric experiments. In: 2013 International Conference of the Biometrics Special Interest Group (BIOSIG), pp. 1–8. IEEE (2013)
10. Komogortsev, O.V., Jayarathna, S., Aragon, C.R., Mahmoud, M.: Biometric identification via an oculomotor plant mathematical model. In: Proceedings of the 2010 Symposium on Eye-Tracking Research & Applications, pp. 57–60. ACM (2010)
11. Komogortsev, O.V., Karpov, A., Price, L.R., Aragon, C.: Biometric authentication via oculomotor plant characteristics. In: 2012 5th IAPR International Conference on Biometrics (ICB), pp. 413–420. IEEE (2012)
12. Komogortsev, O.V., Khan, J.I.: Eye movement prediction by kalman filter with integrated linear horizontal oculomotor plant mechanical model. In: Proceedings of the 2008 Symposium on Eye Tracking Research & Applications, pp. 229–236. ACM (2008)
13. Komogortsev, O.V., Rigas, I.: Bioeye 2015: competition on biometrics via eye movements. In: 2015 IEEE 7th International Conference on Biometrics Theory, Applications and Systems (BTAS), pp. 1–8. IEEE (2015)
14. Rigas, I., Economou, G., Fotopoulos, S.: Biometric identification based on the eye movements and graph matching techniques. Pattern Recog. Lett. **33**(6), 786–792 (2012)
15. Rubner, Y., Tomasi, C., Guibas, L.J.: The earth mover's distance as a metric for image retrieval. Int. J. Comput. Vis. **40**(2), 99–121 (2000)
16. Shen, C., Cai, Z., Guan, X., Du, Y., Maxion, R.A.: User authentication through mouse dynamics. IEEE Trans. Inf. Forensics Secur. **8**(1), 16–30 (2013)

Epigenetically Inspired Modification of Genetic Algorithm and His Efficiency on Biological Sequence Alignment

Kornel Chromiński and Mariusz Boryczka

Abstract In this paper the modification of genetic algorithm inspired by the epigenetic process is presented. The results of the efficiency of the proposed modified algorithm are compared with standard genetic algorithm and a tool which does not use evolutionary processes.

Keywords Genetic algorithm · Epigenetics · Sequence alignment

1 Introduction

The article proposes an original modified genetic algorithm (GA) for comparison of DNA, RNA and protein sequences. Comparison of sequences is of a key importance in modern studies of life on Earth. Sequence comparison is used to determine the common origin of sequences, their phylogenetic classification as well as studies on the functions and structure of a given sequence.

Classification itself, can be also done in a different way, for example with other computational intelligence method: artificial neural networks [5, 10, 11]. Finding the best sequence alignment is not an easy task. The method proposed by the authors was inspired by epigenetic processes occurring in all living organisms, and aims to improve the quality of sequence alignment.

K. Chromiński (✉)
Institute of Technology and Mechatronics, University of Silesia, 12, Żytnia St.,
41-200 Sosnowiec, Poland
e-mail: kornel.chrominski@us.edu.pl

M. Boryczka
Institute of Computer Science, University of Silesia, 39, Będzińska St.,
41-200 Sosnowiec, Poland
e-mail: mariusz.boryczka@us.edu.pl

I. Czarnowski et al. (eds.), *Intelligent Decision Technologies 2016*,
Smart Innovation, Systems and Technologies 57,
DOI 10.1007/978-3-319-39627-9_9

2 Background: Genetic Algorithms

GAs are algorithms inspired by mechanisms occurring in inheritance and evolution processes. Due to their specificity and aspiration to obtain the most well-adjusted individuals, they are usually applied for optimisation purposes. A genetic algorithm operates on a specific population of possible solutions with the use of operations inspired by the evolution of living organisms, such as inheritance, mutation and selection [4, 8].

When solving problems using GA, it is necessary first to design appropriately those functions used for the evaluation of individuals. Individuals are subject to evolution during operation of the algorithm, and an adaptation function is the measure to evaluate the individuals adaptation. A group of individuals creates a so-called population. Each individual of the population is one of possible solutions to the problem with which the genetic algorithm is faced and should be created randomly. Stages of operation of a genetic algorithm are as follows:

1. The operation begins choosing the method of creating the initial population.
2. The adaptation function is calculated for each individual.
3. The next step is **selection**, that is, to choose individuals that will take part in the crossover operation.
4. The next step is the crossover, which involves exchanging the genotype between individuals. The crossover leads to the creation of new individuals.
5. The next operation is the **mutation**. This means that a random change will occur with some probability (usually, a relatively low one) in a part of an individuals sequence from the population.

The steps 2–5 of the GAs operation are repeated until a stopping condition occurs. A specific number of iterations or a specific value of the adaptation function may serve as the stopping condition.

3 Motivation: Alignment of DNA, RNA and Protein Sequences with the Use of Genetic Algorithms

Detecting similarities between sequences is an extremely complicated process. Comparing DNA, RNA and protein sequences is of a special importance for determining biological processes for which a given sequence is responsible and, in the case of proteins, also to determine processes in which they take part [2]. The basic assumption is that biological sequences of a similar construction usually have similar functions. The alignment of sequences is also frequently used in studies on the evolution of living organisms. One of the tools used to find the best sequence alignment are genetic algorithms. The most frequently used one is GAMSA (Genetic Algorithm for

Fig. 1 Creating an individual in the initial population

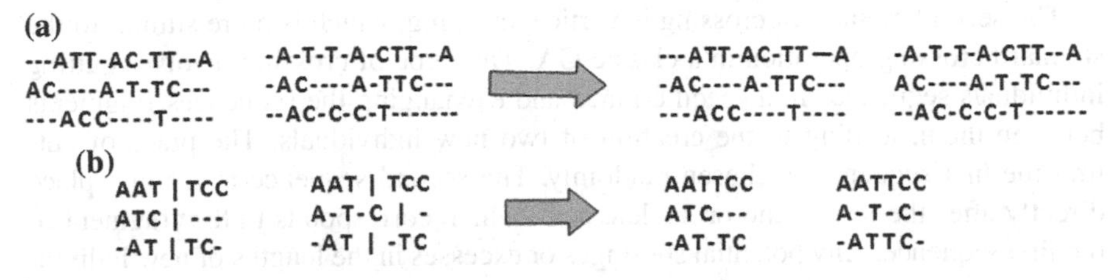

Fig. 2 Horizontal **a** and vertical **b** crossing operation in GAMSA

Multiple Sequence Alignment) [6, 7]. When it comes to operating design, GAMSA does not differ from a standard GA, but what is different, are its operations of population creation, crossing and mutation.

The initial population is created on the basis of sequences to be compared: each individual in the population has all sequences to be compared. In such a case, due to the specificity of the problem solved, it is not possible to change the location of nucleic or amino acids in the sequence as it would change the set-up sequence into a completely different one. In order to make it possible to make genetic operations, encoding of specific individuals in the population is complemented by entering spaces (blank spaces) in the sequence. A limitation concerning the number of spaces entered is such that their number must not be bigger than the length of the sequence to which they are entered. The remaining sequences are completed by spaces inserted in random places in such a number that their length is the same as of the longest sequence containing spaces. Figure 1 features an example of creating an individual in the GAMSA algorithm on the basis of a set-point sequence.

In order to calculate the adaptation function of each individual in the GAMSA algorithm, sequences are compared on the basis of the each for each rule, that is, two sequences are compared at once. One of the methods of calculating the adaptation function encountered in the literature is giving rewards and punishments in accordance to the following rules [7, 9]: +1 for each alignment (the same amino or nucleic acid is present in two sequences on the same position); −1 for non-compliance (there are different amino or nucleic acids on the same position); −2 punishment (for space: if there is an amino or nucleic acid in one of the sequences, and there is space in the other one).

The crossover operation in GAMSA differs from a standard crossover operation done in GA. In GAMSA, there are two methods of crossover and both of them occur simultaneously during operation of the algorithm. The first method of individual crossover is the horizontal approach. During this method of crossing, one entire sequence is exchanged Fig. 2a.

Fig. 3 Mutation operation

The second method of crossing is vertical crossing, which is more similar to the standard crossing operation in a classic GA. This type of crossing involves cutting individuals sequences in a specific place and exchanging the sequences fragments between them, leading to the creation of two new individuals. The place of cutting the first sequence is chosen randomly. The second sequence is cut in a place directly after the last amino or nucleic acid, which corresponds to the fragment of the first sequence. Any potential shortages or excesses in the lengths of new individuals sequences, are eliminated by adding or removing finishing spaces, (see Fig. 2b).

Another operation, that in GAMSA algorithm is different than in a standard GA, is mutation. The first important difference is the probability of mutation. In a classic genetic algorithm mutations occur with a low probability of a very low percent, while in case of GAMSA, the probability of mutation ranges from 50 % even up to 100 %. Another important difference is that the mutation operation in the GAMSA algorithm involves only random addition or deletion of spaces in a sequence (as a consequence of the assumption of consistency of sequences to be compared). The diagram of mutation is shown in Fig. 3.

4 Epigenetically Inspired Modification of the GAMSA

This section presents the modified GAMSA algorithm (named GAMSA_EPI) proposed by us, and its application within a comparison of biological sequences. The proposed modification of the algorithm involves extending the range of changes occurring in individuals of the population in the process of evolution with a change inspired by an epigenetic process. In this section, apart from a short introduction to epigenetics, there is also a detailed description of the implemented modification.

4.1 Epigenetics: Basic Information

Epigenetics is the science that studies non-genetic-inheritance processes as well as the influence of external factors on expression of genes. Expression of genes determines phenotype features of an individual, that is to say, the individuals adaptation to the environment, way of behavior and appearance. For a long time, scientists were wondering why there was a difference in the appearance and behavior of identical twins or why a clone, even though it had identical sequences of genes as the individual that had been cloned, had different colors. Some of those questions were answered thanks to the discovery of mechanisms that are subject of epigenetic studies.

The issue of epigenetics rose at the moment when it was discovered that some changes in the genotype of living organisms are not directly related to the DNA structure and its changes—that is, when scientists started to think about the cause of such changes. Studies discovered numerous particles that influence the way in which genetic code will be read, that is, how the phenotype of an individual will change. It turned out that epigenetic processes play an important role in the diversification of a population and its adaptation to new conditions.

Today we can say that the genotype of living organisms is a repository on which the genetic information is stored and epigenetic processes are a kind of controller responsible for activation of some information and for taking actions on its basis. Epigenetic processes include, e.g., [3] Prion inheritance, Cytosine methylation, Deactivation of a single chromosomes participation in the process of reading a pair of chromosomes. This article describes the modification of the algorithm that imitates activity of prions.

4.2 GAMSA_EPI Algorithm

The proposed modification of the algorithm is based on the epigenetic process of prion inheritance. Prions are protein particles occurring in every living organism. In normal conditions, prions are inactive and they do not influence the inheritance process; however, there may be an external impulse that will result in activation of a prion (transmission from a neutral to an invasive state). Such a transmission usually occurs in every (or in the majority) of representatives of a given population that were subject to the same external factor. In their invasive state, prions may be responsible for diseases (negative changes) or for evolutionary changes, which are positive as they allow an organism to adapt to new conditions. Prions may also have an influence on the process of gene expression and affect the genetic code of an organism by embedding a prion sequence to the inherited sequence.

In the proposed modification, the prion inheritance process in a population subject to one external factor was recreated. In the modified algorithm, GAMSA_EPI, apart from standard processes such as crossing and mutation, there is another modifying process: individuals from a given population are, with some probability, under the risk of being exposed to an external factor. In short, the activation of this factor will result in changes in the code of the individuals, the change will be constant for all individuals and occur in the same sequence place for all individuals, what is presented on Algorithm 1.

Algorithm 4.1: The prion inheritance process in GAMSA_EPI algorithm

Data: population
Result: modified population;
Generate prion from the best individual;
while *the certain percentage of the population are not infected by prion* **do**
  Get a individual;
  if *the individual is not best individual* **then**
    insert prion to individual genotype;
  end
end

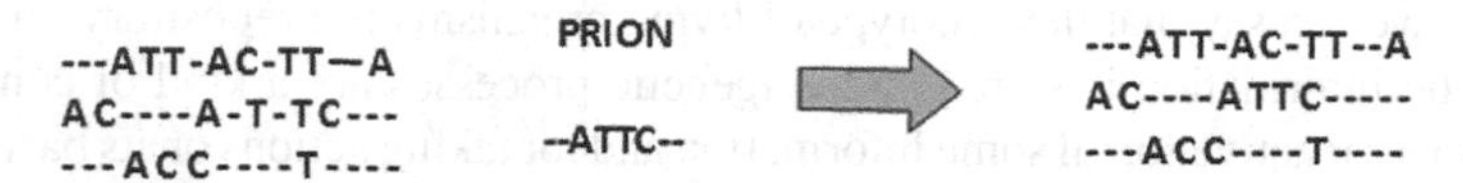

Fig. 4 Prion inheritance process

In GAMSA_EPI algorithm, 5–25 % of the entire code of an individual is subject to epigenetic modification. The prion is extracted in random way and with random length from the individual with the highest value of adaptation function. Prion is a fragment of the sequence of the best individual, prion length comprise 5–25 % length of the sequence of the best individual. Prion's length from a such range, gives the best value of adaptation function in GAMSA_EPI algorithm. This is in order to protect the possibility of obtaining the best solution: the individual with the highest value of the adaptation function is the strongest and will survive the changes in the environment. Figure 4 shows the prion inheritance process. The parameter that decides on the quality of the algorithms operation is the parameter describing the probability of the change resulting from activation of a virtual prion.

5 Experiments

This section discusses experiments results regarding selection of the optimal probability of occurrence of changes resulting from prion activation. The efficiency of the modification, in comparison to the algorithm without the modification, was studied and then compared with the results obtained using a standard sequence comparison algorithm, that is, ClustalW.

A basic algorithm to which modifications were implemented and which was used to compare the efficiency of the modification was GAMSA algorithm. The GAMSA and GAMSA_EPI algorithms involved two methods of crossing individuals occurring with a probability of 100 % and a mutation occurring with a probability of 80 % [1, 6]. Both algorithms were implemented by authors in *R* language.

Table 1 Sequences use in experiment

Group	Sequence	Length (bp)	Description	Organism
Group_1	Seq_1	70	The DNA sequence responsible for coding a MnSOD protein	Homo sapiens
	Seq_2	23		Mus Musculus
	Seq_3	52		Homo sapiens
Group_2	Seq_4	99	Fragment of protein sequence FcRn	Homo sapiens
	Seq_5	88		Bubalus bubalis
	Seq_6	34		Homo sapiens
Group_3	Seq_7	66	The RNA sequence coding protein S8 (A, B, C)	Bacilus stearothermpophilus
	Seq_8	49		
	Seq_9	46		

In the studies, three groups with three sequences (each taken from the databases available on National Center for Biotechnology Information) were used. Descriptions of the sequences are included in Table 1.

The initial population for all groups was created in the same way. The longest sequence was completed by spaces in a random manner. The spaces corresponded to 10 % of the entire length of the sequence and the remaining sequences were completed with the blank spaces (also in random places) so that their length was the same as the length of the longest sequence.

5.1 Selection of the Optimal GAMSA_EPI Parameter

The results of GAs depend on their parameters; this is why the first experiment was aimed at determination of the probability of epigenetic modification in the GAMSA_EPI algorithm at which the modified algorithm will obtain the best results (the best sequence alignment).

Table 2 shows values of the adaptation function for GAMSA_EPI in accordance to the probability of occurrence of the modification as well as the number of the algorithms launches (the number of generations modified with the algorithm). The results for GAMSA_EPI were compared with results obtained using the GAMSA algorithm. The values presented are an average value of the adaptation function for 100 launches of the algorithm. The higher the value of the adaptation function, the better. The best average adaptation function value for the given number of generations in the algorithm are presented in italics.

Table 2 Value of the adaptation function according to the number of generations and the probability of changes in the GAMSA_EPI algorithm

Group_1	GAMSA_Epi						
	Probability of epigenetic modification						GAMSA
Iteration	5 %	10 %	20 %	50 %	70 %	100 %	
5	−226.12	*−225.06*	−226.06	−227.60	−228.86	−227.86	−228.73
10	−213.14	−214.80	*−211.53*	−217.06	−220.40	−214.73	−214.80
15	*−206.48*	−206.60	−207.73	−208.46	−212.20	−207.53	−212.13
20	−204.32	−200.80	*−203.86*	−205.20	−206.06	−204.26	−209.13
30	−203.60	−203.06	−201.86	*−201.66*	−204.00	−204.20	−209.53
50	−203.00	−202, 66	−200, 93	*−199.00*	−201.20	−203.53	−211.66
100	−202.44	−201.53	−199.66	*−196.80*	−198.00	−202.53	−211.66
150	−202.03	−203.33	−199.20	*−193.86*	−194.60	−203.93	−212.20

On the basis of Table 2, it is possible to state that the optimal probability of occurrence of the proposed modification in GAMSA_EPI, in the case of a higher level of iteration, is 50 %. At this probability, the algorithm usually obtains the best average value of the adaptation function, which means better sequence alignment. In comparison to the basic GAMSA algorithm, in the majority of cases, the GAMSA_EPI algorithm obtained a better value of the adaptation function. On the basis of Table 2 we can notice that the modification improved the efficiency of the GA. In the next section, there is a comparison between the original GAMSA_EPI algorithm and the results obtained using the standard GAMSA algorithm and the ClustalW tool.

5.2 *Comparison of Sequences Using GAMSA, GAMSA_EPI and ClustalW*

A study regarding the efficiency of algorithms was conducted on three groups of sequences presented in Table 1. The sequence groups were compared using the GAMSA_EPI algorithm proposed in this article, the original GAMSA algorithm and the ClustalW tool. In the case of ClustalW, the best sequence alignment was evaluated by calculating the adaptation function for an alignment found by the tool in the same way it was used in the case of *GA*. The results obtained by sequences in Group_1, Group_2 and Group_3 are shown in Figs. 5, 6 and 7.

The Figs. 5, 6 and 7 show the average value of the adaptation function for 100 launches of the algorithms according to the number of iterations of the algorithm. The diagrams also include the values of the adaptation functions for the ClustalW tool.

On the basis of Fig. 5, it is possible to assert that even at thirteen iterations in a genetic algorithm (number of populations in the algorithm), they obtain better alignment of compared sequences than the ClustalW tool. In addition, the modified

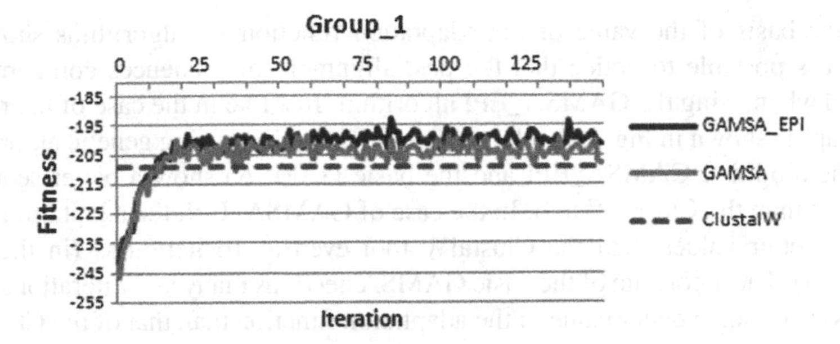

Fig. 5 Average value of the adaptation function for GAMSA_EPI, GAMSA, and ClustalW

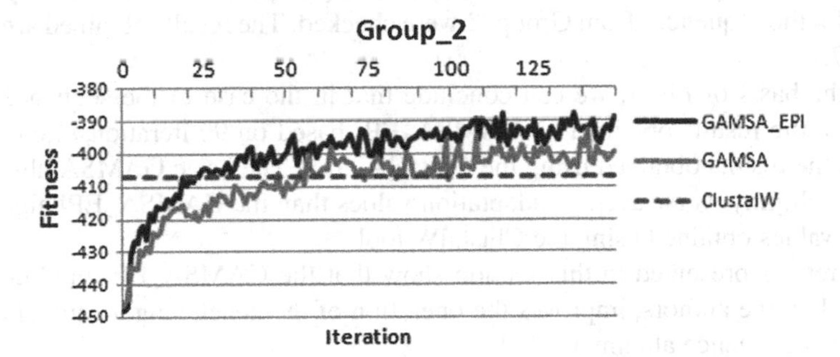

Fig. 6 Values of the adaptation function for GASMA_EPI and GASMA genetic algorithms and the ClustalW tool according to the population number in the algorithm

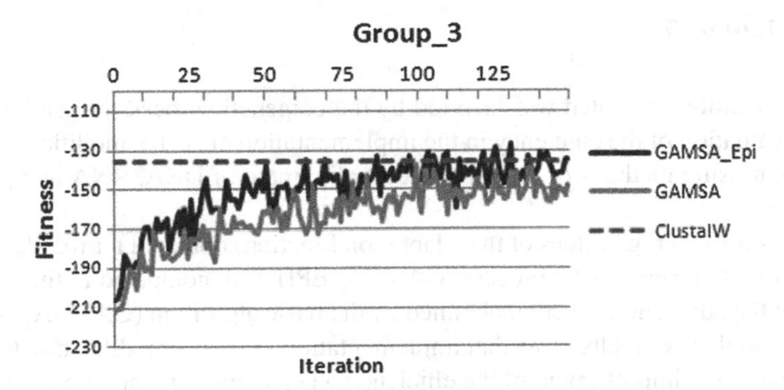

Fig. 7 Average value of the adaptation function of GA according to the iterations number and the value of the adaptation function of the ClustalW tool

GAMSA_EPI algorithm proposed in the article obtained better results of the adaptation function than the basic GAMSA algorithm. Such an approach makes the strongest individual in the population also subject to the mutation operation.

In the next step of the testing process, the efficiency of the GAMSA_EPI algorithm applied to the sequence group from Group_2 was checked. The results of the efficiency evaluation of the modified algorithm are shown in Fig. 6.

On the basis of the value of the adaptation function for algorithms shown in Fig. 6, it is possible to notice that the best alignment of sequences compared are obtained when using the GAMSA_EPI algorithm. Just like in the case of the results for Group_1, shown in Fig. 5, results for Group_2 obtained using genetic algorithms (both the modified GAMSA_EPI and the basic GAMSA) show a better sequence alignment than the ClustalW tool. In the case of GAMSA_EPI, the adaptation function has better values than the ClustalW tool even at 16 iterations (in the 16th generation). The algorithm of the basic GAMSA needs as many as 57 iterations (generations) to obtain a better value of the adaptation function than that of the ClustalW tool.

Just like in the case of the previous sequence groups, results obtained by algorithms for the sequences from Group_3 were checked. The results obtained are show in Fig. 7.

On the basis of Fig. 7, we can conclude that in the case of the sequence from Group_3, the results obtained by GAMSA_EPI based on 90 iterations are comparable to the results obtained using the ClustalW tool. The basic GAMSA algorithm obtained slightly worse average adaptation values than the GAMSA_EPI algorithm and the values obtained using the ClustalW tool.

The results presented in this section show that the GAMSA_EPI modification, proposed by the authors, improves the operation of the genetic algorithm, allowing for a better sequence alignment.

6 Conclusion

The modification presented was inspired by the epigenetic process of prion inheritance. Evaluation of the efficiency in the implementation of such a modification was carried out using methods of finding the best alignment of DNA, RNA and protein sequences.

The results (average values of the adaptation function) obtained for the algorithm extended with the epigenetic process (GAMSA_EPI) were compared to the average values of the adaptation function obtained by the basic algorithm (GAMSA) and the ClustalW tool. The results show that implementation of such a modification to a *GA* may result in the improvement of the efficiency when comparing protein sequences, and that by applying this approach the best sequence alignment can be obtained. On the basis of the results presented, it may be concluded that genetic algorithms (GAMSA and GAMSA_EPI) give better results of sequence comparison than a tool not based on a genetic algorithm, such as ClustalW.

The presented modification may also increase the efficiency of genetic algorithms in other applications, what will be the subject of future researches.

References

1. Agarwal, P., Chauhan, R.: Alignment of multiple sequences using ga method. Int. J. Emerg. Technol. Comput. Appl. Sci. **4**, 411–421 (2013)
2. Anbarasu, A., Narayanasamy, P., Sundararajan, V.: Multiple molecular sequence alignment by island parallel genetic algorithm. Curr. Sci. **78**, 858–863 (2000)
3. Carey, N.: The Epigenetics Revolution: How Modern Biology is Rewriting Our Understanding of Genetics, Disease, and Inheritance. Columbia University Press (2013)
4. Goldberg, D.E.: Genetic Algorithms in Search. Scientific-Technical Publisher, Warsaw (2003). (in Polish)
5. Górny, A., Tkacz, M.A.: Using artificial neural networks for processing data gained via opendap and consolidated from different databases on distributed servers. In: Szczepaniak, P.S., Kacprzyk, J., Niewiadomski, A. (eds.) Advances in Web Intelligence Third International Atlantic Web Intelligence Conference, 2005. LNCS, vol. 3528, pp. 176–182. Springer (2005)
6. Gupta, R., Agarwal, P., Soni, A.: Genetic algorithm based approach for obtaining alignment of multiple sequences. Int. J. Adv. Comput. Sci. Appl. **3**(12), 180–185 (2012)
7. Manning, T., Sleator, R., Walsh, P.: Naturally selecting solutions: the use of genetic algorithms in bioinformatics. Bioengineered **4**(5), 266–278 (2013)
8. Michalewicz, Z.: Genetic Algorithms + Data Structure = Evolutionary Program. Scientific-Technical Publisher, Warsaw (2004). (in Polish)
9. Radenbaugh, A.J.: Applications of Genetic Algorithms in Bioinformatics. San Jose State University, Master Thesis (2008)
10. Tkacz, M.: Artificial neural networks in incomplete data sets processing. In: Kłopotek, M.A., Wierzchoń, S.T., Trojanowski, K. (eds.) IIS: IIPWM'05, pp. 577–584. Advances in Soft Computing, Springer (2005)
11. Tkacz, M.: Artificial neural network resistance to incomplete data. In: Kłopotek, M.A., Wierzchoń, S.T., Trojanowski, K. (eds.) IIS: IIPWM'06, pp. 437–443. Advances in Soft Computing, Springer (2006)

Meta-Bayes Classifier with Markov Model Applied to the Control of Bioprosthetic Hand

Marek Kurzynski and Marcin Majak

Abstract The paper presents an advanced method of recognition of patient's intention to move of multijoint hand prosthesis during the grasping of objects. In the considered decision problem we assume that each prosthesis operation can be divided into sequence of elementary actions and the patient's intention means his will to perform a specific elementary action. A characteristic feature of the explored sequential decision problem is the dependence between its phases at particular instants which should be taken into account in the recognition algorithm. The proposed classification method is based on multiclassifier (MC) system working in sequential fashion, dedicated to EMG and MMG biosignals and with dynamic combining mechanism using the Bayes scheme and Markov model of dependences. The performance of proposed MC system with 3 different types of base classifiers was experimentally compared against 3 sequential classifiers for 1—and 2-instant backward dependence using real data concerning the recognition of six types of grasping movements. The results obtained indicate that use of MC system dedicated to the sequential scheme of recognition process, essentially improves performance of patient's intent classification and that this improvement depends on the type of base classifiers and order of dependence.

Keywords Bioprosthesis · EMG signal · MMG signal · Multiclassifier system · Sequential recognition · Probabilistic model

1 Introduction

The importance of hands in human life cannot be estimated. The loss of even a single hand significantly reduces the human activity. The people who have lost their hands are doomed to permanent care. Restoring to these people even a hand sub-

M. Kurzynski (✉) · M. Majak
Department of Systems and Computer Networks,
Wroclaw University of Technology, Wroclaw, Poland
e-mail: marek.kurzynski@pwr.edu.pl

I. Czarnowski et al. (eds.), *Intelligent Decision Technologies 2016*,
Smart Innovation, Systems and Technologies 57,
DOI 10.1007/978-3-319-39627-9_10

stitute makes their life less onerous. The hand transplantations are still in a medical experiment, mainly due to the necessity of immunosuppression [17]. An alternative is to equip these people with cybernetics prostheses.

The activity of human organism is reflected in characteristic biosignals, which can be measured and next can be applied to the control of the work of technical devices. Electrical potentials accompanying skeleton muscles (called EMG signals) are an example of such biosignals. Through the tensing of these muscles, the disabled person may express his/her intentions as to the workings of the prosthesis [1, 2, 5, 15, 18, 22]. Nevertheless, reliable recognition of intended movement using only the EMG signals analysis is a hard problem hence any attempt to obtain better classification methods and algorithms is fully justified.

According to the author's recent experience [11, 13, 18, 19], increasing the efficiency of the recognition stage may be achieved through the following activities:

1. by introducing the concept of simultaneous analysis of two different types of biosignals, which are the carrier of information about the performed hand movement—the EMG and mechanomiographic (MMG) signals;
2. by using sequential classification scheme which is based on decomposition of hand movement on a sequence of elementary actions with Markov model;
3. through the use of multiclassifier system with base classifiers dedicated to the particular steps of sequential recognition procedure.

The bioprosthesis control system developed in this study includes the above mentioned ideas within a common concept in contrast to the earlier author's works where above suggestions were considered separately. Taking into account above ideas, the paper aims to solve the problem of recognition of the patient's intention to move the multiarticulated prosthetic hand during grasping and manipulating objects in a skillful manner, by measuring and analyzing multimodal signals coming from patient's body. The adopted solution takes into consideration the advantages given by the fusion of the EMG and MMG signals in the original sequential MC system based on the Bayes paradigm and Markov model of dependences among elementary actions.

In the proposed MC system new method of dynamic fusion of base classifiers is developed. The method is dedicated to the sequential recognition scheme with probabilistic model. This specificity of the MC system is visible through the pool of base classifiers which are associated with the particular stages of classification process and the trainable mechanism of fusion based on probabilistic properties of base classifiers.

The paper arrangement is as follows. Chapter 2 includes the concept of prosthesis control system based on the recognition of patient's intention in the sequential scheme and provides an insight into steps of the whole decision control procedure. Chapter 3 presents the key sequential recognition algorithm based on the multiclassifier system with Markov model of trainable combining algorithm. The experiments conducted and the results with discussion are presented in Chap. 4. The paper is concluded in Chap. 5.

2 Bioprosthetic Hand Control System

As mentioned above, the bioprosthesis control is performed by recognizing its intended movement on the base of classification of EMG and MMG signals from user arm stump. This requires the development of three stages: (1) acquisition of signals; (2) reduction of dimensionality of their representation; (3) classification of biosignals (recognition of patient's intention).

Biosignal acquisition and analysis processes influence essentially on the reliability of recognition of prosthesis motion control decisions. The acquisition process should take into account the nature of the measured signals and their measurement conditions [3]. For the needs of experimental research presented in Chap. 4 the special EMG/MMG biosignals measuring system was constructed. The system fully meets the above requirement, mainly due to the use of differential amplifiers, which eliminate interferences in EMG signals and special casing, which isolate the microphone from the external sound sources for MMG signals [14].

After the acquisition stage, the recorded signals have the form of strings of discrete samples. Their size is the product of measurement time and sampling frequency. For a typical motion action, that gives a record of size between 5 and 7 thousand of samples per channel (time of the order of 5–7 s, and 1 kHz sampling). This primary representation of the signals hinders the effective classification and requires the reduction of dimensionality. This reduction leads to a representation in the form of a signal feature vector. In this study, the sequence of autoregressive (AR) model and principal component analysis (PCA) is proposed as a feature extraction and reduction methods, respectively. Former experimental research showed, that both AR and PCA algorithms are effective methods in respect of the recognition error and the calculation costs in the biosignal analysis [7, 11].

In the considered control concept we assume that each prosthesis operation consists of specific sequence of elementary actions, and the patient's intention means his will to perform a specific elementary action [18]. Thus, prosthesis control is a discrete process where at the nth stage ($n = 1, 2, \dots, N$) occurs successively:

1. the measurement of EMG and MMG signal parameters (results of AR and PC analysis) x_n ($x_n \in \mathcal{X} \subseteq \mathcal{R}^d$), that represents patient's will j_n ($j_n \in \mathcal{M} = \{1, 2, \dots, M\}$) (the intention to take a particular action);
2. the recognition of this intention (the result of recognition at the nth stage will be denoted by $i_n \in \mathcal{M}$);
3. the realization of an elementary action $a_n \in \mathcal{A}$, uniquely defined as a recognized intention. This means that there is M number of elementary actions $\mathcal{A} = \{a^{(1)}, a^{(2)}, \dots, a^{(M)}\}$—an exemplary meaning of elementary actions in relation to a dexterous hand prosthesis is defined in Sect. 4.

The assumed character of control decisions (performing an elementary action) means that the task of bioprosthesis control is reduced to the recognition of the patient's intent in successive stages on the basis of the available measurement information. Since the patient's current intention depends on history, the specificity of the investigated classification task reveals in the form of input data, which are not associated only with the direct EMG and MMG signals parameters that manifest the current intention, but comprise up to an extend the historic information that regards the preceding course of control process. In the general case, we suppose that the decision algorithm at the nth instant takes into account the K-instant-backwards-dependence ($K < n$). It means, that decision at the nth instant is made on the base of vector of features

$$\bar{x}_n^{(K)} = (x_{n-K}, x_{n-K+1}, \dots, x_{n-1}, x_n). \tag{1}$$

In consequence, the classification algorithm at the nth instant is of the following form:

$$\Psi_n(\bar{x}_n^{(K)}) = i_n,\ i_n \in \mathcal{M}. \tag{2}$$

Figure 1 shows the block diagram for the complete dynamic process of bioprosthesis control in the explored sequential decision problem. In this study, multiclassifier systems will be applied as classifiers (2) for the particular instances of sequential recognition. In the proposed MC systems, both the pool of base classifiers and the combining mechanism will be constructed using the supervised learning procedure, what leads to the assumption that a learning set $\mathcal{S}$ and a validation set $\mathcal{V}$ are available [8]. In the considered sequential decision problem, the learning set $\mathcal{S}$ consists of m training sequences:

$$\mathcal{S} = \{S_1, S_2, \dots, S_m\}, \tag{3}$$

where a single sequence

$$S_k = ((x_{1,k}, j_{1,k}), (x_{2,k}, j_{2,k}), \dots, (x_{N,k}, j_{N,k})) \tag{4}$$

denotes a single-patient sequence of prosthesis activity that comprises N EMG and MMG signals observation instants, and the patient's intentions.

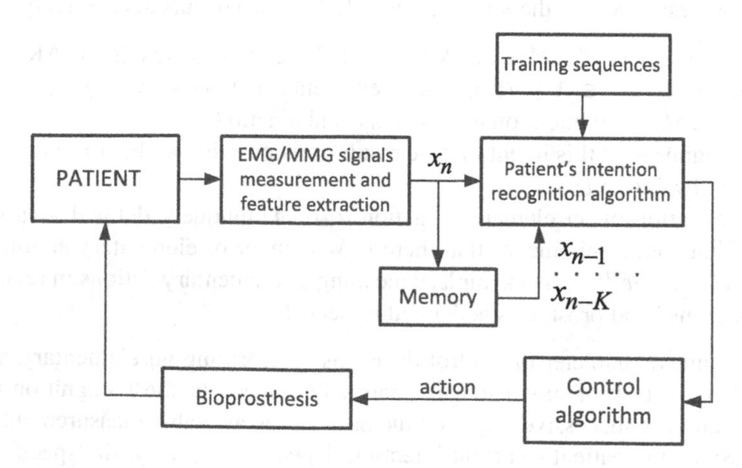

Fig. 1 System of bioprosthesis control via sequential recognition of patient's intentions

Similarly, the validation set V consists of r validation sequences $\mathcal{V} = \{\mathcal{V}_1, \mathcal{V}_2, \dots, \mathcal{V}_r\}$ and a single sequence $\mathcal{V}_k$ has the same form as in (4). The next section, describes the procedure of determining the original MC systems (2) using learning set S and validation set $\mathcal{V}$, in detail.

3 Multiclassifier System

3.1 Preliminaries

The proposed multiclassifier system is built as a combination of the two following probabilistic paradigms:

Markov Model. We will treat the sequential recognition task as a discrete dynamical process, in which the patient's intents in successive stages $j_1, j_2, \dots, j_N$ are observed values of sequence of random variables $\mathbf{J}_1, \mathbf{J}_2, \dots, \mathbf{J}_N$ modeled by first-order Markov chain. The probabilistic formalism for such a dependence is given by the initial probabilities

$$p_{j_1} = P(\mathbf{J}_1 = j_1) \tag{5}$$

and by the transition probabilities

$$p_{j_n j_{n-1}} = P(\mathbf{J}_n = j_n | \mathbf{J}_{n-1} = j_{n-1}). \tag{6}$$

Meta Bayes Classifier. In the concept of Meta Bayes Classifier (MBC), which originally was introduced in [12] we suppose that a base classifier ψ is given, which maps feature space into a set of class numbers, viz.

$$\psi : \mathcal{X} \longrightarrow \mathcal{M}. \tag{7}$$

The MBC ψ^{MBC} constitutes the specific probabilistic generalization of base classifier (7) which has the form of the Bayes scheme built over the classifier ψ. This means, that ψ^{MBC} takes the decision according to the maximum *a posteriori* probability rule:

$$\psi^{MBC}(\psi(x) = k) = i \longleftrightarrow P(i|\psi = k) = \max_{l \in \mathcal{M}} P(l|\psi = k). \tag{8}$$

3.2 Fusion of Base Classifiers

Suppose first, that we have the set of N trained base classifiers:

$$\psi_1(x_1), \psi_2(x_2), \dots, \psi_N(x_N), \tag{9}$$

which classify the patient's intents at the 1st, 2nd, ..., Nth instant, respectively.

The MC system (2) for nth instant is defined as the MBC classifier (8) constructed over the set of base classifiers (9) for nth, $(n-1)$th, ..., $(n-K)$th instants, namely:

$$\Psi_n(\bar{x}_n^{(K)}) = \psi^{MBC}(\psi_{n-K}(x_{n-K}) = i'_{n-K}, \ldots, \psi_{n-1}(x_{n-1}) = i'_{n-1}, \psi_n(x_n) = i'_n). \quad (10)$$

The MC system (10) produces the decision about the patient's intent at the nth instant according to the generalized rule (8):

$$\Psi_n(\bar{x}_n^{(K)}) = i_n \longleftrightarrow P(i_n|\psi_{n-K}(x_{n-K}) = i'_{n-K}, \ldots, \psi_n(x_n) = i'_n) =$$

$$= \max_{l \in \mathcal{M}} P(l|\psi_{n-K}(x_{n-K}) = i'_{n-K}, \ldots, \psi_n(x_n) = i'_n), \quad (11)$$

where:

$$P(i_n|\psi_{n-K} = i'_{n-K}, \ldots, \psi_n = i'_n) = \frac{P(i_n, \psi_{n-K} = i'_{N-k}, \ldots, \psi_n = i'_n)}{P(\psi_{n-K} = i'_{N-k}, \ldots, \psi_n = i'_n)}. \quad (12)$$

Since denominator in (12) has no influence on the classification result of algorithm (11), classifying function of (11) reduces to the nominator, which—assuming that base classifiers (9) are conditionally independent—after simple calculations has the following form:

$$P(i_n, \psi_{n-K} = i'_{n-K}, \ldots, \psi_n = i'_n) = P(\psi_n = i'_n|i_n)\times$$

$$\times \sum_{j_{n-1}} P(\psi_{n-1} = i'_{n-1}|j_{n-1})p_{i_n j_{n-1}} \times \cdots \times \sum_{j_{n-K}} P(\psi_{n-K} = i'_{n-K}|j_{n-K})p(j_{n-K}). \quad (13)$$

The key element in the algorithm (13) presented above is the calculation of probabilities $P(\psi_n = i_n|j_n)$, i.e. class-dependent probabilities of correct classification and misclassification for base classifiers (9).

The proposed method of evaluation of these probabilities is based on the original concept of a hypothetical classifier called Randomized Reference Classifier (RRC) [20]. The RRC is a stochastic classifier defined by a probability distribution which is chosen in such a way, that RRC acts, on average, as an modeled base classifier. It means, that RRC can be considered equivalent to the modeled base classifier, and therefore it is justified to use the class-dependent probabilities of correct classification (misclassification) of RRC as appropriate probabilities for the evaluated base classifier. In the computational procedure, first these probabilities are calculated for validation points and then they are generalized on the whole feature space. Details of the method can be found in [20]. Similarly, initial (5) and transition (6) probabilities in (13) are estimated using validation set $\mathcal{V}$.

4 Experiments

4.1 Experimental Setup

Performance of the MC system developed was evaluated in experiments using real data. The experiments were conducted in the Matlab environment using PRTools 4.1 and Signal Processing Toolbox.

In the control process the grasping of 6 types of objects (a pen, a credit card (standing in a container), a computer mouse, a cell phone (laying on the table), a kettle and a tube (standing on the table)) were considered. Our choice is deliberate one and results from the fact that the control functions of simple bioprosthesis are hand closing/opening and wrist pronation/supination, however for the dexterous hand these functions differ depending on grasped object [2].

In the considered examples, seven steps (elementary actions) can be distinguished in the process of grasping with a hand [18]: a_0—rest position; a_1—grasp preparation; a_2—grasp closing; a_3—grabbing; a_4—maintaining the grasp; a_5—releasing the grasp; a_6—transition to the rest position.

The experiments were carried out on healthy persons. Biosignals were registered using 8 integrated sensors (containing EMG electrode and MMG microphone in one casing) located on a forearm (vide Fig. 2). EMG and MMG signals were registered in specially designed 16-channel biosignals measuring circuit with sampling frequency 1 kHz. On the base of anatomical analysis of forearm muscles [1], for fur-

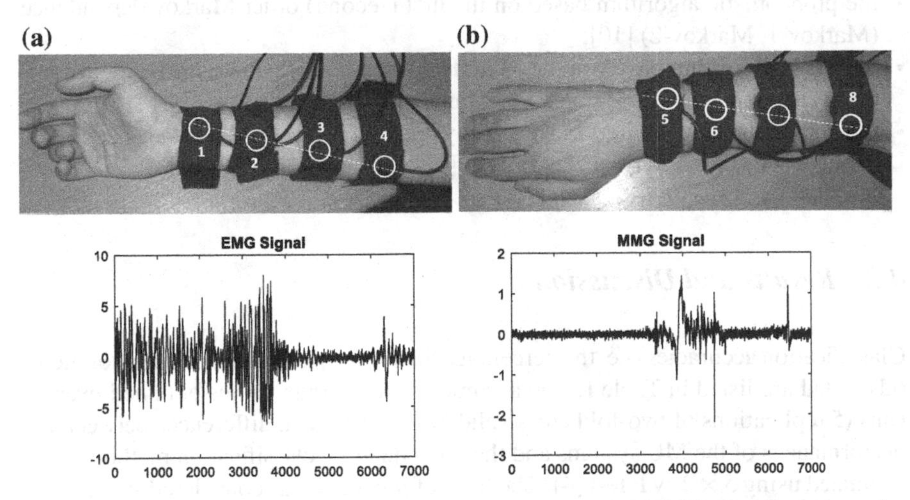

Fig. 2 The layout of the integrated sensors (EMG electrodes and MMG microphones) on the underside **a** and top side **b** of the forearm. Examples of EMG and MMG signals from the channel 2

ther processing the following channels (sensors) were selected: channel 1, channel 2, channel 3, channel 5 and channel 8.

The dataset set used to test of proposed classification method consisted of 2940 measurements, i.e. pairs EMG and MMG signals and segment/movement class forming 420 sequences (4). Each sequence lasted 6 s and was preceded with a 10 s break. The coefficients of AR function for different order of AR model ($p = 20, 30, 50, 80$ per signal and per channel) were considered as primary feature vector. Next, primary features were subjected to the PCA feature extraction procedure with the number of PC's determined by the 95 % of the total variation rule.

The training and testing sets were extracted from each dataset using two-fold cross-validation. For combining the MC system, a two-fold stacked generalization method [21] was used. In this way, the class-dependent probabilities of correct classification/misclassification for base classifiers and initial/transition probabilities of Markov chain are calculated for all objects in the original training set, but the data used for the calculation are unseen during the classifier training.

The experiments were conducted using three different recognition algorithms as base classifiers (4): (LC) Linear classifier based on normal distribution with the same covariance matrix for each class; (*k*-NN) *k*-nearest neighbors classifier (*k* after trials was set to 3); (ANN) feed-forward back-propagation neural network with 1 hidden layer.

The performace of the proposed MC system for K = 1 (MCS-1) and K = 2 (MCS-2) in the sequential scheme was compared against the following six sequential classifiers:

- the probabilistic algorithm based on the first (second) order Markov dependence (Markov-1, Markov-2) [10];
- the fuzzy algorithm based on the Mamdani inference scheme with 1- (2-)instant-backward-dependence (Mamdani-1, Mamdani-2) [18];
- the fuzzy algorithm based on the fuzzy relation with 1- (2-)instant-backward-dependence (FRealtion-1, FRelation-2) [9].

4.2 Results and Discussion

Classification accuracies (i.e. the percentage of correctly classified objects) for methods tested are listed in Table 1. The accuracies are average values obtained over 10 runs (5 replications of two-fold cross validation). Statistical differences between the performances of the MC systems and the six sequential classification methods were evaluated using 5×2 cv F test [4]. The level of $p < 0.05$ was considered statistically significant. In Table 1, statistically significant differences are given under the classification accuracies as indices of the method evaluated, e.g. for the dataset with p = 20, MCS-1(LC) system produced statistically better classification accuracies from the Mamdani-1, FRelation-1 and FRelation-2 methods.

Table 1 Classification accuracies of classifiers compared in the experiment

		The order of AR model/mean accuracy (%)				
No	Classifier	p = 20	p = 30	p = 50	p = 80	Mean
1	MCS-1(LC)	89.1	89.7	92.5	93.4	91.2
		9, 11, 12	11, 12	9, 11, 12	9, 11, 12	
2	MCS-1(3NN)	89.7	90.8	92.9	93.7	91.8
		9, 10, 11, 12	9, 10, 11, 12	9, 11, 12	9, 11, 12	
3	MCS-1(ANN)	92.1	92.7	94.8	95.2	93.7
		9, 10, 11, 12	9,10,11,12	9, 10, 11, 12	7, 9, 10, 11, 12	
4	MCS-2(LC)	90.7	91.5	93.0	93.9	92.3
		9, 10, 11, 12	9,10,11,12	9, 11, 12	9, 11, 12	
5	MCS-2(3NN)	91.3	92.1	93.2	94.6	92.8
		9, 10, 11, 12	9, 10, 11, 12	9, 11, 12	9, 10, 11, 12	
6	MCS-2(ANN)	**92.5**	92.8	**94.9**	**95.8**	94.0
		9, 10, 11, 12	9, 10, 11, 12	9, 10, 11, 12	9, 10, 11, 12	
7	Markov-1	90.8	92.6	93.5	94.2	92.8
8	Markov-2	91.6	**93.2**	94.1	94.8	93.4
9	Mamdani-1	85.9	87.3	89.4	90.2	88.2
10	Mamdani-2	87.1	88.8	90.6	91.1	89.4
11	FRelation-1	78.8	80.3	81.6	82.8	80.9
12	FRelation-2	79.7	80.9	82.6	83.6	81.7

The best score for each dataset is highlighted (*p* denotes the order of AR model)

These results imply the following conclusions: (1) The MC systems produced statistically significant higher scores in 87 out of 144 cases (4 datasets × 6 classifiers compared × 6 MCS's); (2) The MCS-2 system with ANN base classifiers achieved the highest overall classification accuracy averaged over all datasets it outperformed the Markov-1, Markov-2, Mamdani-1, Mamdani-2, FRelation-1, FRelation-2 systems by 1.2, 0.6, 5.8, 4.6, 13.1, 12.3 %, respectively. This results confirm the effectiveness of the use the multiclassifier system in the recognition of patient's intent; (3) There occurs a common effect within each classifier (MC system) type: 1-instant-backwards-dependence is always worse than 2-instant-backwards-dependence. This confirms the effectiveness of the decomposition of decision procedure into sequence of simpler classification tasks; (4) When the order of AR model increases then the accuracy of all methods investigated also increases.

5 Conclusion

The classic methods of analysis of biosignals in the bioprostheses control systems are widely discussed in the literature [6, 7, 12, 16]. However, the classification stage still poses a challenge for researching new solutions enabling the reliable recognition of human intention. In this study a novel method for recognition of sequence of elementary actions of grasping movements is proposed. The method, combining the meta-Bayes concept and Markov model into multiclassifier system and taking into account the K-instant-backwards-dependence among elementary actions, brings new possibilities to biosignal analysis. Results obtained in experimental investigations imply that it is worth trying solution that improves recognition efficiency.

The introduced approach constitutes the general concept of the human-machine interface, that can be applied for the control of a dexterous hand and an agile wheelchair as well as other types of prostheses, exoskeletons, etc. This, however, requires a further study, mainly in the experimental phase, which would allow to assess and verify the effectiveness of the adopted concept.

Acknowledgments This work was supported by the statutory funds of the Dept. of Systems and Computer Networks, Wroclaw Univ. of Technology.

References

1. Boostani, B., Moradi, M.: Evaluation of the forearm EMG signal features for the control of a prosthetic hand Physiolog. Measurement **24**, 309–319 (2003)
2. Carrozza, M., Cappiello, G., et al.: Design of a cybernetic hand for perception and action. Biol. Cybern. **95**, 626–644 (2006)
3. De Luca, C.: Electromyography. Webster, J.G. (ed.) Encyclopedia of Medical Devices and Instrumentation, pp. 98–109. Wiley (2006)
4. Dietterich, T.: Approximate statistical tests for comparing supervised classification learning algorithms. Neural Comput. **10**, 1895–1923 (1998)
5. Englehart, K., Hudgins, B.: A robust, real-time control scheme for multifunction myoelectric control. IEEE Trans. Biomed. Eng. **50**, 848–854 (2003)
6. Kakoty, M., Hazarika, S.: Towards electromyogram-based grasps classification. Int. J. Biomechatronics Biomed. Robot. **3**(2), 6373 (2014)
7. Khushaba, R.: Application of biosignal-driven intelligent systems for multifunction prosthesis control, Ph.D. Thesis, Faculty of Engineering and Information Technology, University of Technology, Sydney (2010)
8. Kuncheva, L.: Combining Pattern Classifiers: Methods and Algorithms. Wiley-Interscience (2004)
9. Kurzynski, M., Zolnierek, A.: Computer-aided sequential diagnosis using fuzzy relations—comparative analysis of methods. Lecture Notes in Bioinformatics, vol. 3745, pp. 242–251 (2005)
10. Kurzynski, M., Puchala, E.: Recognition of EMG signals based on the first-order Markov: model applied to the control of bioprosthetic hand. In: Proceedings of the 5th WSEAS International Conference on Sensors and Signals (SENSIG '12), Malta, pp. 113-118, Sept 2012
11. Kurzynski, M., Trajdos, P. et al.: Two-stage multiclassifier system with correction of competence of base classifiers applied to the control of bioprosthetic hand. In: Proceedings of IEEE 26th International Conference on Tools with Artiffcial Intelligence, pp. 620–626 (2014)
12. Kurzynski, M., Majak, M., Zolnierek, A.: Multiclassifier systems applied to the computer-aided sequential medical diagnosis. J. Healthc. Eng. (to appear)
13. Kurzynski, M., Krysmann, M., et al.: Multiclassifier system with hybrid learning applied to the control of bioprosthetic hand. Comput. Biol. Med. **69**, 286–297 (2016)
14. Luca, G.: Fundamental Concepts in EMG Signal Acquisition. Delsys Inc. (2003)

15. Micera, C., Carpantero, J., Raspopovic, S.: Control of hand prostheses using peripheral information. IEEE Rev. Biomed. Eng. **3**, 48–68 (2010)
16. Oskoei, M., Hu, H.: Support vector machine-based classification scheme for EMG control applied to upper limb. IEEE Trans. Biomed. Eng. **55**, 19561965 (2008)
17. Ravindra, K., Ildstad, S.: Immunosuppressive protocols and immunological challenges related to hand transplantation. Hand Clin. **27**(4), 79–467 (2011)
18. Wolczowski, A., Kurzynski, M.: Human - machine interface in bio-prosthesis control using EMG signal classification. Expert Syst. **27**, 53–70 (2010)
19. Wolczowski, A., Kurzynski, M.: Control of bioprosthetic hand using fusion of information from biosignals and from prosthesis sensors, In: Proceedings of IEEE Asia-Pacific Conference on Computer Aided System Engineering (APCASE), pp. 19–24 (2014)
20. Woloszynski, T., Kurzynski, M.: A probabilistic model of classifier competence for dynamic ensemble selection. Pattern Recognit. **44**, 2656–2668 (2011)
21. Wolpert, D.: Stacked generalization. Neural Netw. **5**, 214–259 (1992)
22. Zecca, M., Micera, S., et al.: Control of multifunctional prosthetic hands by processing the electromyographic signal. Crit. Rev. Biomed. Eng. **30**, 459–485 (2002)

Contextual Modelling Collaborative Recommender System—Real Environment Deployment Results

Urszula Kużelewska

Abstract Nowadays, recommender systems are widely used in many areas as a solution to deal with information overload. There are some popular and effective methods to build a good recommendation system: collaborative filtering, content-based, knowledge-based and hybrid. Another approach, which made a significant progress over the last several years, are context-aware recommenders. There are many additional information related to the context or application area of recommender systems, which can be useful to generate accurate propositions, e.g. user localisation, items categories or attributes, a day of a week or time of a day, weather. Another issue is recommenders evaluation. Usually, they are only assessed with respect to their prediction accuracy (RMSE, MAE). This is good solution, due to possibility of off-line calculation. However, in real environment recommendation lists are finally evaluated by users who take into consideration many various factors, like novelty or diversity of items. In this article a multi-module collaborative filtering recommender system with consideration of context information is presented. The context is included both in post-filtering module as well as in a similarity measure. Evaluation was made off-line with respect to prediction accuracy and on-line, on real shopping platform.

Keywords Collaborative filtering · Contextual recommender systems · Recommender system evaluation

1 Introduction

The amount of information appeared on the Internet increases rapidly. There are many new services, web pages of companies, blogs, shops, music, video, etc. The unconstrained next step to cope with searching them is to use new technologies that can assist us to find resources of interest among the overwhelming available items. One of such tools are recommender systems (RS), which are electronic applications

U. Kużelewska (✉)
Bialystok University of Technology, Wiejska 45a, 15-351 Białystok, Poland
e-mail: u.kuzelewska@pb.edu.pl

I. Czarnowski et al. (eds.), *Intelligent Decision Technologies 2016*,
Smart Innovation, Systems and Technologies 57,
DOI 10.1007/978-3-319-39627-9_11

with the aim to generate for a user a limited list of items from a large items set. In case of personalised RS the list is constructed basing on the active user's and other users' past behaviour. People interact with recommender systems by visiting web sites, listening to the music, rating the items, doing shopping, reading items' description, selecting links from search results. This behaviour is registered as access log files from web servers, or values in databases: direct ratings for items, the numbers of song plays, content of shopping basket, etc. After each action users can see different, adapted to them, recommendation lists depending on their tastes [5, 10, 13].

Context information is additional valuable data, which is worthy to include in recommendation process. To propose a restaurant for a customer the nearest places should be recommended. In shops, before Christmas, the best suggestion are the items, which could be gifts.

In [3] the importance of the contextual information in recommender systems has been mentioned. The authors define it as information, which is known a priori and characterised by additional related to the domain factors having a known hierarchical structure that does not change significantly over time. Due to great attention on this issue and many articles, that have appeared recently [1, 6, 7, 9, 11], finally, the Context-Aware Recommender System (CARS) field has been formed.

The methods, which belong to CARS can be divided into pre-filtering, post-filtering and contextual modelling methods [12]. Pre-filtering algorithms apply contextual knowledge e.g. to remove irrelevant data before recommendations calculation, which is then performed with standard methods. Post-filtering approach uses common algorithms to generate recommendations, as well, then the contextual information is used to adjust recommendation lists. The last type, contextual modelling, use this background data in the process of recommendation generation.

One of the first pre-filtering solutions is *exact pre-filtering* [8], in which the ratings not related to the specific context of interest are removed before recommendation calculation. Another example is *item splitting* and *microprofiling* proposed by Baltrunas and Ricci [4]. They split user profiles into set of overlapping subprofiles representing the given user in a particular context. An example of contextual modelling is RPMF proposed by [14]. The backgound information is encoded in or reflected by the user-specific and item-specific latent factors. Based on this, tree based random partition is applied to split the user-item-rating matrix by grouping users and items with similar contexts, and then apply matrix factorization to the generated sub-matrices. Finally, a framework for building context-aware recommender system was proposed by Hussein et al. in [9].

This paper contains results of experiments on collaborative filtering recommender system *what2buy* with context information included. There were tested both post-filtering and contextual modelling methods. The post-filtering approach is based on collaborative filtering item-based and user-based techniques with standard similarity metrics, whereas the contextual modelling solution involves relationship among item's categories in similarity measure which is used in item-based collaborative filtering module, as well. Quality of prediction (RMSE) and real effectiveness (items from recommendations, which were bought in real environment) was examined in the experiments.

2 Architecture and Description of What2buy Recommender System

The architecture of the system and its individual parts were strictly designed for this particular selling platform. It was created and implemented in re.com.sys Ltd. company. There are over 50 000 products and nearly 1000 categories. There were the following aims to achieve: increase conversion rate, increase average number of items in baskets and propose for loyal customers the items, which are particularly interesting for them.

The conversion rate is an index of users, who bought something during the same visit to the number of users, who were only visiting items. To increase it, new visitors should see interesting products without time-consuming searching. The only information gathered from them were: the visited product pages and the categories the products belong to. It was observed, that the time spent in the store was very short: the visited only several pages from one or two categories.

Relatively high number of items in basket is efficient for both: seller and customer. It reduces the influence of shipping cost. Customers eager to increase the content of their carts, but only if they see an interesting offer.

Last aim is related to increase loyalty of regular customers. The have a great history of transactions, which is a good source of profiles of their preferences. They have favourite products and categories, they know the rules of navigation in the service. For them the related other products from different categories should be proposed.

Taking into considerations the above challenges and objectives, there was proposed a hybrid collaborative filtering system with 2 source of data and 2 types of recommenders. The data were gathered from transactions (*transSData*) and from user pre-transactional behaviour—visits and operations on basket (*visSData*). Transactions data was used as input to user-based collaborative filtering module (*UBR*) which is designed to generate recommendations for regular customers. The data *visSData* is a source to item-based collaborative filtering module (*IBR*) to generate propositions for new customers as well as users, who only wander among the pages. The architecture of the described system with new and regilar users paths is presented in Fig. 1.

It should be noticed importance of contextual information (category of items) in every of the mentioned objectives: in *IBR* the proposed items ought to belong to the same category as the items registered in users' path, whereas in *UBR* it is desirable to recommend items new and surprising from different from well known by user categories. To achieve it there were proposed two approaches: context pre-filtering module or a new similarity measure which process the context information during similarity calculation.

The new measure, $sim_{EuclCtx}$, was proposed on Bialystok University of Technology and tested in what2buy system. It is based on Euclidean similarity, due to the best results of this standard measure with respect to predictive ability. The proposed index takes into consideration relationship among the items as well as among

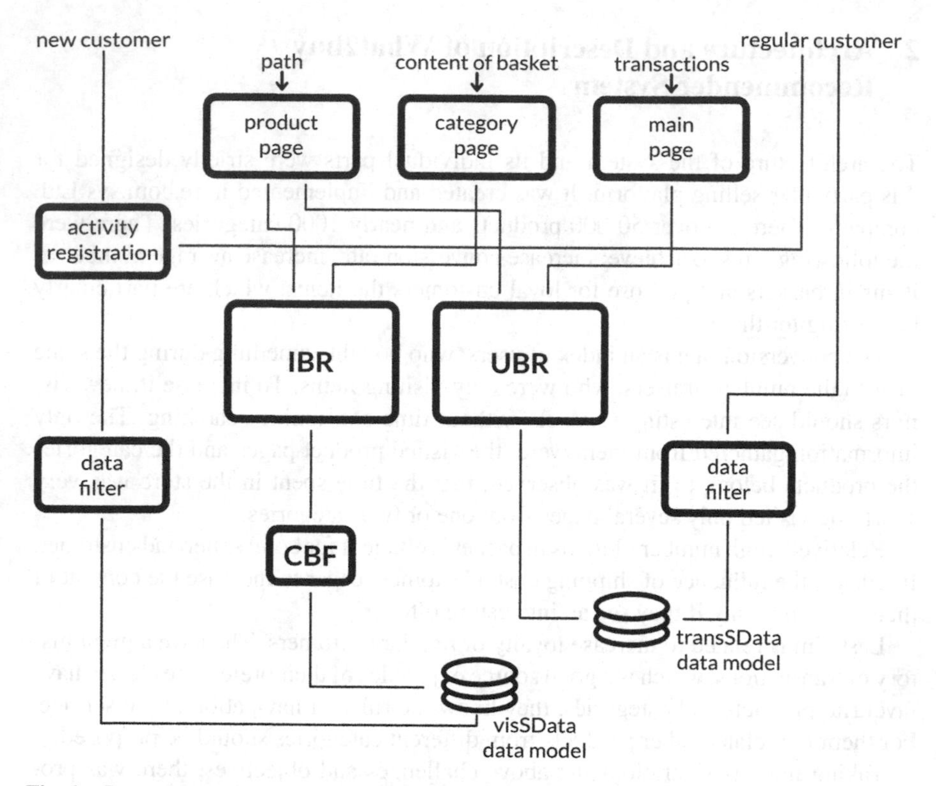

Fig. 1 Components of what2buy recommender system

the categories. In both parts the input data is behavioural information from users. Equation 1 describes the measure.

The final similarity value depends on similarity between items: x_i and x_j and between the categories they belong to: c_a, c_b. High values of this index require strong matching in part of items as well as categories.

$$sim_{EuclCtx}(x_i, x_j) = \frac{sim_{Eucl}(x_i, x_j) + sim_{Eucl}(c_a(x_i), c_b(x_j))}{2} \tag{1}$$

However, in case of cold start, when new items are introduced to the store's offer ($sim_{Eucl}(x_i, x_j) = 0$), it is allows to determine similarity basing on the relationship among the categories.

3 Experiments

The set *transSData* contains data from 13 months (see Table 1), whereas the set *visSData* consists data from eight months (see Table 2). There were saved all transactions data, whereas the in the second set, only successful data (items from paths from the same category and from users who finally added at least one item to the basket).

First of all, it was examined effectiveness of the recommender modules on the source data with respect to RMSE error. There were examined for both modules the following similarity indices: Pearson Correlation (*Pearson*), LogLikehood Similarity (*LogLikehood*), Cosine-based Coefficient (*Cosine*), Euclidean Distance Similarity (*Euclidean*), CityBlock Measure (*CityBlock*) and Euclidean Context-based Similarity (*EuclideanContext*). The results are presented in Table 3 for both *transSData* and *visSData* datasets.

The lowest value of RMSE error was the case with Euclidean Distance Similarity in both modules, however the measure based on context data and Euclidean distance was the following one, slightly only worse: the difference was 2–3 %. It must be mentioned, that for *EuclideanContext* measure on *transSData*, but with the context information taken from *visSData*, the value of RMSE was equal 0.38. It was the lowest error value in all cases.

Table 1 Description of transactions data *transSData*

Total number of users	Number of users versus transaction length (*L*)
154382	L = {1, 2} 98350
	L = [3, 10] 39511
	L ≥ 11 16520
Total number of items	**Number of items versus transaction length (*L*)**
34255	L = {1,2} 12094
	L = [3,10] 10661
	L ≥ 11 11499

Table 2 Description of transactions data *visSData*

Total number of users	Number of users versus path length (*L*)
41664	L = {1, 2} 22725
	L = [3, 10] 15338
	L ≥ 11 3600
Total number of items	**Number of items versus path length (*L*)**
20635	L = {1, 2} 7663
	L = [3, 10] 8045
	L ≥ 11 4926

Table 3 RMSE values for both modules of recommender system: *UBR* and *IBR* with different similarity measures

Pearson	*LogLikehood*	*Cosine*	*Euclidean*	*CityBlock*	*EuclideanContext*
UBR module on *transSData* data					
0.68	0.44	0.45	0.40	0.44	0.41
IBR module on *visSData* data					
0.86	0.57	0.56	0.50	0.59	0.52

Very important measure in recommender systems evaluation is coverage. This index measures filling a recommendation list with required length. This is particularly important in real environment. The recommender modules with all mentioned above similarity measures were examined with respect to coverage recommendation lists of different length. For this experiment only input data of users who had 2 items in their history was taken, because they are the most common in the both datasets. The results are presented in Tables 4 and 5 for both *transSData* and *visSData* datasets respectively.

Table 4 Coverage values [%] for *UBR* module of the examined recommender system with different similarity measures

Pearson (%)	*LogLikehood* (%)	*Cosine* (%)	*Euclidean* (%)	*CityBlock* (%)	*EuclideanContext* (%)
Required recommendation list length = 10					
82	98	98	98	100	100
Required recommendation list length = 20					
78	98	98	98	100	100
Required recommendation list length = 30					
76	98	98	97	100	100

Table 5 Coverage values [%] for *IBR* module of the examined recommender system with different similarity measures

Pearson (%)	*LogLikehood* (%)	*Cosine* (%)	*Euclidean* (%)	*CityBlock* (%)	*EuclideanContext* (%)
Required recommendation list length = 10					
35	89	89	89	99	99
Required recommendation list length = 20					
28	84	84	84	98	98
Required recommendation list length = 30					
17	78	78	78	96	96

Table 6 A level of customers interest [%] of presented recommendations with respect to different similarity measures

Pearson (%)	*LogLikehood* (%)	*Cosine* (%)	*Euclidean* (%)	*CityBlock* (%)	*EuclideanContext* (%)
UBR module on *transSData* data					
0.61	1.5	0.88	0.62	0.61	0.69
IBR module on *visSData* data					
7.15	17.31	4.63	5.45	5.31	8.86

The measures *CityBlock* and proposed in this article *EuclideanContext* generate the longest recommendation lists. The index based on Euclidean distance, which generated the highest quality recommendations in the previous experiment, had worse coverage about 10–20%.

As it was mentioned before, the system is working in real environment on selling platform. It was possible to compare its factual effectiveness evaluated by real customers. It was measured a ratio of the bought items, which were recommended and presented to the customer before, to the total number of items, which were bought that day. The results are presented in Table 6.

It can be seen, that customers the most often select items, which were recommended basing on their path in the service. Surprisingly, the recommendations, which were the most interesting for the customers, were generated by the module *IBR* with LogLikehood similarity measure, which was not the best one in off-line evaluation. The following best results were generated by the system with the context similarity based on Euclidean distance.

4 Conclusions

Internet customers are people, who appreciate their time, convenience and access to wide assortment of products. On the other hand, Internet shops are places, where the customer support can be delivered personally. One of the personalization tools are recommended systems, which identify customer tastes analysing their behaviour on the selling platform.

There are many approaches to personalization in recommender systems: content-based recommenders, collaborative filtering systems, knowledge-based technique. Real environment recommendation services are the most often complex hybrid systems composed of many different method modules. Designing an architecture of such network requires deep analysis of the domain specificity: customers behaviour, frequency of shopping, attachment to favourites products, etc. Often, a very important factor is context information, e.g. categories and dependence among them. Finally, to select and adjust the most optimal approach, it also involves performing and repeating a series of experiments to evaluate overall effectiveness.

In this article a recommender system in real environment was presented. There was described the way of designing and evaluation of its components with taking into account context information, that is dependence among categories of products. A new similarity measure based on Euclidean distance was proposed and evaluated. The final similarity value takes into account users interaction on products as well as on categories the products belong to. Finally, the overall effectiveness of the recommender system, which generates proposition using this measure, was better: customers more often selected the propositions, although in the experiments its RMSE value was not the lowest.

Acknowledgments This work was supported by Rectors of Bialystok University of Technology Grant No. S/WI/5/13.

References

1. Abbas, A., Zhang, L., Khan, S.U.: A survey on context-aware recommender systems based on computational intelligence techniques. Computing **97**, 1–24 (2015)
2. Adomavicius, G., Tuzhilin, A.: Toward the next generation of recommender systems: a survey of the state-of-the-art and possible extensions. IEEE Trans. Knowl. Data Eng. **17**(6), 734–749. (2005)
3. Adomavicius, G., Tuzhilin, A.: Context-aware recommender systems. In: Handbook on Recommender Systems, pp. 217–253. Springer (2011)
4. Baltrunas, L., Ludwig, B., Ricci, F.: Matrix factorization techniques for context aware recommendation. In: 5th ACM Conference on Recommender Systems, pp. 301–304. ACM (2011)
5. Bobadilla, J., Ortega, F., Hernando, A., Gutiérrez, A.: Recommender systems survey. Knowl. Based Syst. **46**, 109–132 (2013)
6. Braunhofer, M., Ricci, F., Lamche, B., Wörndl, W.: A context-aware model for proactive recommender systems in the tourism domain. In: 17th International Conference on Human-Computer Interaction with Mobile Devices and Services Adjunct, pp. 1070–1075. ACM (2015)
7. Datta, S., Gupta, P., Majumder, S.: SCARS: a scalable context-aware recommendation system. In: 3rd International Conference on Computer, Communication, Control and Information Technology, pp. 1–6. IEEE Press (2015)
8. Gorgoglione, M., Panniello, U.: Including context in a transactional recommender system using a prefiltering approach: two real e-commerce applications. In: 23rd IEEE International Conference on Advanced Information Networking and Applications, pp. 667–672. IEEE Press (2009)
9. Hussein, T., Linder, T., Gaulke, W., Ziegler, J.: Hybreed: a software framework for developing context-aware hybrid recommender systems. User Model. User-Adapt. Interact. **24**(1–2), 121–174 (2014)
10. Jannach, D., et al.: Recommender Systems: An Introduction. Cambridge University Press, Cambridge (2010)
11. Liu, H., Zhang, H., Hui, K., He, H.: Overview of context-aware recommender system research. In: 3rd International Conference on Mechatronics, Robotics and Automation, pp. 1218–1221. Atlantis Press (2015)
12. Panniello, U., Tuzhilin, A., Gorgoglione, M.: Comparing context-aware recommender systems in terms of accuracy and diversity. User Model. User-Adapt. Interact. **24**(1–2), 35–65 (2014)
13. Ricci, F., et al.: Recommender Systems Handbook. Springer (2010)
14. Zhong, E., Fan, W., Yang, Q.: Contextual collaborative filtering via hierarchical matrix factorization. In: SIAM International Conference on Data Mining, pp. 744–755 (2012)

Decision Trees on the Foreign Exchange Market

Juszczuk Przemyslaw, Kozak Jan and Trynda Katarzyna

Abstract In this article we present a novel approach to generate a data set directly from real-world forex market data. The data are transformed into a decision table. Every single object in such a table consists of conditional attributes—in this case values of technical analysis indicators as well as of the decision class (BUY, SELL or WAIT). Our second goal was to test the quality of the classification based on two well-known algorithms used for decision tree construction: the CART algorithm and the C4.5 algorithm. All experiments were conducted on three different currency pairs—with 3 data sets for each pair.

Keywords Forex market · Decision tree · CART · C4.5 algorithm

1 Introduction

The forex (foreign exchange market) is a global, decentralised market on which the main instruments are the currencies. It is often described as the most liquid market in the world, and its turnovers reach over billions of dollars every day. On the foreign exchange market, every single instrument is described as a pair of currencies in which two elements can be indicated: the first element is the base currency and the second element is the quote currency. The exchange rate of the currency pair changes dynamically over time and there are many crucial factors that have a high impact on it. In general, every currency pair is the resultant value of demand and supply. The

J. Przemyslaw (✉) · T. Katarzyna
Institute of Computer Science, University of Silesia, Sosnowiec, Poland
e-mail: przemyslaw.juszczuk@us.edu.pl

T. Katarzyna
e-mail: katarzyna.trynda@us.edu.pl

K. Jan
Chair of Knowledge Engineering, Faculty of Informatics and Communication,
University of Economics, Katowice, Poland
e-mail: jan.kozak@us.edu.pl

I. Czarnowski et al. (eds.), *Intelligent Decision Technologies 2016*,
Smart Innovation, Systems and Technologies 57,
DOI 10.1007/978-3-319-39627-9_12

whole market consists of four overlapping sessions located at different time zones (below are the hours given on the basis of GMT+1 time):

- London—(European session)—8 a.m.–16 p.m.;
- New York—1 p.m., 9 p.m.;
- Sydney—10 p.m., 6 a.m.;
- Tokio—midnight–8 a.m.

One of the most important mechanisms applied to the market is leverage. As opposed to regular markets (such as the stock exchange), there is no need on the foreign market to have a considerable amount of money. In its simplest definition, leverage allows to open positions on any currency pair having only partial capital protection. Such an approach is a considerable facilitation for persons with small capital. Moreover, it is also one of the most important features of this market that attracts small, private investors. More details on the forex market can be found, e.g. in [22].

Stock data (including foreign market data) are often considered to be chaotic data without a visible point attractor. In other words, there can be no expectation, that in the longer time horizon some currency pairs' values will converge into a single value (constant price). In general, there are three different attractors: the constant point attractor, the cycle attractor and the chaotic attractor. There have been many articles which indicated that in the first two attractors there is a high possibility to predict future values. As oppose to those, in the chaotic data there is often a need to deal with a large error [8]. On the other hand, there is a large group of mathematical methods known as technical analysis. Such methods can be very helpful in determining the price direction movement as well as the range of such a move. One of the most popular tools based on the mathematical and statistical approach is the moving average. It is a value calculated on the basis of n last price values. High efficiency of such tools was shown for example in [9].

The whole concept of technical analysis is based on three rules: the market discounts everything, prices move on the basis of trends, and history repeats itself. Those assumptions were described in Dow theory, which is often considered as a preliminary to technical analysis. Dow theory existed already at the end of the 19th century but was described in details in the 20th century. Rhea [17] is considered to be one of the most important works on Dow theory. At the end of the 20th century there was still growing interest in technical analysis. In 1992 in [20] one of the first formal studies was presented on the effectiveness of technical analysis. A similar study was proposed once again in [19].

In this article we propose a novel approach to transform forex data into a decision table. We assume that in every time step there are three possibilities: Buy signal—which allows to open a buy position; sell signal—which allows to open a sell position, and wait—which is basically idle. This article may be considered as a preliminary study on the effectiveness of such a classification approach in which there is a visible inequality in the number of elements belonging to the decision classes. Such research will be conducted on the basis of well-known decision tree algorithms. The article is organised as follows: first we provide the background for the problem, then we propose a novel concept of acquiring financial data and transforming the data into decision tables. We also give a brief description of decision tree algorithms. Finally, we present the preliminary experiments and end with some conclusions.

2 Related Works

Automated forex systems hav attracted much attention in the over past few years. The high popularity of such systems is the result of continued growing interest in technical analysis. Besides fundamental analysis, it is one of the major components of modern trading systems. However, the effectiveness of technical analysis is still unclear. In article [6] the authors investigated over 90 different articles correlated with technical analysis, and only 56 of them had positive conclusions. In 20 articles, technical analysis did not generate satisfactory results.

The concept of building an automated forex system based on computer science algorithms is very popular. One of the most often used approaches involves the use of artificial neural networks. The effectiveness of this type of algorithm was recently proven in multiple articles, such as in [21], thus it is natural to use this concept in economy. One of the first articles involving a neural network as a tool for generating trading signals was [5]. The popularity of such an approach has grown with time and many similar concepts have been introduced, e.g. in [14] the authors proposed a web-decision support system based on neural networks. One of the newest articles on the application of neural networks on the forex market is [7].

The second subject of interest in the foreign market is the data prediction problem. One of the most interesting articles dealing with this problem is [10]. In this article the authors propose a transformation of the financial time series into fuzzy grain particle sequences. The main goal of this approach is to use a support vector machine to calculate the upper and lower bounds of those particles. There are also very complex systems based on classical fuzzy time series analysis, e.g. [1]. In this approach the authors propose an algorithm consisting of elements such as voting, statistical analysis and emotional decision-making agents.

Besides the above articles, there are a few approaches involving evolutionary computation, e.g. [3], in which the evolutionary programming mechanism is developed. Genetic algorithms were a point of interest in [12]. The authors tested their approach on the French stock market and the overall results were compared with one of the most well-known strategies: "buy and hold". A method based on the recognition of patterns was proposed in [4]. An interesting fact is that Japanese candlesticks, which are one of the most frequently used elements of technical analysis, have not been a very popular subject in research articles. Modern financial data visualisations, such as Renko or Kagi charts, still remain beyond mainstream research interest.

3 Data Preparation

The data acquisition process is one of the most important elements of the proposed approach. Currently, there are many different concepts that allow to open a position on the forex market. These involve using advanced charts (e.g. Renko or Kagi), predefined sets of technical indicators and even fundamental analysis. The emerging question is as follows: How does one precisely define the price direction on the basis of technical analysis indicator values. In other words, the problem is to define an element of the decision table with a set of attributes, i.e. values of different indicators at some period of time. Such an element should also have a properly assigned value from the 3-element set: buy, sell and wait. We assume that the buy decision may be described as a situation in which the price of a given instrument will rise by some predefined value. The sell decision is a situation in which the price of the instrument drops by some predefined value. Of course, all remaining situations are classified as the wait position. An example fragment of the data set creation process is presented in Fig. 1.

As can be observed, every situation on the market is described by a number of parameters. The decision to buy or sell is set only if significant price movement is observed on the chart. In other cases the value of the decision is set to default. Such an approach ensures that the signal to open the order is generated only on the basis of the price movement and technical analysis indicators, thus the Commodity Channel Index or the Relative Strength Index do not have an affect on the decision. Unfortunately, due to the lack of space we will omit detailed information on the technical analysis indicators used in this paper. More information on technical analysis can be found, e.g. in [11]. The proposed approach points to a significant problem related to the number of objects in every one of the three decision classes. It is understandable that the number of elements belonging to the wait class will be much larger than the size of the two remaining classes. Such a situation raises the question whether

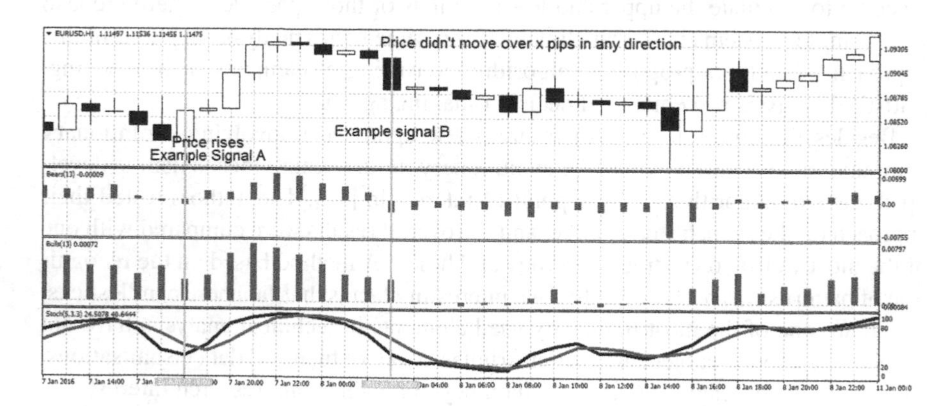

Fig. 1 Data set creation process

classification of objects in such a data set may be effective. We will look closer at this problem in the "Experiments" section.

4 Decision Trees

Readers who want to learn more about decision trees we refer to [18]. Below we give just some brief information and the most important features that will be crucial in the context of using decision trees in financial data. Decision trees have a very simple and intuitive construction and may be described as an acyclic, directed graph in which all vertices are called nodes, the edges are branches, nodes without descendants are leaves, and, finally, the root of the decision tree is a node without a parent. All nodes contain tests on the conditional attributes which are created on the basis of some predefined splitting rule. This splitting rule allows to divide the data according to the values of those attributes. Moreover, every result is represented by branches.

A decision tree is constructed from a training set which consists of objects. Each object is completely described by a set of attributes and a class label. In our problem the set of attributes comprises values of technical analysis indicators and the class label is one of the decisions: "sell", "buy" and "wait". All internal nodes contain splits which test the value of an expression of the attributes. Arcs from an internal node t to its children are labeled with distinct outcomes of the test at t. Finally each leaf node has one class label.

One of the important advantages of using decision trees is the possibility to perform a fast analysis (significantly faster as compared to the classical methods). Moreover, decision trees can easily be stored as decision rules. This feature allows to use them in systems directly correlated with the decision rules. Such an advantage of this structure promotes future development of this technique. In addition, the so-called groups of classifiers presented as decision forests seems especially useful when building decision trees with stochastic methods. In general, the construction of a single decision tree is based on the "divide and conquer" rule, which is recursive data division. Such an approach leads to dividing one single problem into smaller problems.

The task of constructing a tree from the training set has been called either: tree induction, tree building or tree growing. Most tree induction systems use the greedy approach, i.e. trees are induced top-down, where the first attribute (possibily the best) becomes the root of the tree. The child nodes perform division according to the same rule. Mostly, just after decision tree creation, the process of pruning is carried out. Such an approach is used to prevent classificator overtraining. In the literature we can find multiple algorithms used for decision tree construction. Some of the most popular state-of-art algorithms are CART and C4.5. A detailed brief of other similar algorithms can be found in [15].

4.1 The CART Algorithm

The CART method addresses the classification and regression problem by building a binary decision tree according to some splitting rule. Each split is based on a single variable; some variables may be used several times while others may not be used at all. Each sub-leaf is then split further based on independent rules.

Decision trees constructed by the CART algorithm are mostly generated on the Gini split criterium:

$$D_t = \sum_{i \neq j} \cdot p(i|t) \cdot p(j|t) = 1 - \sum_i p(i|t)^2 \quad (1)$$

where: $p(i|t)$ is the probability of decision class i in node t. The Gini splitting rule is based on the Gini index a measure of the random variable concentration. The main goal in this case is uniform division in the child nodes. The procedure is finished when there is no more admissible splitting. In general, the final tree overfits the available data and the prediction error. The main goal is to minimise that prediction error. To achieve this, construction of such a tree involves the pruning procedure. Such methods try to produce a subtree whose expected prediction error is lower than the error value before the pruning procedure.

4.2 The C4.5 Algorithm

The C4.5 algorithm was proposed by Qunlan [16] and is an improvement of his earlier ID3 algorithm. The main improvements concern the splitting criterium so that it is possible to limit the classification error and missing attributes values are no longer a problem. The C4.5 uses the so-called rule of relative profit which is calculated for every node (the node with the highest value is selected):

$$Gain(S,A) = Entropy(S) - I(S,A) = Entropy(S) - \sum_i \cdot \frac{S_i}{S} \cdot Entropy(S_i) \quad (2)$$

where S is a set of examples and i is the number of subsets. Moreover, the pruning method is also used in the C4.5 algorithm. During the learning process it is possible to work with objects with some missing attributes. This algorithm is also prepared to work with continuous attributes.

In our experiments we used two of the algorithms as described above. The CART algorithm is very often considered a state-of-art algorithm. It may also be used to point out the weak elements of the proposed approach. The second choice was C4.5, which is best known and (probably) the most widely used learning algorithm.

5 Experiments

Our goal was to test experimentally if it is possible to present forex market data in the form of rules and to use them in the classification problem. The structure of real-world data was adjusted to the WEKA system. It was used to generate the results of the experiments. All of the results and their analysis are presented below. All experiments were conducted for three algorithms (with default settings): CART algorithm, C4.5 algorithm and Random Tree.

5.1 Data Sets

We prepared 9 real-world data sets which included three currency pairs: *EURUSD* was chosen because it is the most fluent instrument on the currency market. The *GBPUSD* pair is often considered to be one of the most variable pairs. Finally *USDJPY* was chosen as a pair mostly correlated with the technical analysis. For every instrument we prepared 3 sets which depended on the number of pips (1 pips is a 0.0001 change in the price of the instrument). The numbers were equal to 30, 45 and 60, e.g. 45 means that at least 45 pips movements should be observable to describe such a set-up for the *BUY* or *SELL* class. It is also worth noting that every data set consists of three decision classess: *SELL*, *BUY* and *WAIT*. Due to the specifics of these data, a vast majority of the elements belonged to the *WAIT* class. Detailed parameters of the data sets are presented in Table 1.

The data sets were randomly divided (sampling without replacement) into two sets: a training set (9000 objects) and a test set (2998 objects). Every data set consisted of 8 conditional attributes (numeric) and one decision attribute (nominal)). Conditional attributes were selected on the basis of the most often used technical analysis indicators and are given as follows: Alligator, Bulls strength, Bears strength, CCI (Commodity Channel Index), Demarker, RSI (Relative Strength Index), Stochastic Oscillator, and Williams indicator. All indicators where used with their standard parameters.

5.2 Results of Experiments

As can be observed in Table 1, the proposed data sets are very difficult to analyse. The vast majority of elements (86–98 %) in the training set belongs to the *WAIT* class. The same situation is given in the test set, where 86–99 % of the elements are in the *WAIT* class. An even more serious problem is that the random division in many cases leads to a different ratio of objects (in the same class) in the training and testing sets. In many cases that difference is equal to even 50 %, e.g. in *EURUSD* 30p with 6.7 % of objects in the SELL class there are only 2.9 % objects with the same class in the training set.

Table 1 Original parameters in data sets

Dataset	Train set			Test set		
	SELL	BUY	WAIT	SELL	BUY	WAIT
EURUSD 30p.	509(5.6 %)	600(6.7 %)	7891(87.7 %)	87(2.9 %)	83(2.8 %)	2828(94.3 %)
EURUSD 45p.	234(2.6 %)	240(2.7 %)	8526(94.7 %)	35(1.2 %)	31(1.0 %)	2932(97.8 %)
EURUSD 60p.	112(1.3 %)	96(1.1 %)	8792(97.7 %)	15(0.5 %)	13(0.4 %)	2970(99.1 %)
GPBUSD 30p.	660(7.3 %)	607(6.7 %)	7733(85.9 %)	205(6.8 %)	225(7.5 %)	2568(85.7 %)
GPBUSD 45p.	274(3.0 %)	260(2.9 %)	8466(94.1 %)	88(2.9 %)	91(3.0 %)	2819(94.0 %)
GPBUSD 60p.	118(1.3 %)	125(1.4 %)	8757(97.3 %)	46(1.5 %)	41(1.4 %)	2911(97.1 %)
USDJPY 30p.	471(5.2 %)	514(5.7 %)	8015(89.1 %)	131(4.4 %)	95(3.2 %)	2772(92.5 %)
USDJPY 45p.	196(2.2 %)	216(2.4 %)	8588(95.4 %)	50(1.7 %)	28(0.9 %)	2920(97.4 %)
USDJPY 60p.	101(1.1 %)	90(1.0 %)	8809(97.9 %)	18(0.6 %)	13(0.4 %)	2967(99.0 %)

Different measures were used due to the difficulty of the data sets. All preliminary results are presented in Tables 2, 3 and 4. All results should be analysed due to different measures as well as the parameters given in Table 1.

An analysis of the results given in Table 2 (which is the accuracy of the classification and the size of the decision tree) suggests that it is possible to use algorithms for decision tree construction as a tool for financial data prediction. However, it should be pointed out that in some cases the best results (the best according to the accuracy measure) are given by the CART algorithm. This is connected with decision trees with a height equal to 1. As may be expected, such trees classify objects by assigning all of them to the most numerous *WAIT* class.

For a more effective analysis, average values of the remaining measures are given. The set of measures contains: recall, precision, F-measure and ROC area. The results for those measures (Table 3) point out that observation related to the predominance of a single class is correct. Of course, in some examples other objects are properly classified as well. Especially interesting are the results for the Random Tree algorithm in which the ROC area value is frequently better than in the case of either the CART or C4.5 algorithms.

This article is a preliminary study in which we dealt with the possibility of using decision trees as a tool for developing an effective transaction system. Thus the confusion matrix is one of the crucial elements of our experiments. The results for the selected currency pairs and algorithms were presented in Table 4. These are example matrices which are the most frequent in the analysed experiments. The values of these matrices point to the problem related with the *WAIT* class the vast majority of objects is assigned to this class. It is very important to point out that in real-world problems such a situation from the viewpoint of the transaction system is far better than the option of assigning the wrong object to the class of *BUY* or *SELL*. Such an approach coincides with the transaction systems used on the forex market where the most frequent and safest option is to stay out of the market and to not open the

Table 2 Comparative study—accuracy rate

Data set	CART		C4.5		Random tree	
	acc	size	acc	size	acc	size
EURUSD 30	**0.9473**	13	0.9426	191	0.8726	3041
EURUSD 45	**0.9803**	10	0.9793	93	0.9426	1609
EURUSD 60	**0.9703**	13	0.9650	67	0.9503	809
GPBUSD 30	**0.8609**	25	0.8572	309	0.7748	3315
GPBUSD 45	**0.9410**	5	0.9383	113	0.9030	1685
GPBUSD 60	0.9710	9	**0.9720**	19	0.9500	915
USDJPY 30	**0.9250**	5	0.9246	1	0.8482	2785
USDJPY 45	**0.9740**	1	0.9730	21	0.9293	1457
USDJPY 60	**0.9897**	1	0.9877	29	0.9656	721

Abbrev. acc—accuracy rate; size—size of the tree

Table 3 Average values for different measures for currency pairs—italicized numbers are denoted as the best result

Measure	EURUSD											
	EURUSD H1 30 pipsow (weighted avg.)				EURUSD H1 45 pipsow (weighted avg.)				EURUSD H1 60 pipsow (weighted avg.)			
	rec	prec	F1	ROC	rec	prec	F1	ROC	rec	prec	F1	ROC
CART	*0.947*	*0.933*	*0.930*	*0.612*	*0.980*	*0.977*	*0.973*	*0.618*	*0.970*	*0.957*	*0.959*	0.584
C4.5	0.943	0.923	0.927	0.601	0.979	0.974	*0.973*	0.575	0.965	0.955	*0.959*	0.536
RT	0.873	0.909	0.889	0.602	0.943	0.962	0.952	0.571	0.950	0.953	0.952	*0.595*
Measure	GPBUSD											
	GPBUSD H1 30 pipsow (weighted avg.)				GPBUSD H1 45 pipsow (weighted avg.)				GPBUSD H1 60 pipsow (weighted avg.)			
	rec	prec	F1	ROC	rec	prec	F1	ROC	rec	prec	F1	ROC
CART	*0.861*	*0.827*	0.815	*0.632*	*0.941*	0.905	0.915	0.516	*0.974*	*0.965*	0.960	0.531
C4.5	0.857	0.816	*0.819*	0.577	0.938	*0.920*	*0.922*	*0.611*	0.972	0.954	*0.961*	0.533
RT	0.775	0.775	0.775	0.563	0.903	0.904	0.903	0.580	0.950	0.949	0.949	*0.551*
Measure	USDJPY											
	USDJPY H1 30 pipsow (weighted avg.)				USDJPY H1 45 pipsow (weighted avg.)				USDJPY H1 60 pipsow (weighted avg.)			
	rec	prec	F1	ROC	rec	prec	F1	ROC	rec	prec	F1	ROC
CART	*0.925*	*0.874*	*0.891*	0.482	*0.974*	0.949	0.961	0.500	*0.990*	0.979	*0.985*	0.500
C4.5	*0.925*	0.855	0.888	0.500	0.973	*0.958*	*0.962*	*0.565*	0.988	*0.984*	0.984	0.492
RT	0.848	0.869	0.858	*0.542*	0.929	0.949	0.939	0.502	0.966	0.981	0.973	*0.552*

Abbrev. rec—recall (avg.); prec—precision (avg.); F1—F-measure (avg.); ROC—ROC Area (avg.).

Table 4 Confusion matrix for selected pairs

	Classified as		
	SELL	BUY	WAIT
EURUSD H1 30 pipsow—CART SELL—87; BUY—83; WAIT—2828			
SELL	*9*	2	76
BUY	1	*12*	70
WAIT	7	2	*2819*
GPBUSD H1 30 pipsow—CARTSELL—205; BUY—225; WAIT—2568			
SELL	*21*	2	182
BUY	1	*19*	205
WAIT	8	19	*2541*
GBPUSD H1 60 pipsow—RT SELL—18; BUY—13; WAIT—2967			
SELL	*1*	0	17
BUY	0	*3*	10
WAIT	46	30	*2891*
USDJPY H1 60 pipsow—C4.5SELL—18; BUY—13; WAIT—2967			
SELL	*0*	0	18
BUY	0	*1*	12
WAIT	7	0	*2960*

order if the situation is not clear. It is worth noting that in some cases the selected algorithms do surprisingly well even for the two remaining decision classes.

6 Conclusions

Ahe analysis presented in the article confirms that there is a possibility to effectively transform forex financial data into a decision table which may be used to generate decision trees and to perform satisfactory classification. The results given in the "Experiments" section allow to determine that in the future it will be possible to use such algorithms in the process of building forex systems. It should be pointed out that there are different classification measures, because only a general view allows to properly react to such results.

It is worth noting that even in this preliminary study the algorithms did surprisingly well with rash decisions, although it would be beneficial to improve the quality of the classifications, especially for the two remaining classes. The results analysis shows that in such a type of problem, approximate algorithms should be far more effective than the classical approaches. Algorithms in which there is the possibility of optimisation for exact classification measures, such as [13] or ensemble methods [2] should be checked in the near future.

References

1. Bahrepoura, M., Akbarzadeh-T, M.-R., Yaghoobia, M., Naghibi-S, M.-B.: An adaptive ordered fuzzy time series with application to FOREX. Exp. Syst. Appl. **38**(1), 475–485 (2011)
2. Boryczka, U., Kozak, J.: Enhancing the effectiveness of ant colony decision tree algorithms by co-learning. Appl. Soff Comput. **30**, 166–178 (2015)
3. Brabazon, T., ONeill, M.: Trading foreign exchange markets using evolutionary automatic programming. In: GECCO 2002: Proceedings of the Bird of a Feather Workshops, Genetic and Evolutionary Computation Conference (2002)
4. Caginalp, G., Laurent, H.: The predictive power of price patterns. Appl. Math. Financ. **5**(3–4), 181–205 (1998)
5. Cassetti, M.D.: A neural network system for reliable trading signals. Tech. Anal. Stock. Commod. **11**(6), 78–84 (1993)
6. Cheol-Ho, P., Irwin, S.H.: What do we know about the profitability of technical analysis? J. Econ. Surv. **21**(4), 786–826 (2007)
7. Czekalski, P., Niezabitowski, M., Styblinski, R.: ANN for FOREX forecasting and trading. Control Systems and Computer Science (CSCS 2015), pp. 322–328 (2015)
8. Froelich, W., Juszczuk, P.: Predictive capabilities of adaptive and evolutionary fuzzy cognitive Maps—a comparative study. Studies in Computational Intelligence, vol. 252, pp. 153–174. Springer (2009)
9. Gunasekarage, A., Power, D.M.: The profitability of moving average trading rules in South Asian stock markets. Emerg. Markets Rev. **2**, 17–33 (2001)
10. Kim, K.-J.: Financial time series forecasting using support vector machines. Neurocomputing **55**(1–2), 307–319 (2003)
11. Kirkpatrick II, C.D., Dahlquist Julie, R., Technical Analysis. Complete Resource for Financial Market Technicians. FT Press (2010)
12. Korczak, J., Roger, P.: Stock timing using genetic algorithms. Appl. Stoch. Model. Bus. Ind. **18**(2), 121–134 (2002)
13. Kozak, J., Boryczka, U.: Multiple boosting in the ant colony decision forest meta-classifier. Knowl. Based Syst. **75**, 141–151 (2015)
14. Lai, K.K., Yu, L., Wang, S.: A neural network and web-based decision support system for forex forecasting and trading. Data Min. Knowl. Manag. **3327**, 243–253 (2005)
15. Lim, T.-S., Loh, W.-Y., Shih, Y.-S.: A comparison of prediction accuracy, complexity, and training time of thirty-three old and new classification algorithms. Mach. Learn. **40**(3), 203–228 (2000)
16. Quinlan, J.R.: Induction of decision trees. Mach. Learn. **1**(1), 81–106 (1986)
17. Rhea, R.: Dow Theory. Barron's, New York (1932)
18. Safavin, S.R., Landgrebe, D.: A survey of decision tree classifier methodology. IEEE Trans. Syst. **21**(3), 660–674 (1991)
19. Taylor, M.P., Allen, H.: Extended evidence on the usage of technical analysis in foreign exchange. Int. J. Finance Econ. **11**, 327–338 (2006)
20. Taylor, M.P., Allen, H.: The use of technical analysis in the foreign exchange market. J. Int. Money Finance **11**, 304–314 (1992)
21. Tkacz, M.: Artificial neural networks in incomplete data sets processing. In: IntelligentInformation Processing and Web Mining, pp. 577–583 (2005)
22. Wasendorf Sr., R.R.: Foreign Currency Trading: From the Fundamentals to the Fine Points. McGraw-Hill (1997)

On Granular Rough Computing: Covering by Joint and Disjoint Granules in Epsilon Concept Dependent Granulation

Piotr Artiemjew and Jacek Szypulski

Abstract In this work we present the optimization methods of epsilon concept-dependent granulation. We consider two cases of parallel covering and granulation, based on joint and disjoint granules. Additionally we check two variants of majority voting, the first one based on descriptors, which are epsilon-indiscernible with the centers of granules, and the second variant uses all descriptors of respective granules. We verify the effectiveness of our methods on the real data sets from UCI Repository using the SVM classifier. It turned out that disjoint granules versus joint give almost identical results of classification with a significant acceleration of the granulation process. Additionally, the majority voting, based on the epsilon indiscernible descriptors, stabilised the process of granulation in terms of the accuracy of classification. This is a significant result, which lets us to accelerate the process of classification for many popular classifiers at least for k-NN, Naive Bayes, many rough set methods and the SVM classifier, which is supported by our recent works.

Keywords Rough sets · Decision systems · SVM · Granular rough computing · Epsilon concept-dependent granulation · Majority voting

1 Introduction

In recent years the granular computing approach has gained great interest among researchers. The popularity of the approach can be explained by the analogy to natural thinking; it is obvious that we group objects using some similarity measures, by some joint features. We use the granulation of knowledge to resolve problems in ordinary life.

P. Artiemjew (✉) · J. Szypulski
Department of Mathematics and Computer Science,
University of Warmia and Mazury, Sloneczna 54, 10-710 Olsztyn, Poland
e-mail: artem@matman.uwm.edu.pl

J. Szypulski
e-mail: jszypulski@matman.uwm.edu.pl

I. Czarnowski et al. (eds.), *Intelligent Decision Technologies 2016*,
Smart Innovation, Systems and Technologies 57,
DOI 10.1007/978-3-319-39627-9_13

The granular rough computing paradigm was initiated by Professor Lotfi Zadeh in 1979. The paradigm is connected with the Rough Set Theory proposed by Professor Zdzisław Pawlak in 1982. In the rough set theory the granules are defined as indiscernibility classes, where as similarity measures we use rough inclusions. In terms of rough inclusions, an interesting way of granulation of knowledge was proposed by Polkowski in [5, 6]. These methods turned out to be effective in many contexts. Interesting results can be found in the [7, 8]. In this scope of methods, the standard granulation is the basic one—see [5, 6]—this method was extended in joint works of Polkowski and Artiemjew into other variants, among others into granulation in the range of decision classes, and with epsilon variant, considering the indiscernibility ratio of descriptors epsilon (ε)—see [1, 2].

In this paper we have examined a few methods of optimization of the mentioned epsilon concept-dependent granulation. We propose the methods which let us compute the granules and covering of the universe of objects in the parallel way. In the covering process, we use two types of granules, the joint and disjoint one—see the Sects. 2.1 and 2.2. In the process of granular reflection creation, we use two variants of majority voting, with either use of all descriptors in the granules, or only the epsilon indiscernible to centers of granules—see Sect. 2.3. In order to compare mentioned methods we have designed an experimental session on the data from UCI Repository and with the use of the SVM classifier [4].

The rest of the paper is as follows. In Sects. 1.1, 1.2, and 1.3, we have described the methodology, the theoretical introduction to granulation in rough mereology, and the basic information about the used classifier, respectively. In Sect. 2. we have a description of our modifications of epsilon concept-dependent granulation. In Sect. 3 we have an experimental session with the results. In Sect. 4 we have conclusions and future work.

Let us start with a brief description of the methodology.

1.1 Methodology

We use the SVM classifier with RBF kernel as a reference classifier. The motivation to use it arises from our recent experiments [9], which show the effectiveness of this classifier in the context of epsilon concept-dependent granulation. Our modification of the granulation consists of the modification of covering process, which is parallel with the granulation. We have two variants. In the first, the granules are created from redundant indiscernibility classes, and in the second from disjoint indiscernibility classes. In both methods we consider only the central objects, which are new for the covering set. For verification of results we compute the accuracy of classification with the use of the five times Cross Validation 5 method [10].

In the next subsection we show the background information of our methods.

1.2 Granulation in Rough Mereology

Rough mereology is a theory of the predicate $\mu(x, y, r)$ read: "x is a part of y to a degree r", called a *rough inclusion*, see [7].

We recall that an *information system* (a *data table*) is represented as a pair (U, A) where U is a finite set of things and A is a finite set of *attributes*; each attribute $a : U \rightarrow V$ maps the set U into the *value set* V. For an attribute a and a thing v, $a(v)$ is the value of a on v.

We apply a particular form of a rough inclusion defined as follows.

For an attribute a, we let a_{max}, a_{min}, the maximal, resp. the minimal value of the attribute a on objects in the decision system, and then $span(a) = a_{max} - a_{min}$ is the span of a.

Given a parameter ε, defined as the fraction $x \times span(a)$ for $x = 0, 1, 2, \ldots, 100$ percent, we fix a *granulation radius r*.

We call two values $a(u), a(v)$ of the attribute a *ε–similar* if the inequality $\frac{|a(u)-a(v)|}{span(a)} \leq \varepsilon$ holds, in symbol $sim_{\varepsilon}(a(u), a(v))$. For a given object u, we define the *granule about u and of the radius r*, $g_{\varepsilon}(u, r)$ as the set,

$$g_{\varepsilon}(u, r) = \{v : |\{a \in A : sim_{\varepsilon}(a(u), a(v))\}| \geq r\}, \tag{1}$$

where $|.|$ denotes the size of a set.

Having granules defined, we continue with the granulation procedure. We apply the sequential covering method by selecting an object, building a granule around it, removing the granule from the universe of objects, and repeating until all objects are covered. Each of the obtained granules in the covering is factorized by selecting for each attribute the representative value for the granule by majority voting with random tie resolution. By this process, each granule is replaced with a vector of attribute values. The obtained reflection of the original decision system is then subject to classification by means of C–SVC, with the radial basis kernel RBF, see [3, 4].

1.3 The Classifier in a Nutshell

As a reference classifier we use the Support Vector Machine classifier with RBF kernel [3, 4], which turns out to be effective in the context of classification of the granular reflections of data [9]. The training and test data are normalized into the interval [−1, 1], and then the classifier is used on granulated training parts of data sets.

The granulation methods used in the work are as follows.

2 Optimisation of Epsilon Concept Dependent Granulation

2.1 Epsilon Concept Dependent Granulation with Disjoint Granules

This method is the modification of [1, 2]. In this variant we fix the epsilon parameter—the descriptors discernibility ratio—and we compute the indiscernibility classes from the universe of objects. We choose only classes whose central objects are not yet in the covering, and the granules cannot contain any objects from the covering.

The detailed procedure for covering is the following,

(i) from the original decision system (U, A, d), we form the training decision system (TRN) and test decision system (TST),
(ii) $U_{cover} = \emptyset$,
iii we set the granulation radius r_{gran} and the indiscernibility ratio of attributes ε,
(iv) for given $TRN = \{u_1, u_2, ..., u_{|TRN|}\}$, we form the $TRN_{temp} = TRN - U_{cover}$, we get in a random way the object $u \in TRN_{temp}$, and form the granule

$$g^{\varepsilon,cd}_{r_{gran}}(u) = \{v \in TRN_{temp}; \frac{|IND_{\varepsilon}(u, v)|}{|A|} \geq r_{gran} \text{ } and \text{ } d(u) = d(v)\}$$

$$IND_{\varepsilon}(u, v) = \{a \in A : \frac{|a(u) - a(v)|}{span(a)} \leq \varepsilon\}$$

(v) $U_{cover} \leftarrow g^{\varepsilon,cd}_{r_{gran}}(u)$,
(vi) if the U_{cover} is equal TRN, we go to (vii), otherwise to (iv),
(vii) we form the granular reflections of the original TRN system based on the granules from U_{cover} with the use of selected majority voting strategy—see Sect. 2.3.

2.2 Epsilon Concept Dependent Granulation with Joint Granules

This is the modification of the previous method, where during the covering process we use indiscernibility classes computed in the entire TRN set. The central objects of new granules are still on the outside of U_{cover}. The procedure is analogous to the previous one with the exception of the granule definition, which is as follows.

$$g^{\varepsilon,cd}_{r_{gran}}(u) = \{v \in TRN; \frac{|IND_{\varepsilon}(u, v)|}{|A|} \geq r_{gran} \text{ } and \text{ } d(u) = d(v)\}$$

2.3 Majority Voting with Consideration of ε-indiscernible Descriptors

In the paper we consider two variants of granular reflections. We form the granular reflections from granules of covering, based on the descriptors, which are ϵ-indiscernible from the central objects of granules. The covering of the universe U is as follows.

$Cov(U) = \{g^{\varepsilon,cd}_{r_{gran}}(x_i) : \bigcup_{i=1}^{k} g^{\varepsilon,cd}_{r_{gran}}(x_i) = U\}$.

If $|g^{\varepsilon,cd}_{r_{gran}}(x_i)| = n$ then,

$$g^{\varepsilon,cd}_{r_{gran}}(x_i) = \begin{pmatrix} a_1(x_1)\ a_2(x_1)\ ...\ a_m(x_1) \\ a_1(x_2)\ a_2(x_2)\ ...\ a_m(x_2) \\ ...\quad ...\quad ...\quad ... \\ a_1(x_n)\ a_2(x_n)\ ...\ a_m(x_n) \end{pmatrix}$$

Considering the central object of the granule,

$$a_1(x)a_2(x)...a_m(x)$$

we have to perform the following procedure of MV_{type2}.

$$MV_{type2}(a_1(g^{\varepsilon,cd}_{r_{gran}}(x_i))) = Avg\{a_1(x_j) : ||a_1(x) - a_1(x_j)|| \leq \varepsilon \ and \ x_j \in g^{\varepsilon,cd}_{r_{gran}}(x_i)\}$$

$$||a_1(x) - a_1(x_j)|| \leq \varepsilon \ if \ \frac{|a_1(x) - a_1(x_j)|}{span(a_1)} \leq \varepsilon$$

And the granular reflection of $g^{\varepsilon,cd}_{r_{gran}}(x_i)$ looks as follows:

$$MV_{type2}(a_1(g^{\varepsilon,cd}_{r_{gran}}(x_i))), MV_{type2}(a_2(g^{\varepsilon,cd}_{r_{gran}}(x_i))), \ldots, MV_{type2}(a_m(g^{\varepsilon,cd}_{r_{gran}}(x_i)))$$

The MV_{type1} is just the averaging of all of the descriptors in the granule from the covering.

$$MV_{type1}(a_1(g^{\varepsilon,cd}_{r_{gran}}(x_i))) = Avg\{a_1(x_j); x_j \in g^{\varepsilon,cd}_{r_{gran}}(x_i)\}$$

3 Experimental Session

For the experimental session we use the covering with hierarchical choice and the same split of data sets for a better comparison of methods. As a reference classifier we use SVM with the multiple Cross Validation method. The training system of each split was granulated by the respective method. In the experiments we use selected data from UCI Repository, among others, *Wisconsin Diagnostic Breast Cancer*, *Breast Cancer Wisconsin*, *Wisconsin Prognostic Breast Cancer*, *Fertility*, and *Parkinsons*.

On the charts, we have the Average accuracy of classification—this is the average accuracy from 5 times CV-5 test. There is the parameter Radius (r_{gran}) the granulation radius and Epsilon (ε) the indiscernibility ratio of descriptors. These parameters are useful in the process of approximation. The average size of the training set (granular decision system) is the percentage size of the training decision system after the approximation process. These parameters, compared with the accuracy of classification, show us the level of acceleration of the classification process.

3.1 Results of Experiments

Due to lack of space we show the result only for the exemplary data set, but the conclusion of our work is consistent with all of our results.

The results for the chosen *Wisconsin Diagnostic Breast Cancer* data set are as follows. In Fig. 1, we have the accuracy of classification for joint granules and majority voting based on the epsilon indiscernible descriptors. In Fig. 2 we have the analogous result for majority voting with the use of all descriptors of granules. Furthermore, in Figs. 3, and 4, we have the respective results for the disjoint granules. Finally, in Fig. 5 we have the average size of the granulated data set, which lets us see the possible acceleration of classification. We can see the spectra of parameters, which allows us to preserve knowledge from the original training data set.

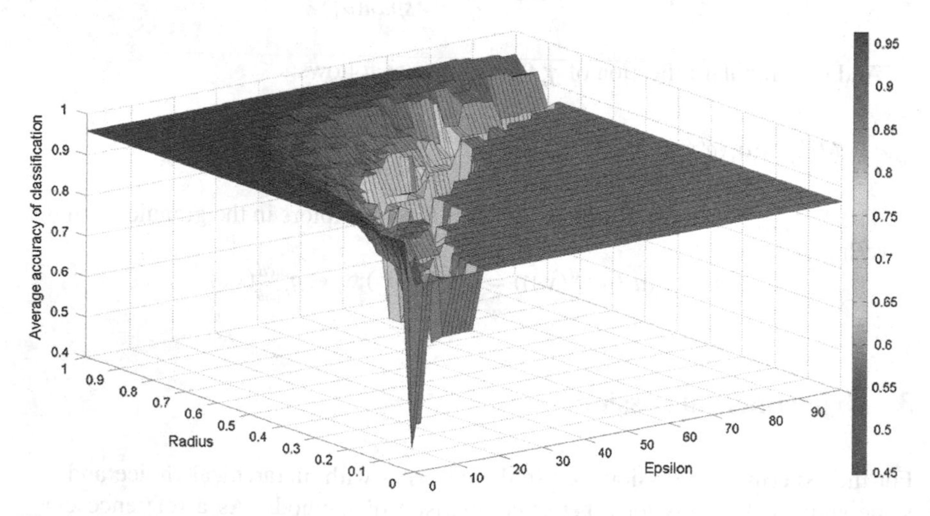

Fig. 1 5xCV5—classification result for WDBC data set for epsilon concept-dependent granulation and granules created from redundant indiscernibility classes; the averaging of values in majority voting with considering descriptors, which are epsilon indiscernible with central objects of granules

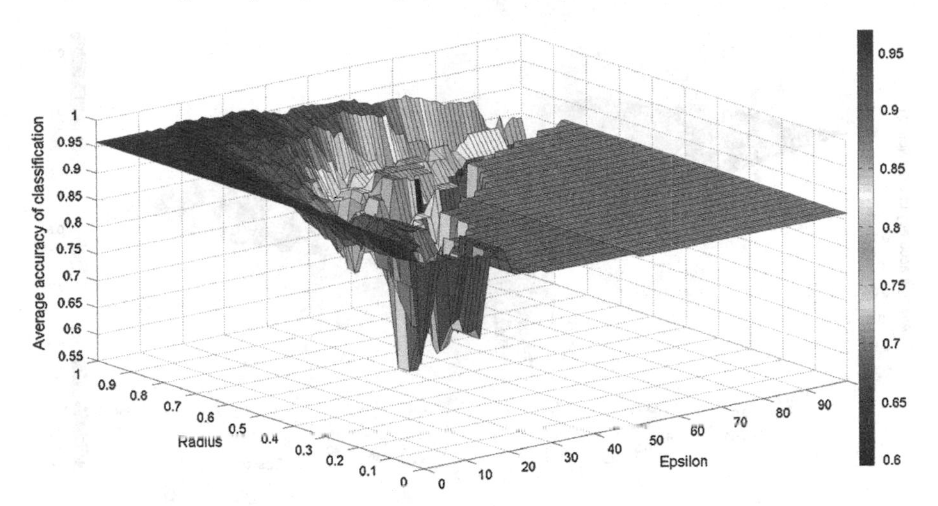

Fig. 2 5xCV5—classification result for WDBC data set for epsilon concept-dependent granulation and granules created from redundant indiscernibility classes; the averaging of values in majority voting with considering all descriptors

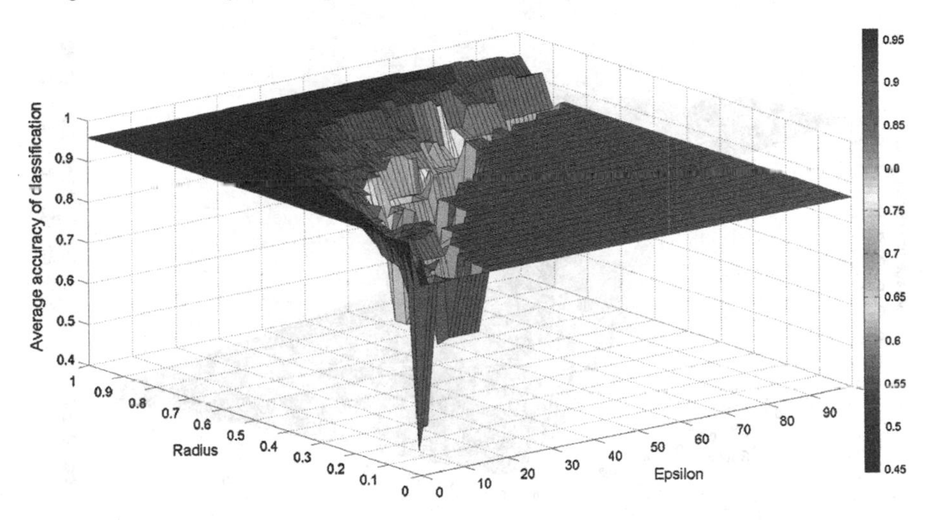

Fig. 3 5xCV5—classification result for WDBC data set for epsilon concept-dependent granulation and granules created from disjoint indiscernibility classes; the averaging of values in majority voting with considering descriptors, which are epsilon indiscernible with central objects of granules

The results of the experiments show the advantage of the method based on the disjoint granules. For both cases of majority voting, the result is comparable with the result for joint granules, but in the first method we have significant acceleration of granulation. Comparing the majority voting strategies, the one based on the epsilon indiscernible descriptors stabilises the classification process.

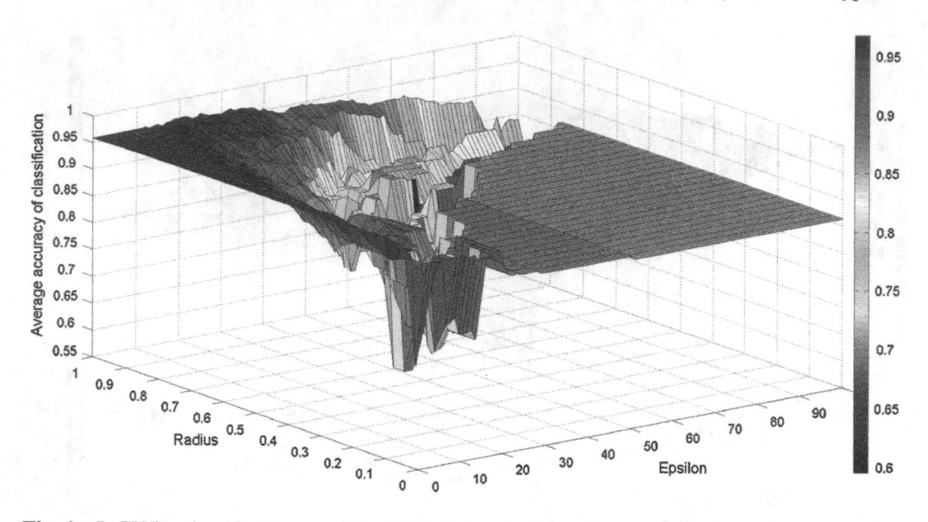

Fig. 4 5xCV5—classification result for WDBC data set for epsilon concept-dependent granulation and granules created from disjoint indiscernibility classes; the averaging of values in majority voting with considering all descriptors

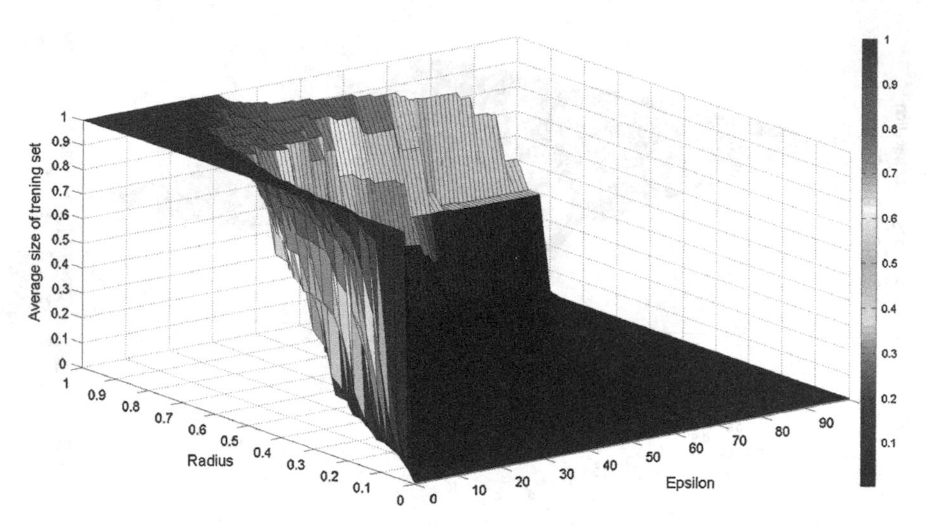

Fig. 5 The average size of granular decision systems for concept-dependent granulation variants with WDBC data set

The best results here are the ones with good enough accuracy and high lowering of the training data set size. In those cases the approximation of the training data set yield significant maintenance of knowledge from the original training data set. This was proven in many previous works, for example: [8, 9].

On the question 'why does the method of granulation based on the disjoint granules work in most cases faster than the one based on the joint granules?' one can answer that indiscernibility classes overlap in many cases (for many parameters), at least for the best parameters, which leads to high approximation of TRN data with quite high effectiveness of classification. Using joint granules we use such overlapping indiscernibility classes one by one during the covering of the universe, and we have to perform redundant operations. Thus using disjoint granules, we use indiscernibility classes with disjoint centers during covering. The process of covering speeds up because convergence occurs more quickly. This is the main source of acceleration, but one can see that the level of acceleration depends on the internal logic of the data sets, and could depend on the density of the data.

4 Conclusions

The basic result of this work is the acceleration of the granulation process by computing disjoint granules in comparison with the joint variant. The result of classification for these methods is almost identical.

Additionally, we have investigated two methods of majority voting for granular reflections creation. The results of the experiments lead us to the conclusion that the majority voting with consideration of only the ε indiscernible descriptors stabilises the granulation in the sense of accuracy of classification. The disadvantage of this method is the need for selection of the mentioned indiscernible descriptors during the granulation process, but the majority voting procedure is accelerated.

An accelerated process of granulation gives us the acceleration of classification for any classifier based on the approximated data set. In particular, in this work we can see the acceleration of the SVM classifier.

In future work we would like to check the other variants of majority voting, especially the voting on decision based on the weights determined by ε indiscernible descriptors.

Acknowledgments The research has been supported by grant 1309-802 from the Ministry of Science and Higher Education of the Republic of Poland.

References

1. Artiemjew, P.: On strategies of knowledge granulation and applications to decision systems. Polkowski, L., Supervisor, Ph.D. Dissertation, Polish Japanese Institute of Information Technology, Warsaw (2009)
2. Artiemjew, P.: A review of the knowledge granulation methods: discrete versus continuous algorithms. In: Rough Sets and Intelligent Systems—Professor Zdzisław Pawlak in Memoriam, vol. 2, pp. 41–59. Springer, Heidelberg (2013)
3. Boser, B.E., Guyon, I.M., Vapnik, V.N.: A training algorithm for optimal margin classifiers. In: Proceedings of the Fifth Annual Workshop on Computational Learning Theory, COLT '92, pp. 144–152. ACM, New York, NY, USA (1992)
4. Chang, C.C., Lin, C.J.: Libsvm: A library for support vector machines. ACM Trans. Intell. Syst. Technol. **2**(3), 27:1–27:27 (2011)

5. Polkowski, L.: Formal granular calculi based on rough inclusions. In: 2005 IEEE International Conference on Granular Computing, vol. 1, pp. 57–69 (2005)
6. Polkowski, L.: Granulation of knowledge in decision systems: the approach based on rough inclusions. The method and its applications. In: Proceedings of International Conference on Rough Sets and Intelligent Systems Paradigms, RSEISP 2007, Warsaw, Poland, 28–30 June 2007, pp. 69–79. Springer, Heidelberg (2007)
7. Polkowski, L.: Approximate Reasoning by Parts: An Introduction to Rough Mereology, 1st edn. Springer Publishing Company (2011)
8. Polkowski, L., Artiemjew, P.: Granular Computing in Decision Approximation: An Application of Rough Mereology. Springer Publishing Company (2015)
9. Szypulski, J., Artiemjew, P.: The rough granular approach to classifier synthesis by means of SVM. In: Proceedings of the 15th International Conference on Rough Sets, Fuzzy Sets, Data Mining, and Granular Computing, RSFDGrC 2015. Tianjin, China, 20–23 Nov 2015, pp. 256–263. Springer International Publishing, Cham (2015)
10. Tibshirani, R.J., Tibshirani, R.: A bias correction for the minimum error rate in cross-validation. Ann. Appl. Stat. **3**(2), 822–829 (2009)

On Approaches to Discretization of Datasets Used for Evaluation of Decision Systems

Grzegorz Baron and Katarzyna Harężlak

Abstract The paper describes research on ways of datasets discretization, when test datasets are used for evaluation of a classifier. Three different approaches of processing for training and test datasets are presented: "independent"—where discretization is performed separately for both sets assuming that the same algorithm parameters are used; "glued"—where both sets are concatenated, discretized, and resulting set is separated to obtain training and test sets, and finally "test on learn"—where test dataset is discretized using ranges obtained from learning data. All methods have been investigated and tested in authorship attribution domain using Naive Bayes classifier.

Keywords Discretization · Decision system · Classification · Naive Bayes classifier · Authorship attribution

1 Introduction

In the area of text analysis and processing very often research focuses on input data preparation methods to improve classification results. A scoring function can be used to evaluate quality of features affecting classification performance [10], a feature scaling method using Naive Bayes classifier can be applied [13], or a feature weighting method and text normalization can be attempted [5]. The paper addresses the issue of the influence of discretization methods applied to datasets used in evaluation of decision systems.

Considering the nature of numerical data, theoretically it can be infinitely dense. In many cases reduction of data density is beneficial or even necessary, and it can be obtained by discretization. Mainly it allows to convert continuous form of data into

G. Baron · K. Harężlak (✉)
Silesian University of Technology, Akademicka 16, 44-100 Gliwice, Poland
e-mail: katarzyna.harezlak@polsl.pl

G. Baron
e-mail: grzegorz.baron@polsl.pl

I. Czarnowski et al. (eds.), *Intelligent Decision Technologies 2016*,
Smart Innovation, Systems and Technologies 57,
DOI 10.1007/978-3-319-39627-9_14

discrete domain, but it can also change the volume of data, delivering the smaller number of continuous values. The former approach is employed when a classifier chosen for further analysis cannot operate on continuous numbers. In other cases such data preprocessing is facultative and can be analyzed in respect to possible benefits, for example improvement of classification performance [1], or more succinct way of expressing knowledge learned from input data.

When evaluation of a classifier performance is executed by using test datasets, the question arises how these sets should be discretized in relation to learning datasets. Three approaches can be employed. The first one relies on independent processing of learning and test datasets. In the second approach test data values are assigned to the bins based on the bins boundaries calculated during the discretization of learning datasets. And thirdly, data from training and test sets can be concatenated together, discretization process performed for such set, and then the resulting dataset splitted back to obtain learning and test sets.

The paper focuses on an analysis of different discretization methods in conjunction with the way of input sets discretization, taking into consideration some most popular discretization algorithms. To determine the influence of data discretization approach on classification quality, the Naive Bayes classifier has been chosen. It is a simple but very useful tool used in various domains, including text analysis. The presented experiments were conducted in an attempt to answer if it is possible to formulate any rules supporting the process of choosing the most suitable discretization method for a specific task.

The task considered as the application domain for described algorithms and procedures is authorship attribution from stylometric analysis of text. It deals with recognition of authorship based on style, in order to determine an author of some anonymous or disputed text, detect plagiarism etc. Statistics or machine learning techniques are mainly used for performing such tasks [7, 11].

The paper is organized as follows. Section 2 presents the theoretical background and methods employed in the research. Section 3 introduces the experimental setup, datasets used and techniques employed. The test results and their discussion are given in Sect. 4, whereas Sect. 5 contains conclusions.

2 Theoretical Background

The background of the presented research includes discretization algorithms, approaches to discretization of test datasets, and Naive Bayes classifiers.

2.1 Discretization

Many machine learning applications operate only on discrete data, whereas the nature of information in real life is often continuous. On the other hand, a number of methods work well with continuous features but perform better in discrete domain. Discretization converts wide spectrum of continuous values into datasets of discrete attributes, constituted by finite sets of intervals. It can be considered as a data reduction method which simplifies information as well as reduces possible information noise. But it is important to notice that in the discretized data always some loss of information occurs, therefore the process must be applied with caution.

Discretization algorithms can be fundamentally divided into two categories: supervised which utilize class information, and unsupervised which omit such information during discretization process. Generally discretization can be considered as four-step process: sorting all values, determining cut-points for splitting (or intervals for merging), performing splitting or merging according to an algorithm criterion, and evaluating the stopping condition of the process. Attribute values from the input set are assigned to one of the evaluated intervals.

Discretization Algorithms. The two most popular unsupervised discretization methods are so-called equal width and equal frequency binning. The former method seeks the minimum and maximum values of an attribute and then divides the whole range into the desired number of discrete intervals of equal width. There is a modification of the algorithm that relies on leave-one-out estimation of entropy [4]. The resulting number of bins is optimized and depends on the nature of input data. The equal frequency algorithm sorts all attribute values in ascending order, evaluates the minimum and maximum values for the discretized attribute, and then divides the range into some required number of intervals so that each part contains the same number of discrete values [8].

For the purpose of the presented research two supervised discretization methods were selected. Both of them utilize the Minimum Description Length principle (MDL). The first one is based on research of Fayyad and Irani [3], whereas the second one uses Kononenko's MDL criterion [6]. Supervised methods are considered as more efficient and delivering better results [2, 8].

Test Datasets Discretization. Application of test datasets is one of the ways of evaluating classifiers. The aim of such approach is to use for that purpose data which was not utilized during the training stage of a decision system building process. In cases of discretized data, it is obvious that learning datasets are discretized applying some parameters like a type of algorithm, number of bins, width of bin, frequency of instances in the bin, class, etc. Similar parameters should be applied for test datasets, but results would be different depending on the type of algorithm, other required parameters, and relationship between discretization processes of training and test datasets, whether they are dependent on each other or not.

2.2 *Bayes Classifiers*

Bayes classifiers are relatively simple but powerful, often used as a reference model for other classification research. The basic Naive Bayes for authorship attribution can be utilized in two versions, depending on the nature of input data. If the features set consists of binary variables that inform only if a word (from the previously selected list) exists in the analyzed text, then the multivariate Naive Bayes classifier can be used. If the information about word occurrences is extracted, the multinomial Naive Bayes algorithm is suitable for problem solving. For big sizes of the vocabulary the second approach is considered better [9].

Bayes classifier is based on Bayes' rule of conditional probability:

$$p(c_j \mid d) = \frac{p(d \mid c_j)p(c_j)}{p(d)} , \tag{1}$$

where: $p(c_j \mid d)$—a'posteriori probability of instance d being in class c_j, $p(d \mid c_j)$—probability of generating instance d given class c_j, $p(c_j)$—a'priori probability of occurrence of class c_j, $p(d)$—probability of instance d occurring, and

$$p(d \mid c_j) = p(d_1 \mid c_j)p(d_2 \mid c_j)\ldots p(d_m \mid c_j) . \tag{2}$$

The MAP (maximum a'posteriori) decision rule is applied to get the result of classification process $NBC(d_1, \ldots, d_n)$:

$$NBC(d_1, \ldots, d_n) = \underset{c}{\operatorname{argmax}} p(C = c) \prod_{i=1}^{n} p(D_i = d_i \mid C = c) . \tag{3}$$

It is commonly assumed that values of numeric attributes are normally distributed, so the probability density function for Gaussian (normal) distribution is utilized. For specific purposes other distributions could be more suitable.

3 Experimental Setup

Processing of datasets during experiments required execution of steps, as follows:

1. preparation of input data,
2. discretization of input data (unsupervised and supervised) using various approaches to learning and test instances,
3. classification using Naive Bayes classifier,
4. classifier evaluation during the test stage.

The following subsections present the main conditions of the performed experiments, the characterization of input datasets, and descriptions of discretization and classification techniques employed.

3.1 Input Datasets

The main condition while creating sets of characteristic features for authorship attribution is that they should uniquely describe all texts of a given author, and at the same time they need to enable distinction from other authors. In the research linguistic descriptors from lexical and syntactic groups were chosen, reflecting frequencies of usage for selected function words and punctuation marks [12]. It is assumed that lexical elements characterize literary style of authors, whereas the style of sentences building is described by syntactic features.

As the base for all experiments texts of two pairs of authors were chosen, male and female [12]. For each author several works were studied. To obtain input data source texts were splitted into blocks of comparable size, and frequencies for selected descriptors were calculated. Each dataset consisted of attributes belonging to one of two classes, corresponding to two recognized authors, and performed classification was binary.

The validation of classification results was performed using test sets. It was important to prepare training and test datasets basing on the disjunctive works of writers. Such approach allows to get objective results. As the result separate training and test datasets were obtained, with balanced classes in each set.

3.2 Approaches to Discretization of Test Datasets

When a quality of a decision system is evaluated by using test sets, and input data needs to be discretized, the relation between discretization procedures for learning and test sets can be considered in three ways:

- "independent" (*Id*)—training and test datasets are discretized separately,
- "glued" (*Gd*)—training and test datasets are concatenated, the resulting set is discretized applying required parameters, and finally data is divided back into learning and test set,
- "test on learn" (*TLd*)—firstly training dataset is discretized using chosen parameters, and then test set is processed using bin's range values calculated for training data.

"Independent" way is the easiest to apply, but intuitively it can be considered as not good, because the way how test set is discretized can be very different from results obtained for training dataset. Since training and test sets are analyzed separately, it is very likely that the bin ranges in both sets are different, and the numbers of bins in both sets may vary. This can possibly lead to the situation, where the same attribute value is assigned to different bins in training and test sets. That seems to be a problem which can degrade the system performance.

"Glued" approach allows to discretize all data in more consistent way. However, test data should be totally independent from training and vice versa, and this assumption is not entirely true because of common processing of both sets. For example minimum and maximum values in test set can be significantly different than in training set. Resulting discretized learning set will be definitely different when compared to the one obtained for data without test set appended.

Discretizing in the "test on learn" manner seems to be more natural and potentially better than previous two, as the ranges of bins found for training dataset are applied for test data. But in this case the possible influence of training information onto the test data exists, which violates the assumption about independence of information used for evaluation of the decision system.

As can be seen all three approaches have possible advantages and disadvantages and deeper experimental investigation is necessary to assess considered solutions. For all three ways of processing several discretization algorithms were employed: unsupervised—equal width, optimized equal width, and equal frequency, and supervised—Fayyad & Irani MDL, and Kononenko MDL.

Some of the properties and relationships described above are illustrated by examples shown in Table 1. For presentation purposes only few instances of bigger datasets being processed are presented. The equal width algorithm was used with number of bins parameter set to 3. Notation used for describing bins reflects the

Table 1 Exemplary results of discretization of input datasets applying equal width algorithm using: "independent" (*Id*), "glued" (*Gd*), and "test on learn" (*TLd*) approach

(a) Bin ranges calculated for training data

Training bin ranges (*Id/TLd*):	(-inf–0.008105], (0.008105–0.009427], (0.009427-inf)
Training bin ranges (*Gd*):	(-inf–0.007057], (0.007057–0.008903], (0.008903-inf)

(b) Input and discretized training data

Training data	Discrete *Id*	Discrete *TLd*	Discrete *Gd*
0.006783	(-inf–0.008105]	(-inf–0.008105]	(-inf–0.007057]
0.006915	(-inf–0.008105]	(-inf–0.008105]	(-inf–0.007057]
0.010151	(0.009427-inf)	(0.009427-inf)	(0.008903-inf)
0.009330	(0.008105–0.009427]	(0.008105–0.009427]	(0.008903-inf)

(c) Bin ranges calculated for test data

Test bin ranges (*Id*):	(-inf–0.006909], (0.006909–0.008608], (0.008608-inf)
Test bin ranges (*TLd*):	the same as training bin ranges (*Id/TLd*)
Test bin ranges (*Gd*):	the same as training bin ranges (*Gd*)

(d) Input and discretized test data

Test data	Discrete *Id*	Discrete *TLd*	Discrete *Gd*
0.009475	(0.008608-inf)	(0.009427-inf)	(0.008903-inf)
0.010135	(0.008608-inf)	(0.009427-inf)	(0.008903-inf)
0.007278	(0.006909–0.008608]	(-inf–0.008105]	(0.007057–0.008903]
0.007493	(0.006909–0.008608]	(-inf–0.008105]	(0.007057–0.008903]

lower and upper boundaries of respective bin (*inf*—infinity used for formal description of first and last intervals). Subtables (a) and (c) show bin ranges calculated for training and test data respectively. Subtable (b) contains input data and outcomes obtained for training data whereas part (d) presents results of test sets discretization.

For unsupervised equal width and equal frequency discretizations the only parameter required was the number of bins. For optimized equal width algorithm the obtained numbers of bins were lower or equal to the parameter value. For both equal width versions the number of bin parameter ranged from 2 to 10 with step 1, and from 10 to 1000 with step 10. For equal frequency the maximum value of this parameter is equal to the number of instances in a discretized dataset. Because of discretizing training and test sets together somehow the maximum value of parameter had to be fitted to lower cardinality of processed datasets. The supervised discretization was applied without any parameters and resulting number of bins depended on nature of data.

It is important to point out that for "independent" discretization of test dataset the resulting number of bins in training and test sets could be different. Such effect caused problems during the classifier evaluation stage, where the numbers of bins in both sets were expected to be equal. To overcome this problem the names of ranges (obtained as strings) were converted to their ordinal numbers, and numerical data type for each attribute was declared. So from a classifier perspective it operated on numerical data, but transformed during discretization. The same conversion was applied to dataset discretized using other methods to unify the experiments.

Naive Bayes classifier can deal with different types of attributes, in particular numeric and nominal ones. During the experiments it operated on numerical data thanks to conversion mentioned above. The normal distribution was used for numeric attributes.

Discretization and classification were performed separately for data based on male and female texts, both groups of results were averaged, and as such were the subject of further analysis. There were also experiments performed for datasets without discretization, to obtain some reference values for comparison.

4 Results and Discussion

To obtain a reference point for discussion, classification for datasets without discretization was performed. For the Naive Bayes classifier the predictive accuracy was 86.94 % (calculated as mean of results obtained for male and female authors separately). This value is indicated in all figures presenting experimental results. Figure 1 gives results for unsupervised methods.

The range of parameters variation for equal width and equal frequency algorithms was initially very wide. Experiments showed that for all ways of discretization the most promising classification results were obtained for relatively small values of a given parameter, typically below 10. For higher values performance was decreasing rapidly. Therefore diagrams presented in Fig. 1 were prepared for number of bins up to 10.

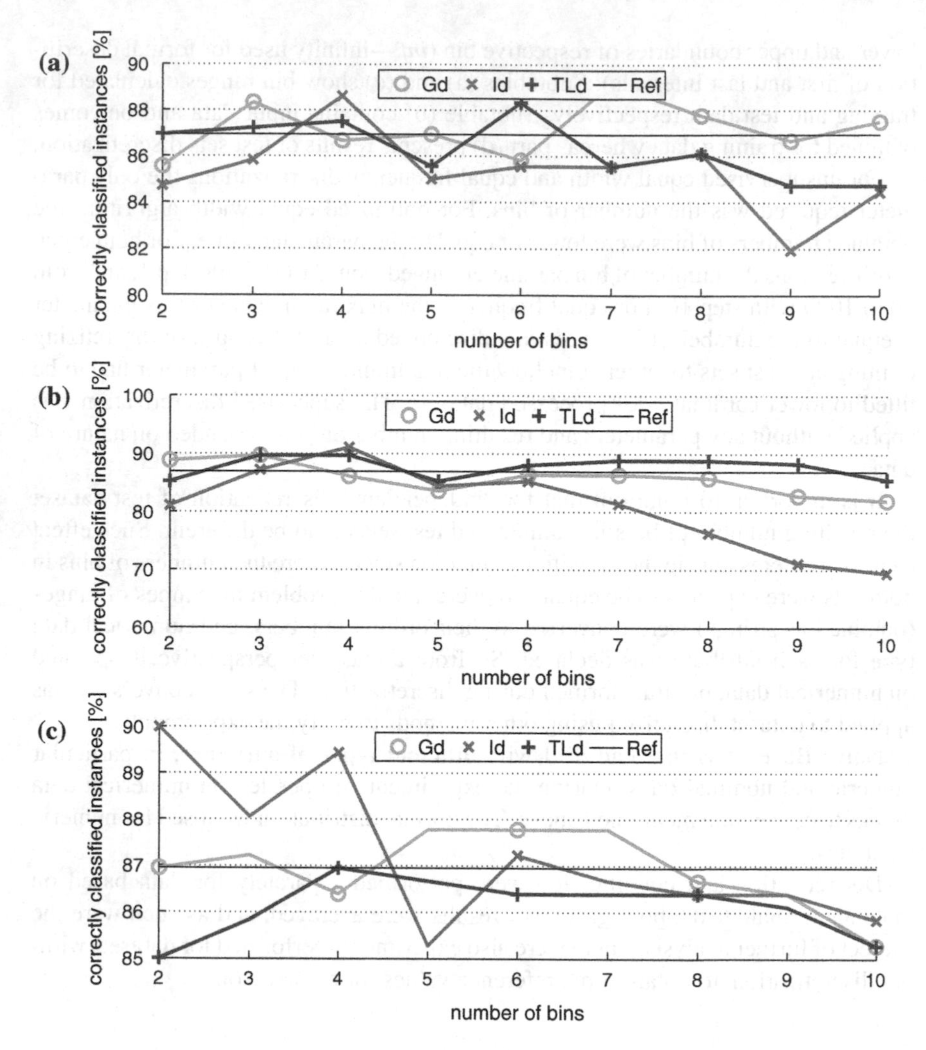

Fig. 1 Classifier performance for: **a** equal width, **b** optimized equal width, **c** equal frequency discretization, with training and test sets discretized using methods: "independent" (*Id*), "glued" (*Gd*), "test on learn" (*TLd*). *Ref* represents reference value obtained for non-discretized data

For equal width discretization (Fig. 1a) all three approaches to test set discretization delivered results better than reference for some values of a number of bins. But only "glued" approach performed well or almost well in the entire analyzed range. The similar observations could be made for optimized equal width discretization algorithm (Fig. 1b), except for the fact that all three discretization procedures delivered more stable results, when compared to that obtained for simple equal width algorithm. The best overall result (considering unsupervised methods) of correctly classified instances was obtained exactly for this algorithm using "independent"

Table 2 Results of Naive Bayes classification for experiments performed using three test datasets discretization approaches for supervised algorithms

Discretization algorithm	Test dataset discretization approach		
	"independent" (%)	"glued" (%)	"test on learn" (%)
Fayyad & Irani	95.56	93.33	91.11
Kononenko	83.33	94.44	91.11

discretization. Also equal frequency binning (Fig. 1c) gave better results for small parameter values. Especially "independent" discretization delivered good results, comparable to the best.

Table 2 presents results obtained for supervised discretization algorithms. Almost all algorithms applied for different approaches perform very well, exceeding the reference level. Only Kononenko MDL in combination with "independent" test datasets discretization delivered worse results.

As aforementioned, the most interesting range of discretization parameters lies below 10. Results changed there dynamically, therefore it was interesting to investigate that area more deeply. The idea was to observe only the range where classifier performance seems to be better than reference. Therefore average value of classifier efficiency for three analyzed ways of discretization in respect to values of algorithm parameters was calculated. Two of three algorithms performed better than reference for parameter equal or lower than 6. A set of boxplot diagrams presenting classification results is presented in Fig. 2, for discretization parameters limited up to 6 (value represents required number of bins).

The main aim of performed research was to find relations between classification accuracy, assessed during the classifier evaluation process, and a method of test datasets discretization. Intuitive analysis could lead to a conclusion that discretization of test sets performed in some reference to training dataset should deliver better

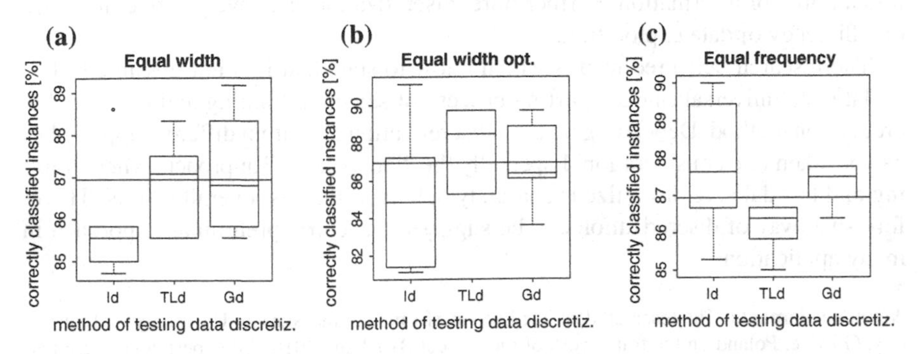

Fig. 2 Selected results of classifier evaluation for: **a** equal width, **b** optimized equal width, **c** equal frequency algorithms, with training set discretized using methods: "independent" (*Id*), "glued" (*Gd*), "test on learn" (*TLd*). Diagrams are based on results obtained for bin number ranged from 2 to 6. Reference level of 86.94 % is indicated

results. One of the reasons was that "independent" discretization outcomes for specific algorithms could have different number of bins in training and test datasets. Yet results presented in Fig. 2 show that a firm hierarchy between tested approaches does not exist. For both equal width algorithms "test on learn" method gives results better than others, whereas it is the worst approach for equal frequency binning. On the other hand, "independent" method, considered as a poor one, performs surprisingly well with this algorithm. It is important to point out that the best results during experiments were obtained for "independent" method for unsupervised as well as supervised algorithms.

Presented research results allow to state that there is no unequivocal rule allowing to select the best approach to test sets discretization. Depending on nature of data and chosen discretization algorithm all analyzed ways can be taken into consideration.

5 Conclusions

The paper presents research on the influence of way of test datasets discretization on results of classifier evaluation. For the executed tests the Naive Bayes was selected and all outcomes were analyzed in comparison with the reference value obtained for non-discretized datasets, using the same classifier. The experiments were binary classification tasks performed in authorship attribution domain.

Study results showed that good quality of decision system was obtained for relatively small number of bins in discretized data. But facts which must be taken into consideration to keep this conclusion valid are as follows: nature of analyzed data—stylometric datasets prepared as aforedescribed; system performing binary classification. Observation of discretized outcomes of supervised algorithms supports prove of such conclusion. Number of bins delivered by these methods, which analyze entropy of data along with its class attribution, were also small what means that such conversion of data (given relatively small number of bins) did not cause significant loss of information. Furthermore, discretization can have positive influence on efficiency of data exploration.

The research delivered results which allow to state that it is not possible to formulate one universal rule supporting process of selecting training and test sets discretization method. Depending on used discretization algorithm different approaches can be taken into consideration. Especially the "independent" approach, where training and test data are discretized separately delivered the best overall results. Therefore such way of discretization can be suggested as entry, preliminary approach in many applications.

Acknowledgments The research described was performed at the Silesian University of Technology, Gliwice, Poland, in the framework of the project BK/RAu2/2016. All experiments were performed using WEKA workbench [4].

References

1. Baron, G.: Influence of data discretization on efficiency of Bayesian Classifier for authorship attribution. Procedia Comput. Sci. **35**, 1112–1121 (2014)
2. Dougherty, J., Kohavi, R., Sahami, M.: Supervised and unsupervised discretization of continuous features. In: Machine Learning: Proceedings of the 12th International Conference, pp. 194–202. Morgan Kaufmann (1995)
3. Fayyad, U.M., Irani, K.B.: Multi-interval discretization of continuous-valued attributes for classification learning. In: Proceedings of the 13th International Joint Conference on Artificial Intelligence (IJCAI), pp. 1022–1029 (1993)
4. Hall, M., Frank, E., Holmes, G., Pfahringer, B., Reutemann, P., Witten, I.H.: The weka data mining software: an update. SIGKDD Explor. **11**(1), 10–18 (2009)
5. Kim, S.B., Han, K.S., Rim, H.C., Myaeng, S.H.: Some effective techniques for Naive Bayes text classification. IEEE Trans. Knowl. Data Eng. **18**(11), 1457–1466 (2006)
6. Kononenko, I.: On biases in estimating multi-valued attributes. In: 14th International Joint Conference on Articial Intelligence, pp. 1034–1040 (1995)
7. Kotsiantis, S.B.: Supervised machine learning: a review of classification techniques. In: Proceedings of the 2007 Conference on Emerging Artificial Intelligence Applications in Computer Engineering: Real Word AI Systems with Applications in eHealth. HCI, Information Retrieval and Pervasive Technologies, pp. 3–24. IOS Press, Amsterdam, The Netherlands (2007)
8. Kotsiantis, S., Kanellopoulos, D.: Discretization techniques: a recent survey. Int. Trans. Comput. Sci. Eng. **1**(32), 47–58 (2006)
9. McCallum, A., Nigam, K.: A comparison of event models for Naive Bayes text classification. In: AAAI-98 Workshop On Learning For Text Categorization, pp. 41–48. AAAI Press (1998)
10. Schneider, K.M.: Techniques for improving the performance of Naive Bayes for text classification. In: Proceedings of 6th International Conference on Intelligent Text Processing and Computational Linguistics (CICLing), pp. 682–693 (2005)
11. Stańczyk, U.: Rule-based approach to computational stylistics. In: Bouvry, P., Kłopotek, M., Marciniak, M., Mykowiecka, A., Rybiński, H. (eds.) Security and Intelligent Information Systems, LNCS (LNAI), vol. 7053, pp. 168–179. Springer, Berlin (2012)
12. Stańczyk, U.: Ranking of characteristic features in combined wrapper approaches to selection. Neural Comput. Appl. **26**(2), 329–344 (2015)
13. Youn, E., Jeong, M.K.: Class dependent feature scaling method using Naive Bayes classifier for text datamining. Pattern Recognit. Lett. **30**(5), 477–485 (2009)

Intelligent Decision Making for Uncertain Unstructured Big Data

A Bayesian Approach to Classify the Music Scores on the Basis of the Music Style

Michele Della Ventura

Abstract This article presents a new version of the algorithm proposed by Della Ventura (12th TELE-INFO International Conference on Recent Researches in Telecommunications, and Informatics, 2013, [1]) to classify the musical scores. Score classification means an automatic process of assignment of the specific score to a certain class or category: baroque, romantic or contemporary music. The algorithm is based on a Bayesian probabilistic model that extends the Naive Bayes classifier by adding a variable tied to the value of the information contained within the. The score is not seen as a single entity, but as a set of subtopics, every single one of which identifies and represents a standard feature of music writing. The classification of the score is done on the basis of its subtopics: an intermediate level of classification is thus introduced, which induces a hierarchical classification. The new algorithm performs equally well on the old dataset, but gives much better results on the new larger and more diverse dataset.

Keywords Categorization • Document classification • Information • Music score • Naive Bayes

1 Introduction

Given the rapid growth of the amount of information available online, information retrieval (IR) is a crucial task for search engines to be able to meet the users' requests. To this end, the classification of the text presents itself as a way to cope with this issue by classifying documents in a fixed and predefined number of categories.

Many classification algorithms were created and used in order to classify SMS [2], newsgroup articles and web pages [3], to understand the users' interests [4], to organize the electronic mail, to find information of interest on the web [5, 6] and to

M.D. Ventura (✉)
Department of Technology, Music Academy "Studio Musica", Treviso, Italy
e-mail: dellaventura.michele@tin.it

I. Czarnowski et al. (eds.), *Intelligent Decision Technologies 2016*,
Smart Innovation, Systems and Technologies 57,
DOI 10.1007/978-3-319-39627-9_15

guide the user's research through the web [7, 8]. As of now, few researches have been performed so as to integrate the search algorithms with IR functions for text music files, i.e. for musical scores: most of the studies refer to audio music files. With respect to scores, various researches have been taken on to classify scores on the basis of the musical genre [9]: classic, pop, rock, jazz. Other studies aim mainly at the segmentation of the symbolic level of the score in order to identify a melody (or a motif), a rhythmic structure or specific features, being essentially based on the musical grammar rules [10, 11] and on the similarity concept [12–15]. Yet, the same segmentation is performed in different ways depending on the musical style of the composition: tonal or atonal music. For instance, in tonal music, unlike atonal music, the segmentation may take into account elements such as the harmonic structure and/or the harmonic functions besides the melodic or rhythmic similarity concept. The very same concept of similarity to compare two musical objects (melodies or motifs) must be applied in different ways [1], based on whether it is tonal or atonal music that is under analysis. It can therefore be inferred that having the possibility to classify the scores would make the operations of segmentation and indexing of a piece of music more rapid and more precise: i.e. the most adequate tools for the specific operation would be used right away.

This article presents a model of analysis of the score that allows the identification of its tonal or atonal style, based on Bayes' Theorem that extends the Naive Bayes classifier by adding a variable tied to the information value (entropy) contained within the score [1]. The Naive Bayes classifier is one of the most used methods for text classification: in spite of its simplicity, in many cases it manages to reach performance levels that are even better than the most known classification algorithms.

This paper is structured as follows. We start by reviewing background and related work in Sect. 2. The Naive Bayesian Classification is described in Sect. 3. The analysis of the musical message is described in Sect. 4. Experiments results are shown in Sect. 5. Finally, Sect. 6 concludes this paper and points out future works.

2 Background and Related Work

In the musical field, the classification concept is rarely, yet interestingly confirmed in the analysis of music files examined on the symbolic level (scores).

Pérez-Sancho et al. [16] investigated whether stochastic language models of harmony including naive Bayes classifiers and 2-, 3- and 4-grams could be used for automatic genre classification on both symbolic and audio data. They reported better classification results when using a richer vocabulary (i.e. including seventh chords), reaching 3-genre classification accuracies on symbolic data of 86 % with naive Bayes models and 87 % using bi-grams [17]. To deal with audio data generated from MIDI they used a chord transcription algorithm and obtain accuracies of 75 % with naive Bayes [17] and 89 % when using bi-grams [18].

Dor and Reich [19] proposed an approach that considers the music scores and classifies composers of classical music based on certain low-level characteristics of their compositions. The proposed composition characteristics are descriptive features derived from the time-ordered sequence of pitches in a composition.

Another interesting entry is the work of Della Ventura [1]. The classification of a score in three different stylistic areas, baroque, romantic and contemporary music, is performed by calculating the entropy value carried by the score (according to Weaver and Shannon's theory). Each stylistic area is identified by a minimum value and a maximum value of entropy, obtained respectively by analyzing many scores of different composers of different historical periods.

This article presents the MSC (Music Scores Classifier) algorithm that, inspired by the above mentioned articles, has the aim of classifying a score in two distinct areas, tonal music and atonal music, using the Naive Bayes classifier and the value of the information obtained through the entropy. Whit this method classification mistake are no possible, that means a score belonging to the tonal period could classified in a atonal period and vice versa.

3 The Naive Bayesian Classification

The Bayesian classification is a statistic technique by means of which one can determine the probability for an element to belong to a certain class. The theorem describes how the opinions while observing *A* are enhanced by having observed event *B* and it describes the relationship existing between the conditional probability of event A given *B* and the reverse conditional probability of event *B* given *A*. The theorem may be generalized by taking into account several events, $p(A_i|B)$. In effect, taking into consideration the space of the events $A_1, A_2, \ldots, A_n$ (called partition of the event space) formula 3.1 is modified as follows:

$$p(A|B) = \frac{p(B|A_i)\ p(A_i)}{p(B)} = \frac{p(B|A_i)\ p(A_i)}{\sum_{j=1}^{n} p(B|A_j)\ p(A_j)} \tag{3.1}$$

A classifier, in statistic analysis (cluster analysis), solves the problem by identifying subpopulations of individuals having certain features within a larger set, with the potential use of a subset of a priori discriminated individuals (training set). Bayesian classifiers are based on the application of Bayes' Theorem. The principle that governs this type of classifiers is based on the fact that some individuals belong to an interest class with a given probability on the basis of certain observations [20, 21]. This probability is calculated assuming that the features observed may be interdependent or independent; in the latter case the Bayesian classifier is called

Naive because it ingenuously assumes that the presence or absence of a certain feature within a given class of interest is not correlated with the presence or absence of other features, significantly simplifying the calculation [22]. With this type of decider, the naive Bayesian classifier acquires the following:

$$\operatorname{argmax}_C p(C) \prod_{i=1}^{n} p(F_i|C) \tag{3.2}$$

4 Analysis of the Musical Message and Training Set

The application of the naive Bayesian classifier demands the use of a collection of experimental observations the class of which must be known. This data set is called a training set and the class of every observation is called target. The target is the learning objective: it is what shall be learned. The classes used to achieve the training set for MSC are derived from an analytic analysis of the music text, excluding a priori those univocal features having a specific compositional style [23]: features belonging only to the tonal period or only to the atonal period. The following have been, therefore, excluded:

- the analysis of the harmonic structure (typical of the tonal style): it is the study of the overlapping of sounds in certain intervals (chords) and their relationships with tonality;
- the analysis of rhythmic accents (typical of the atonal style) Rhythm is defined as a succession of accents [24], where accent means the highest variation of intensity that some sounds have compared to others within the frame of a song or of a musical phrase. Rhythmic accent refers to the division of the piece of music into strong or weak accents organized within the rhythmic cell which is the bar (Fig. 1) [25].

In atonal music, the accent structure as shown in Fig. 1 and typical of the tonal music, is absent, creating a play of effects that often changes within the same piece of music. The features of music composition that may be found in both compositional styles were taken into consideration when creating the training set, i.e.: key signatures, intervallic structure, polyrhythm, information value of the piece of music (entropy).

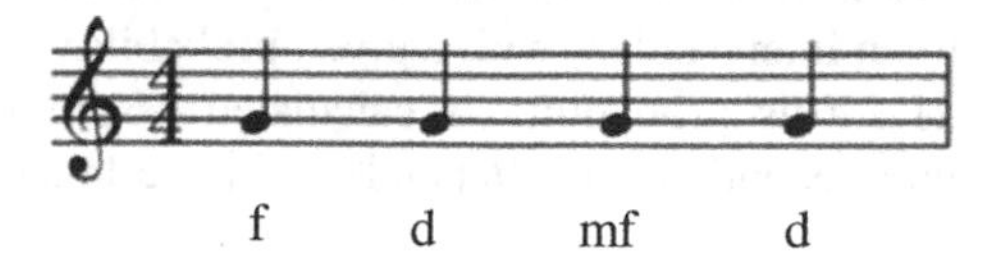

Fig. 1 Representation of the accents, strong/forte (f), weak/debole (d) and half-strong/mezzo-forte (mf)

Fig. 2 Representation of the key signatures

4.1 Key Signatures

Key signatures are alterations shown at the beginning of every stave (Fig. 2), right after the clef, that refer to the tonality of the musical composition, and that are, therefore, typical of the tonal system and missing from the atonal system.

This, which might represent a "false discriminant" in the construction of the training set, actually represents an important class. In effect, in the tonal system there are two tonalities, that of C major and that of A minor, which do not have any key signature. At the same time, it is possible to find scores of the atonal period (albeit in a small number) with key signatures.

4.2 Intervallic Structure

From a macroscopic analysis of the music text of compositions belonging to different historical periods and different composers, there emerges a progressive increase of the interval variety among the various sounds that make up the melody [26]. There is a gradual passage from the exclusive use of consonant intervals (intervals of a third, of a fourth, of a fifth and of a sixth), typical of the Baroque period (Fig. 3a), to the use of dissonant intervals (intervals of a second, of a seventh and all the augmented and diminished intervals), which characterize the Romantic period (Fig. 3b) and the Contemporary music (Fig. 3c). Consonant and dissonant are two music terms that indicate the pleasant or unpleasant quality of an interval and are the subjective evaluation of a sensation. In the training set, the interval is classified as consonant (c), dissonant (d) and dissonant extending beyond the octave (d+): this last interval aims at taking into consideration an aspect that evolved from the Romantic period to the contemporary period and was, therefore, used both in the tonal music and in the atonal music, without it actually being a specific feature of a certain period.

4.3 Polyrhythm

It consists in using different rhythmic combinations within one musical composition. In music one can distinguish between simple irregular rhythmic groups (such as the triplet and the sextuplet) and complex irregular rhythmic groups (such as the quintuplet, the septuplet, the nontuplet, ...). In the training set, the absence of

Fig. 3 Examples of melodies belonging to the time periods

Fig. 4 Examples of different rhythmic structures. **a** Absence of irregular groups. **b** Simple irregular groups. **c** Complex irregular groups

irregular groups (No) (Fig. 4a) was distinguished from the presence of simple irregular groups (Easy) (Fig. 4b) or of simple and complex irregular groups in general (Complex) (Fig. 4c).

4.4 Information Value of the Piece of Music (Entropy)

The analysis based on the information theory considers music to be a linear process supported by its own syntax [27]. However it is not a syntax formulated on the basis of grammar rules, rather on the basis of the occurrence probabilities of every element of the musical message with respect to the preceding element [28]. The fact that the 9events of a composition are organized modularly presents the opportunity to calculate using a formula, or to express by means of an "index", the total information transmitted by a certain musical segment [27]. In a communication occurring through a given alphabet of symbols, the information is associated to every single transmitted symbol. Information may, therefore, be defined as the reduction of uncertainty that might have been present a priori as far as the transmitted symbol is concerned [29]. The wider the message range that the source can transmit (and the bigger the uncertainty of the receiver with respect to the possible message), the larger the quantity of transmitted information and with it its size: the entropy. Shannon [30] produced a definition of the entropy of such a set, identifying it as the information content that the choice of one of the messages will transmit. If every message has the probability pi to be transmitted, the entropy is obtained as the sum on the whole set of functions pi log2 pi, every function being related to a message, i.e.:

$$H(X) = E[I(x_i)] = \sum_{i=1}^{n} I(x_i) \bullet p(x_i) = \sum_{i=1}^{n} p(x_i) \bullet \log_2 \frac{1}{p(x_i)} \quad (4.1)$$

To calculate the entropy and, therefore to quantify the information, it is necessary to refer to a specific alphabet: in the case of music language the classification of the various melodic intervals as symbols of the alphabet was considered [1, 11].

The classification of an interval consists in the *denomination* (generic indication) and in the *qualification* (specific indication) [31]. The *denomination* corresponds to the number of degrees that the interval contains, calculated from the grave (low-pitched) to the acute (high-pitched); it may be of a 2nd, of a 3rd, a 4th, a 5th, etc.; the *qualification* is deduced from the number of tones and semitones that the interval contains; it may be: *just*(G), *major* (M), *minor* (m), *augmented* (A), *diminished* (d), *more-than-augmented* (A+), *more-than-diminished* (d+), *exceeding* (E), *deficient* (df).

This class was introduced in the Training Set as a discriminant in order to reinforce the solidity of the Naive Bayesian classifier. In effect, in the preceding algorithm [1], the music scores were classified in three distinct historical periods, Baroque, Romantic and Contemporary, based on their respective entropy values. This had allowed the identification, with a certain approximation, of indexes (minimum and maximum) of reference for each period. However, some scores, because of their writing features, were classified erroneously: scores of the Classic-Romantic period classified as belonging to the Contemporary period and conversely, scores of the contemporary period classified as belonging to the Romantic period.

5 The Obtained Results

The proposed analysis model was verified by realizing an algorithm (MSC, Music Scores Classifier), the structure of which takes into account every single one of the aspects described above. The objective of the algorithm was to classify the scores considered on a symbolic level, in two distinct periods, the one of tonal music and the one of atonal music, trying to improve the precision of the preceding algorithm [1], based on the quantification of the information of every score (by calculating the entropy). This latter system had highlighted a certain imprecision (even if very reduced) in the classification of certain scores belonging to a precise historical period, yet having writing features typical of another period.

The new algorithm is based on a Bayesian probabilistic model that extends the Naive Bayes classifier by adding a variable tied to the value of the information contained within the score (theory of Shannon and Weaver). The set of data used to construct the training classifier (Table 1) was extracted from a set of 200 labeled scores (belonging to different authors): 100 scores of the tonal period and 100 scores of the atonal period. To perform a first assessment of the performance of the new classifying system a series of experiments was carried out on a set of scores, considering the Naive Bayesian classifier without the data related to entropy but considering only the data presented in paragraph 4: key signatures, intervallic structure and polyrhythm (Table 1a). This data allows the classification of a score in the tonal or atonal period based on indexes (minimum and maximum) identified by the preceding algorithm. A second evaluation of the performance of the new classifying system was performed considering the extended classifier, which also include the "Entropy" class (Table 1b).

Table 1 Datasets used in the experiment

a) Training set

Score	Key signatures	Intervals	Polyrhythm	Tonal?
1	No	c	Easy	Yes
2	No	d	Complex	No
3	No	c	No	Yes
4	No	c	Easy	Yes
5	No	d	Easy	No
6	Yes	d+	Complex	Yes
7	Yes	c	Easy	Yes
8	Yes	d+	No	No
9	Yes	d	No	Yes
10	No	c	No	No

b) Extended training set

Score	Key signatures	Intervals	Polyrhythm	Entropy	Tonal?
1	No	c	Easy	a	Yes
2	No	d	Complex	a	No
3	No	c	No	t	Yes
4	No	c	Easy	t	Yes
5	No	d	Easy	a	No
6	Yes	d+	Complex	t	Yes
7	Yes	c	Easy	t	Yes
8	Yes	d+	No	a	No
9	Yes	d	No	a	Yes
10	No	c	No	t	No

Table 2 Test results

Answer	F. Margola		L. van Beethoven	
	Training set	Extended training set	Training set	Extended training set
Yes	0.0167	0.0111	**0.025**	**0.0083**
No	**0.075**	**0.0188**	0.0063	0.0047
Δ Yes/No	*0.0583*	*0.0077*	*0.0187*	*0.0036*

In both cases the results were positive: the scores under analysis were classified correctly. Below are illustrated the results of the analysis of two compositions:

- the Great Fugue op. 133 of Beethoven, where the composer tries to surpass the concept of Tonality using, therefore, intervals which are typical to the contemporary period, the instance of which to consider for classification is
<Key signatures = Yes, Intervals = d+, Polyrhythm = Easy, Entropy = a>
- first movement of the *sonatina* for pianoforte by Franco Margola, a composition which presents typical features of the tonal period, the instance of which to consider for classification is
<Key signatures = No, Intervals = d, Polyrhythm = no, Entropy = t>

Table 2 illustrates the results of the calculation of the probabilities according to Bayes' theorem, for both scores, distinguishing the values based on the fact that the Training set (without the entropy value) was considered or the Training set (with the entropy value) was considered. The objective was to give an answer to the question: "*is the score tonal?*"

As it can be easily seen, the answers to the question ("*is the score tonal?*") for every score were the same both for the Training Set and for the Extended Training Set. In this latter case, the value related to the difference between the answers (Δ Yes/No) is significantly reduced, but still confirming the correctness of the analysis: in Margola's case that it is an atonal score (Answer ''No'') and in Beethoven's case that it is a tonal score (answer ''Yes'').

Figure 5 shows the data of Table 2, graphically, so as to render its interpretation easy: the upper quadrant refers to the answer ''Yes'' and the lower one to the answer ''No''. The figure also represents the difference between the two answers, so as to render the identification of the answer proposed by the algorithm more immediate.

In case the probabilities related to the answers "Yes" and "No" are equal, there is a case of "*uncertainty*" that does not allow the algorithm to associate the score to a style more than to another style: in such case the classification identified by means of entropy is considered and the algorithm shall ask the user to verify the specific score.

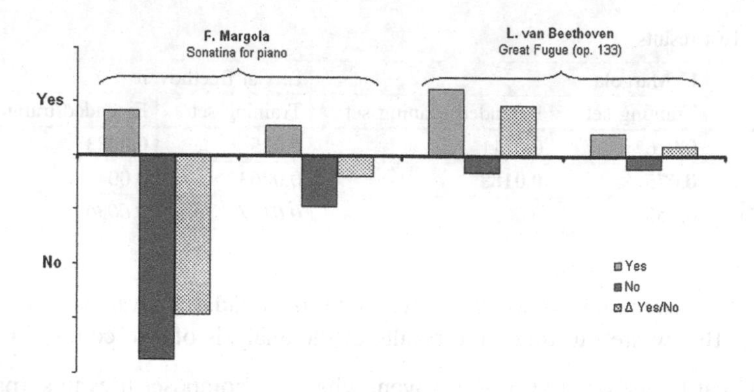

Fig. 5 Graphic representation of the results of Table 2

6 Conclusions

The algorithm proposed for the classification of the scores simultaneously considers the probabilistic theories of Thomas Bayes and the information theory of Warren Weaver and Claude Elwood Shannon. The scores are only classified in two classes, tonal and atonal music, so as to accelerate its analysis and indexation stages by choosing the most adequate tools. The unification of the two different theories allowed the improvement of the precision of the preceding algorithm, eliminating the imprecisions deriving from the musical writing. We have shown in the above preliminary experiment that the proposed algorithm can achieve good performance in diverse musical textures in terms of identifying the historical time period. The increasing amount of documents of a musical nature on the web demands major care in their management (analysis and indexation) in order for the user to obtain more satisfaction. The high degree of complexity of musical phenomena imposes certain forms of achievement that must be adequate and that, for completeness' sake, must cope with the problems under a sufficiently large number of angles. Thus, even from a theoretical—musical point of view, the possibility to integrate different approaches appears as a precursory way of interesting developments. And it is really thanks to the new techniques of artificial intelligence that such forms of integration and verification of the results become achievable.

This method will help students and musicians to find in a very quick way the right scores they need.

Extending this methodology to the harmonic analysis of a musical piece might help us classify scores in different periods: baroque, classic, romantic, post-romantic, contemporary.

References

1. Della Ventura, M.: Detection of historical period in symbolic music data: revisited version. In: 12th TELE-INFO International Conference on Recent Researches in Telecommunications, and Informatics. Baltimore, MD, USA (2013)
2. Ahmed, I., Guan, D., Chung, T.C.: SMS classification based on Naïve Bayes classifier and a priori algorithm frequent item set. Int. J. Mach. Learn. Comput. **4**(2) (2014)
3. Fortuna, B., Mendes Rodrigues, E., Milic-Frayling, N.: Improving the classification of newsgroup messages through social network analysis. In: 16th ACM Conference on Information and Knowledge Management, pp. 877–880. New York (2007)
4. Strapparava, C., Mihalcea, R.: Learning to identify emotions in text. In: 17th ACM Symposium on Applied Computing, pp. 1556–1560. New York (2008)
5. Newman, D.: Visualizing search results and document collections using topic maps. Web Semant. Sci. Serv. Agents World Wide Web **8**(2–3), 169–175 (2010)
6. Kanaka Durga, V., Raja Ramesh, M.R.: Accurate spam mail detection using Bayesian algorithm. Int. J. Adv. Res. Comput. Eng. Technol. **1**(4) (2012)
7. Thorsten, J.: Text categorization with suport vector machines: learning with many relevant features. In: 10th European Conference on Machine Learning, p. 137–142, 21–23 April 1998
8. Wei, Z., Zhang, H., Zhang, Z., Li, W., Miao, D.: A Naive Bayesian multi-label classification algorithm with application to visualize text search results. Int. J. Adv. Intell. **3**(2), 173–188 (2011)
9. Anan, Y., Hatano, K., Bannai, H., Takeda, M.: Music genre classification using similarity functions. In: 12th ISMIR International Society for Music Information Retrieval Conference, pp. 693–698 (2011)
10. Pancini, M.: Problematiche e modelli formali per la segmentazione automatica/interattiva di partiture musicali simboliche. Rapporto Tecnico CNR-PFBC-MUS-TR (2000)
11. Della Ventura, M.: The Influence of the rhythm with the pitch on melodic segmentation. In: 2nd ECC Euro-China Conference on Intelligent Data Analysis and Applications, pp. 191–202. Springer (2015)
12. Cambouropoulos, E., Tsougras, C.: Influence of musical similarity on melodic segmentation: representations and algorithms. In: 1st SMC International Conference on Sound and Music Computing. Paris, France (2004)
13. Cambouropoulos, E.: How similar is similar? Musicae Scientiae Discussion Forum **4B**, 7–24 (2009)
14. Della Ventura, M.: Evaluation of musical similarity on the symbolic level of the musical text. In: 15th ICAI International Conference on Artificial Intelligence. Las Vegas, USA (2013)
15. Della Ventura, M.: Discovering hidden themes in symbolic music text. In: 6th ICSTE International Conference on Software Technology and Engineering. Paris, France (2014)
16. Pérez-Sancho, C., Rizo, D., Inesta, J.M.: Stochastic text models for music categorization. In: da Vitoria Lobo, N., et al. (eds.) SSPR & SPR 2008. LNCS, vol. 5342, pp. 55–64. Springer (2008)
17. Pérez-Sancho, C.: Stochastic language models for music information retrieval. Ph.D. thesis, Universidad de Alicante, Alicante, Spain (2009)
18. Pérez-Sancho, C., Rizo, D., Inesta, J.M., Ponce de León, P.J., Kersten, S., Ramirez, R.: Genre classification of music by tonal harmony. Intell. Data Anal. **14**, 533–545 (2010)
19. Dor, O., Reich, Y.: An evaluation of musical score characteristics for automatic classification of composers. Comput. Music J. **35**(3), 86–97 (2011)
20. Domingos, P., Pazzani, M.: Beyond independence: conditions for the optimality of the simple Bayesian classifier. Mach. Learn. **29**(2/3), 103–130 (1997)
21. Bruntine, W.L., Weigend, A.S.: Bayesian backpropagation. Complex Syst. **5**, 603–643 (1991)
22. Langley, P.W., Thompson, K.: An analysis of Bayesian classifiers. In: 10th National Conference on Artificial Intelligence, pp. 223–228. AAAI Press, Menlo Park, CA (1992)
23. de la Motte, D.: The Study of Harmony. Barenreiter (1976)

24. Fraisse, P.: Psychologie du rythme. Puf. Paris (1974)
25. Fraisse, P.: Les strutctures rythmiques. Erasme, Paris (1958)
26. de la Motte, D.: Il contrappunto. Un libro da leggere e da studiare. Universal Music MGB (1991)
27. Angeleri, E.: Information, meaning and universalit. UTET, Turin (2000)
28. Moles, A.: Teorie de l'information et Perception esthetique. Paris, Flammarion Editeur (1958)
29. Lerdhal, L., Jackendoff, R.: A Grammatical Parallel between Music and Language. Plenum Press, New York (1982)
30. Weaver, W., Shannon, C.: The Mathematical Theory of Information. Illinois Press, Urbana (1964)
31. Coltro, B.: Lezioni di armonia complementare. Zanibon (1979)

A Framework for Extracting Reliable Information from Unstructured Uncertain Big Data

Sanjay Kumar Singh, Neel Mani and Bharat Singh

Abstract Big Data is still in its initial stages and has prompted various basic issues and difficulties to rise, for example, the pace of exchange, information development, and assorted qualities of information and security issues. For example, overseeing and abusing immense measures of information make it more valuable and important has turned into a test driving basic learning for choice making and in picking up an understanding into the general circumstance. Huge information has gotten phenomenal consideration from open and private sectors and in addition from the educated community around the world. In advertising, enormous information is utilized to comprehend the practices and actives of clients. In the experimental fields, huge information can be misused by aiding and taking care of the issues confronting the investigative fields extending from nanotechnology to climatology to geophysics. In the field of law requirement, social administrations and country security, enormous information has exhibited its handiness for government organizations to bolster in their choice making.

Keywords Big data · Unstructured data · Uncertain data · Extracting reliable information

S.K. Singh (✉)
Dell Services, Dublin, Ireland
e-mail: skrs@gmail.com

N. Mani
Adapt Centre, Dublin City University, Dublin, Ireland
e-mail: nmani@computing.dcu.ie

B. Singh
Big Data Labs, Hamburg, Germany
e-mail: bharat.singh@bigdatalabs.org

I. Czarnowski et al. (eds.), *Intelligent Decision Technologies 2016*,
Smart Innovation, Systems and Technologies 57,
DOI 10.1007/978-3-319-39627-9_16

1 Introduction

Big Data' is a term used to portray the ascent in the volume of information that are hard to handle, oversee, dissect, catch and store by utilizing customary database instruments [1]. In the course of recent years, it has been assessed that 90 % of the world information has been made and this implies 2.5 quintillion bytes of information are being made day by day [2]. Huge information originates from numerous sources, for example, exchange records, advanced pictures and recordings, posts on online networking locales, wireless GPS signs and sensors that are utilized to accumulate atmosphere data and these are just to give some examples [3].

Big Data is recognized from conventional database through components, for example, volume which alludes to the measure of information, the speed that speaks to the rate of transmission information through the systems, the assortment that speaks to the heterogeneity of information sorts, and the veracity in the level of dependability connected with the specific sorts of information [4]. This shows there are different sorts of huge information, the information can be ordered into organized, semi-organized and unstructured information [5]. As a rule, 80 % of all accessible information today, are unstructured and semi-structured information and the quickest developing information sort [6].

Unstructured data is partitioned into printed and non-literary. Email messages, Power-Point presentations, Word archives, joint effort programming and texts allude to literary unstructured information. JPEG pictures, streak video records and MP3 sound documents allude to non-literary unstructured information [7]. Propels in data and correspondence advancements have empowered associations to gather and process an assortment of phenomenal measure of information than before in areas, for example, medicinal services, industry, managing an account exchanges, satellite imaging, and online networking [8]. For instance, HP (Hewlett-Packard) creates around 1 trillion occasions for every day [9].

This article presents a framework for extracting reliable information from unstructured uncertain big data. This paper is structured as follows. We start by literature review in Sect. 2. Research objective is described in Sect. 3. Research methodology is described with framework in Sect. 4. Finally, Sect. 5 concludes this paper and points out of future works.

2 Literature Review

It is normal that associations, for example, these will deliver more than 2.5 Exabyte information for each day which is assessed to grow 44-surge by 2020 [10]. This regularly expanding information has introduced a time of huge information, which is thought to be an impetus for advancement, increment in rivalry and efficiency [11]. For instance, as per [12], because of expansions in proficiency, the retail and assembling commercial enterprises are relied upon to increment $325 billion in their GDP, while the medicinal services and taxpayer driven organizations might acquire than $285 billion by 2020. McKinsey states that enormous information might deliver more than $300 billion for the USA government in the field of medicinal services and $250 billion in the organization part for EU [13].

It is assessed that more than 80 % of the information in the organization is unstructured data. There are a wide range of types of unstructured data: video, sound, pictures. Be that as it may, by a long shot the most intriguing and helpful unstructured information is literary data. Literary data is discovered all over the place in the company. Content is found in contracts, email, reports, update, and human asset assessments, et cetera. In a word, literary data makes up the fabric of corporate life, and that is valid for each organization. Unstructured data can be broken into two noteworthy classifications—tedious unstructured information and non-monotonous unstructured information. Figure 1 demonstrates the classes that depict all corporate information.

For an assortment of reasons, by far most of corporate choices are made in view of structured data. There are a few explanations behind this. The essential reason is that organized data is anything but difficult to computerize. Organized information fits actually and ordinarily on standard database innovation. Furthermore, once on database innovation, the information can without much of a stretch be dissected inside the enterprise. It is anything but difficult to peruse and investigations 100,000 records of organized data. There are a lot of systematic apparatuses that can deal with the examination of standard database records.

Figure 1 demonstrates that most corporate choices are made taking into account organized information. Notwithstanding the way that most corporate choices are made on the premise of organized data, there is an abundance of undiscovered potential in the unstructured data of the partnership. The test then is opening that potential.

Figure 2 demonstrates that there is an alternate business esteem recommendation for the diverse sorts of unstructured information. Dull unstructured information has business esteem. Be that as it may, the business esteem in redundant unstructured information is elusive and difficult to open. What's more, as a rule, there just is no business esteem at all in tedious unstructured information.

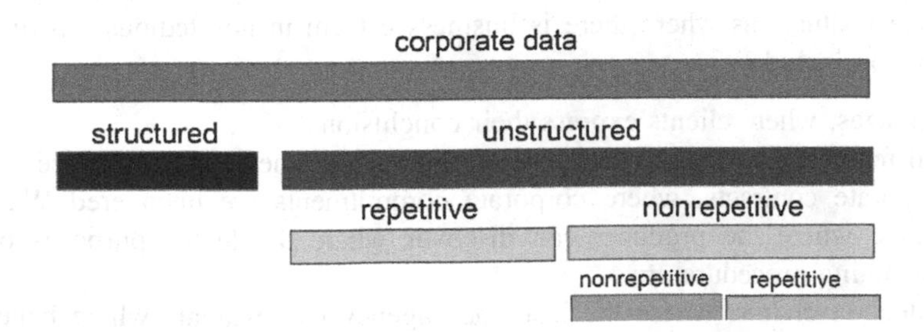

Fig. 1 Big Data type classification

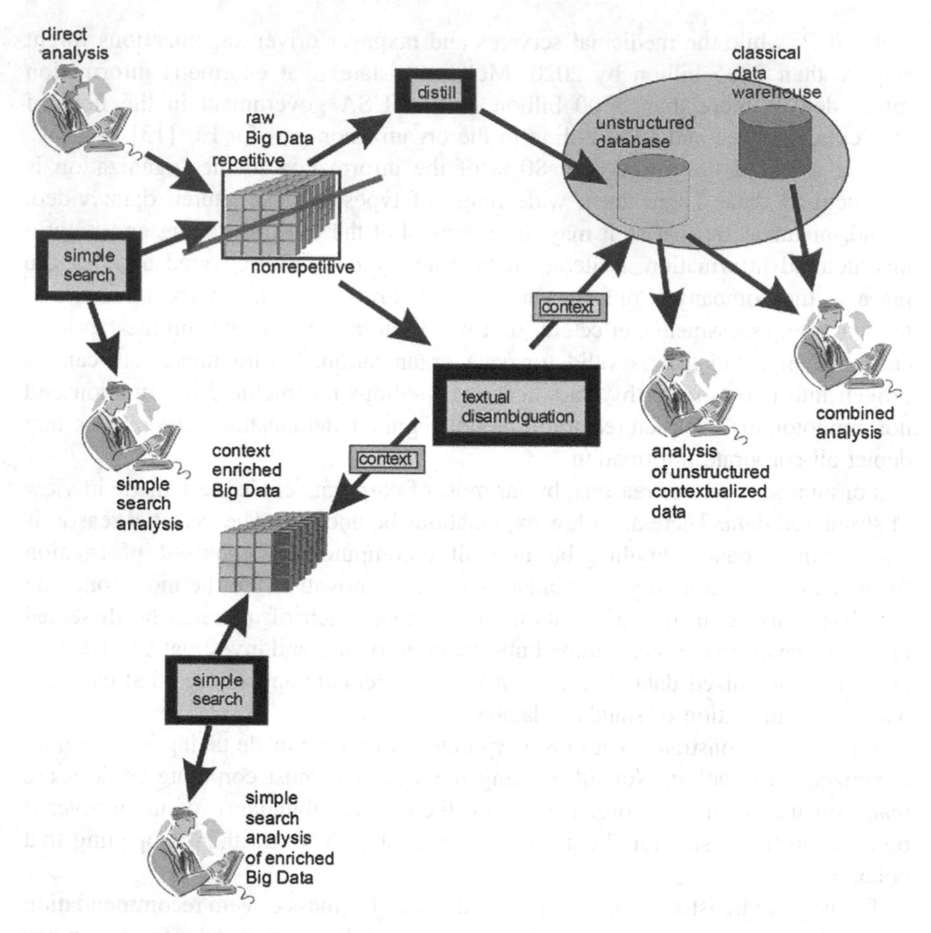

Fig. 2 Unstructured Big Data analysis

It is in non-redundant unstructured information, in any case, where there is gigantic business esteem. There are numerous, numerous situations where the business esteem in non-dull unstructured information is high. A percentage of the more clear situations where there is business esteem in non-tedious unstructured information include:

- Messages, where clients express their conclusions
- Call focus data, where clients have an immediate line to the enterprise
- Corporate contracts, where corporate commitments are uncovered Warranty claims, where the producer can discover where the feeble purposes of the assembling procedure are
- Protection claims, where the insurance agency can evaluate where beneficial business lies
- Showcasing examination organizations, where direct client input can be investigated

'Big data' has gotten remarkable consideration around the world. There are a wide assortment of employments for enormous information, for example, money related exchange, log documents and system movement investigation to distinguish unpredictable and suspicious exercises and these differed utilizes have been beforehand investigated as a part of a number examination [14].

The advantage of utilizing huge information has been augmented well past the business field, for example, in protection administrations, legislative issues, social insurance, guard, sciences, building and numerous different fields. For instance, enormous information has made an upheaval in the investigative exploration [15]. In the natural sciences, now there are numerous useful headway arranges, which expects to store investigative information into an open archive, and make open databases available by different researchers. In Bioinformatics science, there is a need to expand information set for examination information such Next Generation Sequencing information. In the human services field, the treatment of extensive variety of infections and the avoidance of others could be found by enormous information [16].

Big data is described by the accompanying properties:

- Volume: alludes to the information set size of enormous information, as previously stated, it is the primary normal for huge information that is illustrated in Fig. 3. Enormous information originates from various sources that might incorporate exchanges, undertaking data, auto produced information, and social information [25]. Because of every one of these assets, the information is gathered and expanded rapidly.
- Variety: alludes to the information sorts of huge information, Illustrated in Fig. 4. Given that the differing qualities of information is created in parts, for example, interpersonal organization, monetary move, area information (cell phone information and geospatial information), and pictures (satellite and observation), and so on, enormous information are characterized into three principle classifications. They are structure information, unstructured information and semi-structure information.
- Velocity: alludes to the pace of creation, transmission, handling and ingesting of huge information [19], shows as Fig. 5. For instance, the IBM association delineates the speed of huge information by examining 5 million exchange

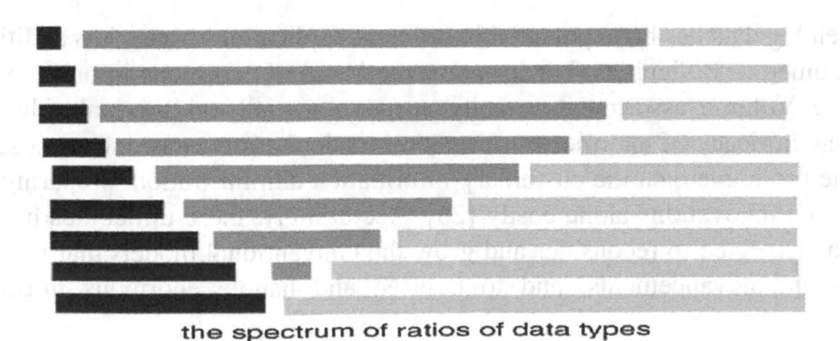

Fig. 3 Volume of Unstructured Big Data

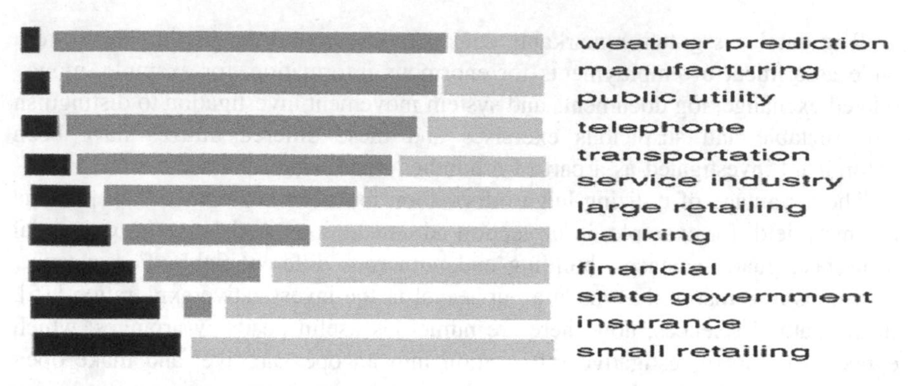

Fig. 4 Variety of Unstructured Big Data

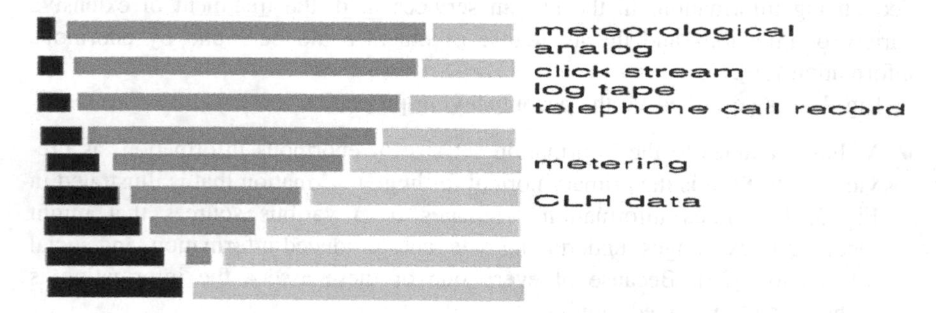

Fig. 5 Velocity of Unstructured Big Data

occasions produced every day to distinguish potential extortion, and breaking down 500 million day by day call subtle element records progressively to anticipate client outline quicker [20].

- Veracity: depicts the validity of information source and the suitability of information for the intended interest group [21]. As the huge information gathered from various assets are from outside control, this implies the outcomes got raises the worry with respect to the precision and culmination of enormous information [22].

Albeit big data has huge potential in different application spaces, it is additionally raises numerous difficulties. For instance, the blend of enormous information attributes (i.e. Volume, assortment, veracity, speed) essentially adds to make challenges [24]. The intricacy of enormous information made the procedure of using so as to extricate profitable data the customary information administration, preparing apparatuses and innovations alone costly [25]. Accordingly, these difficulties have produced an expected to reconsider and grow the conventional models that make new devices and advancements, and to oversee and handle enormous information proficiently. These new apparatuses and advances expect to extricate critical data from enormous information, with ease and short handling time. Enormous information challenges likewise emerge from exercises, for example, looking, sharing, breaking down, information catching and representation [26]. CPU-substantial however I/O-poor is the current test in PC engineering for a very long while [27]. This irregularity is as yet limiting the improvement of the disclosure of Big Data [18].

As indicated by Moore's Law the execution of CPU and circle drive are multiplying every year and a half [19]. In any case, in the course of the most recent decade the rotational velocity of plates has marginally enhanced [20]. Also, the change of data handling techniques is still moderate contrasted and the concurrent expanding rate of data [21]. Huge information applications, for example, Hadoop requires a considerable measure of information development amid its execution [19]. Be that as it may, if the transfer speed interest is not advertisement dressed legitimately, it can influence the rate, unwavering quality and inactivity of the system [23].

3 Research Objective

This study makes a critical commitment in permitting solid choices to be taken in view of unstructured uncertain big data. This paper will research and build up a methodology that will empower endeavors to concentrate esteem from unstructured indeterminate huge information to bolster better business choice making in associations. The proposition will research and create calculations with an abnormal state of unwavering quality to help leaders to take choices for business sector examination, future expectations etc. Additionally, this proposition will build up another structure that gives dependable and solid in-developmental learning from unstructured indeterminate huge information. The particular commitments of this paper are:

- Developing new models for extracting trustworthy and reliable information from unstructured and uncertain big data for supporting decision making.
- Improving reliability of extracted information from unstructured uncertain big data.

4 Research Methodology

Numerous arrangements have been created to address the difficulties postured by enormous information. For instance, Google offered a circulated registering system called MapReduce to address vast scale information adequately [18], and it likewise introduced Big-Table for putting away organized information on a large number of machines [13]. Additionally, the memory stockpiling that has been produced in such a way can offer the huge stockpiling unit to offer it some assistance with storing these information. Notwithstanding, the greater part of the huge information applications, methods and advances can't in a perfect world take care of genuine issues (particularly the constant investigation) [14]. Along these lines, as of recently, we don't totally have fitting instruments to misuse enormous information [15].

As indicated by the report of [16], it can be reasoned that the difficulties emerging while utilizing huge information are:

- Handling unstructured data, and
- Exploiting big data in a timely and cost—effective fashion.

A wide assortment of approaches that incorporate demonstrating, formal model (i.e., hypothetical), exploratory and reenactment have been utilized as a part of software engineering. In this paper, we will utilize both formal demonstrating and exploratory investigation as the examination methodology. A formal model with scientific verification systems, for the calculation outline and investigation is foremost for the way of the exploration being attempted. To handle instability, the examiners need to make connection around the information [17].

There are two approaches to handle this issue [24]. The principal way is utilizing an information combination approach, which joins various less dependable sources to make more helpful, and air conditioning minister information focuses; [25] and the second path is through cutting edge science, for example, fuzzy logic, soft set approaches and robust optimization techniques (Fig. 6).

1. Selection of Unstructured Big Data source
2. Filtering unnecessary data and Pruning Techniques
3. Model for extracting trustworthy and reliable information
4. Make decision based on extracted information

We will execute and check the calculations created on the Hadoop stage to completely use the registering force and capacity limit of distributed computing. Hadoop is a structure that takes into consideration the appropriated preparing of extensive information sets crosswise over bunches of PCs utilizing straightforward programming models [18].

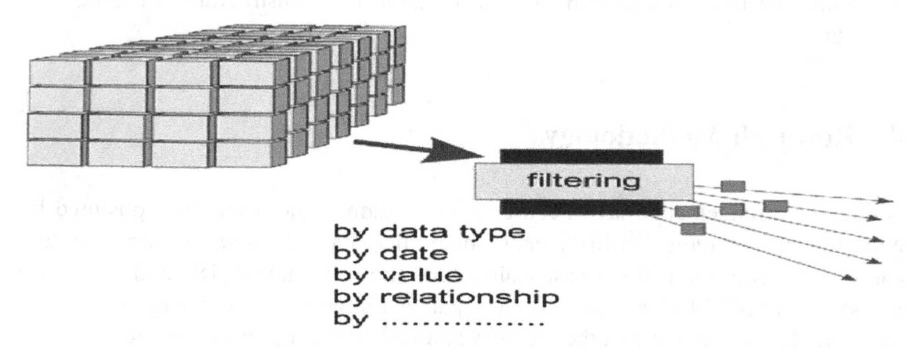

Fig. 6 Extracting reliable information from Unstructured Big Data

We will directed our investigations on five datasets which have distinctive system topologies: PolBooks, Jazz Musician Network, Scale-Free, Random and Small-World. For the trials with genuine datasets, we utilized PolBooks and Jazz Musician Network [23]:

- PolBooks Dataset: This is a system of books about US governmental issues sold by an online store. Edges between books speak to their incessant buy by the same purchasers.
- Jazz Musician Network Dataset: A system of jazz performers who work together in various groups. The vertices speak to the band and edges speak to the performers in like manner.
- Scale-Free Dataset: This is an engineered system which depends on Barabasi's model. This information have a force law vertex degree conveyance which is one of the key legitimate ties showed by interpersonal organizations.
- Random Graph Dataset: This is a manufactured irregular system in view of Er-dos-Renyi system model. The vertices in this net-work are arbitrarily associated taking into account likelihood p.
- Small World Graph Dataset: A little world chart is a kind of diagram in which most vertices can be come to from each other vertex by a little number of jumps. We create little world charts utilizing the same model as a part of.

5 Conclusion

Unstructured uncertain big data is a testing situation confronting the procedure of choice making. In the accompanying paper will examine the writing audit which incorporates a survey of the accompanying: audit of huge information examination, survey of separating dependable and solid data from unstructured and questionable enormous information. Information vulnerability influences the quality and unwavering quality of the ideal choices. Accordingly, unstructured indeterminate enormous information is an extremely basic issue that should be taken a gander at with a specific end goal to touch base to ideal and right choices. Information quality is significant in ensuring a dependable knowledge into the enormous datasets as future work. Along these lines, the assortment and size of instabilities in enormous information requires the advancement of models and techniques, which can evaluate vulnerabilities for a lot of information. Additionally, measuring the effect of instabilities in huge datasets are another test.

References

1. Abawajy, J.: Comprehensive analysis of big data variety landscape. Int. J. Parallel Emergent Distrib. Syst. **30**(1), 5–14 (2015)
2. Abdulhafiz, W.A., Khamis, A.: Handling data uncertainty and inconsistency using multi sensor data fusion. Adv. Artif. Intell. **11** (2013)
3. Aggarwal, C.C., Yu, P.S.: A survey of uncertain data algorithms and applications. IEEE Trans. Knowl. Data Eng. **21**(5), 609–623 (2009)
4. AllegroGraph. AllegroGraph (2015). http://franz.com/agraph/allegrograph/
5. Angelosante, D., Biglieri, E., Lops, M.: Multiuser detection in a dynamic environment: joint user identification and parameter estimation. In: IEEE International Symposium on Information Theory, 2007, ISIT (2007)
6. Bai, Y., Zhuang, H., Wang, D.: Advanced Fuzzy Logic Technologies in Industrial Applications. Springer (2007)
7. Begoli, E., Horey, J.: Design principles for effective knowledge discovery from big data. In: Joint Working IEEE/IFIP Conference on Software Architecture (WICSA) and European Conference on Software Architecture (ECSA) (2012)
8. Brynjolfsson, E., Hitt, L.M., Kim, H.H.: Strength in numbers: how does data-driven decisionmaking affect firm performance? (2011). http://www.SSRN1819486papers.ssrn.com
9. Camacho, J., Macia-Fernandez, G., Diaz-Verdejo, J., Garcia-Teodoro, P.: Tackling the Big Data 4 vs for anomaly detection. In: IEEE Conference on Computer Communications Workshops (INFOCOM WKSHPS) (2014)
10. Chan, J.O.: An architecture for big data analytics. Commun. IIMA **13**(2), 1 (2014)
11. Chang, F., Dean, J., Ghemawat, S., Hsieh, W.C., Wallach, D.A., Burrows, M., Gruber, R.E.: Bigtable: a distributed storage system for structured data. ACM Trans. Comput. Syst. (TOCS) **26**(2), 4
12. Chau, M., Cheng, R., Kao, B., Ng, J.: Uncertain data mining: an example in clustering location data. In: Advances in Knowledge Discovery and Data Mining, pp. 199–204. Springer, Berlin (2006)
13. Chen, C.P., Zhang, C.-Y.: Data-intensive applications, challenges, techniques and technologies: a survey on Big Data. Inf. Sci. **275**, 314–347 (2014)
14. Chen, M., Mao, S., Liu, Y.: Big data: a survey. Mob. Netw. Appl. **19**(2), 171–209 (2014)
15. Cheung, C.F., Lee, W., Wang, Y.: A multi-facet taxonomy system with applications in unstructured knowledge management. J. Knowl. Manag. **9**(6), 76–91 (2005)
16. Chowdhury, M., Stoica, I.: Coflow: a networking abstraction for cluster applications. In: Proceedings of the 11th ACM Workshop on Hot Topics in Networks, pp. 31–36 (2012)
17. Chu, E., Baid, A., Chen, T., Doan, A. Naughton, J.: A relational approach to incrementally extracting and querying structure in unstructured data. In: Proceedings of the 33rd International Conference on Very Large Data Bases (2007)
18. Dean, J., Ghemawat, S.: MapReduce: simplified data processing on large clusters. Commun. ACM **51**(1), 107–113 (2008)
19. Dijcks, J.P.: Oracle: big data for the enterprise. Oracle White Paper (2012)
20. Ding, H., Mao, J., Wei, K., Yang, L.: Fuzzy modeling with unstructured data uncertainty. In: International Conference on Control and Automation, ICCA'05 (2005)
21. Ding, X., Jin, H., Xu, H., Song, W.: Probabilistic skyline queries over uncertain moving objects. Comput. Inform. **32**(5), 987–1012 (2014)
22. Dubois, D., Prade, H.: Rough fuzzy sets and fuzzy rough sets*. Int. J. Gen. Syst. **17**(2–3), 191–209 (1990)
23. Durrant-Whyte, H., Henderson, T.C.: Multisensor Data Fusion Springer Handbook of Robotics, pp. 585–610. Springer (2008)
24. Easton, J.: Carrying out a big data Readiness Assessment (2014)

25. Eswari, T., Sampath, P., Lavanya, S.: Predictive methodology for diabetic data analysis in Big Data. Proc. Comput. Sci. **50**, 203–208 (2015)
26. Feng, L., Li, T., Ruan, D., Gou, S.: A vague-rough set approach for uncertain knowledge acquisition. Knowl.-Based Syst. **24**(6), 837–843 (2011)
27. Florea, M.C., Jousselme, A.-L. Bossé, É.: Fusion of imperfect information in the unified framework of random sets theory: application to target identification: DTIC Document (2007)

Reasoning-Based Intelligent Systems

Visual Builder of Rules for Spacecraft Onboard Real-Time Knowledge Base

Andrey Tyugashev

Abstract Fault tolerance of spacecraft remains one of the most complex problems in space missions. There are several ways to implement the "onboard intelligence allowing the recovery of a spacecraft in case of abnormal situations caused by hardware or software failures. The most common but inflexible way is "to disperse" the recovery logic in the source code of the flight control software. Our approach implies using onboard real-time knowledge base. The rules of the knowledge base could be added or refined from Earth over the radio channel on a timely basis. Currently, the rules of an onboard knowledge base should be specified in a table form, which entails some misunderstandings in the mission team and consequently leads to errors. The improved approach presented in the paper provides special tools–the visualizer and the visual builder of rules. The approach allows space mission operation engineers without special mathematical or programming background to define, visualize and refine knowledge base rules in a very easy manner. Tools prototypes are currently introduced at JSC Information Satellite Systems, Russia.

Keywords Real-Time onboard knowledge base • Visual builder • Spacecraft control system • Spacecraft's fault tolerance feature • Autonomous control

1 Introduction and Background

1.1 Spacecraft Fault Tolerance

The problem of dependability remains one of the most complex problems in modern space missions. A modern spacecraft has a lot of different onboard systems and equipment (motion control system, power supply system, telemetry system,

A. Tyugashev (✉)
Department of Computer Technologies and Control Systems,
ITMO University, 49 Kronverksky Pr., Saint Petersburg 197101, Russia
e-mail: tau@corp.ifmo.ru

I. Czarnowski et al. (eds.), *Intelligent Decision Technologies 2016*,
Smart Innovation, Systems and Technologies 57,
DOI 10.1007/978-3-319-39627-9_17

thermal control system, etc.). Each system, in turn, consists of dozens of devices, sensors, and aggregates. It is well known that, as a rule, a more complex system is less reliable (except in cases where complexity is specially designed for providing dependability). Thus, it is no wonder that there are many faults and failures during space mission operations. The faults and failures may have a different nature, different levels of damage and methods of recovery. Abnormal situations related to such a complex "system of the systems" can be very tricky and require really smart decision support (in many cases we need reasoning with taking causation relationships into account) to be compensated.

The success of a space mission depends on the possibility to recover the spacecraft in case of hardware or software failures. The recovery of conventional machines exploited on the ground usually involves processes of diagnosis and repair accomplished by a human. If there are cosmonauts onboard, they can use their intellect in an abnormal situation to change or fix the failed equipment. The dramatic history of Apollo 13 is well known as opposed to Salyut 7 rescue expedition [1].

In case of an unmanned mission, the situation is more complicated. To ensure the overall success of space missions, it is necessary to ensure the "fault tolerance" feature of a spacecraft. Fault tolerance is one of the most important components of dependability meaning that even in case of failures a spacecraft remains operational (possibly with some losses in quality). The standard level of the fault tolerance of a spacecraft means parrying of the failure of a single device while fully saving the operational mode [2–5].

Actually, we have a lot of impressing examples of the successful usage of the fault tolerance feature of a spacecraft for the recovery of the operational mode of a spacecraft. How can this feature be implemented?

First, it is reasonable to apply the power of human brain. The computers running intelligent software can assist in this. Probably, we can state that technical diagnostics is one of the conventional domains for the application of knowledge bases and expert systems [6–8]. Spacecraft control processes are implemented by the ground flight control together with the onboard control system [2, 9, 10]. The intelligent decision making support systems are widely used both on Earth and onboard [4]. The only way to utilize the human intelligence to "recover" an unmanned spacecraft is the remote control (there was an exception—the repair of the Hubble telescope [11], but this is just an exception). The remote control mode from the ground control center requires the following. The personnel are engaged in the processes of the analysis of the state of a spacecraft, decision making, transmitting of the required commands onboard. All the operations should be undertaken in real-time mode. It is necessary to ensure the continuous work of the ground complex including telemetry receiving stations, radio command channels, and the infrastructure of a flight control center. Of course, it involves the work of well-prepared personnel who have the excellent knowledge of the operational instructions and are ready to make fast and right decisions. The special methods of learning can be required for training these people [12, 13].

But in many practical cases remote control appears to be fully impossible or too late in time. We mean situations when the radio connection with the ground personnel is impossible due to spacecraft orbits (e.g. low-earth-orbit satellites with a short interval of visibility from each of the ground control points, or probes with a long time of radio signal transmission) [2, 14]. If we face with rapid progress of the abnormal situation onboard, the remote control is useless.

The alternative way for unmanned missions is the use of "onboard intelligence" for autonomous control. The term "onboard intelligence" is enclosed in quotes, and this has a reason behind it. The ways of the implementation of "onboard intelligence" are quite different [15].

Fault tolerance should be envisioned at the initial stage of spacecraft design. De-signers and onboard system engineers make a lot of efforts to support the reliability of the spacecraft equipment. The special measures include the introduction of redundancies. Structural redundancy means duplication (some important units could be doubled or tripled). Functional redundancy means that a system remains operational even if a failure happens (with the downshifted quality, i.e. with less precision, etc.). Spacecraft control logic should be specified in the corresponding obligatory documentation both for normal operations and for abnormal situations. The documents include a list of considered failures with the specification of the level of importance, state diagrams and cyclograms representing the needed reaction [3, 4, 14]. If the failure has been diagnosed, the compensating reaction has to be executed. The reaction implies the reconfiguration of the equipment using "hot" or "cold" reserve or the transition of the spacecraft into one of special modes providing safety (these modes exclude catastrophic consequences).

Then it is necessary to implement the control logic for the considered abnormal situations in spacecraft hardware and software. Nowadays, hardware implementation of onboard control is more a historical issue [9, 16]. The only case when hardware fault tolerance makes sense is the compensation of very fast crashing processes (microseconds or less). In other cases, the common approach is using software [2, 14, 17].

In fact, it is impossible to imagine a space mission without application of computers. Computers are used from the design stage through lifting to Space, operations and support till the end of the lifetime of a spacecraft. Currently, all spacecraft, including micro- and nano-satellites, are equipped with onboard control computer systems which combine several onboard computers integrated into network. The control functions are executed by a special sort of software—onboard software (which can also be called "mission-critical flight control software"). Onboard software consists of hundreds of concurrently running programs [2, 14, 18]. Thus, flight control logic both for normal and abnormal functioning of modern space missions is hidden in the onboard software.

1.2 Ways of the Implementation of the Onboard Intelligence

There are several approaches to introducing control logic into software. The most common but inflexible way is to implement it in the program source code (C, Java, an assembler, etc.). In such a case, any change in control logic should entail a very complex, time-consuming and many-staged process of software re-design, coding and testing (including unit testing, integrity testing, system testing, etc.). When we evaluate labor and time consumption and total costs, the typical proportion between hardware and software of the onboard control system can be characterized as 1:10 [5, 16, 19]. Thus, the total cost of the onboard software lifecycle dramatically grows because of required software maintenance efforts [20]. In aerospace projects (as it was noticed decades ago [10, 16]), the processes of design, development and verification of onboard software became a "critical path" of network scheduling, embracing all works connected with designing and manufacturing of a rocket/space system as a whole [16, 19].

There are a lot of examples of successful implementation of software changes and re-uploading them onboard, even when the distance between the Earth and a deep space probe amounts to millions of kilometers [21–23]. The author knows impressing similar examples in Russian space missions, but they are not published enough in English (for example, the recovery of an Earth observation satellite with the serious change of onboard algorithms and software for operations using another onboard system) [3]. The uploading of onboard software becomes a "routine operation" which has already been performed hundreds of times. Let us consider an example. Jim Erickson, Chief Project Manager of Mars Science Laboratory, states that Curiosity is much more reprogrammable than previous missions. He even called it a "software-defined spacecraft" [21]. The flexibility of the software of Curiosity has sometimes been a problem, because it increases the complexity of the mission. Let us cite Erickson's statement: "The more complicated the software, the more likely you'll not get everything perfect. You'll get surprises. Both in development, test and in operations". But this is a situation where flexibility will help, allowing to redesign the way the rover works in response to a potentially mission-ending hazard that was never anticipated.

A very important issue is that software testing even in theory cannot guarantee a total absence of errors. Moreover, onboard software cannot be fully tested for all possible situations related to the real-time mode of functioning and concurrency [3, 4]. This imperfection reduces the overall effectiveness of space missions.

The dominant trend in modern unmanned space missions is the increase of the planned active lifetime (till 10–15 years) [4, 14]. It is known that onboard electronics faces a growing number of faults caused by the long exposure to cosmic hard radiation. In this case, an abnormal situation emerges, and normal spacecraft operations could be impossible. New kinds of abnormal situations can appear caused by unpredictable flight history and history of failures. They cannot be considered at the stage of designing a spacecraft. The changes in control logic related to these situations should be formulated and implemented at the operational stage.

In summary, there is a need in the tools of prompt correction of spacecraft control logic without the necessity of software re-development and upload. The technologies of such re-engineering of space operations in real time entail issues related to the necessity of a timely reaction to an abnormal situation, providing the safety of a spacecraft, and returning a spacecraft to the operational mode without direct access of human personnel [3, 4].

A much more flexible and promising approach than the implementation of control logic in the source code of a program involves the use of some sort of "intelligent software". It can provide flexibility and reduction of labor and total costs.

Intelligent computer software is a term related to a wide spectrum of applications. The known and well-developed approaches include genetic algorithms, fuzzy logic, neural networks, knowledge bases, and reasoning systems [24–27]. The very impressive results in real-time control, presented in public accessed papers, relate to neural networks [28, 29]. But the use of neural networks requires the process of network training. Training process involves the repeated input of patterns to the network (in some cases hundreds or thousands patterns). Unfortunately, there is a non-zero possibility that the network will never become trained enough.

If we consider the problem domain of space mission control, we should take into account the following factors. This is a typical "mission critical" application, with the commercial and/or military customers. In this case, a principle of "trial-and-error" training is inappropriate. Errors at the stage of operation are not permissible. In practice, only pre-flight training of a neural network can be considered. But in this case we still have the insufficient level of confidence because of the lack of well-defined and documented procedure of decision making. This problem is related to the "dispersion" of decision making rules in the network, which leads to the impersonality in responsibility. However, the issue of personal responsibility is of paramount importance in space missions.

Actually, the principles of neural network training do not fully correspond to the problem domain because of the following circumstance. The number of patterns we can use in the training process is limited. But the typical expected lifetime of a modern spacecraft, as it was stated before, is 10–15 years. Thus, there is a non-zero possibility of the emergence of an abnormal situation which was not used as a pattern. In this case, a neural network can make a serious mistake [14, 30, 31]. Probably, this is a reason why, despite the fact that several companies, including AI Solutions, Interface and Control Systems, and Allied Signal Technical Services, started attempts to introduce intelligence into onboard software in the end of 1980s, the manufacturers of spacecraft are skeptical about this idea.

Meanwhile, during designing and manufacturing of a spacecraft, a team of specialists possesses a sum of certain knowledge about a spacecraft as a controlled object. This theoretically gives an opportunity to formulate unambiguous rules for the compensation of each abnormal situation. Thus, rule-based systems such as knowledge bases and reasoning systems look more suitable than fuzzy logic, neural network and genetic algorithms. A very interesting rule-based approach is "situational control", which has been developed in the USSR since the end of 1960s by D. Pospelov, Yu. Klykov, L.S. Zagadskaya, and others [32].

We can state that today the main efforts in the problem domain of intelligent autonomous control of spacecraft should be aimed at the creation of tools allowing knowledge acquisition. These tools must utilize the knowledge of spacecraft designers and system engineers and represent it as well-defined rules, provide means for checking a set of rules for completeness and consistency, allow onboard execution of rules in real-time mode. We need a means that would allow the refinement of onboard control rules during the flight. The specified problem is not simple. Today, the information about rules is usually represented in verbal form. These descriptions have a fragmentary nature, lack causation and structural relationships, and conditions for activation. Some fragments of the knowledge remain in a specialist's brain. Formalization, structurization and checking of these rules are very urgent problems.

2 Framework

This work was performed under the contract with JSC Information Satellite Systems, Krasnoyarsk region, Russia. Consequently, one of the required features is that the methods should support "seamless" incorporation into the customer's existing onboard software lifecycle processes. All data formats must be compatible with the customer's existing programming tools and databases.

To date, such advanced and flexible methodology of autonomous intelligent con-trol has been already implemented at customer site. A special onboard real-time interpreter of rules is used for autonomous integrated control of a spacecraft. The interpreter is periodically started by the dispatcher of the onboard operating system at fixed time intervals. The rules are incorporated in the so-called "DKD program" (DKD is the acronym for "Duty Control and Diagnosis" in Russian) [33]. The main functions of DKD autonomous control program are the detection of abnormal situations and the execution of the corresponding set of actions needed to eliminate a failure. Abnormal situations are associated with the patterns of spacecraft state vectors. A state vector consists of elementary conditions reflecting the current onboard situation. We can consider a "general" state vector combining the parameters of all onboard systems (not used in practice), and particular vectors checked at fixed time intervals (for example, a particular state vector can include parameters important in the current spacecraft operation mode).

The presence of particular state vectors allows the implementation of a certain kind of "reasoning". It means that we go from one proposition (predicate, etc.) to another. An additional important feature is that during the processing of one particular state vector we can "activate" checking of another one, if it is necessary.

The DKD program is organized as a set of rules. Each rule combines a state vector and the required actions. Each recognizable abnormal situation is associated with the pattern of a particular state vector of a satellite. First, we should diagnose the presence of a certain abnormal situation.

Let $A = \{AS_q\}$, $q = 1..S$ is a set of recognized Abnormal Situations.

Let $L = \{\alpha_i\}$, $i = 1..M$ is a set of conditions reflecting current onboard situation.

For example, α_{101} can mean "main gyrodynes are in the operational mode", α_{254}—"the second solar panel is not fully opened", etc.

The full set of conditions forms a "general" satellite state vector. As it was stated above, in practice a more effective way is to use a set *SV* of particular state vectors (subsets of the general state vector).

Then $SV \rightarrow A$ will be a mapping between patterns of state vectors (in other words, complex conditions like $\alpha_{101} \wedge \neg\alpha_{254} \wedge \alpha_{120}$) and Abnormal Situations.

Secondly, a diagnostic program should execute the required set of actions (sup-ported both by onboard equipment and software modules). Suppose $F = K\ U\ P = \{f_j\}$, $j = 1..N$ is a set of actions, which can be executed onboard. *F* unites set *K* of commands executed by onboard equipment and the set *P* of onboard programs.

Similarly, $A \rightarrow 2^F$ will be a mapping between Abnormal Situations and the powerset of Actions (a set of all possible subsets).

But this model is not fully adequate. More precisely, we often need not a single action or just a straight step-by-step consequence of actions, but a "cyclogram" (commonly used term in the aerospace domain), containing pairs (f_j, t_j) where t_j is a time of f_j execution. In other words, a cyclogram represents coordinated synchronized operations.

The specially designed domain-specific language (DSL) is currently used to specify the rules. The language is specially designed to be easily understood by non-programmers and differs significantly from C, Fortran or Java. The rule building is an interactive process supported by a special "REAL" programming system presented in Fig. 1.

Actually, a designer of the control logic fills up the fields of the tables in a special database. The tables are logically connected to represent a structure of the rules. There are a table of onboard parameters, a table of abnormal situations, and a table of available onboard actions. The designer chooses a specific action to fill up the table of "recommendations" associated with the specified abnormal situation.

First, all the parameters checked in the conditions contained in state vectors should be specified, as well as the base of executing actions. Then the user forms a particular state vector and associates a set of actions with it. Then the special programming tool converts the database and rules into compact onboard structures.

The verification of rules is performed using a special testing complex. The main part of this complex is the software for the simulation of functioning of all onboard systems (both for normal modes and for predicted failures) and the physical parameters of the flight of a spacecraft. This testing complex is also used for system testing of all onboard software and for training of spacecraft personnel. The checked rules are saved into the memory of the onboard control system during spacecraft manufacturing. The most important feature is that these data can also be transmitted onboard on a timely basis. Thus, it is possible to change the control logic without the necessity of full re-development of the software. We can see that

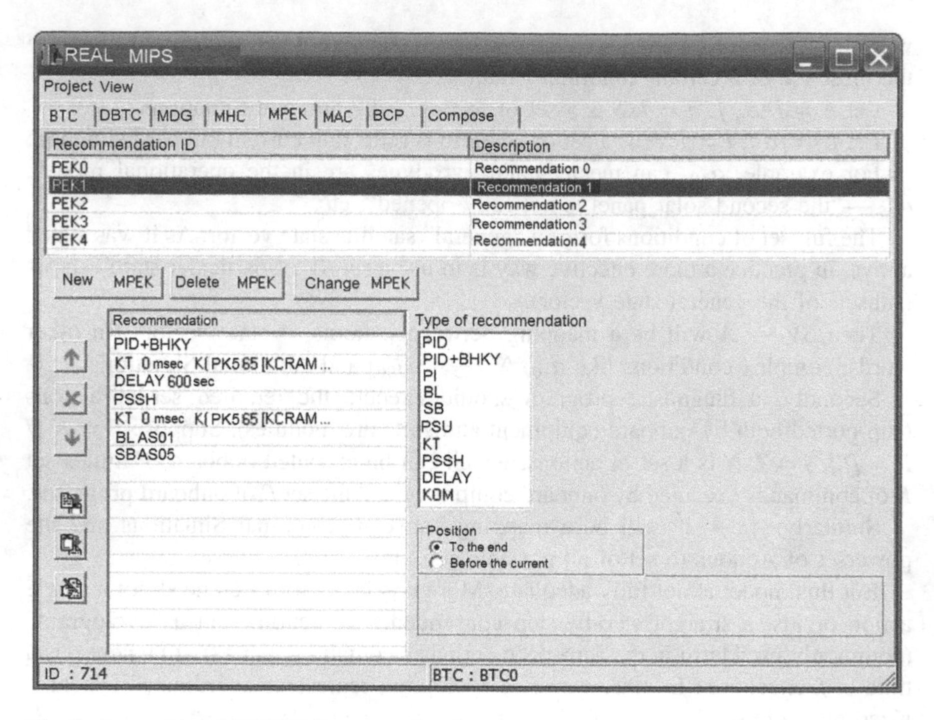

Fig. 1 Screenshot *of "REAL" Programming System*

the onboard rule interpreter is a kind of real-time knowledge base. By the moment of the launch of a spacecraft several hundred rules are usually specified. At the operational stage, this number usually increases by 20–30 % [14]. The opportunity of the specification and updating of rules by non-programmers makes intelligent autonomous control programs the main and most effective tool for spacecraft "remote repair" during its lifetime.

Onboard DKD interpreter supports real-time logical inference (reasoning) about the required actions in forward direction—from the signature of an abnormal situation (the pattern of a particular state vector) to the conclusion about the truthfulness of the hypothesis. Activation of inference is performed in cycle with the individual period for each particular vector, or from one the set of actions activated by one state vector to another. The result of the interactive construction of the autonomous control program is a set of tables printed in a dedicated program document (see Fig. 2).

Despite the fact that the table form of information representation is understood better than a plain text, it is not the best form ensuring clear and fast understanding.

AS #	ABNORMAL SITUATION	PARAMETER	Value	Name of Set of Actions	Actions
0	AS04	CMEHA	0	PEK 04	**OFF** AS01, AS03, AS05, AS06 **PIP** (+DBTC1)
1	AS05	CMEHA	1	PEK 05	**PIP** (+DBTC1) **BLOCK** AS00, AS01, AS02, AS03 **KT** 5265 (KLBATC=1), 1,0100 (Start YFCA), 1;5266 (KLBATC=0), 1;5267 (CMEHA=0), **delay** 30 sec. **OFF** AS04, AS06 **PCY** **BLOCK** AC04 (c №14) **KT** 5265 (KLBATC =1), 1;0124 (Start YFCB), 1;5266 (KLBATC =0), 1;5267 (CMEHA=0), **delay** 30 sec. **OFF** AS04, AS06 **PCY** **BLOCK** AS04 (c №14) **KT** 5265 (KLBATC=1), 1;0100 (Start YFCA), 1;5266 (KLBATC=0), 1;5267 (CMEHA=0), **delay** 30 sec **OFF** AS04, AS06 **PCY** **PID+BHKY** (+DBTC1) **OFF** AS01, AS03, AS04, AS06

Fig. 2 Fragment *of DKD Autonomous Control Program specified by existing DSL*

3 Methods

3.1 Advantages of Visual Form of Information Representation

The main idea of the proposed approach is to combine the flexibility of autonomous satellite control, based on the use of the real-time onboard rule interpreter, and the advantages of the visual form of representation.

There are many reasons for the choice of the visual form of representation of information. Visual form provides simultaneous perception as opposed to textual representation limited by the successive impression. The graphical form of representation is applied to ensure the better reaction time and accuracy of operations of railroad station dispatchers, atomic power station personnel, pilots (the head-up indicator is a good example). The form of information representation is important not only for human-computer interaction, but for human-human interaction, too. In the area of knowledge representation, convenience and clearness for a human is of prime importance. A real-time onboard knowledge base is an example of mission critical systems where the cost of any error or inaccuracy is unacceptably high. The set of rules should be complete, consistent and well structured. The used language makes a “footprint” on the results of thinking. The language should contribute to clear, correct and fast reasoning; the language can be considered as a tool for knowledge [34, 35].

It is a well-known fact that one of the most serious problems of knowledge bases is knowledge acquisition [15, 36–39]. Frequently, a specialist possessing the knowledge is not a mathematician or IT professional. Consequently, he or she faces the problem of the formal representation of the knowledge required by the computer system. A knowledge engineer could help in such a situation, but we cannot fully exclude the "broken phone" effect. There is misunderstanding between the participants of the process. A number of approaches have been proposed for eliminating this problem, for example, the use of a restricted subset of natural language [32, 35]. Another way is an interactive mode of introducing rules. An expert system provides an opportunity to ask clarifying questions. But even in this case we cannot guarantee the absence of inaccuracies and errors. With regard to knowledge, it is reasonable to pay attention to the graphical form of representation.

One can say that the human culture is visually oriented. When we want fast, clear and unambiguous representation, we use graphical form—charts, diagrams, drawing [34]. The best (or even the only) way of representing an enormous amount of knowledge is visual communication. Time tables, bargraphs, maps, even pictorial icons figure prominently in our routine activities [40]. Additionally, the nature of control programs (analysis of logical conditions → detection of the situation → actions) quite corresponds to the graphical form of representation (as opposed to computational programs). Of course, considering the autonomous control domain, we need a means to describe not just descriptive knowledge but also procedural knowledge with "active nature" [15]. The graphics matches the requirements of specification and design stages.

In practice, as a rule, the requirements to the logic of spacecraft functioning are represented in the textual form (or, at best—using tables). Consequently, there are some ambiguities and inconsistencies in the corresponding documents.

Thus, a visual notation and tools for the visual building of onboard real-time knowledge base rules have been proposed.

3.2 Notation

There are a lot of studies devoted to the problem of visual knowledge representation [41–46]. But the performed analysis has shown that any known approach should be very seriously customized if we plan to use it within space missions [35].

The Visual languages for various purposes are being actively developed and used in Russian Aerospace Industry. The examples known to author include Mars Design Bureau, Moscow, Arsenal Design Bureau, Saint Petersburg, Progress Rocket and Space Center, Samara. Unfortunately, the results are practically not published because of many reasons (including security and other issues, see [47]. The very advanced methodology "GRAFIT-FLOKS" with the considering of fundamentals in Human understanding and impression issues was developed and successfully used for years at Academician Pilyugin Center, Moscow [48]. The Visual Notation presented in the paper, substantially based on Parondzhanov ideas.

The notation is not the same, but in some aspects is similar to notation developed at Academician Pilyugin Center. Visual language consists of the following elements:

- Graphical primitives and lines;
- Text labels located inside or outside of graphical primitives.

Actually, the visual notation is based on commonly-used standard flowcharts. The actions are represented by rectangles; the primitives for logical conditions also are intuitively recognizable. But the structure of the flowchart is optimized from the prospective of ergonomics, clear and precise understanding by a human in accordance with the ideas of Parondzhanov [34]. For example, line-crossing is strictly prohibited. The control flow is directed only from top to bottom and from left to right. Straight bottom line from the conditional primitive always corresponds to the "true" branch. These features made the language more concise and thus more intuitive and understandable in comparison with conventional flowcharts.

Some updates have been made in relation to the notation designed by Parondzhanov. First, a flowchart represents one particular state vector (mapping 1:1). A flow-chart consists of several vertical branches which are executed concurrently. The branches correspond to abnormal situations (or it can be said that one branch is one visual rule). Each branch contains exactly one logical condition (complex condition, as a rule), and a set of executed actions. "False" parts are empty. Simple actions are represented by "regular" rectangles. Actions corresponding to satellite control commands with the complex internal structure have the code name displayed in dedicated field, and the comment in other field. Special "KT" block is used to represent a fragment of a "cyclogram" where special fields for specifying of the time of actions are added (the examples can be found in Fig. 3). Delays are

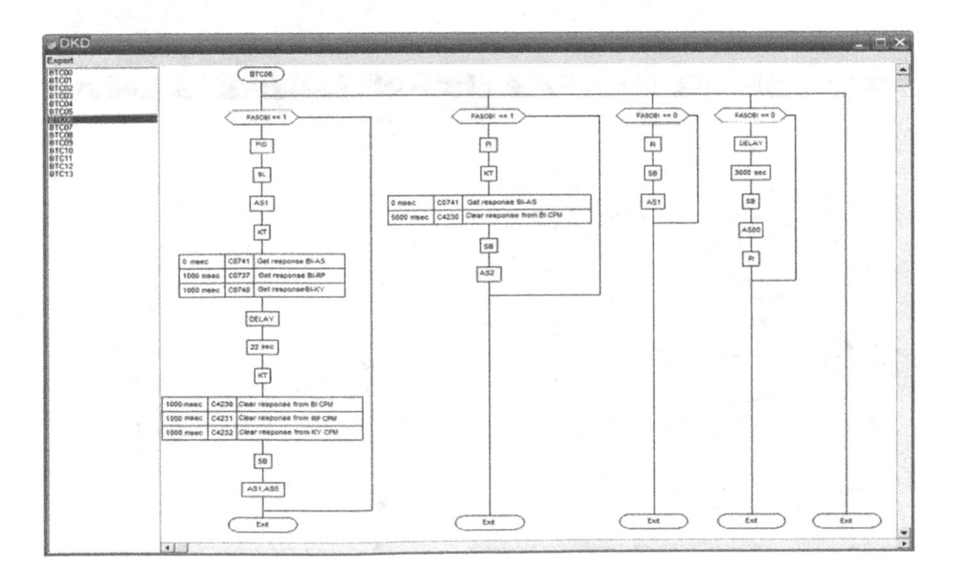

Fig. 3 Screenshot of the Rule's visualization tool

represented by the sequence of two rectangles: first marked as “ПАУЗА” (“Delay” in Russian), and the following rectangle displays the time interval.

4 Practical Results, Discussion and Future Work

One of the causes of errors in mission critical software is the complexity of the development process itself. Misunderstanding between onboard system specialists, de-signers of the satellite control logic, programmers and testers leads to the bugs. In fact, proposed method allows excluding programmers from the development process. This makes it possible to eliminate one type of errors. In practice, we use “programming without programmers” [49]. But other errors caused by the inaccuracies and incompleteness of the rules still preserve and can influence the success of a space mission. Visual verification method [50] is widely used for checking and technical diagnostic of machines and equipment. The structure of the used rules can be visually checked by all the participants of the space mission project. The method of visual rule checking was successfully introduced at the customer site. The developed tools allow:

- Visualization and analysis of previously designed Rules
- Visual building of newly introducing rules.

The screenshot of the Visualization tool is presented in Fig. 3.

As the logical dependencies are allowed between the rules (allowing step-by-step ‘reasoning’), the special feature of the visualizer has been added. We can see and check these dependencies in graphical form as well (see Fig. 4).

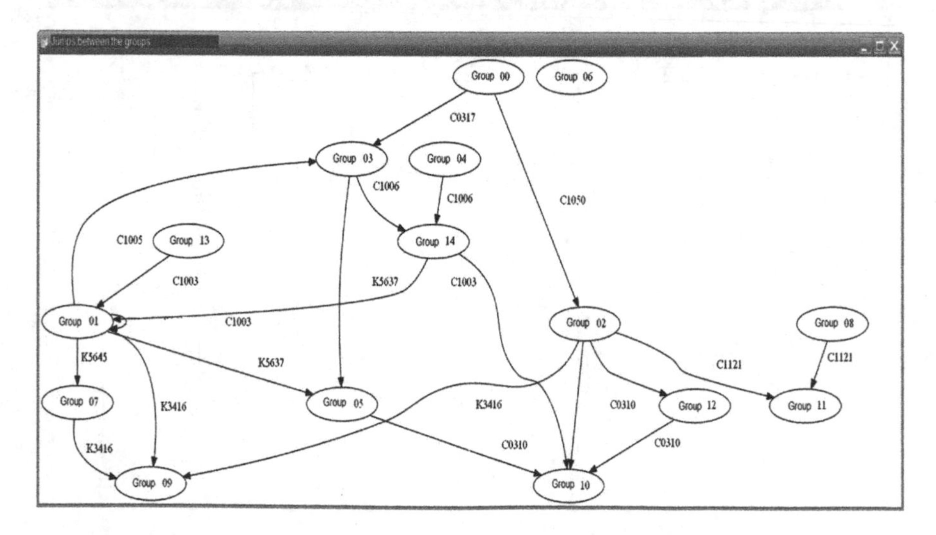

Fig. 4 The screenshot displays the graph of logical dependencies between the rules

The DKD program is represented by the graph; the nodes correspond to rules (state vectors), while the edges display the logical dependencies appearing when during the implementation of one set of actions we find the action that assumes checking of another particular state vector.

Since the designers of the satellite control logic took part in design and discussion of the notation of the visual domain specific language, they enjoy opportunities given by it.

The graphical construction tool supports the creation of an autonomous diagnosis program "from scratch". Initially, the "blank pre-form" of a rule appeared. The user needs to specify the parameters which should be checked in a particular state vector. Then the user can introduce a new abnormal situation and a corresponding set of actions in graphical manner.

As of today, the prototypes of visualization and graphical construction tools have been successfully accepted by the customer. All the tests both at university site and at customer site were executed using real "DKD" programs developed for real satellites which are in use now.

These tools were implemented using C++ programming languages and Graphviz library.

The method and the toolset were documented in accordance with strict industrial level standards applicable to the development of real-time embedded software including DO178B. The presented tools can be classified, by some attributes, as the real-time knowledge base or procedural oriented expert system. Of course, a fully-functional expert system requires additional features to be introduced. It is necessary to support the hierarchy of the rules and introduce more automation into the process of rules verification [14, 51–54].

The presented methods and prototypes of the visualization and graphical construction tools are just first steps in the planned development. The scope of the contract with the JSC Information Satellite Systems covers both the improvement of the developed prototypes and the development of other tools. The planned improvement includes visual language and visual tools for other kinds of autonomous control programs (not diagnostic). Another important aspect is to introduce a feature which will support the detection of incompleteness in the rules of the DKD diagnostic.

The following additional tools are under development now:

- Verification Tool for Satellite Autonomous Integrated Control Programs;
- Documentation Generation Tool providing an automated template-based generation and version control system with a guarantee of the strict correspondence between versions of the program and documentation;
- Networked Integrated Development Environment providing unified access point for designers of Satellite Autonomous Control Programs with saving of user workspaces.

All tools together should form "SIPR MP" (Russian acronym for "System for Intellectual Support for Design and Verification of Integrated Control Programs").

The "SIPR MP" is intended to be used as a complex software engineering toolset. A special stage of the development of "SIPR MP" is dedicated to the collection of notes and comments from users during the introduction of the system at the customer's site. The found errors should be removed, and the requested improvement of methods and the software engineering toolset should be provided.

5 Conclusions

The paper presents a flexible approach to the fault tolerance control of satellites based on an onboard knowledge base and a real-time interpreter of rules. The architecture of the onboard knowledge base supports the possibility of changing the rules from Earth on a timely basis over a radio channel. The domain specific visual language was introduced for knowledge representation. The visual rule builder provides a clear, user-friendly and unambiguous notation, developed by the designers of the satellite control logic. The process of satellite control is simplified by excluding the necessity of coding the control logic in programming languages and the associated long-term and labor-consuming multi-stage redevelopment cycle of the software. The prototypes of the developed tools were successfully accepted by the customer—JSC Information Satellite Systems, Krasnoyarsk region, Russia (the manufacturer of two-thirds of Russian spacecraft). In future, it is planned to implement additional tools including a verification tool, an automated documentation generation tool, and integrated development environment.

References

1. Ars Technica Information Portal, http://arstechnica.com/science/2014/09/the-little-known-soviet-mission-to-rescue-a-dead-space-station
2. Kozlov, D.I., Anshakov, G.P., Mostovoy, Y.A.: Upravlenie kosmicheskymi apparatami zondirovaniya Zemly: Komputerniye tekhnologii (In Russian). Mashinostroenie, Moscow (1998)
3. Kirilin, A.N., Akhmetov, R.N., Sollogub, A.V., Makarov, V.P.: Metody obespecheniya zhivuchesty nizkoorbitalnykh avtomaticheskykh KA zondirovaniya Zemly (In Russian). Mashinostroenie, Moscow (2010)
4. Akhmetov, R.N., Makarov, V.P., Sollogub, A.V.: Principles of the earth observation satellites control in contingencies. Inf. Control Syst. **1**, 16–22 (2012)
5. Eickhoff, J.: Onboard Computers. Onboard Software and Satellite Operations. An Introduction. Springer, Berlin (2012)
6. Luger, G.F., Stubblefield, W.A.: Artificial Intelligence and the Design of Expert Systems. Benjamin/Cummings Publishing Co, Redwood City, CA (1989)
7. Watanabe, S.: Knowing and Guessing. Wiley, New York (1969)
8. Lambert-Torres, G., Abe, J.M., et al. (eds.): Advances in Technological Applications of Logical and Intelligent Systems: Selected Papers from the Sixth Congress on Logic Applied to Technology. Series Frontiers in Artificial Intelligence and Applications, vol. 186. IOS Press (2008)

9. Tomayko, J.E.: Computers in Space: Journeys with NASA. Alpha Books, Indianapolis, Indiana (1994)
10. Tomayko, J.E.: Computers Take Flight: A History of NASA's Pioneering Digital Fly-By-Wire Project. NASA History Office, Washington, D.C. (2000)
11. Space.com News Portal, http://www.space.com/23640-hubble-space-telescope-repair-anniversary.html
12. Lisitsyna, L., Lyamin, A., Skshidlevsky, A.: Estimation of Student Functional State in Learning Management System by Heart Rate Variability Method. In: Neves-Silva, R., Tsihrintzis, G.A., Uskov, V., Howlett, R.J., Jain, L.C. (eds.) Smart Digital Futures 2014, vol. 262, pp. 726–731. IOS Press (2014)
13. Lisitsyna, L., Lyamin, A.: Approach to Development of Effective E-Learning Courses. In: Neves-Silva, R., Tsihrintzis, G.A., Uskov, V., Howlett, R.J., Jain, L.C. (eds.) Smart Digital Futures 2014, vol. 262, pp. 732–738. IOS Press (2014)
14. Khartov, V.V.: Autonomnoe upravlenie kosmicheskymi apparatami svyazi, retranslyacii i navigacii (In Russian). Aviakosmicheskoe priborostroenie (Aerospace Instrument-Making), 6, 12–23 (2006)
15. Smith, R.K., Muscettola, N.: Knowledge Acquisition for the Onboard Planner of an Autonomous Spacecraft. Technical Report, American Association for Artificial Intelligence WS98-03 (1998)
16. Koczela, L.I., Burnett, G.I.: Advanced Space Missions and Computer Systems. IEEE Trans Aerosp. Electron. Syst. **AES-4**(3), 456–467 (1968)
17. Sghairi, M., de Bonneval, A.: Challenges in building fault-tolerant flight control system for a civil aircraft. IAENG Int. J. Comput. Sci. **35**(4), 120–125 (2008)
18. Koltashev, A.A.: Effectivnaya technologiya upravleniya cyclom zhizni bortovogo programmnogo obespechenia sputnikov svyazi i navigacii (In Russian). Aviakosmicheskoe priborostroenie (Aerospace Instrument-Making), 12, 20–25 (2006)
19. Tyugashev, A.A., Ermakov, I.E., Ilyin, I.I.: Ways to Get More Reliable and Safe Software in Aerospace Industry. In: Program Semantics, Specification and Verification: Theory and Applications (PSSV 2012), pp. 121–129. Nizhni Novgorod, Russia (2012)
20. Kransner, S., Bernard, D.E.: Integrating autonomy technologies into an embedded spacecraft system-flight software system engineering for new millennium. In: IEEE Aerospace Conference, vol. 2, pp. 409–420. IEEE Press, Snowmass (1997)
21. Planetary Society: http://www.planetary.org/blogs/emily-lakdawalla/2014/08190630-curiosity-wheel-damage.html
22. Space.com Information Portal: http://www.space.com/17034-mars-rover-curiosity-software-upgrade.html
23. Planetary Society: http://www.planetary.org/explore/space-topics/space-missions/mer-updates/2004/04-09-mer-update.html
24. Hayes-Roth, B.: An Architecture for Adaptive Intelligent Systems. Artif. Intell. **72**, 329–365 (1995)
25. Nakamatsu, K., Abe, J.M. (eds.): Advances in Logic Based Intelligent Systems: Selected Papers of LAPTEC 2005. IOS Press (2005)
26. Grabot, B., Geneste, L., Dupeux, A.: Experimental design, expert system and neural network approaches: comparison for the choice of parameters. In: International Conference on Systems, Man and Cybernetics 'Systems Engineering in the Service of Humans', vol. 4, pp. 15–20. Le Touquet, France (1993)
27. Nakamatsu, K., Jain, L.C. (eds.): The Handbook on Reasoning-Based Intelligent Systems. World Scientific (2013)
28. Bianchini, M., Maggini, M., Scarselli, F. (eds.): Innovations in Neural Information Paradigms and Applications. Springer-Verlag Berlin Heidelberg (2009)
29. Bianchini, M., Maggini, M., Sarti, L., Scarselli, F.: Recursive neural networks learn to localize faces. Pattern Recognit. Lett. **26–12**, 1885–1895 (2005)
30. Hartmann, G.L.: Fault Tolerant Hardware/Software Architectures for Flight Critical Functions. Introduction/Overview. In: Fault Tolerant Hardware/Software Architectures for Flight Critical

functions. AGARD Lecture Series No.143. NATO Advisory Group for Aerospace Research and Development. Laughton, Essex, UK (1985)
31. Lemos, J.M., Neves-Silva, R., Igreja, J.M.: Adaptive Control of Solar Energy Collector Systems. Springer International Publishing (2014)
32. Pospelov, D.A.: Situational Control: Theory and Practice. Batelle Memorial Institute, Columbus, OH (1986)
33. Kochura, E.V.: Razrabotka macroprogramm integralnogo upravleniya KA (In Russian). Vestnik SibAU **1**, 105–107 (2011)
34. Parondzhanov, V.D.: Druzhelyubnye algoritmy, ponyatnye kazhdomu. Kak uluchshit' rabotu uma bez lishnih hlopot (In Russian). DMK Press, Moscow (2010)
35. Tyugashev, A.A.: Graficheskiye yazyki programmirovania i ih primenenie v sisitemah upravlenia realnogo vremeni (In Russian). Russian Academy of Sciences, Samara, Russia (2009)
36. Shadbolt, N., Schreiber, G. (eds).: Advances in Knowledge Acquisition: 9th European Knowledge Acquisition Workshop, EKAW '96, Springer, New York (1996)
37. Ruiz, P.P., Foguem, B.K., Grabot, B.: Generating knowledge in maintenance from Experience Feedback. Knowl. Based Syst **68**, 4–20 (2014)
38. Osipov, G.S.: Priobretenie znaniy intellectualnymi systemami (In Russian). Nauka, Moscow (1997)
39. Chassiakos, A.P., Vagiotas, P.: A knowledge-based system for maintenance planning of highway concrete bridges. Adv. Eng. Softw. **36**(11–12), 740–749 (2005)
40. Drucker, J.: Graphesis: Visual Forms of Knowledge Production. Harvard University Press, Boston (2014)
41. Chein, M., Mugnier, M.L.: Graph-Based Knowledge Representation: Computational Foundations of Conceptual Graphs. Springer, Berlin (2008)
42. Lengler, R., Eppler, M.: Towards a periodic table of visualization methods for management. In: IASTED Proceedings of the Conference on Graphics and Visualization in Engineering (GVE 2007), pp. 83–88. ACTA Press, USA (2007)
43. Eppler, M.J., Burkhard, R.A.: Visual representations in knowledge management: framework and cases. J. Knowl. Manag. **11**(4), 112–122 (2007)
44. Nobécourt, J., Biébow, B.: Mdws: A modeling language to build a formal ontology in either description logics or conceptual graphs. In: Knowledge Engineering and Knowledge Management Methods, Models, and Tools. LNCS, vol. 1937, pp. 57–64. Springer, Heidelberg (2002)
45. Pfeiffer, H.D., Hartley, H.D.: Visual CP representation of knowledge. In: Bernhard Ganter, Guy W. Mineau (eds). In: 8th International Conference on Conceptual Structures, ICCS 2000. LNCS, vol. 1867, pp.1211–1237. Springer, Heidelberg (2000)
46. Travers, M.: A visual representation for knowledge structures. In: HYPERTEXT'89, pp. 147–158. ACM, NY (1989)
47. Finn, A.: Legal issues for military intelligent decision-making technologies. In: Knowledge-Based Intelligent Information and Engineering Systems. 12th International Conference KES2008. LNCS, vol. 5177, Part I, pp. 14–15. Springer-Verlag Germany (2008)
48. Parondzhanov, V.D., Trunov, Y.V.: Systema upravlenia razgonnogo blocka Fregat (In Russian). Vestnik NPO imeni S.A. Lavochkina (NPO Lavochkina Bulletin), 1 (22), 16–25 (2014)
49. Martin, J.: Application Development without Programmers. Prentice-Hall, PTR Upper Saddle River, NJ, USA (1982)
50. Kalentyev, A.A., Tyugashev, A.A., Bogatov, A.V., Shulyndin, A.V.: Visual toolset for real-time onboard programs verification support. In: Program Semantics, Specification and Verification: Theory and Applications (PSSV 2011), Saint Petersburg, pp. 120–127. Yaroslavl State University, Russia (2011)
51. Sullivan, G.A.: A knowledge-based control architecture with interactive reasoning functions. IEEE Trans. Knowl. Data Eng. **8**(1), 179–183 (1996)

52. Nakamatsu, K., Abe, J.M., Akama, S.: A logical reasoning system of process before-after relation based on a paraconsistent annotated logic program bf-EVALPSN. Int. J. Knowl. Based Intell. Eng. Syst. **15**(3), 145–163 (2011)
53. Giurca, A., Gasevic, D.: Handbook of Research on Emerging Rule-Based Languages and Technologies: Open Solutions and Approaches. Information Science Reference. Hershey, New York (2008)
54. Neves-Silva, R., Rato, L.M., Lemos, J.M.: Time Scaling Internal State Predictive Control of a Solar Plant. IFAC Control Eng. Pract. (Special Issue on IFAC-B'02 Prize Winning Applications), **11**(12), pp. 1459–1467 (2003)

A Generic Document Retrieval Framework Based on UMLS Similarity for Biomedical Question Answering System

Mourad Sarrouti and Said Ouatik El Alaoui

Abstract Biomedical document retrieval systems play a vital role in biomedical question answering systems. The performance of the latter depends directly on the performance of its biomedical document retrieval section. Indeed, the main goal of biomedical document retrieval is to find a set of citations that have high probability to contain the answers. In this paper, we propose a biomedical document retrieval framework to retrieve the relevant documents for the biomedical questions (queries) from the users. In our framework, we first use GoPubMed search engine to find the top-K results. Then, we re-rank the top-K results by computing the semantic similarity between questions and the title of each document using UMLS similarity. Our proposed framework is evaluated on the BioASQ 2014 task datasets. The experimental results show that our proposed framework has the best performance (MAP@100) compared to the existing state-of-the-art related document retrieval systems.

Keywords Information retrieval · Biomedical question answering system · Gopubmed · Unified modeling language system · Semantic similarity

1 Introduction

By the rapidly increasing of knowledge in the biomedical domain, it becomes very difficult even for experts to absorb all the relevant information in their field of interest. Information Retrieval (IR) systems present a list of document that might have the associated information, but the majority of them leave it to the user to find and extract the required information [8]. For example, the biomedical question "Is the PTPN22 gene a biomarker for Rheumatoid Arthritis?" from BioAsk training datasets, should get back the response "Yes", but instead the user is presented with a large number of

M. Sarrouti (✉) · S.O. El Alaoui
Laboratory of Computer Science and Modeling, FSDM, Sidi Mohammed Ben Abdellah University, Fes, Morocco
e-mail: mourad.sarrouti@usmba.ac.ma

S.O. El Alaoui
e-mail: said.ouatikelalaoui@usmba.ac.ma

I. Czarnowski et al. (eds.), *Intelligent Decision Technologies 2016*,
Smart Innovation, Systems and Technologies 57,
DOI 10.1007/978-3-319-39627-9_18

documents that are potentially relevant to explore in the quest of an accurate answer. Unlike IR systems, Question Answering (QA) systems aim to provide inquirers with direct and precise answers to their questions, by employing Information Extraction (IE) and Natural Language Processing (NLP) methods [3]. In other words, QA systems allow to quickly get precise answers to user's questions with the least amount of reading required.

Typically an automated QA system consists of three main elements, which independently can be studied and developed, [3, 9, 11, 14]: Questions Processing, Documents Processing and Answers Processing. Figure 1 illustrates the generic architecture of a biomedical QA system. For a given biomedical question written in natural language, the Question Processing phase aims to analyze the question and create IR query, identifying the type of question as well [16] . Indeed, the first task is called Query Reformation and the second is called Question Classification [11, 16]. The document Processing phase allows to process the returned documents by an IR system and provides candidate passages which could probably contain the answer. Finally, the type of question identified in the question processing and the candidates answers generated in the document processing are used in the Answer Processing phase in order to extract the final answer.

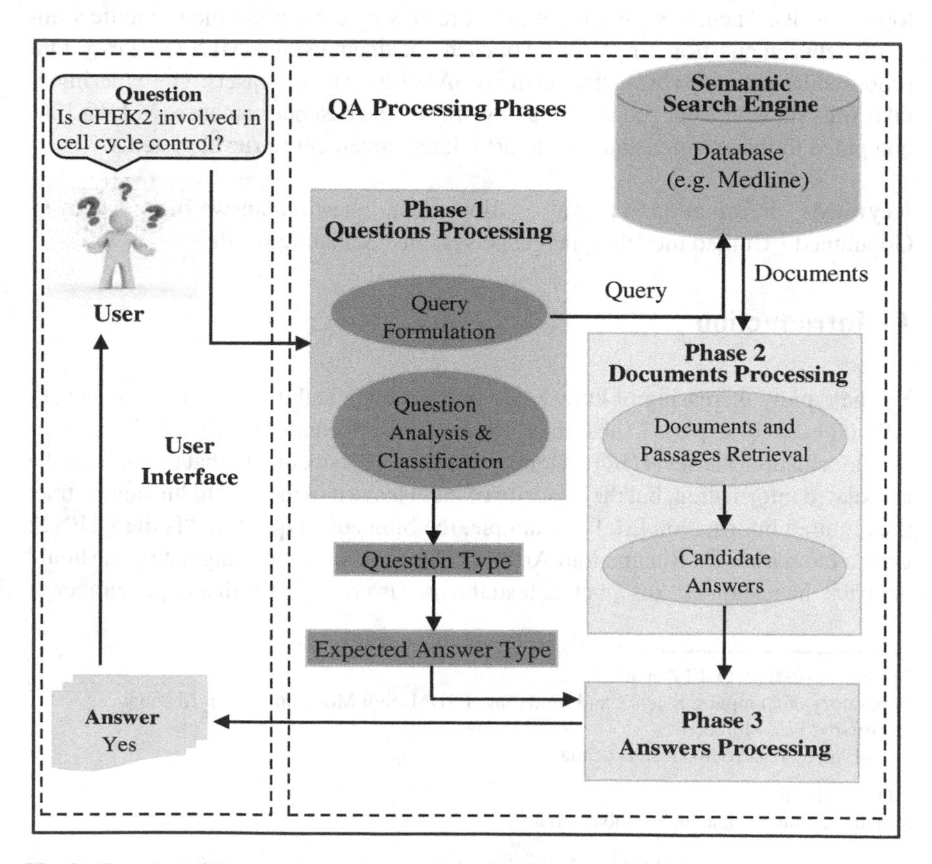

Fig. 1 Generic architecture of questions answering system

Previously, we have addressed the problem of biomedical question classification [16] and in this paper we are interested on biomedical document retrieval, which is an important component of biomedical QA systems. As we mentioned earlier, the task of biomedical document retrieval is to find a list of relevant documents that are likely to contain the answer. In other words, if a list of relevant documents is determined correctly, it can be useful for finding the location of the answer. Therefore, the task of document retrieval has a significant impact on the overall performance of the biomedical QA system.

In light of this, GoPubMed [7] is considered one of the information retrieval tools most widely used in biomedical QA systems to access the MEDLINE database. MEDLINE is a major biomedical literature database repository, which is supported by the U.S. National Library of Medicine (NLM). In other words, the goal of GoPubMed, like all other search engines, is to retrieve documents considered relevant to a user query. Researches have done great effort to optimize retrieval result rankings, hoping to place the most relevant ones at the top of the ranking list. Nevertheless, no ranking solution is perfect, due to the inherent complexity of ranking search results. For instance, in [13], GoPubMed has been used as biomedical document retrieval. The query reformulation component includes sentence splitting, tokenization, part-of-speech tagging and chunking using the Stanford CoreNLP. They have also kept the top 100 documents returned by GoPubMed.

In this paper, we propose a novel biomedical document retrieval framework. Comparing with the previous works, our proposed method has the following contributions:

1. The novel framework is based on GoPubMed semantic search engine and our query reformulation method.
2. In our knowledge, this is for the first time that UMLS similarity has been used for re-ranking the top-K citations and keep the top-k ones which have high probability to contain the answers.

The remainder of the paper is organized as follows. Section 2 introduces related work and discussion about the main biomedical QA approaches. Section 3 describes the overall architecture of the proposed framework. Section 4 presents our experiments on a benchmark dataset and the results of our biomedical document retrieval framework. Finally, conclusion and future work are made in Sect. 5.

2 Related Work

Although research on QA systems has boomed in recent years, document retrieval has been a large part in the research community of text mining after the introduction of QA Track in the Text REtrieval Conference (TREC[1]) in 1999 as well as the presentation of biomedical QA in the BioASK[2] [18].

---

[1]http://trec.nist.gov/.

[2]http://bioasq.org/.

However, QA system has been a well studied research area [15]. Biomedical QA system has its own challenges such as the presence of complex technical terms, compound words, domain specific semantic ontologies, domain-specific format and typology of questions [3].

MedQA [10] is a biomedical QA system which generates paragraph-level answers from the MEDLINE collection. The system consists of information retrieval, extraction, and summarization techniques to automatically generate paragraph-level answers for definitional questions. For query formulation and document retrieval, the system use a shallow syntactic parser and a standard IR engine.

Abacha and Zweigenbaum [1] have described their implemented medical QA system called MEANS. The system consists of three main steps: corpora annotation, question analysis, and answer search. The authors have exploited natural language processing techniques as well as biomedical and semantic resources (e.g. UMLS) to build RDF annotations of the source documents and SPARQL queries representing the users questions.

In [6], SNUMedinfo team has leased 2014 MEDLINE/PubMed Journal Citations and used Indri search engine [17]. In fact, they have experimented with semantic concept-enriched dependence model and sequential dependence model. They have also shown that the semantic concept enriched dependence model showed significant improvement over baseline.

In [19], Weissenborn et al. have presented a biomedical QA system which is composed of three subsystems: question analysis, document retrieval, and answer extraction. They have used GoPubMed search engine in order to find the relevant documents to the question. in addition, the authors have completely depended on GoPubMed ranking of documents.

Continuing, Neves [13] has presented a biomedical document retrieval system. The latter includes sentence splitting, tokenization, part-of-speech tagging and chunking using the Stanford CoreNLP package for query reformulation. The approach is based on GoPubMed search engine. Neves has also completely depended on GoPubMed ranking of documents.

To our knowledge, all the above methods have not taken into account the semantic similarity between question and each title of document returned by a search engine in order to re-rank these documents again. In this paper, in order to improve the performance of document retrieval of biomedical QA systems, we propose a generic biomedical document framework based on UMLS similarity.

3 Proposed Method

In this section, we describe in detail our proposed document retrieval for biomedical QA system. Indeed, the main goal of this method is to find the high informative documents for a given biomedical question from PubMed articles.

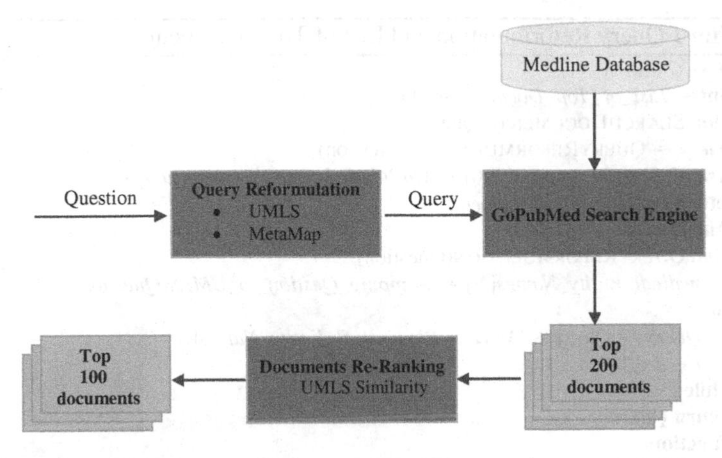

Fig. 2 Flowchart of our biomedical documents retrieval framework

Table 1 Example of mapping question to UMLS concepts

Question	UMLS concepts	CUI
Which are the	Cardiac	C0018787
Cardiac manifestation of	Manifestation of	C1280464
Marfan syndrome	Marfan syndrome	C0024796

CUI indicates Concept Unique Identifier

In order to achieve this goal, we construct a biomedical document retrieval framework to solve the semantic search by reformulating the query and using UMLS Similarity [12] to rank the returned biomedical documents. Moreover, we have proposed two algorithms. The first one allows to reformulate the query and search the top 200 documents using GoPubmed[3] web service [7]. The second one aims to re-rank the top 200 documents and keep only a set of 100 PubMed documents which have high probability to contain the answer. The flowchart of our biomedical document retrieval framework is presented in Fig. 2 and below are the various steps described in details.

1. **Query Reformulation**: in this step, we process the biomedical question, written in natural language, to make it efficient and optimized for searching. Indeed, We have used MetaMap [2] for mapping terms in questions to Unified Medical Language System (UMLS) in order to extract the Biomedical Entity Names (BENs) and connect them with the "AND" operator. The UMLS [5] is a repository developed by the US National Library of Medicine, integrating over 2 million names for some 900 000 concepts from more than 60 families of biomedical vocabularies as well as 12 million relations among these. Table 1 illustrates an example of mapping question to UMLS using Metamap. Besides, based on the above definition, the query reformulation method is defined in the proposed Algorithm 1.

[3]http://gopubmed.org/web/gopubmedbeta/bioasq/pubmedmedline.

Algorithm 1 Query Reformulation and List of Top Documents

```
1: Input ← Question
2: Output ← List_of_Top_Documents(200)
3: function SEARCHDOCUMENTS(Question)
4:     Query ← QUERYREFORMULATION(Question)
5:     List_of_Top_Documents[200] ← GoPubMed_Web_Service(Query)
6:     return List_of_Top_Documents
7: end function
8: function QUERYREFORMULATION(Question)
9:     Biomedical_Entity_Names[N] ← Mapping_Qestion_to_UMLS(Question)
10:    do
11:        Query ← Query + "AND" + Biomedical_Entity_Names[i]
12:        i ← i + 1
13:    while i < N
14:    return Query
15: end function
```

2. **Pubmed Document Retrieval Using GoPubMed**: As it was shown in Algorithm 1, the query generated in the query reformulation phase will be fired to GoPubMed semantic search engine [7] in order to find the top 200 documents.
3. **Biomedical Document Re-Ranking**: the document re-ranking is the main and important step in the proposed framework. Indeed, we do not completely depend on GoPubMed ranking of documents. So we re-rank the obtained 200 documents again by computing the similarity between a given question and the title of each document. We have used UMLS similarity package[4] [12] to obtain similarity between biomedical concepts of a question and the concepts of document title. In fact, we have used path length as similarity measure where the similarity score is inversely proportional to the number of nodes along the shortest path between the concepts. Moreover, our proposed biomedical document Re-Ranking is presented in Algorithm 2.

4 Experimental Results and Discussion

In this section, we conduct our experiments on benchmark dataset to show the performance of our biomedical document retrieval framework. We first describe the dataset, then we present the experimental results, and finally discuss the results.

4.1 Datasets

To demonstrate the efficiency of the proposed framework, we perform experiments on benchmark dataset provided by biomedical experts. Actually, The experimental dataset comes from the official dataset of biomedical semantic QA Taskb phase

[4]http://maraca.d.umn.edu/cgi-bin/umls_similarity.cgi.

Algorithm 2 Biomedical Document Re-Ranking

```
1: Input ← Question, List_of_Top_Documents(200)
2: Output ← Top_Documents(100)
3: Q ← Question
4: relDocs ← List_of_Top_Documents(200)
5: function RANKDOCUMENTS(Q, relDocs)
6:     scores ← {}
7:     do
8:         T[i] ← relDocs[i].title
9:         scores[i] ← COMPUTESIMILARITY(Q, T[i])
10:        i ← i + 1
11:    while i < N
12:    scores, TopDocument[100] ← SortScores(scores, relDocs)
13:    return TopDocument[100]
14: end function
15: function COMPUTESIMILARITY(Q, docTitle)
16:    question_concepts_CUI[N] ← Mapping_Question_to_UMLS(Question)
17:    docTitle_concepts_CUI[M] ← Mapping_Title_to_UMLS(docTitle)
18:    similarity ← 0
19:    sumSimilarity ← 0
20:    do
21:        QCUI ← Question_Concepts_CUI[i]
22:        do
23:            TCUI ← docTitle_Concepts_CUI[j]
24:            similarity ← UMLS :: Similarity(QCUI, TCUI)
25:            if similarity ≠ −1 then
26:                sumSimilarity ← sumSimilarity + similarity
27:            end if
28:            j ← j + 1
29:        while j < M
30:        i ← i + 1
31:    while i < N
32:    return sumSimilarity
33: end function
```

A [18]. In the dataset, there are five batches of questions in the testing set where includes 100 questions in each batch.

4.2 Evaluation Metrics

The typical evaluation measures used in IR are: mean precision, mean recall, mean F-measure and mean average precision (MAP) [18]. In fact, MAP is our main evaluation measure. For the test in 2014, the first 100 documents from the resulting list are permitted to be submitted.

Table 2 TOP 10 MAP@100 results of document retrieval systems [4] on batch 1 of BIOASQ 2014 and the results of our proposed framework

System	Mean precision	Mean recall	Mean F-measure	MAP
SNUMedinfo1	0.0457	0.5958	0.0826	0.2612
SNUMedinfo3	0.0457	0.5947	0.0826	0.2587
SNUMedinfo2	0.0451	0.5862	0.0815	0.2547
SNUMedinfo4	0.0457	0.5941	0.0826	0.2493
SNUMedinfo5	0.0459	0.5947	0.0829	0.2410
Top 100 baseline	0.2274	0.4342	0.2280	0.1911
Top 50 baseline	0.2290	0.3998	0.2296	0.1888
Main system	0.0413	0.2625	0.0678	0.1168
Biomedical text ming	0.2279	0.2068	0.1665	0.1101
Wishart-S2	0.1040	0.1210	0.0793	0.0591
Our system	**0.2331**	**0.3644**	**0.2253**	**0.2758**

4.3 Results and Discussion

To conduct the experiments, we have used the batch 1 of testing datasets (Benchmark dataset) of BioASQ 2014 task [18]. We first have applied MetaMap [2] to extract the Biomedical Entity Names of a given biomedical question and connect them with the "AND" operator in order to construct the query. Then, the latter will be fired to GoPubmed[5] semantic search engine [7] in order to find the top 200 documents (see Algorithm 1). After that, as we have not depended on GoPubMed ranking of documents, the proposed Algorithm 2 has been used in order to re-rank the 200 documents and keep only the top 100 documents. Table 2 presents the comparison between our results and the top 10 results on batch 1 of testing datasets in BioASQ 2014.

Overall, from Table 2, it can be seen clearly that the results of our proposed framework have an absolute competitiveness with the top 10 results in term of MAP. Indeed, the performance of our system was 0.2758 of MAP. Moreover, Our proposed framework significantly outperforms the baseline system (Top 100 Baseline) by a wide margin in term of mean average precision (0.0847 MAP).

In addition, Table 3 shows the results obtained by GoPubMed document ranking and our proposed algorithm (see Algorithm 2) for biomedical document re-ranking. We can see that when using GoPubMed document ranking, the mean average precision was 0.1439. While Algorithm 2 increased the performance to 0.2758 of MAP and the improvement is statistically significant. Hence, the proposed algorithm for document re-ranking plays a vital role on the overall performance of our framework.

Studies have shown that the biomedical document retrieval systems can improve the performance of biomedical QA systems, because the answers extraction is

[5]http://gopubmed.org/web/gopubmedbeta/bioasq/pubmedmedline.

Table 3 Results obtained by GoPubMed document ranking and our proposed algorithm 2 on batch 1 of BIOASQ 2014

System	Mean precision	Mean recall	Mean F-measure	MAP
Our query reformulation and GuPubMed document ranking	0.2253	0.3111	0.1913	0.1439
Proposed framework	**0.2331**	**0.3644**	**0.2253**	**0.2758**

based on the documents returned by document retrieval systems which have high probability to contain answers. Therefore, our proposed biomedical document retrieval framework can be used in order to find relevant documents to the biomedical question with high mean average precision. Moreover, the importance of our results using the proposed framework thus lies both in their generality and their relative ease of application to biomedical QA systems.

5 Conclusion and Future Work

In this paper, we have tackled an original biomedical document retrieval framework. First, we have used Metamap to extract biomedical named entities and connect them in order to generate queries. Then, the top 200 relevant documents are retrieved by GoPubMed search engine. Next, we have kept only the top 100 documents after re-ranking the top 200 documents by computing the semantic similarity between question and documents title. Finally, the experiments on the BioASQ 2014/2015 document retrieval task have demonstrated that our proposed framework is proved to be effective and competitive for biomedical documents retrieval compared to several state-of-the-art systems.

In our future work, we will focus on integrating our biomedical document retrieval framework in a biomedical QA system.

Acknowledgments The authors would like to thank BioASK challenges [18] for providing us with benchmark datasets.

References

1. Abacha, A.B., Zweigenbaum, P.: Means: a medical question-answering system combining nlp techniques and semantic web technologies. Inf. Process. Manag. **51**(5), 570–594 (2015)
2. Aronson, A.R.: Effective mapping of biomedical text to the umls metathesaurus: the metamap program. In: Proceedings of the AMIA Symposium, p. 17. American Medical Informatics Association (2001)

3. Athenikos, S.J., Han, H.: Biomedical question answering: a survey. Comput. Methods Programs Biomed. **99**(1), 1–24 (2010)
4. Balikas, G., Partalas, I., Ngomo, A.C.N., Krithara, A., Gaussier, E., Paliouras, G.: Results of the bioasq track of the question answering lab at clef 2014. Results of the BioASQ Track of the Question Answering Lab at CLEF 2014, 1181–1193 (2014)
5. Bodenreider, O.: The unified medical language system (umls): integrating biomedical terminology. Nucl. Acids Res. **32**(suppl 1), D267–D270 (2004)
6. Choi, S., Choi, J.: Classification and retrieval of biomedical literatures: Snumedinfo at clef qa track bioasq 2014. In: Proceedings of Question Answering Lab at CLEF (2014)
7. Doms, A., Schroeder, M.: Gopubmed: exploring pubmed with the gene ontology. Nucl. Acids Res. **33**(suppl 2), W783–W786 (2005)
8. Dwivedi, S.K., Singh, V.: Research and reviews in question answering system. Procedia Technol. **10**, 417–424 (2013)
9. Gupta, P., Gupta, V.: A survey of text question answering techniques. Int. J. Comput. Appl. **53**(4), 1–8 (2012)
10. Lee, M., Cimino, J., Zhu, H.R., Sable, C., Shanker, V., Ely, J., Yu, H.: Beyond information retrieval medical question answering. In: AMIA Annual Symposium Proceedings, vol. 2006, p. 469. American Medical Informatics Association (2006)
11. Loni, B.: A Survey of State-of-the-Art Methods on Question Classification, pp. 01–40. Delft University of Technology, Delft (2011)
12. McInnes, B.T., Pedersen, T., Pakhomov, S.V.: Umls-interface and umls-similarity: open source software for measuring paths and semantic similarity. In: AMIA Annual Symposium Proceedings, vol. 2009, p. 431. American Medical Informatics Association (2009)
13. Neves, M.: Hpi in-memory-based database system in task 2b of bioasq. In: Proceedings of Question Answering Lab at CLEF (2014)
14. Neves, M., Leser, U.: Question answering for biology. Methods **74**, 36–46 (2015)
15. Ryu, P.M., Jang, M.G., Kim, H.K.: Open domain question answering using wikipedia-based knowledge model. Inf. Process. Manag. **50**(5), 683–692 (2014)
16. Sarrouti, M., Lachkar, A., Ouatik, S.E.: Biomedical question types classification using syntactic and rule based approach. In: Proceedings of the 7th International Joint Conference on Knowledge Discovery, Knowledge Engineering and Knowledge Management, pp. 265–272 (2015)
17. Strohman, T., Metzler, D., Turtle, H., Croft, W.B.: Indri: A language model-based search engine for complex queries. In: Proceedings of the International Conference on Intelligent Analysis, vol. 2, pp. 2–6. Citeseer (2005)
18. Tsatsaronis, G., Balikas, G., Malakasiotis, P., Partalas, I., Zschunke, M., Alvers, M.R., Weissenborn, D., Krithara, A., Petridis, S., Polychronopoulos, D., et al.: An overview of the bioasq large-scale biomedical semantic indexing and question answering competition. BMC Bioinform. **16**(1), 138 (2015)
19. Weissenborn, D., Tsatsaronis, G., Schroeder, M.: Answering factoid questions in the biomedical domain. BioASQ@ CLEF 1094 (2013)

An IoA Cloud-Based Farmer Support System "AgriMieru"

Alireza Ahrary, Masayoshi Inada and Yoshitaka Yamashita

Abstract In recent years, the aging of agricultural workers has progressed rapidly, successor problem is becoming more serious. Under such circumstances are coming out also new farmers that will beginner to agriculture. However, the establishment of farming technology has become a major management challenge for new farmers. In this study, we focus our efforts to solve the management issues of the farming acquisition of technology and to develop an Internet of Agriculture (IoA) cloud-based designed farmer support system, called "AgriMieru". In details, we will describe the water management system and Machine to Machine (M2M) system, cloud system design and interface, those are fundamental keystones of this newly proposed system.

Keywords Internet of Agriculture (IoA) · M2M · Soil moisture · pF measurement

1 Introduction

In recent years, there is a rapid reduction and aging of the Japanese agriculture workers, and there are no clear solutions for their replacement in the short term. According to the Ministry of Agriculture, Forestry and Fisheries in 1960 the total agriculture worker's population was approximately 14.54 million people. Meanwhile in the 2013 statistics, the total population of agriculture workers is 2.39 million people, where 1.478 million people are 65 years old or older, which represents the 60 %. This leads to the conclusion that the situation has been getting

A. Ahrary (✉)
Faculty of Computer and Information Sciences, SOJO University, Kumamoto, Japan
e-mail: ahrary@cis.sojo-u.ac.jp

M. Inada
FusionTech Inc., Kumamoto, Japan

Y. Yamashita
Sanwa Hi-Tech Co. Ltd., Kumamoto, Japan

I. Czarnowski et al. (eds.), *Intelligent Decision Technologies 2016*,
Smart Innovation, Systems and Technologies 57,
DOI 10.1007/978-3-319-39627-9_19

worse in a fast manner [1–3]. Under such circumstances, we observed that around 50,000 people enter as new agriculture workers during the year 2013.

However, according to the survey results related to farming situation of new farmers done by the National Agricultural Chamber of the National new farming consultation center, securing farmland for new farmers, securing funds, and the acquisition of farming technology is a major management issue. In particular, for the acquisition of farming technology, it is pointed out that 55.5 % of people who answered the questionnaire experienced a hard time doing this process.

For the farming technology, important factors related to the healthy growth of the crop are: temperature management, and water management, this study is focused in the water and moisture management issues. Documents related to irrigation information or reference materials published by the Ministry of Agriculture, Forestry and Fisheries were not updated regularly and were not related to the different soil conditions presented in lands around the different Japanese prefectures or their related water characteristics. Having information related to the soil moisture is relevant for the farmer since it is vital for the healthy crop growth. Therefore, it is important the development of such system and the related apparatus to grasp the different moisture levels in the soil [4, 5].

In this study, we focus our efforts to solve the management issues of the farming acquisition of technology and to develop an Internet of Agriculture (IoA) cloud-based designed farmer support system, called "AgriMieru". In details, we will describe the water management system and Machine to Machine (M2M) system, cloud system design and interface, those are fundamental keystones of this newly proposed system.

2 IoA Cloud-Based System Construction

We designed a system to collect the various information of the farming management in the real environment trough Machine to Machine (M2M) interface automatically. M2M interface is a microcomputer with various I/O to transmit the measured data to cloud system through the 3G network. Some related technologies in order to provide the better solutions in this field are described in follow.

i. Sensor networks
   The use of different specific-application designed sensor in a network array connected to a high-speed network, allow information exchange among different business partners. This sensor network is the core of multiple application layers that will be built over the information provided.
ii. Cloud computing
   Many current applications, from storage to Software as a Service (ASA) are using this common affordable platform in order to reduce their hardware

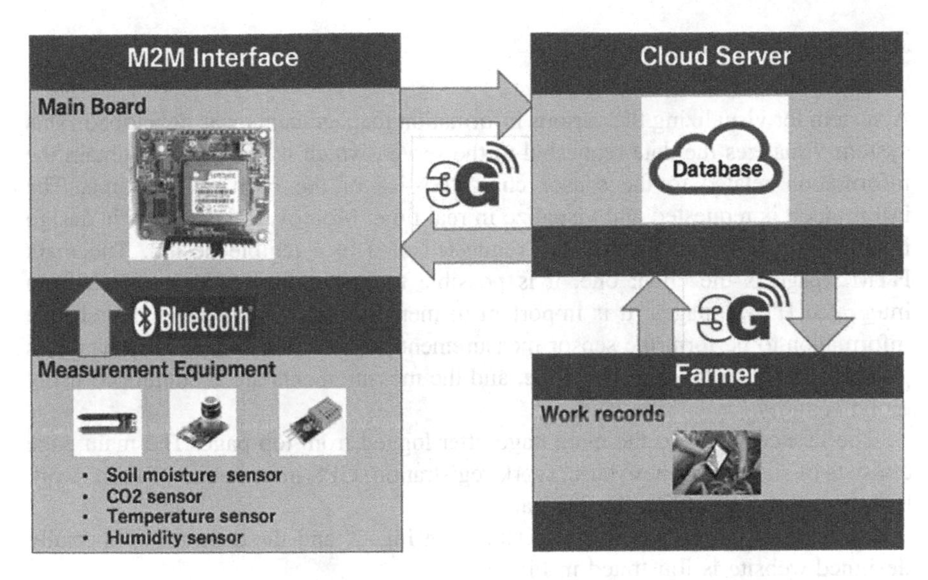

Fig. 1 System construction of "AgriMieru"

expenses. The information retrieved from the sensor networks could be processed or stored in the cloud for future purposes.

iii. Control Area Network (CAN)
The data generated by the different agriculture machines, e.g. yield of pesticides, production, etc., can be stored in the cloud as well.

The structure of the M2M interface for "AgriMieru" is illustrated as shown in Fig. 1. The information from measuring sensors are collected in main board and transfer through the specially designed 3G shield to the cloud database.

The specification of specially designed M2M interface is shown in Table 1. The analog or digital signals can be transformed by M2M interface. And these signals will be transformed into the cloud database by 3G network.

Table 1 Specification of specially designed M2M interface

Item	Specification
Microcontroller	ATmega2560
Operating voltage	5 V
Input voltage	7–12 V
Digital I/O	54
Analog I/O	16
Flash memory	256 KB
Network	3G

3 User Interface

A system for visualizing the various information managements was developed. This system visualizes the data requested to the server, which is dedicated to obtain the information related to the sensor characteristics of the farm in real time. The information is requested and visualize in real time. Moreover, the system is design in order to correctly visualize the requested data in a legible design. The main HTML page is the client one, it is possible to access the M2M interface from integrated HTML pages. It is important to mention that coding pages contain the information to perform the sensor measurement requests, the sensor information is presented through the cloud website, and the measurements are accumulated using cloud services.

The farmer access to the main page after logged from top page. The main page consists of sign up for new farm, work registration, GPS information of farm, work history, graph and online display parts.

Example of the user interface is shown in Fig. 2, and the diagram of specially designed website is illustrated in Fig. 3.

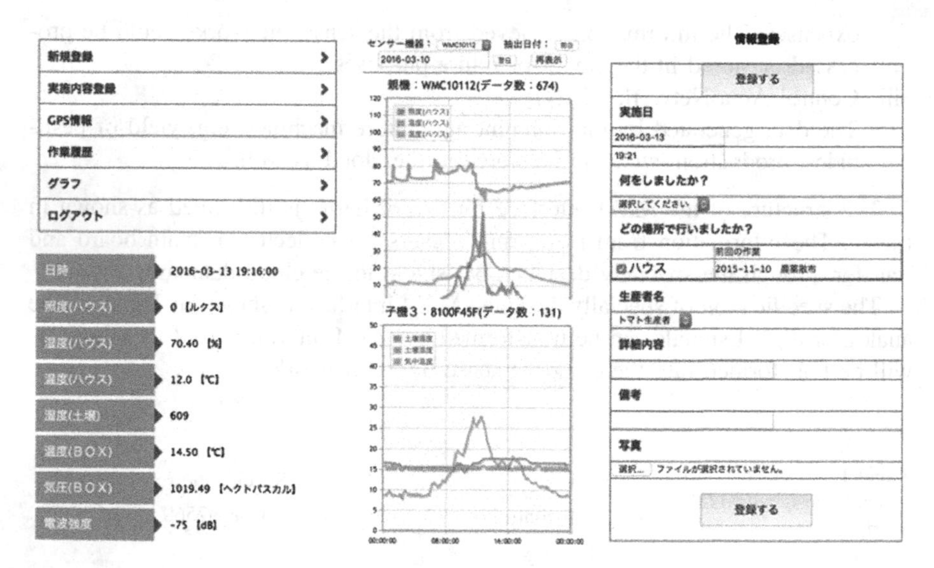

Fig. 2 User interface

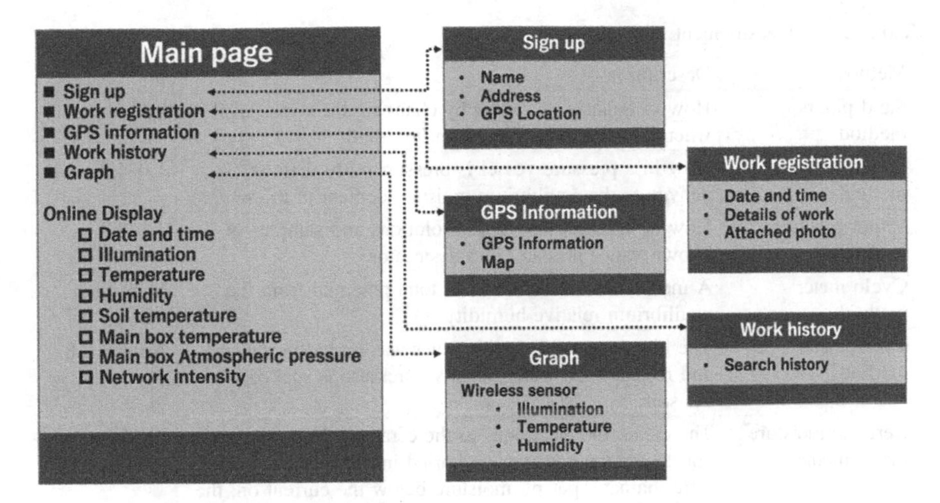

Fig. 3 Diagram of specially designed website

4 pF Measurement Methods

Moisture is strongly related to the different characteristics of the soil and its various forces. In the sand and clay, the water-holding capacity is different as well as the soli farming capabilities. Rather than show the amount of water presented in the soil in percentage (%), the different devices that shows the moisture amount are directly related to the force needed to extract water from the soil. Therefore, the water-content meter readings were correlated significantly with tensiometer pF values is widely used.

The moisture tension represents the degree of strength to extract water from the soil. It is also representing the wetness of the soil, generally used as a unit of pressure. A full pF value indicates a low amount of water presented in the soil, as it is shown when roots extract the moisture of the soil. The different pF Measurement Methods are presented in Table 2, depending on the measurement range interest. In this study, using a tensiometer in order to get the pF value due to a low cost consideration.

Figure 4 shows the measurement instrument used in this study. The instrument is a tensiometer which its principle is to measure the pressure force required to extract moisture from the porous cap presented in the soil. If the soil is drying, will contain the water in the bolus cup, therefore the pF value is low.

In details, a Grove Moisture Sensor was used in this study. The sensor has an analog output directly related to the soil resistance change due to the water content. The principle resides in the change of the resistance value due to the water content of the soil between the transistor-based circuits shown in Fig. 5 related to the change of the base current. The base current change is amplified and the output signal is emitted to R_2 as a voltage change.

Table 2 pF measurement methods

Method	Description	pF
Sand pillars method	How to balance the water by changing the water level when a sample was placed on fine sand	Less than 1.0
Pressurization method	How in a pressure vessel is pressurized to thereby dehydrate the equilibrium moisture content of the sample	1.0–4.2
Vapor pressure method	How to balance the various solutions and samples of known vapor pressure in a desiccator	4.2–7.0
Cyclo-meter method	A method of determining the total potential from the equilibrium relative humidity	More than 3.2
Tensiometer method	The bisque cup filled with water was buried in the soil and measure the water negative pressure is sucked into the soil	Around 2.8
Ceramic moisture meter method	The closer the electrodes to the ceramic pieces having a constant pore diameter are buried in the soil Pore diameter per pF moisture below the current on, the above measures to be turned off	1.0–3.0
Piezometer method	A pipe takes out the pressure of the water and measuring the water column height	Positive pressure range

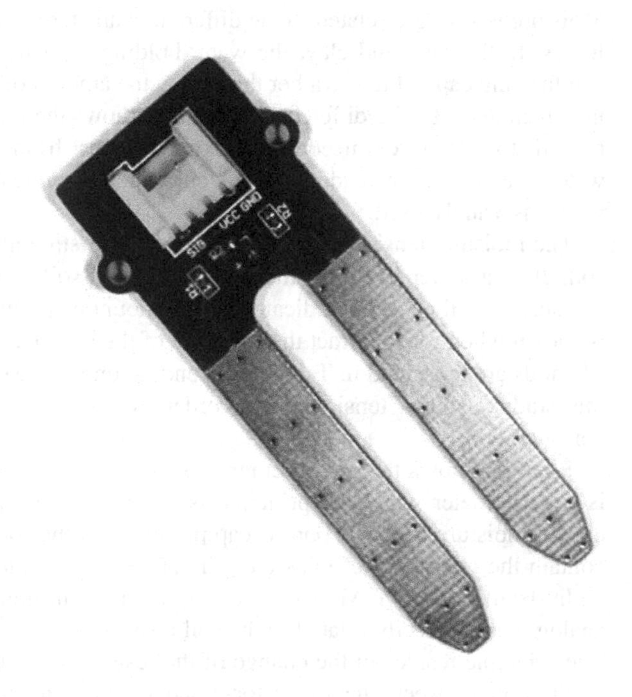

Fig. 4 Tensiometer

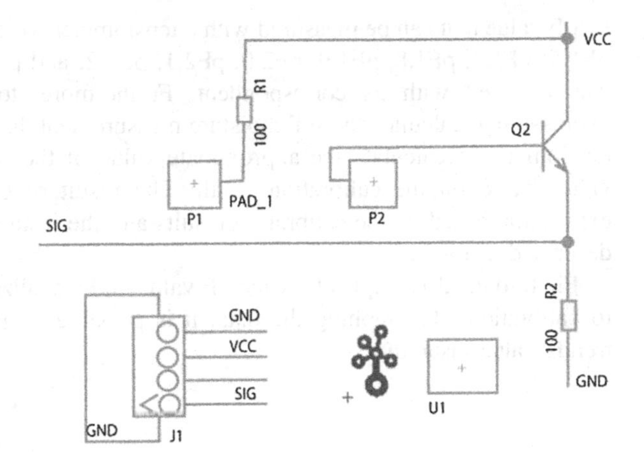

Fig. 5 The electronic schematic of tensiometer

5 Experimental Results

This experiment is to obtain data of the soil moisture conditions in the soil moisture measurement system, in order to have clear values of the measured moisture, the system needs to be calibrated.

Moreover, to calculate the soil moisture measurement for the pF value that could not be measured by a tension meter that was used this time, an approximate expression is obtained based on the calibration result. The calibration process carried out is to adjust the measured value based on the measured value acquired by the measurement instrument used as a reference.

Using the soil moisture measurement system for soil for the experiment that was used in previous session, and acquires the data of the soil moisture conditions. At the same we set up a tensiometer in the same sample soil in order to get the reference value (Fig. 6).

Fig. 6 Tensiometer used for experiment

pF value that can be measured with a tensiometer were divided in nine as, pF1.0, pF1.5, pF1.7, pF1.8, pF1.9, pF2.0, pF2.1, pF2.2, and pF2.3. Each reference value was adjusted with its correspondent. Furthermore, to obtain the approximate expression to calculate the soil moisture measurement that can not be measured by a tensiometer to calculate the approximate value of the soil moisture measurement value based on the calibration result. The result of obtaining the approximate expression based on the calibration results and the results from the tensiometer are described in Fig. 7.

First, to get 30 samples for each pF value and visualize them in order to proceed to calibration. By graphing the data, it is possible to grasp the comparisons and trends value instantly.

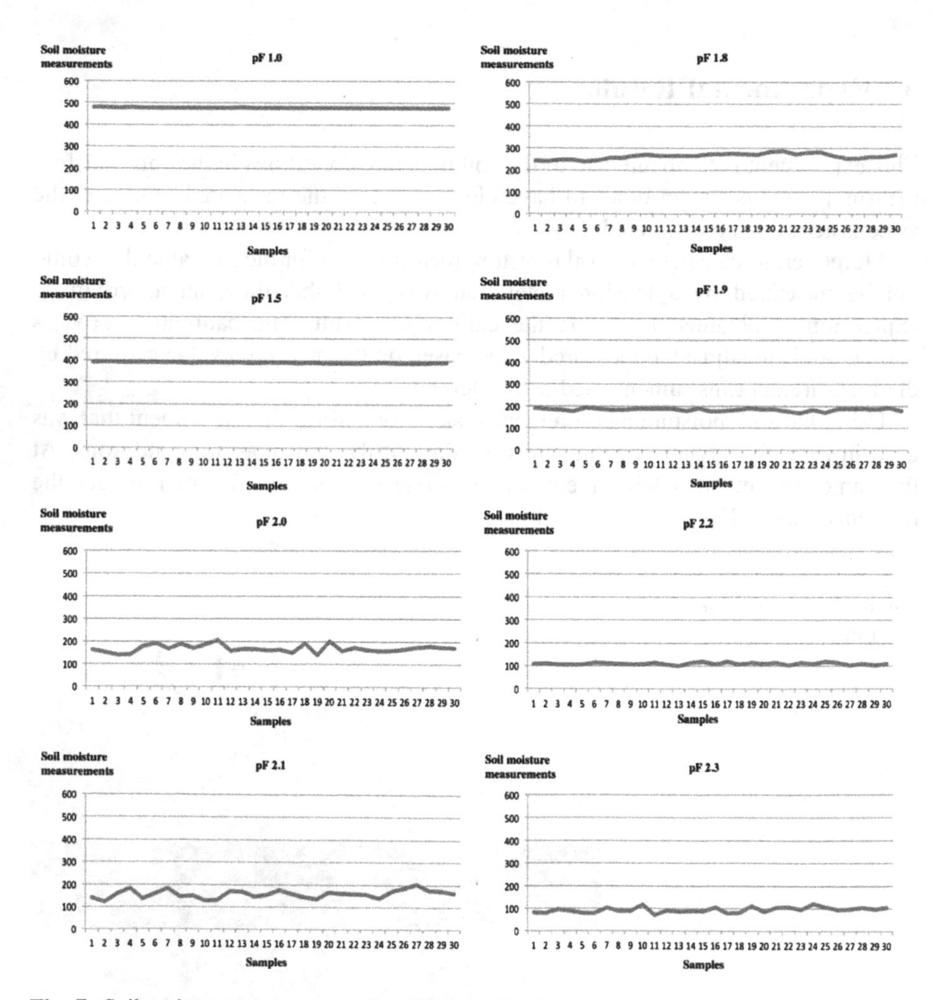

Fig. 7 Soil moisture measurements for different pF values

In this experiment, a soil moisture measurement system was implemented. Experiments of soil moisture measurement system were accord to the soil moisture condition and proper state of the plant to be the experimental subjects.

6 Conclusions

Due to social factors like the reduction of the agriculture worker's population and the reduce number of new workers it is necessary the development of automated systems to cover simple but intrinsically important functions related to the healthy crops growth. In this study, we focus our efforts to solve the management issues of the farming acquisition of technology and to develop an Internet of Agriculture (IoA) cloud-based designed farmer support system, called "AgriMieru".

One main factor related to the healthy crops growth is the presence/absence of moisture or amount of water present in the soil. Moisture is measured based on the strength necessary to obtain water from any given soil. This process could be expensive. For this study purposes it is necessary to develop an affordable system that could perform real time moisture measures in real time. The solution relies in the use of a Grove moisture sensor which is output signal is transmitted to a main board which process it and relies the data to a cloud server. At the same time the measured data will be stored in a cloud service for future or further.

The system needs to be calibrated in order to provide accurate measures to the client, therefore, a sample data is taken in controlled environment and the system is calibrated accordingly. The calibration process is indispensable and important for the purposes of the system but at the same time its efficiency relies in the well known measures of the type of soil to be used.

Acknowledgments This work was supported by the social and system-related industry commercialization support funded by Kumamoto prefecture.

References

1. Seneviratne, S.I., Corti, T., Davin, E.L., Hirschi, M., Jaeger, E.B., Lehner, I., Orlowsky, B., Teuling, A.J.: Investigating soil moisture–climate interactions in a changing climate: a review. Earth Sci. Rev. **99**, 125–161 (2010)
2. Vereecken, H., Huisman, J.A., Bogena, H., Vanderborght, J., Vrugt, J.A., Hopmans, J.W.: On the value of soil moisture measurements in vadose zone hydrology: a review. Water Resour. Res. **44** (2008)
3. Sueyoshi, T.: DEA non-parametric ranking test and index measurement: slack-adjusted DEA and an application to Japanese agriculture cooperatives. Omega **27**(3), 315–326 (1999)
4. van der Meer, C., Yamada, S.: Japanese Agriculture: A Comparative Economic Analysis. Routledge (2005)
5. Yukumoto, O.: Japanese agriculture and agricultural machinery current status and problems. AMA-Agric. Mech. Asia Afr. Latin Am. **40**(1), 72–75 (2011)

Review Paper: Paraconsistent Process Order Control

Kazumi Nakamatsu, Jair Minoro Abe and Seiki Akama

Abstract We have already proposed the paraconsistent process order control method based on an annotated logic program bf-EVALPSN. Bf-EVALPSN can deal with before-after relations between two processes (time intervals) in its annotations, and its reasoning system consists of two kinds of inference rules called the basic bf-inference rule and the transitive bf-inference rule. In this paper, we review how bf-EVALPSN can be applied to process order control with a simple example.

Keywords Paraconsistent annotated logic program · Bf-EVALPSN · Process order control

1 Introduction

We have already proposed the paraconsistent process order control method based on an annotated logic program bf-EVALPSN [4–6]. Bf-EVALPSN can deal with before-after relations between two processes(time intervals) in its annotations, and its reasoning system consists of two kinds of inference rules called the basic bf-inference rule and the transitive bf-inference rule. In this paper, we review how bf-EVALPSN can be applied to process order control with a simple example.

K. Nakamatsu (✉)
University of Hyogo, Himeji, Japan
e-mail: nakamatu@shse.u-hyogo.ac.jp

J.M. Abe
Paulista University, Sao Paulo, Brazil
e-mail: jairabe@uol.com.br

S. Akama
C-republic, Kawasaki, Japan
e-mail: akama@jcom.home.ne.jp

I. Czarnowski et al. (eds.), *Intelligent Decision Technologies 2016*,
Smart Innovation, Systems and Technologies 57,
DOI 10.1007/978-3-319-39627-9_20

2 Annotated Logic Program bf-EVALPSN

In this section a special version of EVALPSN, bf-EVALPSN [4, 5] that can deal with before-after relation between two processes are reviewed briefly. The details of EVALPSN can be found in [3, 4].

In bf-EVALPSN, a special annotated literal $R(p_m, p_n, t) : [(i,j), \mu]$ called *bf-literal* whose non-negative integer vector annotation (i,j) represents the before-after relation between processes Pr_m and Pr_n at time t is introduced. The integer components i and j of the vector annotation (i,j) represent the after and before degrees between processes $Pr_m(p_m)$ and $Pr_n(p_n)$, respectively, and before-after relations are represented paraconsistently in vector annotations.

Definition 1 An extended vector annotated literal, $R(p_i, p_j, t) : [(i,j), \mu]$ is called a bf-EVALP literal or a bf-literal for short, where (i,j) is a vector annotation and $\mu \in \{\alpha, \beta, \gamma\}$. If an EVALPSN clause contains bf-EVALP literals, it is called a bf-EVALPSN clause or just a bf-EVALP clause if it contains no strong negation. A *bf-EVALPSN* is a finite set of bf-EVALPSN clauses.

We provide a paraconsistent before-after interpretation for vector annotations representing bf-relations in bf-EVALPSN, and such a vector annotation is called a *bf-annotation*. Exactly speaking, there are fifteen kinds of bf-relation according to before-after order between four start/finish times of two processes.

Before (`be`)/**After** (`af`) are defined according to the bf-relation between each start time of two processes. If one process has started before/after another one starts, then the bf-relations between them are defined as "before/after", which are represented as the left figure in Fig. 1.

Other kinds of bf-relations are shown as follows.

Disjoint Before (`db`)/**After** (`da`) are defined as there is a time lag between the earlier process finish time and the later one start time, which are described as the right figure in Fig. 1.

Immediate Before (`mb`)/**After** (`ma`) are defined as there is no time lag between the earlier process finish time and the later one start time, which are described as the left figure in Fig. 2.

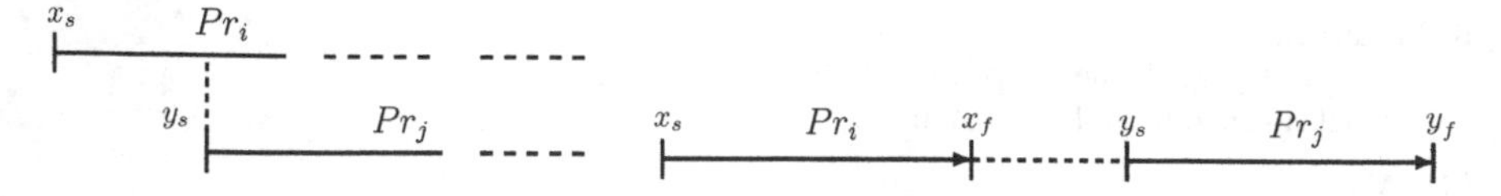

Fig. 1 Before (`be`)/After (`af`) and Disjoint Before (`db`)/After (`da`)

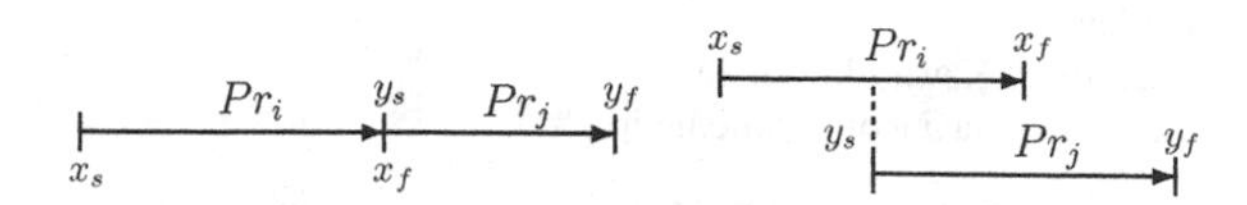

Fig. 2 Immediate Before (`mb`)/After (`ma`) and Joint Before (`jb`)/After (`ja`)

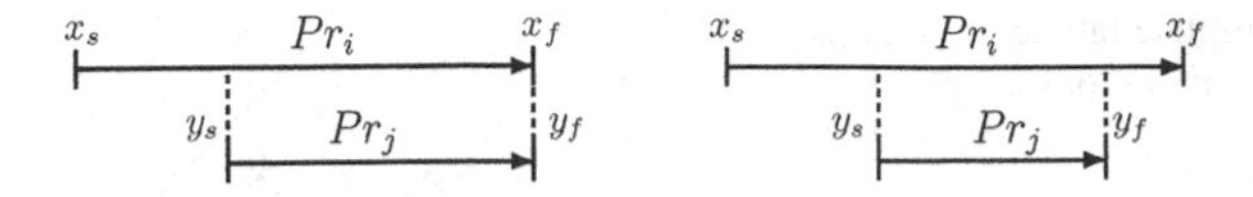

Fig. 3 S-included Before (sb)/After (sa) and Included Before (ib)/After (ia)

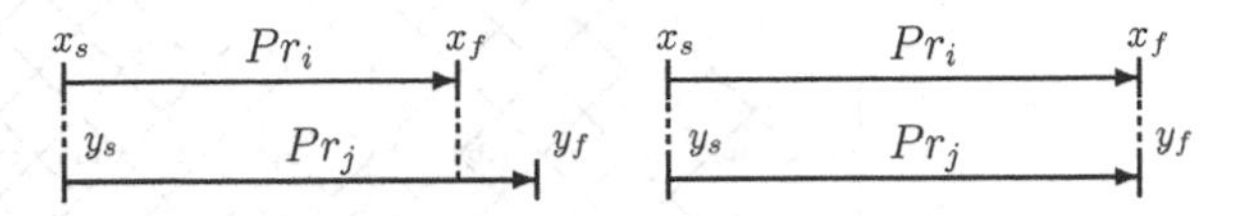

Fig. 4 F-included Before (fb)/After (fa) and Paraconsistent Before-after (pba)

Joint Before (jb)/**After** (ja) are defined as two processes overlap and the earlier process had finished before the later one finished, which are described as the right figure in Fig. 2.

S-included Before (sb), **S-included After** (sa) are defined as one process had started before another one started and they have finished at the same time, which are described as the left figure in Fig. 3.

Included Before (ib)/**After** (ia) are defined as one process had started/finished before/after another one started/finished, which are described as the right figure in Fig. 3.

F-included Before (fb)/**After** (fa) are defined as the two processes have started at the same time and one process had finished before another one finished, which are described as the left figure in Fig. 4.

Paraconsistent Before-after (pba) is defined as two processes have started at the same time and also finished at the same time, which is described as the right figure in Fig. 4.

The epistemic negation over bf-annotations, be, af, db, da, mb, ma, jb, ja, ib, ia, sb, sa, fb, fa, pba is defined and the complete lattice of bf-annotations is shown in Fig. 5.

Definition 2 The epistemic negation $\neg_1$ over the bf-annotations is obviously defined as the following mappings:

$$\neg_1(\texttt{af}) = \texttt{be}, \quad \neg_1(\texttt{be}) = \texttt{af}, \quad \neg_1(\texttt{da}) = \texttt{db}, \quad \neg_1(\texttt{db}) = \texttt{da}, \quad \neg_1(\texttt{ma}) = \texttt{mb},$$
$$\neg_1(\texttt{mb}) = \texttt{ma}, \quad \neg_1(\texttt{ja}) = \texttt{jb}, \quad \neg_1(\texttt{jb}) = \texttt{ja}, \quad \neg_1(\texttt{sa}) = \texttt{sb}, \quad \neg_1(\texttt{sb}) = \texttt{sa},$$
$$\neg_1(\texttt{ia}) = \texttt{ib}, \quad \neg_1(\texttt{ib}) = \texttt{ia}, \quad \neg_1(\texttt{fa}) = \texttt{fb}, \quad \neg_1(\texttt{fb}) = \texttt{fa}, \quad \neg_1(\texttt{pba}) = \texttt{pba}.$$

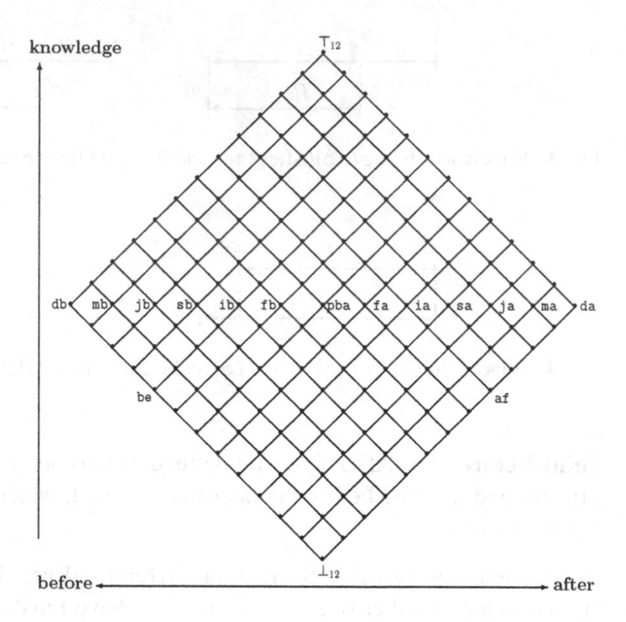

Fig. 5 The complete lattice $\mathcal{T}_v(12)_{bf}$ for Bf-annotations

3 Reasoning Systems in bf-EVALPSN

In order to represent the *basic bf-inference rule* two literals are introduced: $st(p_i, t)$: "process Pr_i starts at time t", and $fi(p_i, t)$: "process Pr_i finishes at time t". Those literals are used for expressing process start/finish information and may have one of the vector annotations, $\{\bot(0,0), \mathtt{t}(1,0), \mathtt{f}(0,1), \top(1,1)\}$.

A group of basic bf-inference rules named (0, 0)-*rules* to be applied at the initial stage (time t_0) are introduced.

(0, 0)-rules Suppose that no process has started yet and the vector annotation of bf-literal $R(p_i, p_j, t)$ is $(0, 0)$, which shows that there is no knowledge in terms of the bf-relation between processes Pr_i and Pr_j, then the following two basic bf-inference rules are applied at the initial stage.

(0, 0)-rule-1 If process Pr_i started before process Pr_j starts, then the vector annotation $(0, 0)$ of bf-literal $R(p_i, p_j, t)$ should turn to $\mathtt{be}(0, 8)$, which is the greatest lower bound of $\{\mathtt{db}(0, 12), \mathtt{mb}(1, 11), \mathtt{jb}(2, 10), \mathtt{sb}(3, 9), \mathtt{ib}(4, 8)\}$.

(0, 0)-rule-2 If both processes Pr_i and Pr_j have started at the same time, then it is reasonably anticipated that the bf-relation between processes Pr_i and Pr_j will be one of the bf-annotations, $\{\mathtt{fb}(5, 7), \mathtt{pba}(6, 6), \mathtt{fa}(7, 5)\}$ whose greatest lower bound is $(5, 5)$ (refer to Fig. 5). Therefore, the vector annotation $(0, 0)$ of bf-literal $R(p_i, p_j, t)$ should turn to $(5, 5)$.

$(0,0)$-rule-1 and $(0,0)$-rule-2 are translated into the bf-EVALPSN,

$$R(p_i,p_j,t):[(0,0),\alpha]\wedge st(p_i,t):[\mathtt{t},\alpha]\wedge\sim st(p_j,t):[\mathtt{t},\alpha]\rightarrow R(p_i,p_j,t):[(0,8),\alpha]$$
$$R(p_i,p_j,t):[(0,0),\alpha]\wedge st(p_i,t):[\mathtt{t},\alpha]\wedge st(p_j,t):[\mathtt{t},\alpha]\rightarrow R(p_i,p_j,t):[(5,5),\alpha]$$

Suppose that $(0,0)$-rule-1 or 2 has been applied, then the vector annotation of bf-literal $R(p_i,p_j,t)$ should be one of $(0,8)$ or $(5,5)$. Therefore, we need to consider two groups of basic bf-inference rules to be applied for following $(0,0)$-rule-1 and 2, which are named $(0,8)$-*rules* and $(5,5)$-*rules*, respectively.

(0, 8)-rules Suppose that process Pr_i has started before process Pr_j, then the vector annotation of bf-literal $R(p_i,p_j,t)$ should be $(0,8)$. We obtain the following inference rules to be applied for $(0,0)$-rule-1.

(0, 8)-rule-1 If process Pr_i has finished before process Pr_j starts, and process Pr_j starts immediately after process Pr_i finished, then the vector annotation $(0,8)$ of bf-literal $R(p_i,p_j,t)$ should turn to $\mathtt{mb}(1,11)$.

(0, 8)-rule-2 If process Pr_i has finished before process Pr_j starts, and process Pr_j has not started immediately after process Pr_i finished, then the vector annotation $(0,8)$ of bf-literal $R(p_i,p_j,t)$ should turn to $\mathtt{db}(0,12)$.

(0, 8)-rule-3 If process Pr_j starts before process Pr_i finishes, then the vector annotation $(0,8)$ of bf-literal $R(p_i,p_j,t)$ should turn to $(2,8)$ that is the greatest lower bound of the set, $\{\mathtt{jb}(2,10),\ \mathtt{sb}(3,9),\ \mathtt{ib}(4,8)\}$.

$(0,8)$-rule-1, 2 and 3 are translated into the bf-EVALPSN,

$$R(p_i,p_j,t):[(0,8),\alpha]\wedge fi(p_i,t):[\mathtt{t},\alpha]\wedge st(p_j,t):[\mathtt{t},\alpha]\rightarrow R(p_i,p_j,t):[(1,11),\alpha]$$
$$R(p_i,p_j,t):[(0,8),\alpha]\wedge fi(p_i,t):[\mathtt{t},\alpha]\wedge\sim st(p_j,t):[\mathtt{t},\alpha]\rightarrow R(p_i,p_j,t):[(0,12),\alpha]$$
$$R(p_i,p_j,t):[(0,8),\alpha]\wedge\sim fi(p_i,t):[\mathtt{t},\alpha]\wedge st(p_j,t):[\mathtt{t},\alpha]\rightarrow R(p_i,p_j,t):[(2,8),\alpha]$$

(5, 5)-rules Suppose that both processes Pr_i and Pr_j have already started at the same time, then the vector annotation of bf-literal $R(p_i,p_j,t)$ should be $(5,5)$. We have the following inference rules to be applied for following $(0,0)$-rule-2.

(5, 5)-rule-1 If process Pr_i has finished before process Pr_j finishes, then the vector annotation $(5,5)$ of bf-literal $R(p_i,p_j,t)$ should turn to $\mathtt{sb}(5,7)$.

(5, 5)-rule-2 If both processes Pr_i and Pr_j have finished at the same time, then the vector annotation $(5,5)$ of bf-literal $R(p_i,p_j,t)$ should turn to $\mathtt{pba}(6,6)$.

(5, 5)-rule-3 If process Pr_j has finished before process Pr_i finishes, then the vector annotation $(5,5)$ of bf-literal $R(p_i,p_j,t)$ should turn to $\mathtt{sa}(7,5)$.

Basic bf-inference rules $(5,5)$-rule-1, 2 and 3 are translated into the following bf-EVALPSN,

$$R(p_i, p_j, t) : [(5,5), \alpha] \wedge fi(p_i, t) : [\mathtt{t}, \alpha] \wedge \sim fi(p_j, t) : [\mathtt{t}, \alpha] \rightarrow R(p_i, p_j, t) : [(5,7), \alpha]$$
$$R(p_i, p_j, t) : [(5,5), \alpha] \wedge fi(p_i, t) : [\mathtt{t}, \alpha] \wedge fi(p_j, t) : [\mathtt{t}, \alpha] \rightarrow R(p_i, p_j, t) : [(6,6), \alpha]$$
$$R(p_i, p_j, t) : [(5,5), \alpha] \wedge \sim fi(p_i, t) : [\mathtt{t}, \alpha] \wedge fi(p_j, t) : [\mathtt{t}, \alpha] \rightarrow R(p_i, p_j, t) : [(7,5), \alpha]$$

If one of $(0, 8)$-rule-1, 2, $(5, 5)$-rule-1, 2 and 3 has been applied, a final bf-annotation such as $\mathtt{jb}(2, 10)$ between two processes should be derived. However, even if $(0, 8)$-rule-3 has been applied, no bf-annotation could be derived. Therefore, a group of basic bf-inference rules named $(2, 8)$-rules should be considered for following $(0, 8)$-rule-3.

(2, 8)-rules Suppose that process Pr_i has started before process Pr_j starts and process Pr_j has started before process Pr_i finishes, then the vector annotation of bf-literal $R(p_i, p_j, t)$ should be $(2, 8)$ and the following three rules should be considered.

(2, 8)-rule-1 If process Pr_i finished before process Pr_j finishes, then the vector annotation $(2, 8)$ of bf-literal $R(p_i, p_j, t)$ should turn to $\mathtt{jb}(2, 10)$.
(2, 8)-rule-2 If both processes Pr_i and Pr_j have finished at the same time, then the vector annotation $(2, 8)$ of bf-literal $R(p_i, p_j, t)$ should turn to $\mathtt{fb}(3, 9)$.
(2, 8)-rule-3 If process Pr_j has finished before Pr_i finishes, then the vector annotation $(2, 8)$ of bf-literal $R(p_i, p_j, t)$ should turn to $\mathtt{ib}(4, 8)$.

Basic bf-inference rules $(2, 8)$-rule-1, 2 and 3 are translated into the bf-EVALPSN,

$$R(p_i, p_j, t) : [(2,8), \alpha] \wedge fi(p_i, t) : [\mathtt{t}, \alpha] \wedge \sim fi(p_j, t) : [\mathtt{t}, \alpha] \rightarrow R(p_i, p_j, t) : [(2,10), \alpha]$$
$$R(p_i, p_j, t) : [(2,8), \alpha] \wedge fi(p_i, t) : [\mathtt{t}, \alpha] \wedge fi(p_j, t) : [\mathtt{t}, \alpha] \rightarrow R(p_i, p_j, t) : [(3,9), \alpha]$$
$$R(p_i, p_j, t) : [(2,8), \alpha] \wedge \sim fi(p_i, t) : [\mathtt{t}, \alpha] \wedge fi(p_j, t) : [\mathtt{t}, \alpha] \rightarrow R(p_i, p_j, t) : [(4,8), \alpha]$$

The application orders of all basic bf-inference rules are summarized in Table 1.

Suppose that there are three processes Pr_i,Pr_j and Pr_k starting sequentially, then we consider to derive the vector annotation of bf-literal $R(p_i, p_k, t)$ from those of

Table 1 Application orders of basic Bf-inference rules

Vector annotation	Rule	Vector annotation	Rule	Vector annotation	Rule	Vector annotation
$(0,0)$	rule-1	$(0,8)$	rule-1	$(0,12)$		
			rule-2	$(1,11)$		
			rule-3	$(2,8)$	rule-1	$(2,10)$
					rule-2	$(3,9)$
					rule-3	$(4,8)$
	rule-2	$(5,5)$	rule-1	$(5,7)$		
			rule-2	$(6,6)$		
			rule-3	$(7,5)$		

bf-literals $R(p_i, p_j, t)$ and $R(p_j, p_k, t)$ transitively. We describe the rules by vector annotations.

Transitive Bf-inference Rules

$\mathbf{TR0}\quad (0,0) \wedge (0,0) \rightarrow (0,0)$

$\mathbf{TR1}\quad (0,8) \wedge (0,0) \rightarrow (0,8)$

$\mathbf{TR1-1}\quad (0,12) \wedge (0,0) \rightarrow (0,12)$

$\mathbf{TR1-2}\quad (1,11) \wedge (0,8) \rightarrow (0,12)$

$\mathbf{TR1-3}\quad (1,11) \wedge (5,5) \rightarrow (1,11)$

$\mathbf{TR1-4}\quad (2,8) \wedge (0,8) \rightarrow (0,8)$

$\mathbf{TR1-4-1}\quad (2,10) \wedge (0,8) \rightarrow (0,12)$

$\mathbf{TR1-4-2}\quad (4,8) \wedge (0,12) \rightarrow (0,8)$

$\mathbf{TR1-4-3}\quad (2,8) \wedge (2,8) \rightarrow (2,8)$

$\mathbf{TR1-4-3-1}\quad (2,10) \wedge (2,8) \rightarrow (2,10)$

$\mathbf{TR1-4-3-2}\quad (4,8) \wedge (2,10) \rightarrow (2,8)$

$\mathbf{TR1-4-3-3}\quad (2,8) \wedge (4,8) \rightarrow (4,8)$

$\mathbf{TR1-4-3-4}\quad (3,9) \wedge (2,10) \rightarrow (2,10)$

$\mathbf{TR1-4-3-5}\quad (2,10) \wedge (4,8) \rightarrow (3,9)$

$\mathbf{TR1-4-3-6}\quad (4,8) \wedge (3,9) \rightarrow (4,8)$

$\mathbf{TR1-4-3-7}\quad (3,9) \wedge (3,9) \rightarrow (3,9)$

$\mathbf{TR1-4-4}\quad (3,9) \wedge (0,12) \rightarrow (0,12)$

$\mathbf{TR1-4-5}\quad (2,10) \wedge (2,8) \rightarrow (1,11)$

$\mathbf{TR1-4-6}\quad (4,8) \wedge (1,11) \rightarrow (2,8)$

$\mathbf{TR1-4-7}\quad (3,9) \wedge (1,11) \rightarrow (1,11)$

$\mathbf{TR1-5}\quad (2,8) \wedge (5,5) \rightarrow (2,8)$

$\mathbf{TR1-5-1}\quad (4,8) \wedge (5,7) \rightarrow (2,8)$

$\mathbf{TR1-5-2}\quad (2,8) \wedge (7,5) \rightarrow (4,8)$

$\mathbf{TR1-5-3}\quad (3,9) \wedge (5,7) \rightarrow (2,10)$

$\mathbf{TR1-5-4}\quad (2,10) \wedge (7,5) \rightarrow (3,9)$

$\mathbf{TR2}\quad (5,5) \wedge (0,8) \rightarrow (0,8)$

$\mathbf{TR2-1}\quad (5,7) \wedge (0,8) \rightarrow (0,12)$

$\mathbf{TR2-2}\quad (7,5) \wedge (0,12) \rightarrow (0,8)$

$\mathbf{TR2-3}\quad (5,5) \wedge (2,8) \rightarrow (2,8)$

$\mathbf{TR2-3-1}\quad (5,7) \wedge (2,8) \rightarrow (2,10)$

$\mathbf{TR2-3-2}\quad (7,5) \wedge (2,10) \rightarrow (2,8)$

$\mathbf{TR2-3-3}\quad (5,5) \wedge (4,8) \rightarrow (4,8)$

$\mathbf{TR2-3-4}\quad (7,5) \wedge (3,9) \rightarrow (4,8)$

$\mathbf{TR2-4}\quad (5,7) \wedge (2,8) \rightarrow (1,11)$

$\mathbf{TR2-5}\quad (7,5) \wedge (1,11) \rightarrow (2,8)$

$\mathbf{TR3}\quad (5,5) \wedge (5,5) \rightarrow (5,5)$

$\mathbf{TR3-1}\quad (7,5) \wedge (5,7) \rightarrow (5,5)$

$\mathbf{TR3-2}\quad (5,7) \wedge (7,5) \rightarrow (6,6)$

4 Process Order Control in Bf-EVALPSN

In this section, a simple example of the process order control is shown. The process order control method has the following steps: **step 1**, translate the safety properties of the process order control system into bf-EVALPSN; **step 2**, verify if permission for starting the process can be derived from the bf-EVALPSN in **step1** by the basic bf-inference rule and the transitive bf-inference rule or not.

We assume a pipeline system consists of two pipelines, PIPELINE-1 and 2, which deal with pipeline processes Pr_0, Pr_1, Pr_2 and Pr_3. The process schedule of those processes are shown in Fig. 6. Moreover, we assume that the pipeline system has four safety properties $SPR-i(i=0,1,2,3)$.

SPR–0 process Pr_0 must start before any other processes, and process Pr_0 must finish before process Pr_2 finishes,
SPR–1 process Pr_1 must start after process Pr_0 starts,
SPR–2 process Pr_2 must start immediately after process Pr_1 finishes,
SPR–3 process Pr_3 must start immediately after processes Pr_0 and Pr_2 finish.

Step 1. All safety properties $SPR-i(i=0,1,2,3)$ can be translated into the following bf-EVALPSN clauses.

$$SPR-1$$

$$\sim R(p_0,p_1,t):[(0,8),\alpha] \rightarrow st(p_1,t):[\mathtt{f},\beta], \quad (1)$$

$$\sim R(p_0,p_2,t):[(0,8),\alpha] \rightarrow st(p_2,t):[\mathtt{f},\beta], \quad (2)$$

$$\sim R(p_0,p_3,t):[(0,8),\alpha] \rightarrow st(p_3,t):[\mathtt{f},\beta], \quad (3)$$

$$st(p_1,t):[\mathtt{f},\beta] \wedge st(p_2,t):[\mathtt{f},\beta] \wedge st(p_3,t):[\mathtt{f},\beta] \rightarrow st(p_0,t):[\mathtt{f},\gamma], \quad (4)$$

$$\sim fi(p_0,t):[\mathtt{f},\beta] \rightarrow fi(p_0,t):[\mathtt{f},\gamma]. \quad (5)$$

$$SPR-1$$

$$\sim st(p_1,t):[\mathtt{f},\beta] \rightarrow st(p_1,t):[\mathtt{f},\gamma], \quad (6)$$

$$\sim fi(p_1,t):[\mathtt{f},\beta] \rightarrow fi(p_1,t):[\mathtt{f},\gamma]. \quad (7)$$

$$SPR-2$$

$$\sim R(p_2,p_1,t):[(11,0),\alpha] \rightarrow st(p_2,t):[\mathtt{f},\beta], \quad (8)$$

$$\sim st(p_2,t):[\mathtt{f},\beta] \rightarrow st(p_2,t):[\mathtt{f},\gamma], \quad (9)$$

$$\sim R(p_2,p_0,t):[(10,2),\alpha] \rightarrow fi(p_2,t):[\mathtt{f},\beta], \quad (10)$$

$$\sim fi(p_2,t):[\mathtt{f},\beta] \rightarrow fi(p_2,t):[\mathtt{f},\gamma]. \quad (11)$$

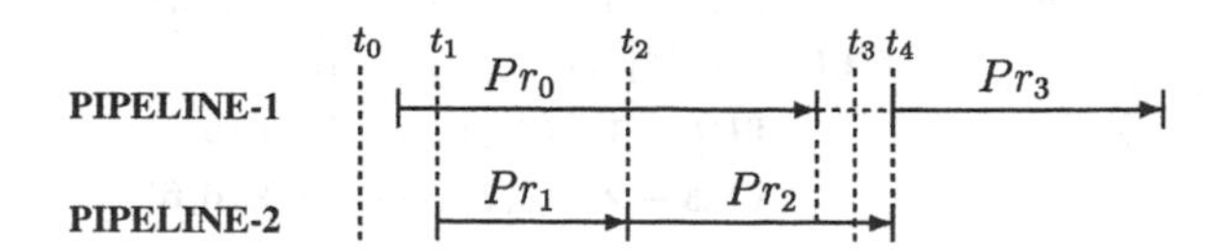

Fig. 6 Pipeline process schedule

$$SPR-3$$

$$\sim R(p_3, p_0, t) : [(11, 0), \alpha] \rightarrow st(p_3, t) : [\mathtt{f}, \beta], \quad (12)$$

$$\sim R(p_3, p_1, t) : [(11, 0), \alpha] \rightarrow st(p_3, t) : [\mathtt{f}, \beta], \quad (13)$$

$$\sim R(p_3, p_2, t) : [(11, 0), \alpha] \rightarrow st(p_3, t) : [\mathtt{f}, \beta], \quad (14)$$

$$\sim st(p_3, t) : [\mathtt{f}, \beta] \rightarrow st(p_3, t) : [\mathtt{f}, \gamma], \quad (15)$$

$$\sim fi(p_3, t) : [\mathtt{f}, \beta] \rightarrow fi(p_3, t) : [\mathtt{f}, \gamma]. \quad (16)$$

Step 2. Here, we show how the bf-EVALPSN process order safety verification is carried out at five time points, t_0, t_1, t_2 and t_3 in the process schedule (Fig. 6). We consider five bf-relations between processes Pr_0, Pr_1, Pr_2 and Pr_3 represented by the vector annotations of bf-literals,

$$R(p_0, p_1, t), \quad R(p_0, p_2, t), \quad R(p_0, p_3, t), \quad R(p_1, p_2, t), \quad R(p_2, p_3, t)$$

which should be verified based on safety properties $SPR-0$, 1, 2 and 3 in real-time.

Initial Stage (at time t_0) no process has started at time t_0, thus, the bf-EVALP clauses,

$$R(p_i, p_j, t_0) : [(0, 0), \alpha], \quad \text{where } i = 0, 1, 2, j = 1, 2, 3 \quad (17)$$

are obtained by transitive bf-inference rule **TR0**; then, bf-EVALP clauses (17) satisfy each body of bf-EVALPSN clauses (1), (2) and (3), respectively, therefore, the forbiddance,

$$st(p_1, t_0) : [\mathtt{f}, \beta], \quad (18)$$

from starting each process $Pr_i (i = 1, 2, 3)$ is derived; moreover, since bf-EVALP clauses (18) satisfy the body of bf-EVALPSN clause (4), the permission for starting process Pr_0, $st(p_0, t_0) : [\mathtt{f}, \gamma]$ is derived; therefore, process Pr_0 is permitted for starting at time t_0.

2nd Stage (at time t_1) process Pr_0 has already started but all other processes $Pr_i (i = 1, 2, 3)$ have not started yet; then the bf-EVALP clauses,

$$R(p_0, p_1, t_1) : [(0, 8), \alpha], \quad (19)$$

are obtained, where the bf-EVALP clause (19) is derived by basic bf-inference rule (0, 0)-rule-1; moreover, the bf-EVALP clauses,

$$R(p_0, p_2, t_1) : [(0, 8), \alpha], \qquad R(p_0, p_3, t_1) : [(0, 8), \alpha] \quad (20)$$

are obtained by transitive bf-inference rule TR1; as bf-EVALP clause (19) does not satisfy the body of bf-EVALPSN clause (1), the forbiddance from starting process Pr_1, $st(p_1, t_1) : [\mathtt{f}, \beta]$ cannot be derived; then, since there does not exist the forbiddance, the body of bf-EVALPSN clause (6) is satisfied, and the permission for starting process Pr_1, $st(p_1, t_1) : [\mathtt{f}, \gamma]$ is derived; on the other hand, since bf-EVALP clauses (20) satisfy the body of bf-EVALPSN clauses (8) and (12) respectively, the forbiddance from starting both processes Pr_2 and Pr_3,

$$st(p_2, t_1) : [\mathtt{f}, \beta], \quad st(p_3, t_1) : [\mathtt{f}, \beta] \tag{21}$$

are derived; therefore, process Pr_1 is permitted for starting at time t_1.

3rd Stage (at time t_2) process Pr_1 has just finished and process Pr_0 has not finished yet; then, the bf-EVALP clauses,

$$R(p_0, p_1, t_2) : [(4, 8), \alpha], \quad R(p_1, p_2, t_2) : [(1, 11), \alpha], \quad R(p_2, p_3, t_2) : [(0, 8), \alpha] \tag{22}$$

are derived by basic bf-inference rules (2, 8)-rule-3, (0, 8)-rule-2 and (0, 0)-rule-1, respectively; moreover, the bf-EVALP clauses,

$$R(p_0, p_2, t_2) : [(2, 8), \alpha], \qquad R(p_0, p_3, t_2) : [(0, 12), \alpha] \tag{23}$$

are obtained by transitive bf-inference rules **TR1-4-6** and **TR1-2**, respectively; then, since bf-EVALP clause (22) does not satisfy the body of bf-EVALPSN clause (8), the forbiddance from starting process Pr_2, $st(p_2, t_2) : [\mathtt{f}, \beta]$ cannot be derived; since there does not exist the forbiddance, the body of bf-EVALPSN clause (9) is satisfied, and the permission for starting process Pr_2, $st(p_2, t_2) : [\mathtt{f}, \gamma]$ is derived; on the other hand, since bf-EVALP clauses (23) satisfy the body of bf-EVALPSN clause (12), the forbiddance from starting process Pr_3, $st(p_3, t_2) : [\mathtt{f}, \beta]$ is derived; therefore, process Pr_2 is permitted for starting, however process Pr_3 is still forbidden from starting at time t_2.

4th Stage (at the t_3) process Pr_0 has finished, process Pr_2 has not finished yet, and process Pr_3 has not started yet; then, the bf-EVALP clauses,

$$R(p_0, p_1, t_3) : [(4, 8), \alpha], \quad R(p_1, p_2, t_3) : [(1, 11), \alpha], \quad R(p_2, p_3, t_3) : [(0, 8), \alpha]$$

have been already reasoned at the previous stage; moreover, the bf-EVALP clauses,

$$R(p_0, p_2, t_3) : [(2, 10), \alpha], \qquad R(p_0, p_3, t_3) : [(0, 12), \alpha] \tag{24}$$

are obtained by basic bf-inference rule (2, 8)-rule-1; then, bf-EVALP clause (24) satisfies the body of bf-EVALP clause (14), and the forbiddance from starting process Pr_3, $S(p_3, t_3) : [\mathtt{f}, \beta]$ is derived; therefore, process Pr_3 is still forbidden to start because process Pr_2 has not finished yet at time t_3.

5 Concluding Remarks

In this paper, we have reviewed the process order control method based on a paraconsistent annotated logic program bf-EVALPSN, which can deal with before-after relation between processes with a small process order safety verification example.

References

1. Blair, H.A., Subrahmanian, V.S.: Paraconsistent logic programming. Theoret. Comput. Sci. **68**, 135–154 (1989)
2. da Costa, N.C.A., et al.: The Paraconsistent logics $P\mathcal{T}$. Zeitschrift für Mathematische Logic und Grundlangen der Mathematik **37**, 139–148 (1989)
3. Nakamatsu, K., et al.: Annotated Semantics for Defeasible Deontic Reasoning. LNAI, vol. **2005**, pp. 432–440. Springer (2001)
4. Nakamatsu, K., Abe, J.M.: The development of paraconsistent annotated logic program. Int. J. Reasoning-Based Intell. Syst. **1**, 92–112 (2009)
5. Nakamatsu, K., Abe, J.M., Akama, S.: A logical reasoning system of process before-after relation based on a paraconsistent annotated logic program bf-EVALPSN. J. Knowl. Based Intell. Eng. Syst. **15**, 145–163 (2011)
6. Nakamatsu, K., Abe, J.M.: The paraconsistent process order control method. Vietnam J. Comput. Sci. **1**, 29–37 (2014)

Recent Advances in Fuzzy Systems

Modeling and Forecasting of Well-Being Using Fuzzy Cognitive Maps

Tatiana Penkova and Wojciech Froelich

Abstract In this paper we address the problem of modeling and forecasting of well-being. First, we apply a graph-based model of a Fuzzy cognitive map to discover cause-and-effect relationships among indicators of well-being. Second, the discovered model is applied to forecast the future state of well-being. The model is constructed using historical multivariate time series containing six consolidated indexes that represent well-being on the considered territory. Experiments with real-world data provided evidence for the usefulness of the proposed approach. Moreover, the interpretation of the obtained FCM graph led to the discovery of unknown dependencies within the data. The analysis of the unknown dependencies requires further research.

1 Introduction

Estimation of well-being is an important problem and a key factor supporting decision-making processes in territory management. The estimation of well-being has raised the active interest of several researchers [3–5, 7, 14, 27, 28]. The estimation of a synthetic well-being index was proposed in [21] where the weighting of the partial indicators was used. The creation of the territory well-being standard includes the following: identification of hierarchy of indicators (i.e., the set of primary indicators and levels of their aggregation), identification of significance coefficients of indicators and identification of normative values of indicators in form of range. This process is performed by experts using historical data based on territory characteristics and specifications [27]. The method of estimation of the territory well-being level is an improvement on the approach to estimation of complex socio-economic

T. Penkova (✉)
Institute of Computational Modelling SB RAS, Krasnoyarsk, Russia
e-mail: penkova_t@icm.krasn.ru

W. Froelich
The University of Silesia, ul. Bedzinska 39, Sosnowiec, Poland
e-mail: wojciech.froelich@us.edu.pl

I. Czarnowski et al. (eds.), *Intelligent Decision Technologies 2016*,
Smart Innovation, Systems and Technologies 57,
DOI 10.1007/978-3-319-39627-9_21

objects [4]. The theory and estimation of individual and social welfare measures was proposed in [3]. The social context of well-being was investigated in [14]. The constriction of well-being indexes has also been investigated in [5, 7].

The approach for the modeling and forecasting of well-being proposed in this paper is based on the application of the soft-computing model of Fuzzy cognitive maps. A Fuzzy cognitive map is a knowledge representation tool inheriting different aspects of fuzzy sets and neural networks [8, 18]. FCMs model knowledge through fuzzy concepts represented as nodes and relationships between them represented as weighted arcs. The causal relationships among concepts are either determined by experts knowledge or by learning when historical data are available. There is a growing interest in FCMs, especially in the fields of control [30], medicine [29], computer science [20], time series forecasting [10, 11, 15, 23, 24], decision support [31], and machine learning [17]. A review of FCM research is given in [26].

In this paper we address the problem of modeling and forecasting of well-being. The well-being indexes are changing over time and constitute the considered time series. First, we apply a Fuzzy cognitive map to discover cause-and-effect relationships among indicators of well-being. Second, the discovered model is applied to forecast the future course of well-being. The forecasted multivariate time series contain six comprehensive indexes representing well-being in the considered territory. Experiments with real-world data provide evidence for the usefulness of the proposed approach. Moreover, the interpretation of the obtained FCM graph led to the discovery of unknown dependencies within the data.

The rest of this paper is organized as follows. In Sect. 2 the theoretical background on the estimation of well-being is presented. Theoretical background related to Fuzzy cognitive maps is given in Sect. 3. The contribution of this study, i.e., the application of FCM to the modeling and forecasting of well-being, is presented in Sect. 4. Experiments using real-world data are described in Sect. 5. Section 6 concludes the paper.

2 Estimation of Well-Being

Let us assume that $P_k \in \Re$, $k = 1, 2, \ldots, n$ denotes real valued indicators related to well-being (e.g., 'Alcoholism', 'Drug addiction', etc.) where is the number of all considered indicators. The estimation of well-being is based on the values of indicators and consists of the following two steps:

1. Estimation of primary indicators by the calculation of individual well-being indexes i_k. The value of an individual index demonstrates significant improvement if $i_k > 1$. If $i_k < 1$, then it demonstrates significant degradation of the indicator.
2. Estimation of comprehensive indicators by the calculation of consolidated well-being indexes I applying previously calculated individual indexes. The value of

the consolidated index is $I > 1$ or $I < 1$, which demonstrates an improvement in or the degradation of the well-being level, respectively.

Every individual index of the kth indicator is calculated by Formula (1):

$$i_k = 1 + \Delta P_k \cdot S_k, \tag{1}$$

where: ΔP_k is the compliance coefficient of actual values of the kth indicator with standard; $S_k = \pm 1$ is the coefficient which characterizes the 'polarity' of kth indicator, where: $S_k = 1$, when the change of indicator is proportional to the index and $S_k = -1$, when the change of indicator is inversely proportional to the index. The compliance coefficient ΔP_k is calculated using Formula (2):

$$\Delta P_k = \begin{cases} 0, & \text{for: } P_k \in [N_k, Z_k], \\ \frac{P_k - Z_k}{Z_k - N_k}, & \text{for: } P_k > Z_k, \\ \frac{P_k - N_k}{Z_k - N_k}, & \text{for: } P_k < N_k, \end{cases} \tag{2}$$

where: P_k is actual value of the kth indicator; $[N_k, Z_k]$ is the range of normative values of kth indicator, N_k is the lower limit of the range, and Z_k is the upper limit of the range. In cases where the value of the indicator P_k falls within the range of normative values, i.e., $P_k \in [N_k, Z_k]$, the compliance coefficient $\Delta P_k = 0$. In cases where actual value of indicator is above the upper limit of range, the compliance coefficient has a positive value $\Delta P_k > 0$. In cases where the actual value of the indicator is below the lower limit of range, the compliance coefficient has a negative value $\Delta P_k < 0$.

After the calculation of all individual indexes i_k, the consolidated well-being index I is calculated by Formula (3):

$$I = \sum_{k=1}^{n} u_k \cdot i_k, \tag{3}$$

where: $u_k > 0$ is a significance coefficient of kth indicator, where it is assumed that $\sum u_k = 1$, and n is the number of all individual indicators.

To illustrate the described calculation procedure, we provide a numerical example. Table 1 presents individual indicators used to calculate the comprehensive indicator of 'psycho-emotional tension'.

Table 1 Individual indexes for the consolidated index 'Psycho-emotional tension'

k	Indicator	P_k	N_k	Z_k	ΔP_k	i_k	u_k	S_k
1	Children's drug addiction	7.16	4.20	7.20	0	1.00	0.21	−1
2	Teenage drug addiction	180.56	120.00	150.00	1.02	−0.02	0.20	−1
3	Drug addiction	265.38	200.00	250.00	0.31	0.69	0.15	−1
4	Alcoholism	1333.63	1100.00	1300.00	0.17	0.83	0.15	−1

First, according to Formula (2), for all primary indicators P_k, we calculate compliance coefficients ΔP_k. For example, as can be noted in Table 1, the actual value of 'Children's drug addiction' falls within the normative range, therefore the coefficient of compliance is identified as: $\Delta P = 0$. The actual values of other indicators fall above the upper limit of the range; therefore, the compliance coefficient is calculated according to the second condition in Formula (2).

Second, we calculate the individual well-being indexes i_k according to Formula (1). Taking into account the negative polarity S_k for 'Teenage drug addiction', the corresponding individual index is calculated as: $i_2 = 1 + 1.02 \cdot (-1) = -0.02$.

In the last column of Table 1 we placed significance indexes corresponding to the individual indexes. Using the values of individual indexes and applying Formula 3, we calculate the consolidated well-being index. (I) for 'Psycho-emotional tension' as $I = 0.21 \cdot 1.00 + 0.20 \cdot (-0.02) + 0.17 + 0.15 \cdot 0.69 + 0.15 \cdot 0.83 = 0.23426$.

This method provides the estimation of well-being by assessing the changes in the indicators values relative to their normative values.

3 Introduction to Fuzzy Cognitive Maps

Let us assume we observe real-valued variables $v_1, v_2, \ldots, v_n \in V$, where: V is the set of the considered well-being indicators. Let C denotes a set of fuzzy sets, where every set $c \in C$ is a node of the FCM. At time step $t \in [0, 1, \ldots, t_e], t_e \in \aleph$, the value of $v_i(t)$ is mapped by the fuzzification function to the state of the corresponding concept $c_i(t) = \mu_i(v_i(t))$. The value of $c_i(t)$ is the degree in which $v_i(t)$ belongs to the fuzzy set c_i. The fuzzification is usually simplified as a normalization: $c_i(t) = \frac{v_i(t) - \min(v_i)}{\max(v_i) - \min(v_i)}$.

The FCM is defined as an ordered pair $< C, W >$, where C is the set of concepts and W is the connection matrix that stores the weights $w_{ij} \in [-1, 1]$ assigned to the pairs of concepts. The value $w_{ij} = 1$ expresses full positive and $w_{ij} = -1$ full negative impact of the ith causal concept on the jth effect concept respectively. The intermediate values of weights refer to partial causality [8].

The FCM model can be exploited for the prediction of a concept's states $c_i'(t)$ and, after their defuzzification, the corresponding values of varibles $v_i'(t)$. The prediction is carried out using Eq. (4):

$$c_j'(t) = f\left(\sum_{i=1, i \neq j}^{n} w_{ij} c_i(t-1)\right), \tag{4}$$

where $n = card(C)$ is the cardinality of set $C, f(x)$ is the transformation function. The transformation function restricts the weighted sum of concepts states into the interval $[0, 1]$. For the purpose of this study, we use the logistic transformation: $f(x) = \frac{1}{1+e^{-gx}}$, where $g > 0$ is the parameter that determines the gain of the transformation.

After performing the prediction of the concepts state, to obtain the predicted values of variables $v'_j(t)$, denormalization is performed by using the formula: $v'_j(t) = c'_j(t)(\max(v_j) - \min(v_j)) + \min(v_j)$.

For the purpose of this paper we decided to apply Mean Absolute Error (MAE), the simplest approach to the calculation of forecasting errors, which is given by the following formula:

$$e = \frac{1}{n \cdot card(T)} \sum_{t}^{card(T)} \sum_{j=1}^{n} |v'_j(t) - v_j(t)|, \quad (5)$$

where T denotes the considered (learning or testing) period in which the errors were accumulated, $card(T)$ is the length of the considered period of time calculated in time steps, and n is the number of variables. $v'_j(t)$, $v_j(t)$ denotes the predicted and actual values of the time series, respectively.

The set of concepts C is provided by an expert, and only the matrix W is learned, using historical data. There are two known approaches to learning FCMs: adaptive and population-based. Adaptive algorithms are based on the idea of Hebbian learning borrowed from the theory of artificial neural networks. The adaptive learning methods involve: DHL [19], BDA [16], AHL [22] and other algorithms. The population-based approaches for learning FCMs are: RCGA (real coded genetic algorithm) [32], PSO-based algorithm (applies particle swarm optimization method) [25], simulated annealing optimization-based algorithm [12], and differential evolution-based algorithm [17].

As reported in the literature [9], the RCGA is one of the most competitive among the population-based. For that reason, it has been selected to be used in this study.

4 An FCM-Based Model of Well-Being

The first goal of this study is to create an FCM model representing the dependencies among consolidated indexes of well-being. The resulting FCM model is applied as a decision support tool for policy makers but also as a predictive model allowing the forecasting of well-being. To accomplish the aforementioned objective we map the considered consolidated indexes to the concepts of Fuzzy cognitive maps. The mapping is shown in Table 2.

Table 2 Mapping between consolidated indexes and the concepts of FCM

Index	Concept
Population structure	c_1
Labour market	c_2
Housing facilities	c_3
Standard of living	c_4
Psycho-emotional tension	c_5
Medical provision	c_6

As stated in Sect. 2, all consolidated indexes assume values in the range [0, 1]; therefore, the normalization of the original and denormalization of the forecasted time series are not required.

For the purpose of this paper, the evolutionary approach based on the RCGA is applied. The RCGA creates the population of genotypes; each of them is a vector of weights of a candidate FCM. The goal of the evolutionary algorithm is to optimize the matrix W with respect to the predictive capability of the FCM. The applied RCGA algorithm relies on the template of a genetic algorithm (Algorithm 1).

Algorithm 1: Genetic Learning of FCM.

Input: Multivariate time series $\{V(1), V(2), \ldots, V(t_e)\}$.
Output: Optimized matrix: W.

Initialize randomly the first population $P_k, k = 1$ of genotypes;
While (stopping-criterion is not satisfied) {
  $Evaluation(P_k)$;
  $P_{k+1} \leftarrow Selection(P_k)$;
  $Mutation(P_{k+1})$;
  $Crossover(P_{k+1})$;
  $k \leftarrow k + 1$;
}
return $p_{best} \in P_k$ - the genotype with the highest fitness value;

Index k denotes the number of generations. The constituents of the algorithm are the following:

- **Genotype**. Every genotype $p \in P$ includes the vector of numbers coming from the matrix W of the candidate FCM. Subsequent rows of W are placed linearly one after the other into the vector of genotype. The elements on the diagonal of the matrix W are omitted, as they do not take part in reasoning.
- **Evaluation of Genotypes**. To use the RCGA for learning FCMs, we defined the fitness function as $fitness(FCM) = -e$, where e is the accumulated forecasting error calculated for the learning period.
- **Selection**. During the selection process, a new population P_{k+1} of genotypes is produced. The newly created population is later supplemented using the operators of mutation and crossover. For the purpose of this paper the elite selection is applied [13].
- **Mutation and crossover**. To supplement the population, the offspring of the elite genotypes are produced using standard probabilistic mutation and one-point crossover. The probabilities of mutation and crossover are the parameters of the evolution.
- **Stopping-criteria**. The algorithm stops when at least one of the following conditions holds:

1. no improvement in the best fitness value has been recognized after k_{run} consecutive generations, k_{run} is the parameter,
2. the maximum number of generations k_{max} has been reached, k_{max} is the parameter.

5 Experiments

For the validation of the proposed approach, we use real-world data gathered for an industrial city in Siberia (Novokuznetsk, Russia). The data, given in Table 3, are publicly available [1, 2, 6]. For the purpose of our experiments the data has been divided into learning and testing parts. The learning part contained data from five years (2005–2010). Testing was performed for three years (2011–2013).

For the learning FCM we used the RCGA algorithm with the following parameters: cardinality of the population = 100, maximal number of iterations k_{max} = 500, number of iterations without the change of fitness k_{run} = 10, probability of mutation = 0.1, and probability of crossover = 0.8. After numerous trials we set up the gain of the transformation function as $g = 0.5$, and the cardinality of the elite population to 20 %. In Table 4 we present the obtained values of FCM weights.

As can be noted in Table 4, high bidirectional dependency has been found for the concepts $c_1 - c_2, c_1 - c_3, c_1 - c_4$. This means that the population structure is highly positively related to the labor market, housing facilities and standard of living. On the other hand, fairly high negative weights have been recognized for the dependencies related to the concept of psycho-emotional tension. Increased standard of living may lead to decreased psycho-emotional tension. w_{45}: 'Standard of living' (c_4) $->$ 'Psycho-emotional tension' (c_5). Moreover, decreased psycho-emotional tension improve the indicator related to the population structure. w_{51}: 'Psycho-emotional tension' (c_5) $->$ 'Population structure' (c_1).

Further analysis of the obtained dependecies is a challenge for future research and should be made by the domain specialists. To present the obtained FCM to the domain experts we illustrate it in Fig. 1. We show only the weights $w_{ij} \geq 0.8$ and $w_{ij} < -0.2$

Table 3 Well-being indexes of the comprehensive indicators (Novokuznetsk, 2005–2013)

Indicator	2005	2006	2007	2008	2009	2010	2011	2012	2013
Population structure	0.58	0.78	0.84	0.95	0.85	0.75	0.85	0.8	0.82
Labour market	0.8	0.84	0.82	0.8	0.73	0.25	0.6	0.74	0.76
Housing facilities	0.75	0.81	0.81	0.75	0.74	0.7	0.85	0.87	0.8
Standard of living	0.67	0.71	0.74	0.82	0.76	0.68	0.74	0.75	0.73
Psycho-emotional tension	0.6	0.54	0.32	0.56	0.46	0.46	0.64	0.74	0.76
Medical provision	0.67	0.8	0.87	0.95	0.91	0.7	0.85	0.87	0.86

Table 4 Causal dependencies among comprehensive indicators

Concept	c_1	c_2	c_3	c_4	c_5	c_6
Population structure c_1	–	0.90	0.96	0.82	0.57	0.87
Labour market c_2	0.84	–	0.66	0.29	0.58	−0.03
Housing facilities c_3	0.87	0.67	–	0.59	−0.03	−0.06
Standard of living c_4	0.80	0.87	0.15	–	−0.27	0.30
Psycho-emotional tension c_5	−0.28	0.06	0.03	0.07	–	−0.05
Medical provision c_6	−0.07	0.14	−0.04	0.9	0.04	–

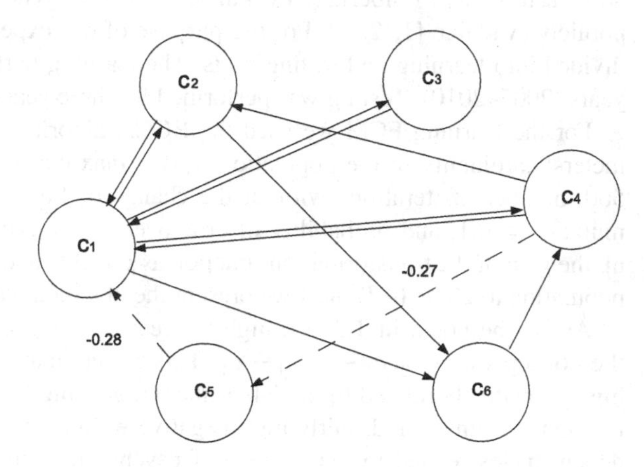

Fig. 1 FCM model of well-being

Table 5 Mean absolute errors

2011	2012	2013
0.0389	0.0510	0.0382

The obtained model has been applied to forecasting future state of well-being. The forecasting accuracies calculated as MAPE for every year within the testing period are given in Table 5.

Taking into account that the obtained errors are accumulated over six variables of the multivariate time series, they can be evaluated as very low. They are also satisfactory from the considered domain of application. This provides evidence that the applied FCM model can be effectively used for modeling and forecasting of well-being.

6 Conclusions

In this study we proposed a new FCM-based model of well-being. The model has been constructed using real-world data. The discovered model graphically illustrates the dependencies between the consolidated well-being indexes and as such can be a valuable decision support tool for policy makers. Moreover, the obtained model has been applied to the forecasting of the future course of well-being. The obtained results are very encouraging and thus motivate our further research in the considered domain. The limitation of the demonstrated approach is a small amount of available data. In spite of that, the paper proposes a general approach that can be easily scaled up in the future.

Acknowledgments The reported study was funded by RFBR according to the research project No. 16-37-00014-mol_a.

References

1. Collection of Statistical Reports for 2005-2013 Years: Socio-economic situation of the city of Novokuznetsk. Federal State Statistic Service of Kemerovo region, Department of State Statistics of Novokuznetsk. Novokuznetsk (in Russian)
2. Federal State Statistic Service of Russian Federation. http://www.gks.ru
3. Becht, M.: The theory and estimation of individual and social welfare measures. J. Econ. Surv. **9**(1), 53–87 (1995)
4. Borisova, E.: Index method of comprehensive quantitative estimation of the quality of complex object. In: II International Conference: Mathematics, Computer, Education, Izhevsk, Russia, vol. 1, pp. 249–259 (2005)
5. Brenger, V., Verdier-Chouchane, A.: Multidimensional measures of well-being: standard of living and quality of life across countries. World Dev. **35**(7), 1259–1276 (2007)
6. Chechenin, G.: State of Public Health and Environment of Novokuznetsk in 2000-2013. In: Collective Monograph State of Public Health and Environment of Novokuznetsk. Automated System Socio-Hygienic Monitoring, Novokuznetsk State Institute of Postgraduate Medicine (2015) (in Russian)
7. Diamantopoulos, A., Winklhofer, H.M.: Index construction with formative indicators: an alternative to scale development. J. Mark. Res. **38**(2), 269–277 (2001)
8. Dickerson, J., Kosko, B.: Virtual worlds as fuzzy cognitive maps. Presence **3**(2), 173–189 (1994)
9. Froelich, W., Juszczuk, P.: Predictive capabilities of adaptive and evolutionary fuzzy cognitive maps—a comparative study. In: Nguyen, N.T., Szczerbicki, E. (eds.) Intelligent Systems for Knowledge Management, Studies in Computational Intelligence, vol. 252, pp. 153–174. Springer (2009)
10. Froelich, W., Papageorgiou, E.I., Samarinas, M., Skriapas, K.: Application of evolutionary fuzzy cognitive maps to the long-term prediction of prostate cancer. Appl. Soft Comput. **12**(12), 3810–3817 (2012)
11. Froelich, W., Salmeron, J.L.: Evolutionary learning of fuzzy grey cognitive maps for the forecasting of multivariate, interval-valued time series. Int. J. Approx. Reason. **55**(6), 1319–1335 (2014)
12. Ghazanfari, M., Alizadeh, S.: Learning FCM with Simulated Annealing (2008)
13. Goldberg, D.: Genetic Algorithms. Pearson Publishing (2013)
14. Helliwell, J.F., Putnam, R.D.: The social context of well-being. Philos. Trans. R. Soc. Lond. B: Biol. Sci. **359**(1449), 1435–1446 (2004)
15. Homenda, W., Jastrzebska, A., Pedrycz, W.: Modeling time series with fuzzy cognitive maps. In: FUZZ-IEEE 2014, Beijing, China, pp. 2055–2062 (2014)
16. Huerga, A.V.: A balanced differential learning algorithm in fuzzy cognitive maps. In: Proceedings of the 16th International Workshop on Qualitative Reasoning, pp. 1–7 (2002)

17. Juszczuk, P., Froelich, W.: Learning fuzzy cognitive maps using a differential evolution algorithm. Pol. J. Environ. Stud. **12**(3B), 108–112 (2009)
18. Kosko, B.: Fuzzy cognitive maps. Int. J. Man-Mach. Stud. **24**, 65–75 (2010)
19. Kosko, B.: Differential Hebbian learning. In: Neural Networks for Computing, April, pp. 277–282. American Institute of Physics (1986)
20. Lopez, C., Salmeron, J.L.: Modeling maintenance projects risk effects on ERP performance. Comput. Stand. Interfaces **36**(3), 545–553 (2014)
21. Murias, P., Martinez, F., De Miguel, C.: An economic wellbeing index for the Spanish provinces: a data envelopment analysis approach. Soc. Indic. Res. **77**(3), 395–417 (2006)
22. Papageorgiou, E., Stylios, C.D., Groumpos, P.P.: Active Hebbian learning algorithm to train fuzzy cognitive maps. Int. J. Approx. Reason. **37**(3), 219–249 (2004)
23. Papageorgiou, E.I., Froelich, W.: Application of evolutionary fuzzy cognitive maps for prediction of pulmonary infections. IEEE Trans. Inf. Technol. Biomed. **16**(1), 143–149 (2012)
24. Papageorgiou, E.I., Froelich, W.: Multi-step prediction of pulmonary infection with the use of evolutionary fuzzy cognitive maps. Neurocomputing **92**, 28–35 (2012)
25. Papageorgiou, E.I., Parsopoulos, K.E., Stylios, C.D., Groumpos, P.P., Vrahatis, M.N.: Fuzzy cognitive maps learning using particle swarm optimization. J. Intell. Inf. Syst. **25**, 95–121 (2005)
26. Papageorgiou, E.I., Salmeron, J.L.: A review of fuzzy cognitive maps research during the last decade. IEEE Trans. Fuzzy Syst. **21**(1), 66–79 (2013)
27. Penkova, T.: Method of wellbeing estimation in territory management. In: Murgante, B., Misra, S., Rocha, A., Torre, C., Rocha, J., Falco, M., Taniar, D., Apduhan, B., Gervasi, O. (eds.) Computational Science and Its Applications ICCSA 2014, Lecture Notes in Computer Science, vol. 8582, pp. 57–68. Springer International Publishing (2014)
28. Penkova, T.: Decision making support technique based on territory wellbeing estimation. In: Neves-Silva, R., Jain, L.C., Howlett, R.J. (eds.) Intelligent Decision Technologies, Smart Innovation, Systems and Technologies, vol. 39, pp. 513–523. Springer International Publishing (2015)
29. Salmeron, J.L., Papageorgiou, E.I.: A fuzzy grey cognitive maps-based decision support system for radiotherapy treatment planning. Knowl.-Based Syst. **30**(1), 151–160 (2012)
30. Salmeron, J.L., Papageorgiou, E.I.: Fuzzy grey cognitive maps and nonlinear Hebbian learning in proccess control. Appl. Intell. **41**(1), 223–234 (2014)
31. Salmeron, J.L., Vidal, R., Mena, A.: Ranking fuzzy cognitive maps based scenarios with topsis. Exp. Syst. Appl. **39**(3), 2443–2450 (2012)
32. Stach, W., Kurgan, L., Pedrycz, W., Reformat, M.: Genetic learning of fuzzy cognitive maps. Fuzzy Sets Syst. **153**(3), 371–401 (2005)

Embedded Dynamic Fuzzy Cognitive Maps for Controller in Industrial Mixer

Márcio Mendonça, Flávio Neves Jr., Lúcia V.R. de Arruda, Ivan Rossato Chrun and Elpiniki I. Papageorgiou

Abstract This paper presents the application of certain intelligent techniques to control an industrial mixer. Control design is based on Hebbian modification of Fuzzy Cognitive Maps learning. This research study develops a Dynamic Fuzzy Cognitive Map (DFCM) based on Hebbian Learning algorithms. It was used Fuzzy Classic Controller to help validate simulation results of an industrial mixer of DFCM. Experimental analysis of simulations in this control problem was conducted. Additionally, the results were embedded using efficient algorithms into the Arduino platform in order to acknowledge the performance of the codes reported in this paper.

Keywords Fuzzy Cognitive Maps · Hebbian Learning · Arduino microcontroller · Process control · Fuzzy logic

M. Mendonça (✉) · F. Neves Jr. · L.V.R. de Arruda · I.R. Chrun
Paraná Federal Technological University CPGEI, Curitiba, Brazil
e-mail: mendonca@utfpr.edu.br

F. Neves Jr.
e-mail: nevesjr@utfpr.edu.br

L.V.R. de Arruda
e-mail: lvrarruda@utfpr.edu.br

I.R. Chrun
e-mail: ivanchrun@gmail.com

E.I. Papageorgiou
Department of Computer Engineering, Technological Education Institute/University of Applied Sciences of Central Greece, Lamia, Greece
e-mail: epapageorgiou@teiste.gr

I. Czarnowski et al. (eds.), *Intelligent Decision Technologies 2016*,
Smart Innovation, Systems and Technologies 57,
DOI 10.1007/978-3-319-39627-9_22

1 Introduction

Developments in optimal control theory, robust control and adaptive control have expanded significantly the automation concept and the feasibility of automatic control in practice. However, these techniques rely on mathematical models for the plant and controllers.

On the other hand, intelligent control techniques try to control actions without resorting to complete or partial mathematical models. Otherwise, the ability of a human to find solutions to a particular problem is known as human intelligence. In short, human beings are able to deal with complicated processes based on inaccurate and/or approximate information. The strategy adopted by them is also of imprecise nature and usually capable of being expressed in linguistic terms. Thus, by means of Fuzzy Logic concepts, it is possible to model this type of information [1].

Artificial Intelligence (AI) has applications in various areas of knowledge, such as mathematical biology, neuroscience, computer science and others. The research area of intelligent computational systems aims to develop methods that try to mimic or approach the capabilities of humans to solve problems. These new methods seek to emulate human's abilities to cope with very complex processes, based on inaccurate and/or approximated information.

In general, Fuzzy Cognitive Map (FCM) is a tool for modeling the human knowledge. It can be obtained through linguistic terms, inherent to Fuzzy Systems, but with a structure similar to the Artificial Neural Networks (ANN), which facilitates data processing, and has capabilities for training and adaptation. FCM is a technique based on the knowledge that inherits characteristics of Cognitive Maps and Artificial Neural Networks [2–4], with applications in different areas of knowledge and applications [5–10]. Besides the advantages and characteristics inherited from these primary techniques, FCM was originally proposed as a tool to build models or cognitive maps in various fields of knowledge. It makes the tool easier to abstract the information necessary for modeling complex systems, which are similar in the construction to the human reasoning. Dynamic Fuzzy Cognitive Maps (DFCM), need to be developed to model and manage behaviors of non-linear time-dependent system and often in real time. Examples of different DFCMs can be found in the recent literature, e.g., [10–14].

This paper has two objectives. First objective is the development of two controllers using an acyclic DFCM with same knowledge as this of Fuzzy Controller, and with similar heuristic, thus producing comparable simulated results. Second aim is to show an embedded DFCM in the low cost and processing microcontroller Arduino.

To succeed the goals, we initially use the similar DFCM proposed in [11] to control an industrial mixing tank. In contrary to [11], we use a Hebbian algorithm to dynamically adapt the DFCM weights. In order to validate our DFCM controller, we compared its performance with a Fuzzy Logic Controller. This comparison is carried out with simulated data.

2 Development

To demonstrate the evolution of the proposed technique (DFCM) we will use a case study well known in the literature as seen in [3, 15, 16] and others. This case was selected to illustrate the need for refinement of a model based on FCM built exclusively with knowledge. The process shown in Fig. 1 consists of a tank with two inlet valves for different liquids, a mixer, and an outlet valve for removal of liquid produced by mixing and specific gravity meter that measures the specific gravity of the produced liquid. In this research, to illustrate and exemplify the operation of the industrial mixer, the liquids are water with specific gravity 1 and soybean oil with a specific gravity of about 0.9.

Valves (V_1) and (V_2) insert two different liquids (specific gravities) in the tank. During the reaction of the two liquids, a new liquid characterized by it's a new specific gravity value is produced. At this time the valve (V_3) empties the tank in accordance with a campaign output flow, but the liquid mixture should be in the specified levels of the volume and specific gravity.

Although relatively simple, this process is a TITO (Two Inputs Two Outputs) type with coupled variables. To establish the quality of the control system of the produced fluid, a weighting machine placed in the tank measures the (specific gravity) produced liquid.

When the value of the measured variable G (liquid mass) reaches the range of values between the maximum and minimum [G_{min}, G_{max}] specified, the desired mixed liquid is ready. The removal of liquid is only possible when the volume (V) is in a specified range between the values [V_{min} and V_{max}]. The control consists to keep these two variables in their operating ranges, as:

$$V_{min} < V < V_{max} \tag{1}$$

$$G_{min} < G < G_{max}. \tag{2}$$

In this study was tried to limit these values from approximately the range of 800–850 mg for the mass and approximately the range of 850–900 ml for the volume. The initial values for mass and volume are 800 mg and 850 ml respectively. According to Papageorgiou et al. [18], through the observation and analysis of operation of the process is possible for experts to define a list of key concepts related to physical quantities involved. The concepts and cognitive model are:

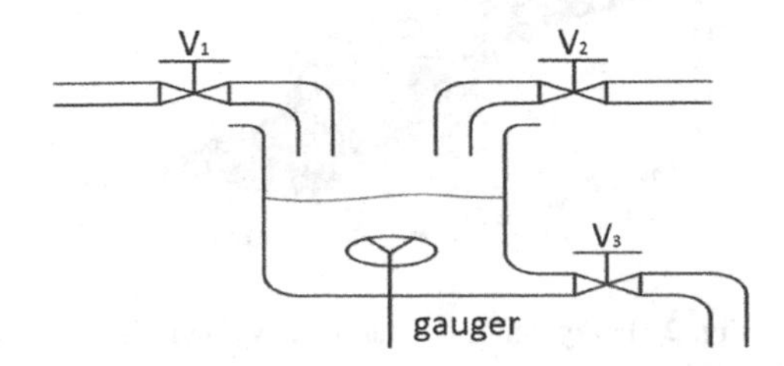

Fig. 1 Mixer Tank (*Source* adapted from [17])

- Concept 1—State of the valve 1 (closed, open or partially open).
- Concept 2—State of the valve 2 (closed, open or partially open).
- Concept 3—State of the valve 3 (closed, open or partially open).
- Concept 4—quantity of mixture (volume) in the tank, which depends on the operational state of the valves V_1, V_2 and V_3.
- Concept 5—value measured by the G sensor for the specific gravity of the liquid.

Considering the initial proposed evolution for FCM we use a DFCM to control the mixer which should maintain levels of volume and mass within specified limits.

The process model uses the mass conservation principle to derive a set of differential equations representing the process used to test the DFCM controller. As a result the tank volume is the volume over the initial input flow of the intake valves V_1 and V_2 minus the outflow valve V_3, this valve V_3 and the output campaign was introduced in this work for increase the complexity original process [17]. Similarly, the mass of the tank follows the same principle as shown below. The values used for m_{e1} and m_{e2} were 1.0 and 0.9, respectively.

$$V_{tank} = V_i + V_1 + V_2 - V_3 \tag{3}$$

$$Weight_{tank} = M_i + (V_1 m_{e1}) + (V_2 m_{e2}) - M_{out} \tag{4}$$

3 Fuzzy Controller Development

To establish a correlation and a future comparison between techniques, a Fuzzy Controller was also developed. The Fuzzy rule base uses the same heuristic control strategy and conditions.

Fuzzy Control is a technique used for decades, especially in process control [1]. It is a motivation for DFCM validation, and this study uses a same knowledge for two controllers, with two different formalisms. It isn't scope discuss the development of the Fuzzy Controller. But, some details of structure are pertinent functions are triangles and trapezoidal and 6 rules are considered in its base. The surface of this controller is showed in Fig. 2. Moreover, the rules are symmetric and similar by

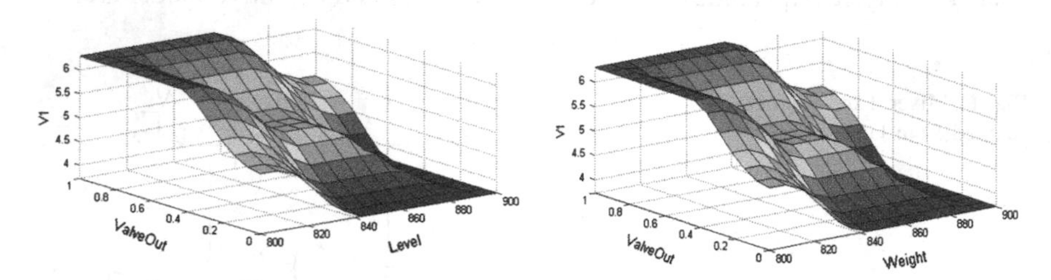

Fig. 2 Fuzzy controller surfaces, V_1 and V_2

two output valves; in this specific case the surface of valve 2 are same as in valve 1. The base rules are:

1. If (Level is low) or (Valve Out is high) then (V_1 is high) and (V_2 is high)
2. If (Level is medium) or (Valve Out is medium) then (V_1 is medium) and (V_2 is medium)
3. If (Level is high) or (Valve Out is low) then (V_1 is low) and (V_2 is low)
4. If (Weight is low) or (Valve Out is high) then (V_1 is high) and (V_2 is high)
5. If (Weight is medium) or (Valve Out is medium) then (V_1 is medium) and (V_2 is medium)
6. If (Weight is high) or (Valve Out is low) then (V_1 is low) and (V_2 is low)

4 DFCM Development

The structure of the DFCM controllers is similar to the developed Fuzzy Controllers, using same heuristics, e.g., in case the output valve (V_3, in accordance to Fig. 1) increases, the inputs valves (V_1 and V_2) increase too. In other hand, in case volume and weights of the mixture increase, the input valves decrease. For example, the relationships W54 and W53, in the DFCM, are similar in effects or control actions of the Fuzzy Controller's base rules.

The development of the DFCM is made through three distinct stages. First, the DFCM is developed as structure, concepts and causal relationships, similar to a classic FCM, where concepts and causal relationships are identified through sensors and actuators of the process. The concepts can be variables and/or control actions, as already mentioned. The output valve is defined by a positive relationship, i.e., when the campaign increases, the output (V_3) flow also increases, similarly, the input valves increase too. Moreover, when Volume Mixture and Weight Mixture increase, V_1 and V_2 decrease. In both cases the flow of the valves increases or decreases proportionally. The second development stage is the benchmark well-known Genetic Algorithm [19]. The Fig. 3 shows the schematic graph of a DFCM controller.

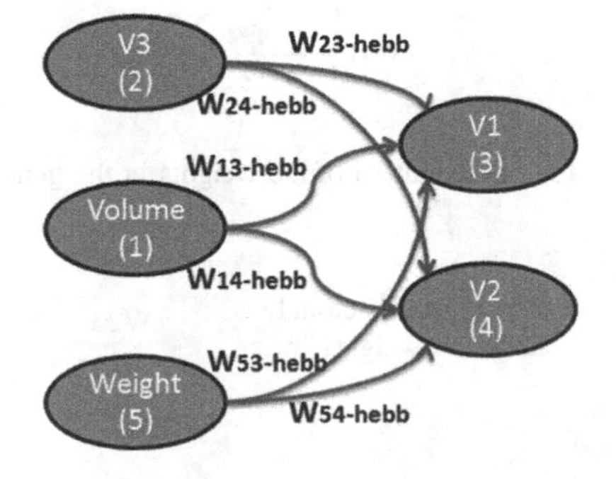

Fig. 3 DFCM controller

In this research, details of the G.A. are: the initial values of causal relationships are determined through genetic algorithms. The genetic algorithm used is a conventional one, with a population of 20 individuals, simple crossing and approximately 1 % of mutation. The chromosomes were generated by real numbers with all the DFCM weights, individuals were random and the initial method of classification was the tournament method with 3 individuals. Finally, the fitness function for simplicity considers the overall error of the two desired outputs.

Figure 4 shows an evolution; with 60 generations of the Genetic Algorithm proposed, it stabilizes and reaches the initial solution for the opening of the valves, approximately 42 %. It was obtained different generation values as the results, e.g., 45, 57, and 60 (which was the chosen value in this work); due G.A. being a classic random search method, used in this work. Table 1 shows initial values of the DFCM weighs. Different proposals and variations of this method applied in tuning FCM can be found [13].

The third stage of the DFCM development concerns the-line tuning or refinement of the model for dynamic response of the controller. In this case, when a change of output set-point in the campaign occurs, the weights of the causal relationships are dynamically tuned. To perform this function a new kind of concept and relation was included in the cognitive model.

In order to dynamically adapt the DFCM weights we used the Hebbian Learning algorithm for FCM that is an adaptation of the classic Hebbian method [2]. Different proposals and variations of this method applied in tuning or in learning for

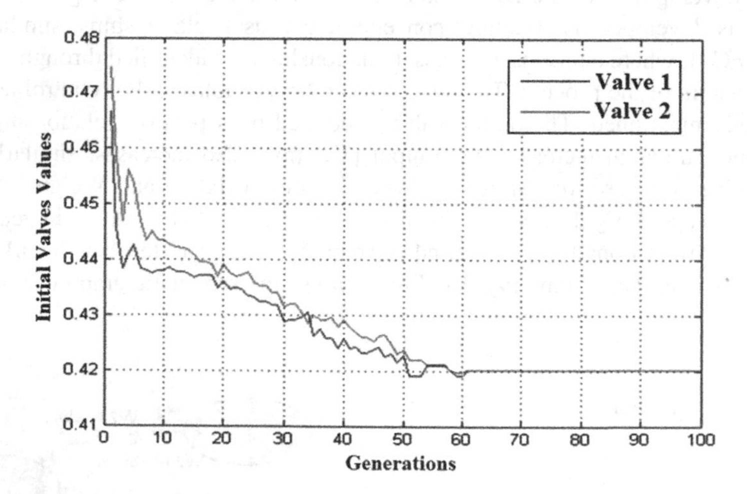

Fig. 4 Evolution of the weights in the genetic algorithm

Table 1 Initials casual relationship weights

W23	W24	W13	W14	W53	W54
−0.23	−0.26	−0.26	−0.26	0.23	0.15

FCM are known in the literature [12]. In this paper, the method is used to update the intensity of causal relationships in a deterministic way according to the variation or error in the intensity of the concept or input variable. Specifically, the application of Hebb learning provides control actions as follows: if the weight or volume of the liquid mix increases, the intake valves have a causal relationship negatively intensified and tend to close more quickly. Conversely, if the volume or weight mixture decreases, the intake valves have a causal relationship positively intensified. The mathematical equation is presented in (5).

$$W_i(k) = W_{ij}(k-1) \pm \gamma \Delta A_i \tag{5}$$

where: ΔA_i is the concept variation resulting from causal relationship, and it is given by $A_i = A_i(k) - A_i(k-1)$, γ is the learning rate at iteration k.

This version of the Hebbian algorithm is an evolution of the two proposals of Matsumoto et al. [20].

Causal relationships that have negative causality has negative sign and similarly to positive causal relationships. The equations applied in this work are adapted of the original version.

$$W_i(k) = k_p * (W_{ij}(k-1) - \gamma * \Delta A_i) \tag{6}$$

where: $\boldsymbol{\gamma} = 1$ for all, and $\boldsymbol{k_p}$ is different for every weights pairs. It is had their assigned values empirically by observing the dynamics of process performance, recursive method, $\boldsymbol{k_p} = 40$ for (W14; W23), $\boldsymbol{k_p} = 18$ for (W13; W24) and $\boldsymbol{k_p} = 2.5$ for (W53; W54), with normalized values.

The DFCM inference is similar to Classic FCM [2], and the inference equations are shown below (Eqs. 7 and 8).

$$A_i = \int \left(\sum_{\substack{j=1 \\ j \neq i}}^{n} (A_j \times W_{ji}) \right) + A_i^{previous} \tag{7}$$

$$f(x) = \frac{1}{1 + e^{-\lambda x}} \tag{8}$$

Figure 5 shows the results of Hebbian Learning Algorithm for FCM considering the variations ΔAi of the concepts concerning volume, weight and outlet valve, while in the is displayed the weights of the causal relationship in the process. This figure also shows the evolution of the weights of the causal relationships during the process into a range FCM definition [−1, 1].

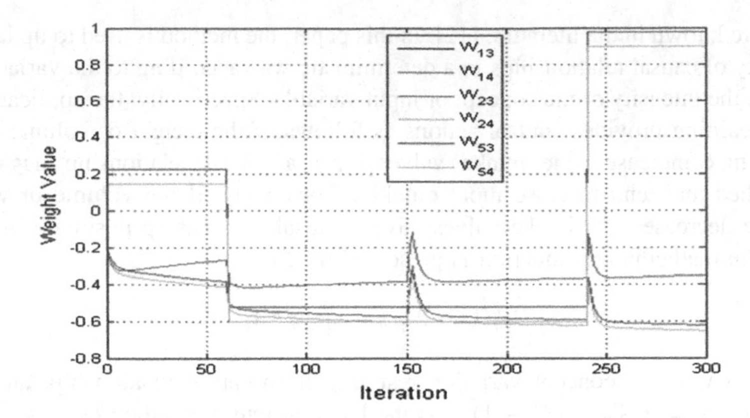

Fig. 5 Evolution of the weights in the Hebbian Learning

5 Simulated Experimental Results

The results of DFCM are shown in Fig. 6, which show the behavior of the controlled variables within the predetermined range for the volume and the weight of the mixture. It is noteworthy that the controller keeps the variables in the control range and pursues a trajectory according to a campaign, where the output flow is also predetermined. In this initial experiment, a campaign with a sequence of values ranging from 7, 5 and 11 ml/min can be seen as a set-point output flow (outlet valve). Similarly, the results of the Fuzzy Controller are shown in Fig. 7.

Table 2 shows that the simulated numeric results of the DFCM controller had a similar performance compared to the conventional Fuzzy Logic Controller, and

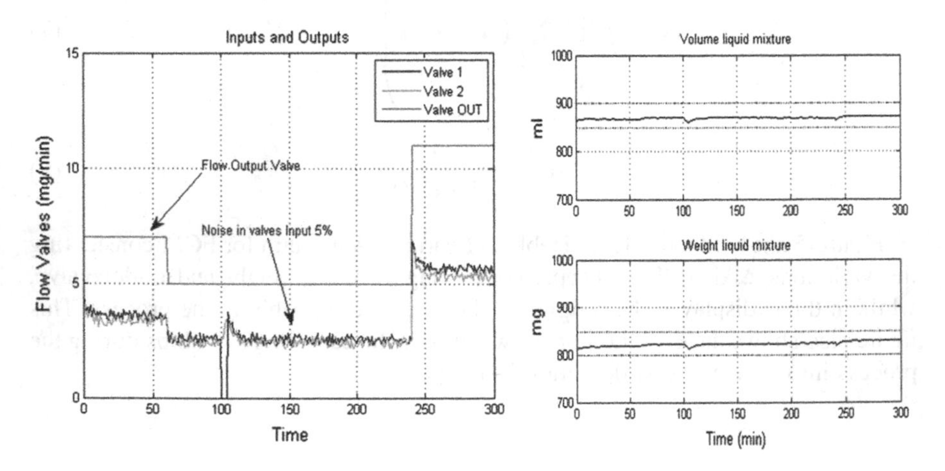

Fig. 6 Valves and results of the DFCM controller

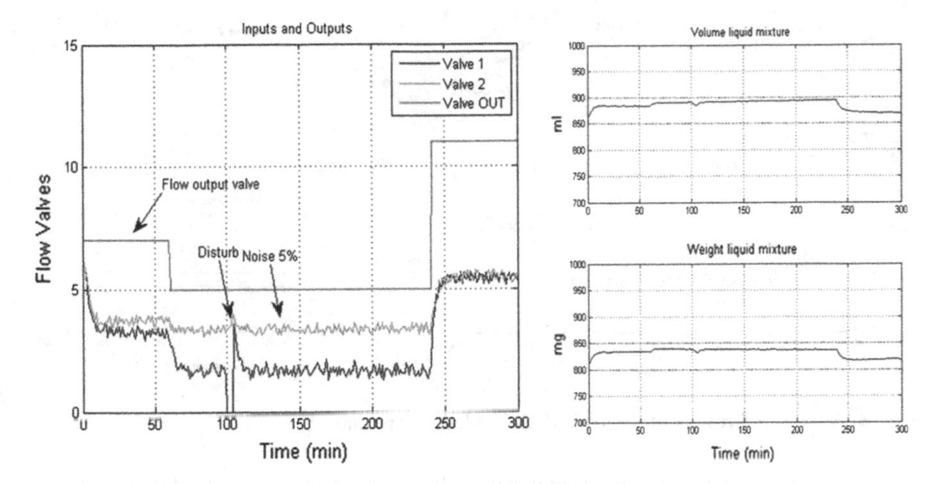

Fig. 7 Valves and results of the fuzzy controller

Table 2 Quantitative results

	DFCM	DFCM-Arduino	Fuzzy logic
	Max-Min	Max-Min	Max-Min
Volume mix (ml)	13.6	16.9	31.2
Weight mix (mg)	17.1	14.1	27.5
	Max value	Max value	Max value
Volume mix (ml)	873.2	864.4	894.2
Weight mix (mg)	829.7	816.7	840

DFCM embedded in Arduino with low difference in same conditions, with noise and with disturb (instant locked valve).

In order to extend the applicability of this work, the developed DFCM controller is embedded into an Arduino platform which ensures the portability of the FCM generated code. Arduino is an open-source electronics prototyping platform which uses ATMega® series microcontrollers.

The equations for level and weight are calculated by Matlab simulating the process. Through a serial communication established with Arduino, Matlab sends the current values of Volume, weight and output valve to Arduino that receives these data, calculates the values of the concept 1 (valve 1) and concept 2 (valve 2) and then returns these data to Matlab. After this, new values of Volume and Weight are recalculated. Details on how this technique can be used are presented in Matlab Tutorial and Arduino codes, by accessing the link [21]. The cycle of communication between Arduino to Matlab can be checked in [20].

Figure 8 shows the results obtained with the Arduino platform providing data of the actuators, Valve 1 and Valve 2, with Matlab performing data acquisition. The algorithm switches the sets of causal relations that operate similarly to a DFCM simulated with noise and disturb in the valve 1. The noise in Fig. 8 is the sum of the

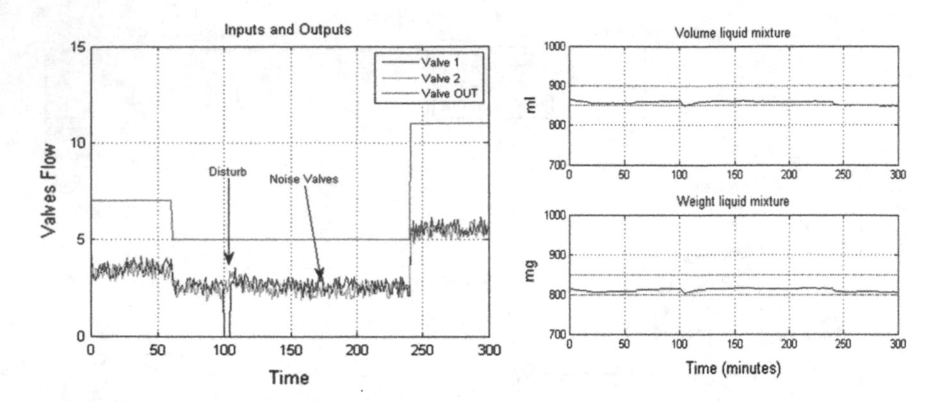

Fig. 8 Valves and results of the DFCM embedded in Arduino

real noise, observed in data transference between Arduino and Matlab, plus a simulated white noise. Equation (9) shows the composition of the experiment noise.

$$Noise_{Experiment} = Noise_{Simulated} + Noise_{Arduino-Matlab} \tag{9}$$

6 Conclusions

The contribution of this study focuses on the introduction of Fuzzy Cognitive Maps in the embedded control area. In simulated data, the results are similar for the two controllers, with small advantage for DFCM with or without Arduino, observed that DFCM Controller is adaptive. From the data obtained from Arduino Microcontroller, based on the variations of the DFCM embedded in the platform, it is observed that the controlled variables were in well-behaved ranges, which suggests that the DFCM codes have low computational complexity due to the simplicity of its inference mathematical processing. Thus, we can emphasize the portability and the possibility of developing DFCM controllers on low cost platforms. In short, this work showed that DFCM can be embedded.

Future studies will address more simulations with different campaigns, for a more general conclusion. Other methods of dynamic adaptation can be used for comparison ground, e.g., Reinforcement Learning for DFCM.

References

1. Zadeh, L.A.: An Introduction to Fuzzy Logic Applications in Intelligent Systems. Kluwer Academic Publisher, Boston (1992)
2. Kosko, B.: Fuzzy cognitive maps. Int. J. Man-Mach. Stud. **24**(1), 65–75 (1986)

3. Glykas, M.: Fuzzy Cognitive Maps: Advances in Theory, Methodologies. Tools and Applications. Springer, Berlin, Heidelberg (2010)
4. Kosko, B.: Neural Networks and Fuzzy Systems: A Dynamical Systems Approach to Machine Intelligence. Prentice Hall, New York (1992)
5. Dickerson, J.A., Kosko, B.: Virtual worlds as fuzzy cognitive maps. Presence **3**(2), 173–189 (1994)
6. Lee, K.C., Lee, S.: A cognitive map simulation approach to adjusting the design factors of the electronic commerce web sites. Expert Syst. Appl. **24**(1), 1–11 (2003)
7. Papageorgiou, E., Stylios, C., Groumpos, P.: Novel for supporting medical decision making of different data types based on Fuzzy Cognitive Map Framework. In: Proceedings of the 29th Annual International Conference of the IEEE embs cité internationale, Lyon, France, August, pp. 23–26 (2007)
8. Papageorgiou, E., Stylios, C., Groumpos, P.A.: Combined fuzzy cognitive map and decision trees model for medical decision making. In: Annual International Conference of the IEEE Engineering in Medicine and Biology Society. IEEE Engineering in Medicine and Biology Society, vol. 1, pp. 6117–6120 (2006)
9. Huang, Y.C., Wang, X.Z.: Application of fuzzy causal networks to waste water treatment plants. Chem. Eng. Sci. **54**(13/14), 2731–2738 (1999)
10. Papageorgiou, E.I.: Fuzzy Cognitive Maps for Applied Sciences and Engineering from Fundamentals to Extensions and Learning Algorithms. Springer (2014)
11. Mendonça, M., Angélico, B., Arruda, L.V.R., Neves, F.: A dynamic fuzzy cognitive map applied to chemical process supervision. Eng. Appl. Artif. Intell. **26**, 1199–1210 (2013)
12. Miao, Y., Liu, Z.Q., Siew, C.K., Miao, C.Y.: Transformation of cognitive maps. IEEE Trans. Fuzzy Syst. **18**(1), 114–124 (2010)
13. Papageorgiou, E.: Learning algorithms for fuzzy cognitive maps. IEEE Trans. Syst. Cybern. Part C: Appl. Rev. **42**, 150–163 (2012)
14. Mendonça, M., Arruda, L.V.R.: A Contribution to the Intelligent Systems Development Using DCN. OmniScriptum GmbH & Co, KG (2015)
15. Miao, Y., Liu, Z.Q., Siew, C.K., Miao, C.Y.: Dynamical cognitive network—an extension of fuzzy cognitive. IEEE Trans. Fuzzy Syst. **9**(5), 760–770 (2001)
16. Axelrod, R.: Structure of Decision: The Cognitive Maps of Political Elites. Princenton University Press, New Jersey (1976)
17. Stylios, C.D., Groumpos, P.P., Georgopoulos, V.C.: An fuzzy cognitive maps approach to process control systems. J. Adv. Comput. Intell. **5**, 1–9 (1999)
18. Papageorgiou, E.I., Parsopoulos, K.E., Stylios, C.S., Groumpos, P.P., Vrahatis, M.N.: Fuzzy cognitive maps learning using particle swarm optimization. J. Intell. Inf. Syst. **25**, 95–121 (2005)
19. Goldberg, D.E.: Genetic Algorithms in Search Optimization and Machine Learning. Addison-Wesley, Mass (1989)
20. Matsumoto, D.E., Mendonça, M., Arruda, L.V.R., Papageorgiou, E.: Embedded Dynamic fuzzy cognitive maps applied to the control of industrial mixer. In: Simpósio Brasileiro de Automação Inteligente – XI SBAI (2013)
21. Tutorial Matlab-Arduino. http://epapageorgiou.com/index.php/fcm-research-group

Analyzing Cloud Business Services with Choquet Fuzzy Integrals and Support Vector Machines

Jose L. Salmeron and Pedro Palos

Abstract Cloud computing poses both opportunities and challenges for companies and IT professionals. Some of these are technical challenges that can be solved over time, while others are related to uncertainties arising from the commitment to a recent innovation. The objective of this research is to identify some of the uncertainties that IT professionals may have and can discourage them from adopting cloud computing. In fact, this paper is focused on predicting the perceived easy-of-use of cloud business services. For that purpose, we use Choquet Fuzzy Integral and Support Vector Machines.

Keywords Cloud services · Choquet fuzzy integrals · Support vector machines

1 Introduction

Cloud computing poses both opportunities and challenges for companies and IT professionals. Some of these are technical challenges that can be solved over time, while others are related to uncertainties arising from the commitment to a recent innovation. The objective of this research is to identify some of the uncertainties that IT professionals may have and can discourage them from adopting cloud computing.

An innovation is an idea, practice or object that is perceived as new [17]. Although the newness of cloud computing is certainly debatable, there is no doubt that its introduction challenges our conventional understanding of the location and management of IT infrastructure, the nature of products and services, business processes and practice of its services (both for IT professionals and consumers).

The factors which are potentially affecting the intention of IT professionals in the use of cloud computing to deliver products and services to their customers are taken mainly from the theory of diffusion of innovation [17].

J.L. Salmeron (✉) · P. Palos
Data Science Lab, Universidad Pablo de Olavide, Km. 1 Utrera Road, 41013 Seville, Spain
e-mail: salmeron@acm.org

I. Czarnowski et al. (eds.), *Intelligent Decision Technologies 2016*,
Smart Innovation, Systems and Technologies 57,
DOI 10.1007/978-3-319-39627-9_23

The theory identifies five variables that have a profound influence on the rate of adoption of innovation including: perceived attributes of innovation, the type of decision in innovation, communication channels, the nature of the social system, and changing the promotional efforts of agents. The perceived attributes of innovation are an important predictor of intent in adopting innovations [17].

There are many benefits in taking the services offered by a cloud provider. Its application will depend on the nature, size and needs of the company. The decision to choose the option that best meets customer needs is a complex task due to the appearance of different suppliers.

Each provider has its own pricing policy, a degree of flexibility in offering services and a technical support appropriate to the service that he thinks he should supply. The offer revolves around these three important pillars. Many companies like Amazon, Google, Microsoft and Salesforce have become cloud computing providers.

The open source community is also present as a provider within the business model that offers, looking very active in the area of cloud computing with numerous contributions, especially in virtualization technologies [15].

Virtualization is a key technology for the cloud which allows a more efficient and flexible use of resources. Virtualization is a key element of the cloud for its advantages such as flexibility, isolation and utilization rate of resources. Building a cloud environment often initially involves choosing a management solution for the cloud. Often, this decision is difficult, because each solution has its specific characteristics [6].

2 Theoretical Background

There are many variables that can influence the adoption of cloud computing. Research in this field is still scarce due to its recent emergence and adoption by businesses. Often, technological literature focuses on addressing issues and challenges related to adoption, such as service availability, performance, lack of interoperability standards and difficulty of integration and customization [7, 8, 10].

Moreover it is possible to find several studies that emphasize the importance of confidence, both in the adoption of cloud technology, as in the privacy in data storage. This paper focuses on the importance of various especially significant aspects in the literature of adoption of technological innovations, like Cooperation, Complexity Technology, Training, Top Management Support and Communication [16, 21].

- Cooperation, either in its internal or external aspects, provides synergies, it reassures users and helps achieve expectations. Internal cooperation is the exerted between the different functional areas of a company [12, 13]. Furthermore, external cooperation refers to the links that the organization maintains with the cloud provider.

- Top Management Support. It is defined as the active involvement of those responsible for the management in the successful implementation of technology [2]. This active participation materializes mainly through leadership and continuous contact with those who are directly linked to the IT planning [19]. Through these measures, users tend to assimilate the expectations of management, in addition to perceive that those responsible for an organization support its implementation. This increases the employees' favourable attitudes regarding IT.
- Training. Training is described as the degree to which a company instructs its employees in the use of a tool in terms of quality and quantity. In a complex information system such as cloud computing, the organization needs to train employees and develop skills for effective use in the future [9]. This reduces the potential stress of staff and provides greater motivation and a better understanding of the benefits of the cloud.
- Communication. Quality communication occurs when members employ a certain amount of time exchanging information and views, either formally or informally [11]. This type of communication increases the distribution of ideas and improves knowledge transfer, especially when the information transmitted is credible and from reliable sources.
- Technological complexity. It is defined as the degree by which an innovation is perceived as relatively difficult to understand and use. In the case of cloud computing, the aspects related to the complexity could be the time needed for the development of tasks, application integration with cloud infrastructure, interface design or efficiency in data transfer, etc. [9]. In short, the technological complexity can influence the adoption of a cloud solution.

3 Methodological Framework

3.1 Choquet Fuzzy Integral

Non-additive measures are known in literature as fuzzy measures, monotonic measures and capacities between others. Some additive operators such as simple weighted average, ordered weighted average, quasi arithmetic means, weighted min and weighted max are usually used for aggregation purpose. These operators assume that the attributes are always independent between them. This assumption is not correct in scenarios where in many cases, the attributes are strongly interrelated. According to this, aggregation should not be always carried out using common additive operators instead, Fuzzy Choquet Integrals are useful to aggregate with interrelated attributes [5].

Let's define a fuzzy measure μ on a finite set $N = \{1, 2, \ldots, n\}$ as a function $\mu : P(N) \rightarrow [0, 1]$ (where $P(N)$ is the power set of N) satisfying the following conditions:

$$\mu(\phi) = 0 \tag{1}$$

$$\mu(N) = 1 \tag{2}$$

$$A \subseteq B \text{ implies that } \mu(A) \leq \mu(B) \tag{3}$$

The third condition allows measures that do not satisfy the strong condition of additivity. For our purposes, this means that we can model systems where the high value of an attribute of the system in itself does not indicate deviations unless a set of other attributes show deviations from their usual values at the same time [14].

Fuzzy Choquet Integral is one of the most general formulations when using monotone measures as the basis of aggregation [18]. To formulate the definition, we assume n attributes measures $(c_1, \ldots, c_n)$ that it generate the corresponding $(s_1, \ldots, s_n)$ values after an evaluation performed.

The basic properties of the operator are determined by the monotone measure, such as symmetry, additivity and linearity. A discrete Fuzzy Choquet Integral with respect to a monotone measure μ is defined as

$$C_\mu(s_1, \ldots, s_n) = \sum_{i=1}^{n} (s_{(i)} - s_{(i-1)})\mu(C_{(i)}) \tag{4}$$

where $s(i)$ denotes a permutation of the s_i values such that $s_{(1)} \leq s_{(2)} \leq \ldots \leq s_{(n)}$ and $C_{(i)} = \{c_{(i)}, c_{(i+1)}, \ldots, c_{(n)}\}$.

3.2 Support Vector Machines

Let us start from a binary classification

$$\{x_i, y_i\}, i = 1, ..., n, y_i \in \{-1, 1\}, x_i \in R^n \tag{5}$$

where x_i are data points, and y_i are labels. The data points are separated with a hyperplane given by $w^T x + b = 0$, where w is a n-dimensional coefficient vector that is normal to the hyperplane, and b is the offset from the origin (Fig. 1).

The linear SVM obtains an optimal separating margin by solving an optimization problem [20] as follows

$$\begin{aligned} \min(w, \xi) &= \tfrac{1}{2}||\mathbf{w}||^2 + C \textstyle\sum_{i=1}^{n} \xi_i \\ \text{s.t. } y_i(\mathbf{w}^T \mathbf{x}_i + b) &\geq 1 - \xi_i, \xi_i \geq 0 \end{aligned} \tag{6}$$

where $\xi_i = \max(0, 1 - y_i(w \cdot x_i + b))$ if and only if ξ_i is the smallest non-negative value satisfying $y_i(w \cdot x_i + b) \geq 1 - \xi_i$.

Figure 2 shows an example of a linear SVM, where the solid line h in the figure is the final SVM solution.

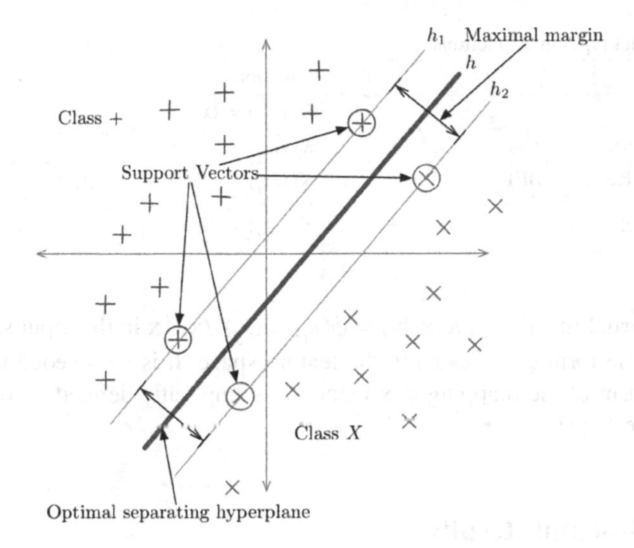

Fig. 1 A linear support vector machine for binary classification

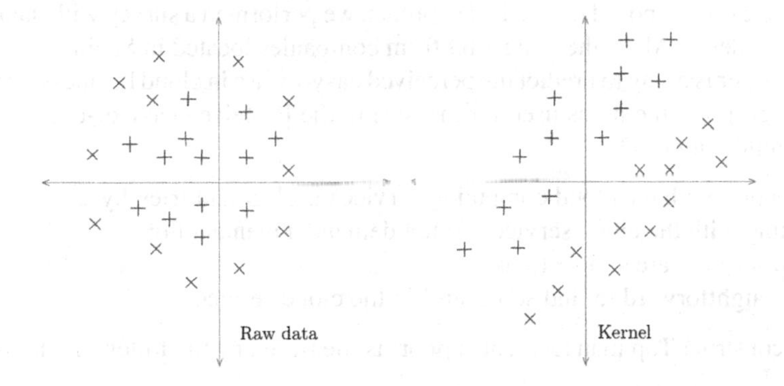

Fig. 2 From raw data to higher dimensional with kernels

Usually, the classification cannot be done linearly. In order for the linear classification to work well in non-linear data, kernels are introduced [20]. The original input space can be mapped into some higher-dimensional feature space where the training set is linearly separable [4]. With this kind of mapping, the decision function can be expressed as

$$g(\mathbf{x}) = \operatorname{sgn}\left(\sum_{i=1}^{n} \alpha_i \cdot y_i \cdot K(\mathbf{x}_i, \mathbf{x}) + b \right) \tag{7}$$

Table 1 Kernel types and functions

Kernels	Functions
Linear kernel	$K(x, x_i) = (x^T x_i)$
Polynomial kernel	$K(x, x_i) = \left((x^T x_i) + 1\right)^n$
Radial Based Kernel (RBF)	$K(x, x_i) = exp(-\gamma \|\|x - x_i\|\|^2)$
Sigmoid kernel	$K(x, x_i) = \frac{e^{2\cdot(x^T x_i)+b} - 1}{e^{2\cdot(x^T x_i)+b} + 1}$

where the kernel function is $K(\mathbf{x}_i, \mathbf{x}) = \phi(\mathbf{x}_i)^T \phi(\mathbf{x}_j)$, $(\mathbf{x}_i)^T \mathbf{x}$ in the input space is represented as the form $\phi(\mathbf{x}_i)^T \phi(\mathbf{x}_j)$ in the feature space. It is not needed to know the functional form of the mapping $\phi(\mathbf{x}_i)$ since it is implicitly defined by one selected kernel (Table 1) [1].

4 Experimental Results

To evaluate our proposed theoretical approach we performed a survey with data from 128 respondents. All of them are CIO from companies located in Spain.

This paper is going to predict the perceived easy-of-use in cloud business services. For that purpose, the items used for measuring the Perceived easy-of-use construct are the following [2, 3]:

- Interaction with the cloud computing services is clear and friendly
- Working with the cloud services do not demand a mental effort
- Cloud services are simple to use
- It is straightforward to find some stuff in the cloud service

The construct Top management support is measured by the following items:

- Top management is interested in cloud services
- Top management understands the importance of cloud services
- Top management sponsors cloud services
- Top management understands the opportunities of cloud services

The items used for measuring the Technological complexity construct are the following:

- It is hard to understand what cloud services is doing
- Working with cloud services takes too long
- Working with cloud services needs a hard training
- In general terms, working with cloud services is so hard

The construct Communication is measured by the following items:

- Communication about the cloud services is fluid
- There is not constraints about cloud services' communication
- The information about cloud services is correct

Table 2 Error measurement

Measures	Results
MSE	0.08317709
RMSE	0.28840439
SMAPE	1.83722632

After run the proposal hybrid methodology we checked the results with three common error measures. The first one is the Mean Squared Error (MSE). It measures the average of the squares of the errors and is computed as follows

$$MSE = \frac{1}{n} \cdot \sum_{i=1}^{n} (\hat{x}_i - x_i)^2 \quad (8)$$

where n is the number of experiments, x_i is the real value and $\hat{x}_i$ is the estimated one. The second error measure computed is the Root-Mean-Square Error (RMSE). It measures the differences between the estimated values and the real observed values. It is computed as follows

$$RMSE = \sqrt{\frac{1}{n} \cdot \sum_{i=1}^{n} (\hat{x}_i - x_i)^2} \quad (9)$$

The third error measures is the Symmetric Mean Absolute Percentage Error (SMAPE) is based on relative errors. It is usually defined as follows

$$SMAPE = \frac{1}{n} \cdot \sum_{i=1}^{n} \frac{|\hat{x}_i - x_i|}{\left(|\hat{x}_i| + |x_i|\right)/2} \cdot 100 \quad (10)$$

The results are detailed in Table 2.

We consider that the results confirm that our proposal is a worthy endeavour.

5 Conclusions

Cloud computing is challenging for companies and IT professionals. The research proposes a hybrid machine learning methodology for predicting the perceived easy-of-use of cloud business services.

For that purpose, we use Choquet Fuzzy Integral and Support Vector Machines. The results confirm that this proposal is a worthy endeavour.

References

1. Akay, M.F.: Support vector machines combined with feature selection for breast cancer diagnosis. Expert Syst. Appl. **36**(2), 3240–3247 (2009)
2. Bueno, S., Salmeron, J.L.: Fuzzy modeling enterprise resource planning tool selection. Comput. Stan. Interfaces **30**(3), 137–147 (2008)
3. Burda, D., Teuteberg, F.: Exploring consumer preferences in cloud archiving—a student's perspective. Behav. Inf. Technol. **35**(2), 89–105 (2016)
4. Chen, H.L.: A new hybrid method based on local fisher discriminant analysis and support vector machines for hepatitis disease diagnosis. Expert Syst. Appl. **38**(9), 11796–11803 (2011)
5. Choquet, G.: Theory of capacities. Ann. Inst. Stat. Fourier **5**, 131–295 (1953/54)
6. Endo, P.T., Gonçalves, G.E., Kelner, J., Sadok, D.: A survey on open-source cloud computing solutions. In: Brazilian Symposium on Computer Networks and Distributed Systems, pp. 3–16 (2010)
7. Feuerlicht, G., Govardhan, S.: Impact of cloud computing: beyond a technology trend. Syst. Integr. **20010**, 2 (2010)
8. Feuerlicht, G., Burkon, L., Sebesta, M.: Cloud computing adoption: what are the issues. Systémová integrace, pp. 187–192 (2011)
9. Gangwar, H., Date, H., Ramaswamy, R.: Understanding determinants of cloud computing adoption using an integrated TAM-TOE model. J. Enterp. Inf. Manage. **28**(1), 107–130 (2015)
10. Géczy, P., Izumi, N., Hasida, K.: Cloudsourcing: managing cloud adoption. Glob. J. Bus. Res. **6**(2), 57–70 (2012)
11. Hsu, J. S.-C., Shih, S.P., Chiang, J.C., Liu, J.Y.: The impact of transactive memory systems on IS development teams' coordination, communication, and performance. Int. J. Proj. Manage. **30**(3), 329–340 (2012)
12. Joo, J., Sang, Y.: Exploring Koreans smartphone usage: an integrated model of the technology acceptance model and uses and gratifications theory. Comput. Hum. Behav. **29**(6), 2512–2518 (2013)
13. Lee, D.Y., Lehto, M.R.: User acceptance of YouTube for procedural learning: an extension of the technology acceptance model. Comput. Educ. **61**, 193–208 (2013)
14. Llamazares, B.: Constructing choquet integral-based operators that generalize weighted means and OWA operators. Inf. Fusion **23**, 131–138 (2015)
15. Mahjoub, M., Mdhaffar, A., Halima, R.B., Jmaiel, M.: A comparative study of the current cloud computing technologies and offers. In: 2011 First International Symposium on Network Cloud Computing and Applications (NCCA), pp. 131–134. IEEE (2011)
16. Pearson, S.: Toward accountability in the cloud. IEEE Internet Comput. **15**(4), 64 (2011)
17. Rogers, E.M.: Diffusion of Innovations: modifications of a model for telecommunications. In: *Die Diffusion von Innovationen in der Telekommunikation*, vol. 17, pp. 25–38. Springer, Berlin Heidelberg (1995)
18. Sanz, J., Lopez-Molina, C., Cerrn, J., Mesiar, R., Bustince, H.: A new fuzzy reasoning method based on the use of the Choquet integral. In: Proceedings of EUSFLAT, pp. 691–698 (2013)
19. Son, H., Park, Y., Kim, C., Chou, J.-S.: Toward an understanding of construction professionals' acceptance of mobile computing devices in South Korea: an extension of the technology acceptance model. Autom. Constr. **28**, 82–90 (2012)
20. Vapnik, V.N.: The Nature of Statistical Learning Theory. Springer, New York (1995)
21. Walterbusch, M., Martens, B., Teuteberg, F.: Evaluating cloud computing services from a total cost of ownership perspective. Manage. Res. Rev. **36** (6), 613–638 (2013)

Intelligent Methods for Eye Movement Data Processing and Analysis

Time-Preserving Visual Attention Maps

Michael Burch

Abstract Exploring the visual attention paid to a static scene can be done by visual analysis in form of attention maps also referred to as heat maps. Such maps can be computed by aggregating eye movement data to a density field which is later color coded to support the rapid identification of hot spots. Although many attempts have been made to enhance such visual attention maps, they typically do not integrate the time-varying visual attention in the static map. In this paper we investigate the problem of incorporating the dynamics of the visual attention paid to a static scene in a corresponding attention map. To reach this goal we first compute time-weighted Voronoi-based density fields for each eye-tracked person which are aggregated or averaged for a group of those people. These density values are then smoothed by a box reconstruction filter to generate aesthetically pleasing diagrams. To achieve better readability of the similar color values in the maps we enrich them by interactively adaptable isolines indicating the borders of hot spot regions of different density values. We illustrate the usefulness of our time-preserving visual attention maps in an application example investigating the analysis of visual attention in a formerly conducted eye tracking study for solving route finding tasks in public transport maps.

1 Introduction

The analysis and visualization of eye movement data is important in marketing, psychology, human-computer interaction, but also in visualization where visual stimuli are presented and a task is given to be solved by paying visual attention to the stimulus [12]. Not only the recording of the data is a challenging issue, but also the extraction of useful insights from the vast amount of recorded spatio-temporal eye movements of a large number of study participants.

Many metrics can be recorded in eye movement studies such as completion times, error rates, fixation durations, saccade lengths, pupil dilations and the like [11] but

M. Burch (✉)
VISUS, University of Stuttgart, Stuttgart, Germany
e-mail: michael.burch@visus.uni-stuttgart.de

I. Czarnowski et al. (eds.), *Intelligent Decision Technologies 2016*,
Smart Innovation, Systems and Technologies 57,
DOI 10.1007/978-3-319-39627-9_24

we found the participants' scanpaths very crucial to derive meaningful patterns about their visual task solution strategies. Although a visual analysis of these scanpaths based on line-based diagrams is useful we found that visual scalability is a challenging issue, in particular when it comes to a visualization with gaze plots [15]. The many line crossings are responsible for a high degree of visual clutter [14], "a state in which excess items or their disorganization lead to a degradation of performance at some task". Consequently, we would like to design a visual representation of the eye movement data based on visual attention maps that do not suffer from visual clutter problems but strongly aggregate the data. Neither individual participants nor time-varying information can be extracted from such a standard visual attention map (without the explicit use of interaction techniques).

But, fortunately, those maps give a great overview about visual attention paid to a static scene or stimulus in a user study which can be a starting point for further explorations with additional sophisticated visualization and analysis methods. Typically, visual attention maps do not encode the information about the order of visits in a stimulus which we will focus on in this research paper.

To reach this goal, we first map the scanpath of each study participant to a sequence of fixations to the 2D stimulus. We use these fixation points as the centroids of Voronoi cells spanning the entire stimulus which are attached by weight values proportional to the occurrence index in the sequence. Doing this for all participants (i.e., for all scanpaths), aggregating or averaging those values, interpreting them as density fields, applying partially randomized reconstruction filters, and finally, color coding them and augmenting them by isolines gives the desired results of time-preserving visual attention maps. Our technique supports several interaction techniques like switching to a standard visual attention map, selecting and highlighting Areas of Interest (AOIs), adapting color coding, or using circles of varying radii (with Euclidian distances) instead of Voronoi-based space subdivision for the generation of the density fields.

We illustrate the usefulness of our approach by applying it to scanpath data from a formerly conducted eye tracking study investigating a route finding task in public transport maps [5, 10], i.e., people were told start and destination stations and they had to follow a suitable path with their eyes. There are two scenarios, with explicitly highlighted stations and without highlights meaning people had to perform a search task first before doing the route finding task.

2 Related Work

Eye tracking has become an emerging strategy in the fields of information visualization, visual analytics, and human-computer interaction focusing on the evaluation of time-varying visual attention paid by people to a displayed static or dynamic stimulus [1, 12]. Although it got quite easy to reliably track people's eye movements when they are solving a task by inspecting a visual stimulus, we can observe a trend towards more and more challenging becoming analysis requirements for such vast amounts of spatio-temporal data [3].

One reason for that is the complex structure of the data of a large number of participants who apply different visual task solution strategies to find the answer of the given task which is investigated by [5, 9] using stimuli from public transportation maps, from node-link tree diagrams [6, 7], and from trajectory representations [13]. Visualization techniques are generally a good strategy to explore eye movement data of which plenty have been developed, implemented, and designed in the past which is surveyed by [2].

However, many of the techniques do not allow to visually explore the time-varying visual attention of a large group of study participants. The attention map [4, 16], for example, aggregates the data of all participants over time in one static color coded density field which only allows to have a look at the hot spots of visual attention paid to a stimulus. Neither time nor individual participants can be observed by this static map. A gaze plot [15] on the other side can be used to display time-varying eye movements for each individual participant by overplotting color coded polylines, but on the negative side, long fixation sequences and many people make such a display suffer from visual clutter [14].

Burch et al. [8] designed the AOI rivers technique which shows time-varying eye gaze frequencies of many study participants but the spatial information of the stimulus cannot be derived that easily due to the fact that the rivers are mapped to a horizontal time axis pointing from left to right. Space is given by applying categorical color coding to the individual rivers which represent Areas of Interest (AOIs) in the static stimulus. Our approach does not explicitly define AOIs but works more on fixations-to-points data which lets visually see all of the possible AOIs that might later by selected for further analysis.

Consequently, we do not argue that this approach solves the challenges that traditional visual attention maps have, but it gives support to analysts that would like to get a hint about time-varying visual attention paid to a static stimulus to get a starting point for further explorations.

3 Data and Visual Design

To compute a time-preserving visual attention map, the eye movement trajectories (scanpaths) must first be transformed into time-weighted areas of visual attention by taking into account the sequence of those fixations. The weighted areas are then aggregated and averaged resulting in a scalar two-dimensional density field which is then used as input for the visualization approach. The area computation can be based on either Voronoi cells spanning the complete 2D stimulus (see Fig. 1) or on circular regions with a user-defined radius. Both options can be selected interactively and are illustrated in Algorithm 1.

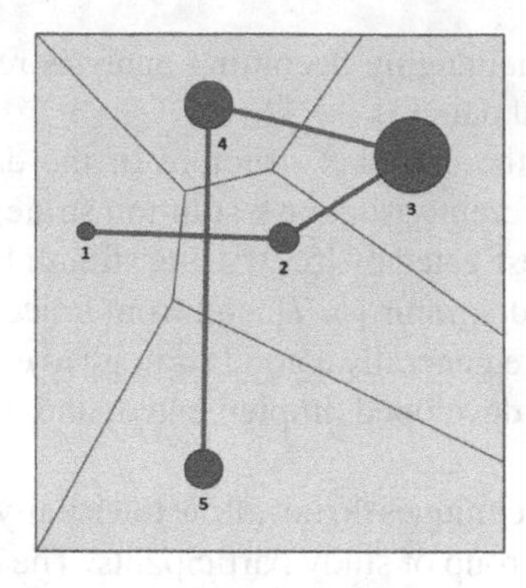

Fig. 1 A scanpath consisting of 5 fixations and 4 saccades where the fixations build the centroids of Voronoi cells. The *circle* sizes proportionally indicate the fixation durations that accumulate to the completion time for answering a given task. The progressing time or the sequential order of the fixations is used for computing density values

---

Algorithm 1 Generation of a Time-Preserving Visual Attention Map

---

TPVAM($S := \{S_1, \dots, S_n\}, r, c$):
- S; // n scanpaths of different lengths
- r; // Rendering radius (if Voronoi cell size has to be reduced)
- c; // User-defined color scale
- w; // 2D array for storing the time-preserving weights
- A; // 2D array for storing current density field values

for $i := 1$ **to** n **do**
- $S_i := (p_{i,1}, \dots, p_{i,m_i})$;
- **for** $j := 1$ **to** m_i **do**
  - $w_{i,j} := \frac{j}{m_i}$;
- **end for**
- Compute Voronoi cells from points of S_i;
- Add density values $w_{i,j}$ up to distance r from centroid to array A;

end for

Divide array values in A by n;

Smoothing step by applying reconstruction filter to array A;
Compute isolines in array A;
Apply color coding to the density values in array A;

3.1 Data Model and Transformations

Eye movements can be modeled as a set of individual scanpaths (Fig. 1)

$$S := \{S_1, \dots, S_n\}$$

which are sequences of fixations, i.e., each

$$S_i := (p_{i,1}, \ldots, p_{i,m_i}),\ 1 \leq i \leq n.$$

Each fixation to a point $p_{i,j}$, $1 \leq j \leq m_i$ is attached by a duration $t_d(i,j)$ which is the time span between the point in time the eye enters fixation point $p_{i,j}$ and the point in time it leaves it again. The movement of the eye from one fixation point to the next one in the sequence is called a saccade, i.e., a rapid eye movement where no visual attention is paid.

We map a time-based real-valued weight $w_{i,j}$ from the interval [0, 1] to each of the fixations. This weight represents either the occurrence time or number of the corresponding fixation in the sequence. If the number of the sequence is used for computing the weight it is just given by $w_{i,j} := \frac{j}{m_i}$.

Due to the fact that each fixation $p_{i,j}$ builds the centroid of a Voronoi region, all positions in a two-dimensional plane are mapped to the same weight as their corresponding centroid value $w_{i,j}$ (see Fig. 1). Instead of using the sequential order of the fixations in the sequence also the time can be used as a mapping to a weighted Voronoi model which demands to adapt the weight function $w_{i,j}$ to the time-based scenario. Both concepts can be treated similarly for the visual design of the time-preserving visual attention maps. This procedure is done for each of the scanpaths giving a weight-aggregated but average time-preserving Voronoi-based scanpath visualization. The resulting two-dimensional density array is the input data structure for our visualization approach.

As an extension to the pure Voronoi-based model, a radius r can be used to reduce the size of the resulting Voronoi regions. This concept does not give undesirably too much attention to the borders of the stimulus, but as a negative consequence, not all points in the stimulus are equally covered as in the Voronoi model. This bounding circle avoids longish Voronoi cells and does not take stimulus positions into account which are far away from a centroid.

To get rid of the many hard line intersections which make the resulting diagrams hard to analyze we apply a smoothing step in form of a reconstruction filter. The resulting curved and aesthetically appealing shapes still reflect the data in the visualization, i.e., they do not introduce a lie factor in the diagrams as described by Tufte [17]. Algorithm 1 explains how the data-to-visualization mapping works in detail.

3.2 *Time-Varying Density Fields and Isoline Enhancement*

Isolines can be computed by taking into account the formerly computed density value differences. The number of isolines is dependent on the number of value categorization also referred to as classing, i.e., the more value categories the more isolines will be visible in the final diagram. The isolines help to indicate where the hot spots are located and how the density values vary. This is in particular useful for neighbored regions in scenarios where the color codings are difficult to perceptually separate due to limitations of the human's perceptual abilities [18].

Finally, the user has to select a color coding which maps real-valued numbers between 0 and 1 from the computed density field A to color values from a predefined color palette (parameter c in Algorithm 1). This results in a time-varying density field for the temporal visual attention paid to a visual stimulus by a certain number (or all) eye tracked participants.

3.3 Interaction Techniques

Apart from having a look at the static visual attention maps we integrated several interaction techniques to navigate in the data and to get details-on-demand information which cannot be directly observed from the visualization. For example, by hovering over a color coded region all involved study participants' fixations are displayed in a textual form. The number of isolines can be adapted to a user-defined scenario. Also the selected color palette can be exchanged on user's demand while also logarithmic color coding might be an option to better distinguish high values from low ones.

We use Voronoi subdivisions into cells but also circle based subdivisions of space can be selected which typically do not cover the complete display space, i.e., different visual representations are the result. Apart from only aggregating over all participants and computing the average time in the scanpaths the user can also decide to visually encode only the maximum or minimum value in time, i.e., the first or last visit. Many more additional time-varying metrics such as fixation duration, absolute time, or order of fixation sequence can be displayed.

The user can also decide interactively to switch to a standard visual attention map. This can support him to compare time-based visual attention to a more frequency-based visual attention. Moreover, the visual stimulus can be inspected without any overlay of color coded density fields which is important to understand the semantics of the displayed scene for deriving visual task solution strategies of one or more study participants.

4 Application Example

We illustrate our approach with data from a metro map study. 40 participants were asked to find a route from a start to a destination station in several metro map scenarios, i.e., either with explicit station highlighting or not [5, 9]. The data was recorded by a Tobii T60 eye tracking device. 24 metro maps were displayed to the study participants showing public transport systems from all over the world. Before running the experiment we asked all of our participants about prior knowledge about the metro systems. Nobody indicated that he has ever used the metro system in Barcelona (see Fig. 2).

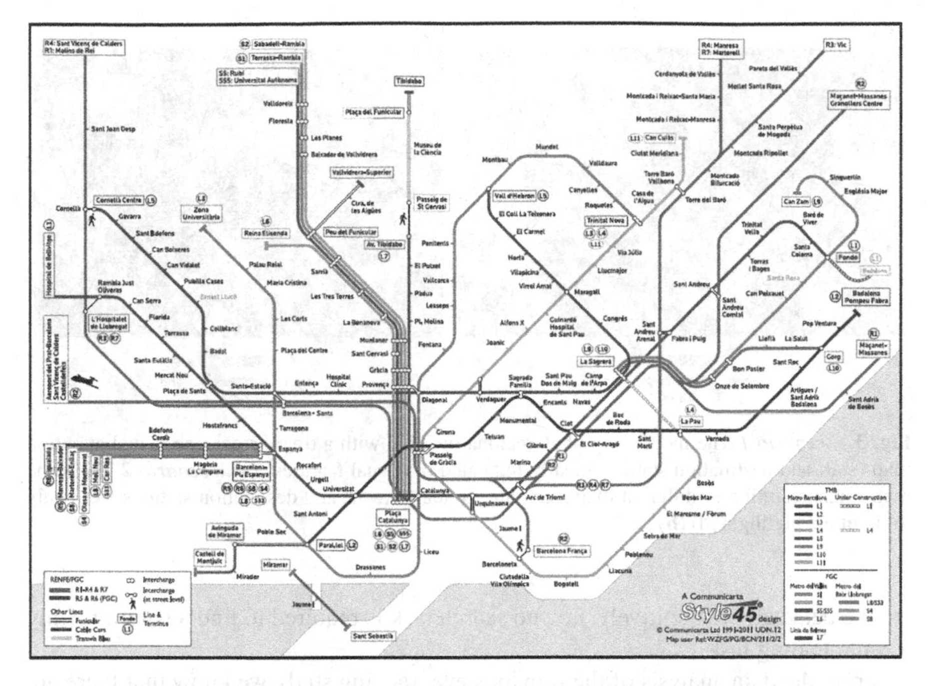

Fig. 2 The metro map of Barcelona shown as 2D static stimulus in the user study as it was designed by Communicarta Ltd. In the eye tracking study we had different scenarios (independent variables), i.e., with explicit station highlights and without. This may result in a longer search task prior to the route finding task [5]

In the scenarios in Fig. 3a, b we showed the metro map of Barcelona (same stimulus) and two different route finding tasks to show the differently applied time-varying strategies depending on the independent variable settings. We used a partially randomized box filter to achieve aesthetically pleasing visual attention maps. The maps are visually augmented by isolines. The topographic color palette is selected which maps earlier visited regions to white and grayish colors, later ones to reddish, and finally, yellow color codings (see color legend).

4.1 Scenario 1: Stations Highlighted

In this specific scenario, the study participants were told to find a route from the airport in Barcelona to the station Badalona Pompeu Fabra. Both stations are far apart from each other, i.e., the airport is located at the left hand side of the map and the destination station at the right hand side (nearly vertically centered). Metro lines had to be changed at least once to reach the destination when starting at the airport. In scenario 1 both stations were highlighted by clearly visible visual features that

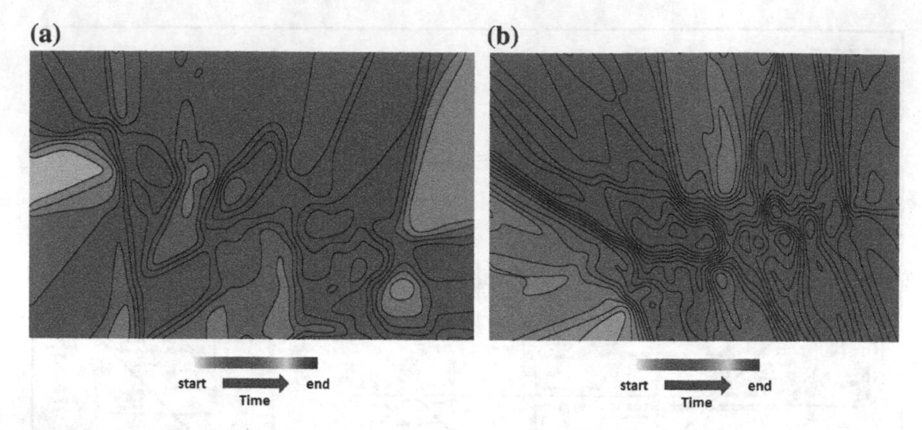

Fig. 3 *Scenario 1* The metro map of Barcelona overlaid with a time-preserving visual attention map (start and destination stations clearly indicated by visual features) (**a**). *Scenario 2* The same metro map resulting in different map scanning strategies (start and destination stations were only told but not highlighted) (**b**)

can be detected preattentively, i.e., no search task is required to find a solution, only a route finding task.

From the data analysis of the previous eye tracking study we know that there are several stages in which the route finding task can be subdivided. People followed the metro lines with their eyes and solved the route finding task partially by first finding possible candidates for interchange points. This means the complete route is subdivided into subroutes, each located between the start station, the interchange points, and the destination station. Some people also solved the route finding task vice versa, i.e., starting from the destination station going back to the start station.

This phenomenon is also reflected in Fig. 3a by the large gray colored regions located to the left and to the right. These indicate that most of the participants started somewhere there. The small gray colored region at the bottom right reflects that many of the participants first had a look at the provided map legend. For map readers this is important although we already gave the participants detailed instructions about how to read the map before running the eye tracking study. This was done in a test experiment phase which is important to check if the participants are all able to understand the task and read the stimuli reliably.

From Fig. 3a we can also see the intermediate fixation areas and also where most of the people finished the route finding task. The orange/yellow regions indicate that those were fixated close to the end, meaning people typically do not finish the task at the start or destination stations but much more often at the interchange points. This is a natural phenomenon and is inline with our previous results where people found the route but crosschecked the solution by combining the interchange points and the subroutes to the complete route.

4.2 Scenario 2: Stations Not Highlighted

Figure 3b shows a different scenario where people were asked the same task, i.e., finding a route from the airport in Barcelona to the station Badalona Pompeu Fabra. To avoid learning effects we used a between-subjects study design while we also permutated and randomized the shown stimuli in the study to avoid those learning and fatigue effects.

The difference to the scenario 1 is that the start and destination stations were not highlighted at all. This means people had to find both stations first before finding a route between them. Consequently, the main task was split into a search task and a route finding task meaning that the time-varying visual attention is much more distributed over the stimulus than in scenario 1. This insight can also be detected in the corresponding time-preserving visual attention map displayed in Fig. 3b. Here, we cannot see any gray colored regions, a phenomenon for which the formerly solved search task is to blame.

A possible explanation for the fact that the airport is fixated at later times may be a result of finding the airport first due to the additional airport sign or the prior knowledge of the study participants about the airport location. Then people tend to look much longer for finding the destination station. Once they found the destination station, they start solving the task backwards, i.e., beginning with the destination station and going back to the airport step-by-step taking into account possible interchange points.

In Fig. 4 we can see the application of the isoline interaction feature. The user interactively decides to explore the density values on a more fine-granular basis. This has the benefit that smaller differences between the values are visible but negatively, also many more isolines are introduced meaning much more overdraw of the visual attention map. Since we are not able to decide this algorithmically we leave this decision for the user.

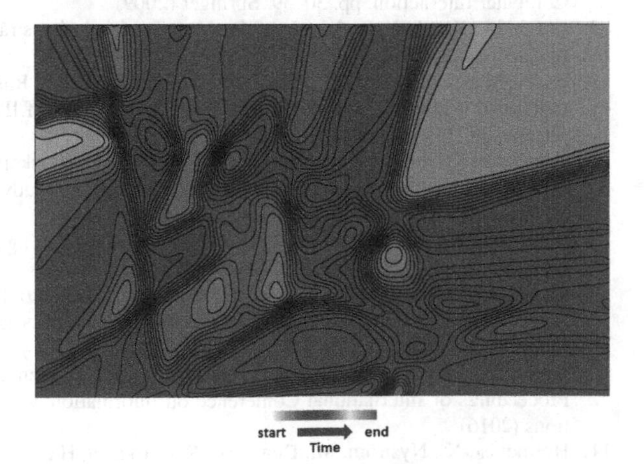

Fig. 4 *Isoline interaction* The isoline number is increased to have a look at more fine-granular density value differences (other task than in Fig. 3a, b)

5 Conclusion and Future Work

In this paper we illustrated an approach which is capable of incorporating the time-varying information of eye movements into a visual attention map. We used a Voronoi-based separation of the stimulus into subregions which are aggregated over all study participants and averaged. To achieve more aesthetically appealing results we used a smoothing step by applying reconstruction filters and finally, augment the maps by isolines. We illustrated our idea by means of a small application example where route finding tasks had to be solved by 40 participants in a formerly conducted eye tracking study. Interaction techniques are useful to further explore the data while the static time-preserving visual attention map served as a good starting point. For future work we plan to explore different stimuli and study tasks meaning the semantics can be a deciding factor for the visual appearance of the diagrams. Based on this we plan to evaluate the usefulness of our technique for certain parameter settings.

Acknowledgments I would like to thank Robin Woods from Communicarta Ldt to provide some metro maps for showing them in our eye tracking studies. Without those maps this work would not have been possible.

References

1. Andrienko, G.L., Andrienko, N.V., Burch, M., Weiskopf, D.: Visual analytics methodology for eye movement studies. IEEE Trans. Vis. Comput. Graphics **18**(12), 2889–2898 (2012)
2. Blascheck, T., Kurzhals, K., Raschke, M., Burch, M., Weiskopf, D., Ertl, T.: State-of-the-art of visualization for eye tracking data. In: EuroVis—STARs, pp. 63–82 (2014)
3. Blascheck, T., Burch, M., Raschke, M., Weiskopf, D.: Challenges and perspectives in big eye-movement data visual analytics. In: Proceedings of the 1st International Symposium on Big Data Visual Analytics, pp. 17–24 (2015)
4. Bojko, A.: Informative or misleading? Heatmaps deconstructed. In: Proceedings of Human-Computer Interaction. pp. 30–39. Springer (2009)
5. Burch, M., Kurzhals, K., Weiskopf, D.: Visual task solution strategies in public transport maps. In: Proceedings of ET4S@GISCIENCE, pp. 32–36 (2014)
6. Burch, M., Andrienko, G.L., Andrienko, N.V., Höferlin, M., Raschke, M., Weiskopf, D.: Visual task solution strategies in tree diagrams. In: Proceedings of IEEE Pacific Visualization Symposium, pp. 169–176 (2013)
7. Burch, M., Konevtsova, N., Heinrich, J., Höferlin, M., Weiskopf, D.: Evaluation of traditional, orthogonal, and radial tree diagrams by an eye tracking study. IEEE Trans. Visual. Comput. Graphics **17**(12), 2440–2448 (2011)
8. Burch, M., Kull, A., Weiskopf, D.: AOI rivers for visualizing dynamic eye gaze frequencies. Comput. Graphics Forum **32**(3), 281–290 (2013)
9. Burch, M., Raschke, M., Blascheck, T., Kurzhals, K., Weiskopf, D.: How do people read metro maps? An eye tracking study. In: Proceedings of the 1st International Workshop on Schematic Mapping (Schematics) (2014)
10. Burch, M., Woods, R., Netzel, R., Weiskopf, D.: The challenges of designing metro maps. In: Proceedings of International Conference on Information Visualization Theory and Applications (2016)
11. Holmqvist, K., Nyström, M., Dewhurst, R., Jarodzka, H., van de Weijer, R.: Eye Tracking: A Comprehensive Guide to Methods and Measures. Oxford University Press (2011)

12. Kurzhals, K., Fisher, B.D., Burch, M., Weiskopf, D.: Evaluating visual analytics with eye tracking. In: Proceedings of the Fifth Workshop on Beyond Time and Errors: Novel Evaluation Methods for Visualization, BELIV, pp. 61–69 (2014)
13. Netzel, R., Burch, M., Weiskopf, D.: Comparative eye tracking study on node-link visualizations of trajectories. IEEE Trans. Vis. Comput. Graphics **20**(12), 2221–2230 (2014)
14. Rosenholtz, R., Li, Y., Mansfield, J., Jin, Z.: Feature congestion: a measure of display clutter. In: Proceedings of the Conference on Human Factors in Computing Systems (CHI), pp. 761–770 (2005)
15. Scinto, L.F., Pillalamarri, R., Karsh, R.: Cognitive strategies for visual search. Acta Psychol. **62**(3), 263–292 (1986)
16. Spakov, O., Miniotas, D.: Visualization of eye gaze data using heat maps. Electron. Electr. Eng. **2**(74), 55–58 (2007)
17. Tufte, E.R.: The Visual Display of Quantitative Information. Graphics Press (1992)
18. Ware, C.: Visual Thinking: for Design. Morgan Kaufmann Series in Interactive Technologies, Paperback (2008)

Eye Movements in Reading the Texts of Different Functional Styles: Evidence from Russian

Tatiana Petrova

Abstract This study is one of the first eye-tracking experiment on Russian language material, checking out if the functional text style is among the readability categories and if it influences the effect of reading perspective. In Experiment participants (30 native speakers of Russian) read three texts on different topics each written in a different functional style. Questionnaires and retelling the texts were additionally used to collect data on text comprehension and accessibility. We suggest that the following eye-tracking data can be informative when we need to evaluate text readability: amplitude of saccades, number of regressions, fixation duration while searching for an answer in the text. The results indicate that the text on the same topic is easier read if it is written in a publicistic style than in a scientific style. There were no significant differences in eye-tracking data between texts written in publicistic style and colloquial style.

Keywords Text · Functional style · Readability · Eye-tracking · Process of comprehension

1 Introduction

Mechanisms of understanding of a text, which are the basis of text processing while reading, are among the main problems of modern psycholinguistics. Large number of experiments has shown that the understanding of the text (written or oral) is a complicated and multi-step process. Thus, understanding of a written text includes a plurality of additional variables, such as, recognition of letters, drawings and structural components. In modern linguistics, the question of an influence of a text type on text perception and comprehension has been studied extensively [1–6].

T. Petrova (✉)
St. Petersburg State University, Universitetskaya emb., 11,
199034 Saint Petersburg, Russia
e-mail: tatianapetrova4386@gmail.com

I. Czarnowski et al. (eds.), *Intelligent Decision Technologies 2016*,
Smart Innovation, Systems and Technologies 57,
DOI 10.1007/978-3-319-39627-9_25

M-A.Laane identifies the following variables affecting the "readability" of the text: content (abstracts, organization, coordination), style (semantic and syntactic elements), the format (design, layout, illustrations), organization (paragraphs, structure, headings) [3]. William H. DuBay notes that the text style can be experimentally verified. According to his opinion in the study of different texts there are following indicators of readability: the average length of sentences (in words), the percentage of "simple" words, the number of different "complicated" words, the number of pronouns (1, 2 and 3 person), the minimum length of a sentence (in syllables), the maximum length of a sentence (in syllables), the average sentence length (in syllables), the percentage of monosyllabic words, the percentage of polysyllabic words [4].

Mortimer Adler in his work "How to Read Books" made a very important conclusion, saying that reading is an active process [7]. At first glance, you may divide all speech activity on two processes—active and passive. In such opposition it is quite obvious, that speech production (writing and speaking) is an active process, and reception (reading and listening) is a passive process. But it is impossible to assert that while processing a text (written or oral) the communicant is passive and doesn't act. Communicant makes a number of actions while reading or listening a text: receives information, decodes it, interprets and comprehends it in a particular way. Reading it is not an easy task, as we naively believe. It consists of different stages, on each of these stages you can improve and hone your skills [7]. Thus, reading it is an active verbal and cogitative process consisting of several stages. The text is a complex and multifaceted object of perception, "it can be more or less understood holistically, from a minimum of assimilation to the full disclosure of the author's intention." The degree of integrity of text comprehension, respectively, depends on the degree of activity and involvement of a reader in the reading process. However, the type of a text will also affect the level of reading comprehension. There are two types of reading: "for information" and "for understanding" as M. Adler suggests. While reading for information we look for the facts in the text, get some knowledge and accept it. While reading for understanding we look for the meanings, concepts and implications and try to combine them with our own worldview. Reading for information is a necessary stage of studying at the starter levels, when demonstration of language units and their functioning is a subject of training and is a main goal of using of the text. However, when, skills are developed and the language material is acquired, the teacher should pay more attention to motivation of students to understand the text, but not just to extract the information from it. Thus, differentiating process of reading on two types, we can also speak about two types of the texts: that don't demand long processing and efforts for understanding and which main content lies on a surface, and also that which are acquired not at once and demand more time for perception and comprehension [7].

2 Research

2.1 Hypothesis

Detection the features of processing and understanding of texts of different functional styles in reading process was the purpose of the experiment described in this article. The hypothesis is based on the assumption that the processes of perception, including reading, processing and reproduction, in texts of different functional styles are different.

2.2 Material

For a pilot study three functional styles were chosen: scientific, publicistic and colloquial, as representing the greatest interest as a material at the lessons of Russian language and Russian for foreigners. Thus, in order that ease or complexity of the text wasn't caused by thematic features, and also to output some average result on functional style, but not on concrete unit, 9 texts on three different topics were taken, and each topic was presented by three texts of different functional styles. The choice of a subject was caused by existence of texts of these topics in all selected functional styles, and also by a potential possibility of using these texts at lessons of Russian and Russian as a foreign language. 3 topics connected with daily human life were chosen: thunder-storm, relations of children and parents, cars. Thus, the material of the research contained 9 texts:

- 3 texts about a thunder-storm (scientific, publicistic and colloquial);
- 3 texts about the relations between parents and children (scientific, publicistic and colloquial);
- 3 texts about cars (scientific, publicistic and colloquial).

The functional styles of these texts were defined by experts (Table 1). Seven respondents (4 women, 3 men)—teachers of Russian language and literature—were

Table 1 Results of an expert evaluation of a functional style of the texts

Text expert	Text 1[a]	Text 2	Text 3	Text 4	Text 5	Text 6	Text 7	Text 8	Text 9
1	scient	publ	publ	scient	publ	publ	publ	publ	coll
2	scient	publ	coll	scient	publ	coll	scient	publ	coll
3	scient	publ	coll	scient	publ	coll	scient	publ	coll
4	scient	publ	coll	scient	publ	coll	scient	publ	coll
5	scient	publ	coll	scient	publ	publ	scient	publ	coll
6	scient	publ	coll	scient	publ	coll	scient	publ	coll
7	scienti	publ	coll	scient	publ	publ	scient	publ	coll

[a]Experimental texts can be found in the Appendix 1

asked to define to which of the functional styles (scientific/publicistic/colloquial) this or that text belonged.

There was a unanimous expert decision in determining a functional style for six of nine texts. The greatest interest is the results in an assessment of text 6. Two experts defined this text as belonging to a publicistic style, four of seven—to scientific, one of experts dropped a hint of doubt, choosing between these two styles, but as a result stopped on the publicistic. This text represents an article from the encyclopedic dictionary, respectively, if to judge on a resource, the text is scientific.

For the further analysis of the texts Text Internet project was used [http://ru.readability.io]. This allowed us to calculate a number of indicators of the texts (are presented in Table 2, from the line "number of signs"), and also on their basis to calculate indexes of readability for each text (on the formulas adapted for Russian).

It is possible to notice that indicators in each group (scientific, publicistic and colloquial) are different. The texts considerably differ from each other according to quantitative characteristics.

The analysis of readability parameters of the texts revealed that scientific texts are allocated against texts of publicistic and colloquial style from the formal point of view. Their peculiarities are the following:

- contain bigger number of the "compound words" demanding bigger amount of time for text processing;
- contain less number of sentences at an equal number of words in the text, that indicates the complexity of syntax used in scientific texts;
- are focused on more adult (from the point of view of education) audience—students, whereas texts of two other styles can be apprehended by children.

However, it should be noted that these indicators of "readability" are especially formal and consider only quantitative characteristics, ignoring the qualitative.

The main objective of the research was a checkout if the functional text style is among the readability categories and if it influences the effect of reading perspective. This technique gives an opportunity to see the processes of thinking and language behavior of the person in their interaction and is the most exact for research of reading processes as it allows not only to identify difficulties when reading, but also to find when and where specifically these difficulties arise and are resolved by the reader [8]. Today it is one of the main methods of the simultaneous analysis of a large number of dependent variables for language handling research.

2.3 Subjects

30 native speakers of Russian aged from 17 to 30 years participated in eye-tracking experiment. All subjects had the higher or incomplete higher education.

Table 2 The analysis of texts on the category of "readability"

Text Index	Scientific			Publicistic			Colloquial		
	Storm Text 1	Car Text 6	Children Text 7	Storm Text 5	Car Text 8	Children Text 2	Storm Text 9	Car Text 3	Children Text 4
Flesch-Kincaid	18.99	17.5 6	22.21	6.34	8.93	10.31	7.44	5.44	3.17
Coleman-Liau	17.22	16.0 7	22.14	7.56	10	10.7	6.67	5.21	2.63
Dale-Chale	14.41	13.0 6	20.75	5.1	7.84	10.16	7.42	5.28	4.37
ARI	19.05	17.1 4	24.02	7.48	10.1 7	11.02	6.72	4.81	2.47
Audience	4–6 grade	1–3 grade	postgraduate students	7–9 grade	7–9 grade	10–11 grade	4–6 grade	4–6 grade	7–9 grade
Number of signs	1104	1119	1265	1014	1015	1054	939	895	883
Number of gaps	142	143	143	148	151	146	152	143	144
Number of letters	961	975	1122	863	862	906	787	751	739
Number of words	143	144	144	149	152	147	153	144	145
Number of phrases	7	9	9	10	11	12	16	10	7
Number of more than 4-syllabic words	22	23	41	11	12	18	6	6	7
Number of less than 4-syllabic words	121	121	103	138	140	129	147	138	138
Average number of words in a phrase	18.57	16	16	14.9	13.8	12.25	9.5	14.4	20.7
Percentage of compound words from the total number of words	15.3	15.9	28.4	7.2	7.8	12.2	3.9	4.1	4.8

2.4 Procedure

After the calibration procedure each subject received the following instruction: "Hello, dear participant! Thanks for readiness to take part in experiment. Please, attentively study the text for further retelling". Then the record of experiment began. Participants of experiment weren't limited in time and could study each text as much as they wanted for adequate perception of the text with its further reproduction. Eye movements were monitored with an SR Eyelink 1000 plus eye tracker (SR Research Ltd., Kanata, Ontario, Canada) sampling at 250 Hz. The system was configured in 'desktop mode' and equipped with a chin rest. While subjects read binocularly, only one eye (the right eye by default) was tracked. Viewing distance was approximately 60 cm. Approximately 2.4 characters were encompassed by 1° of visual angle. We used a 22' LCD monitor (HP Compaq LA2205 wg) with a refresh rate of 60 Hz (screen resolution 1680 × 1050). When the subject reported about end of this stage of experiment, he (or she) was asked to retell the text. All the retellings were recorded on a dictophone. After each retelling the subject was asked to answer 7 questions checking the degree of text comprehension. Each subject read three texts, according to the number of texts in the protocol (Appendix 2). Each protocol contained three different texts of three different styles. So, if the subject read a text "Car" written in scientific style, on the second step the same subject read text "Children" written in colloquial style, and the third step—the text "Storm" in publicistic style. The texts were randomised among the subjects. We used drift correction each time next text was presented. And we recalibrated in case we saw that it was necessary. It took about 20 min for each subject to participate in the experiment.

3 Results and Discussion

As a result of experiment 90 records of eye movements during reading the texts were made. The averaged values for each text type are in Appendix 3.

In the analysis of the results the following indicators in each text were considered: an average duration of fixations, an average amplitude of saccades, a number of fixations, a quantity of saccades, a reading time, a percentage of regressions (returns to the earlier read fragment) to the total number of saccades.

For statistical data processing Mann-Whitney's U-criterion was used from SPSS Statistics. Statistically significant difference between indicators of reading of the text was revealed for texts "Storm" and "Children", namely: reading time, number of saccades and fixations was significantly higher in the texts of scientific style than in texts of colloquial and publicistic styles; and also in reading scientific text about the Storm the percentage of regressions was higher than in colloquial and publicistic texts (See Tables 3, 4 and 5). Bold print in the tables indicates statistically significant differences. We connect the lack of a significant difference in indicators

Table 3 Results of comparison of eye-movement indicators while reading the texts of different functional styles on the topic "Car" (p-level)

	Colloquial ⇔ Scientific	Scientific ⇔ Publicistic	Colloquial ⇔ Publicistic
Duration of fixations	0.436	0.28	0.739
Amplitude of saccades	0.631	0.353	0.912
Reading time	0.123	0.853	0.19
Number of fixations	0.165	0.971	0.218
Quantity of saccades	0.165	0.971	0.218
Percentage of regressions to the total number of saccades	0.684	0.353	0.075

Table 4 Results of comparison of eye-movement indicators while reading the texts of different functional styles on the topic "Storm" (p-level)

	Colloquial ⇔ Scientific	Scientific ⇔ Publicistic	Colloquial ⇔ Publicistic
Duration of fixations	0.971	0.971	0.912
Amplitude of saccades	0.393	0.912	0.481
Reading time	**0.043**	**0.011**	0.353
Number of fixations	**0.043**	**0.019**	0.436
Quantity of saccades	**0.043**	**0.019**	0.436
Percentage of regressions to the total number of saccades	**0.003**	**0.05**	0.579

Table 5 Results of comparison of eye-movement indicators while reading the texts of different functional styles on the topic "Children" (p-level)

	Colloquial ⇔ Scientific	Scientific ⇔ Publicistic	Colloquial ⇔ Publicistic
Duration of fixations	0.315	0.971	0.739
Amplitude of saccades	0.853	0.796	0.796
Reading time	**0.002**	**0.002**	0.971
Number of fixations	**0.004**	**0.001**	0.912
Quantity of saccades	**0.004**	**0.001**	0.912
Percentage of regressions to the total number of saccades	0.853	0.28	0.218

of reading the text about cars with features of Text 6 which is intermediate (see an expert assessment of the text style).

The analysis revealed that the process of reading the texts of scientific style opposes the process of reading the publicistic and colloquial texts. There were significant differences in the processing of publicistic and colloquial texts.

The analysis of answers to the questions after each text allowed to conclude that subjects are capable to apprehend the main ideas of the texts of different functional styles, but the degree of understanding is different.

Texts of scientific style were more difficult for comprehension. There were significantly more mistakes in the answers after scientific text reading in comparison with the answers after publicistic or colloquial texts. Texts of scientific style demand more time for comprehension and we suppose that degree of their comprehension directly depends on the volume of working memory of a subject. Whereas texts of publicistic and colloquial style stimulate the reader to create his own text and to identify himself with the author of the text [9].

The mistakes made in answers to the questions on comprehension of such texts are connected first of all with a replacement of the read information by an actual information from the reader's background (Table 6).

The analysis of retellings confirms the conclusions drawn above (Table 7).

Texts of colloquial and publicistic style are reproduced easier: subjects make less mistakes at a statement of the main content of a text, they generalize the received information a little, draw conclusions on its basis and transform, "personalize" the text.

Some retellings contained subject's comments such as: *"as I understand"*, *"to my mind"*, *"I think so"*, *"It seems to me"* etc. (Table 8).

Table 6 Percentage of correct answers after the text questions

	Storm (%)	Children (%)	Car (%)
Scientific	81	89	73
Publicistic	83	90	87
Colloqiual	89	79	83

Table 7 Percentage of retellings that contain factual mistakes in comparison with the original text

	Storm (%)	Children (%)	Car (%)
Scientific	37.5	20	67
Publicistic	37.5	0	20
Colloqiual	30	12.5	0

Table 8 Percentage of retellings containing subject's comments

	Storm (%)	Children (%)	Car (%)
Scientific	75	40	67
Publicistic	37.5	17	30
Colloqiual	0	25	44

4 Conclusion

Generalizing all received results, it is possible to conclude that strategies of reading texts of different functional styles are different.

We suggest that the following eye-tracking data can be informative when we need to evaluate text readability: amplitude of saccades, number of regressions, fixation duration while searching for an answer in the text. The readability level of the text influences the effect of reading perspective. The results of checking the texts by readability formulas (Table 2) are correlated with eye tracking data and retellings.

The results indicate that the text on the same topic is easier read if it is written in a publicistic style than in a scientific style. There were no significant differences in eye tracking data between texts written in publicistic style and colloquial style. But the publicistic text is read, processed and reproduced better, than scientific.

These conclusions are important from the point of view of methodology of teaching languages (at least Russian). Scientific text is more difficult for reading and understanding. So, if the teacher selects it for studying in a classroom more pre-text work will be needed. It's not recommended to take scientific texts for demonstrating grammar or syntactic constructions as it will require much time for understanding the meaning of the phrase or the text, while in colloquial and publicistic texts it's easier to concentrate on studying grammar.

There are the following perspectives for further research:

- how do foreign students learning Russian language read texts of different functional styles, are there the same trends as in reading texts by native speakers;
- how does the instruction affect the reading and comprehension of a text;
- is there any difference in reading texts with pre-text work and without it;
- what types of pre-text work are better for text reading and comprehension.

Acknowledgments The research was supported by grant No 14-04-00586 "An experimental study of context-related factors influencing different stages of written and oral speech perception and comprehension" from Russian Foundation for Humanities.

Appendix

1. **Texts** (in Russian)

Text 1 (scientific)

Groza′ - atmosfernoye yavleniye, pri kotorom vnutri oblakov ili mezhdu oblakom i zemnoy poverkhnost'yu voznikayut elektricheskiye razryady — molnii, soprovozhdayemyye gromom. Kak pravilo, groza obrazuyetsya v moshchnykh kuchevo-dozhdevykh oblakakh i svyazana s livnevym dozhdem, gradom i shkval'nym usileniyem vetra. Po usloviyam razvitiya grozy razdelyayutsya na vnutrimassovyye i frontal'nyye. Vnutrimassovyye grozy nad materikom voznikayut v rezul'tate mestnogo progrevaniya vozdukha ot zemnoy poverkhnosti, chto privodit k razvitiyu v nèm voskhodyashchikh tokov mestnoy konvektsii i k obrazovaniyu moshchnykh kuchevo-dozhdevykh oblakov. Poetomu vnutrimassovyye grozy nad sushey razvivayutsya preimushchestvenno v poslepoludennyye chasy. Nad morem grozy voznikayut vsledstviye razvitiya konvektsii v kholodnykh vozdushnykh massakh, kotoryye dvizhutsya v nizkiye shiroty nad tèploy vodnoy poverkhnost'yu, poetomu v dannom sluchaye nekotoryy maksimum v sutochnom khode imeyet mesto v nochnyye chasy. Pri groze sostoyaniye atmosfery neustoychivo, poetomu mozhno ozhidat' poyavleniya effektivnykh metodov upravleniya grozami, osnovannykh na regulirovanii toy ili inoy gruppy protsessov, formiruyushchikh grozy (Tables 9 and 10).

Text 2 (publicistic)

Detki rastut ochen' bystro. I vot uzhe laskovaya devochka-printsessa, obnimavshaya
mamu po 20 raz v den', nachinayet khamit', krichat' i stanovitsya skrytnoy. Ne vse mamy byvayut gotovy k takomu rezkomu izmeneniyu v lyubimykh detyakh. V otnosheniyakh mezhdu nimi voznikayet khaos, negativ i neponimaniye. Vozmozhno li sokhranit' khoroshiye otnosheniya s rebenkom-podrostkom? Osnovnoye, chto stoit ponyat' i prinyat' mame: podrostkovyy period — eto otrezok vremeni, kogda

Table 9 Distribution of texts according to the protocols

Style	Topic		
	Protocol 1	Protocol 2	Protocol 3
Scientific	Storm	Car	Children
Publicistic	Children	Storm	Car
Colloquial	Car	Children	Storm

Table 10 Eye movement averaged values of the texts

Text style	Topic	Duration of fixations	Amplitude of saccade	Reading time (ms)	Number of fixations	Percentage of regressions
Scient	Car	242.954	4.993	121650	427	30.3
	Storm	230.005	5.004	174641	615	33.8
	Children	225.616	5.07	211772	752	28
Publ	Car	224.324	4.834	119419	415	27.4
	Storm	230.613	4.835	92871	337	27.1
	Children	225.665	5.163	86702	300	32
Coll	Car	228.636	4.88	82688	293	32.8
	Storm	233.21	4.67	108284	375	25.7
	Children	233.711	4.86	87617	319	27.8

rebenok nachinayet vzroslet', no vzroslym yeshche ne yavlyayetsya. I, kak lyuboy drugoy promezhutok vremeni, on zakonchitsya i proydet bezvozvratno. V podrostkovyy period yest' svoi polozhitel'nyye momenty. Eto vremya, kogda mozhno nauchit' svoikh detey byt' vzroslymi: v povedenii, myshlenii, vospriyatii okruzhayushchey deystvitel'nosti. Da, oni soprotivlyayutsya nravoucheniyam, no zato khorosho berut primer so svoikh roditeley. Kak nikogda, dlya vzroslykh stanovitsya aktual'nym sovet: « Ne vospityvayte detey, vospityvayte sebya. Vse ravno oni budut pokhozhimi na vas » . Samokontrol' i samodistsiplina — vot o chem, v pervuyu ochered', dolzhna pozabotit'sya mama.

Text 3 (colloquial)

Avtomobil' - eto chlen sem'i. Yego nuzhno kormit', poit', obuvat', odevat', myt', ukhazhivat', lechit', sledit', okhranyat' i t.d., i t.p. A on, kak lyubimoye domashneye zhivotnoye, budet sluzhit' vam veroy i pravdoy, poka vy yego ne prodadite ili ne razob'yete. Funtsionala u avto tozhe massa: i tebe vnedorozhnik, legkovushka bol'shaya i malen'kaya, moshchnaya i ne ochen', sportivnyye i dazhe raketa "zemlya-vozdukh". YA inogda assotsiiruyu raznyye avtomobili s porodami sobak. Tak vot, inogda mashina i byvayet skoreye dvornyagoy, no ona gotova otklikat'sya na lyuboy tvoy impul's nastol'ko, naskol'ko mozhno ot neè etogo ozhidat'. A byvayet, chto samyy dorogoy i « porodistyy » chetyrekhkolesyy drug sdykhayet na pervoy yame. YA v silu vozrasta i temperamenta vybral Opel' Astra OPS, dlya goroda - vpolne. Vizhu yeye i nikak naradovat'sya ne mogu, kak rebenok na lyubimuyu igrushku. A dlya derevni - u otchima Patriot, svoyu zhalko po etim dorogam dolbit'.

Text 4 (colloquial)

Ya rodila syna, kogda mne bylo 24 goda. Ya schitayu, chto toropit'sya ne stoit, mozhno pozhit' dlya sebya, razvlekat'sya, letom otdykhat' na more - yesli pozvolyayut finansy. Za neskol'ko let mozhno nakopit' opredelennuyu summu na prilichnoye pridannoye rebenku, vozmozhno, platnyye rody i platnoye nablyudeniye rebenka do goda - ne nado begat' v polikliniku i stoyat' v ocheredyakh, vrach sam priyedet na dom: polovinu infektsiy, v tom chisle i takuyu., chto my chut' ne popali v bol'nitsu, poluchili v poliklinike. Chto kasayetsya ucheby, ya by vnachale zakonchila institut, zachem sozdavat' sebe problemy - a vdrug gestoz ili yeshche kakiye-to problemy, i kak uchit'sya? Da i kogda roditsya rebenok, problem men'she ne budet. A tak, zakonchite institut, ustroites' na rabotu, porabotayete, a tam mozhno i v dekret ukhodit'. Vazhny yeshche i otnosheniya s Vashim drugom, ya vnachale by raspisalas', a uzhe potom beremenela, zachem potom nuzhny problemy.

Text 5 (publicistic)

Prezhde chem otpravit'sya na progulku za gorod, uznayte prognoz pogody. Yesli, soglasno etomu prognozu, ozhidayetsya groza, voz'mite s soboy zont ili plashch iz nepromokayemogo materiala. Dukhota — eto vernyy priznak priblizhayushcheysya grozy. Samoye opasnoye v grozu — nakhodit'sya ryadom s metallicheskimi ogradami, na smotrovykh bashnyakh i okhotnich'ikh vyshkakh. Vo vremya grozy ni v koyem sluchaye ne prikasaytes' k lyubym metallicheskim predmetam i ne lozhites' na zemlyu! Pri pervykh priznakakh nachinayushcheysya grozy postaraytes' spryatat'sya v mashine ili v lyubom ukrytii: palatke, shalashe, izbe, zdanii s molniyeotvodom. Yesli takoy vozmozhnosti u vas net, i vy vynuzhdeny ostavat'sya na otkrytoy mestnosti, spryach'tes' v kanave ili lozhbine, skrestiv nogi. Izbav'tes' ot lyubykh metallicheskikh predmetov: perelozhite ikh v ryukzak i otodvin'te yego v storonu.Yesli groza zastala vas v lesu, izbegayte otdel'no stoyashchikh i ochen' vysokikh derev'yev. Luchshe vsego pryatat'sya v kustakh ili pod nevysokimi berèzami i klènami. Sosna, topol', yel' i dub — « lyubimyye » derev'ya molniy.

Text 6 (scientific)

Avtomobil' - sredstvo peredvizheniya, kotoroye vpervyye poyavilos' v XIX-m v. Pervyye avtomobili privodilis' v dvizheniye parom, no eta konstruktsiya ne imela uspekha. Po suti, vek avtomobilya nachalsya s vvedeniya v deystviye ekipazhey s benzinovymi dvigatelyami GotlibaDaymlera i Karla Bentsa (1885-86). Dvigateli vnutrennego sgoraniya dlya etikh ekipazhey byli izobreteny yeshche ran'she neskol'kimi inzhenerami, prezhde vsego, Nikolausom Otto (1876). Do nastoyashchego vremeni osnovnyye komponenty avtomobilya ostalis' neizmennymi. Imeyetsya korpus (rama), k kotoromu prikreplyayutsya takiye komponenty, kak dvigatel' ili istochnik energii, sistema peredach, privodyashchaya v dvizheniye kolesa, rul', tormoza, i sistema podveski, kotoryye sluzhat dlya upravleniya avtomobilem, ostanovok i podderzhki korpusa. Pervyye avtomobili sobiralis' nemnogimi znatokami-mekhanikami, no sovremennoye massovoye proizvodstvo nachalos' v nachale 1900-kh gg. s deyatel'nosti Genri Forda i R. E. Oldsa v SSHA. Na bol'shinstve sovremennykh avtomobil'nykh zavodov vse chasti konstruktsii soyedinyayutsya vmeste na sborochnykh konveyyerakh. Pered otpravkoy na prodazhu gotovyy avtomobil' podvergayetsya ispytaniyu.

Text 7 (scientific)

V psikhologicheskoy literature polemika vokrug problemy « materinskogo instinkta » (opyat' zhe bez analiza samogo ponyatiya « instinkt ») razgorelas' vo vtoroy polovine XX stoletiya. Odni issledovateli utverzhdali primat sotsial'nykh faktorov v formirovanii materinskogo otnosheniya, drugiye priderzhivalas' ubezhdeniya, chto materinskaya privyazannost' podchinyayetsya vo mnogom tem zhe vrozhdennym mekhanizmam, kotoryye rodnyat chelovecheskiy vid s zhivotnymi. Rol' biologicheskikh faktorov v formirovanii materinskogo otnosheniya obsuzhdayetsya v etologicheskikh issledovaniyakh. Imprinting i privyazannost' pervonachal'no rassmatrivalis' kak prisposobitel'nyy mekhanizm vida, uvelichivayushchiy shansy vyzhivaniya. Deystvitel'no dlya rebenka ustanovleniye i podderzhaniye kontakta s mater'yu yavlyayetsya vital'noy zadachey. Issledovaniya pokazyvayut, chto psikhosomaticheskoye ravnovesiye rebenka tesno svyazano s vzaimodeystviyem rebenka i materi. Khronicheskaya nekhvatka privyazannosti privodit u rebenka k nervnoy anoreksii, rvote, bessonnitse, chastomu srygivaniyu, oslableniyu immunnoy sistemy. Naprotiv, tesnyy telesnyy kontakt sposobstvuyet chuvstvu bezopasnosti i privodit k umen'sheniyu strakha i trevogi. Dzh. Boulbi schital privyazannost' pervichno spetsificheskoy sistemoy, smysl kotoroy v podderzhanii vzaimodeystviya mezhdu mater'yu i mladentsem.

Text 8 (publicistic)
Avtomobil'. Kakoye eto priyatnoye slovo, tak kak ono oznachayet sredstvo peredvizheniya, kotoroye uprostilo chelovecheskuyu zhizn'. A ved' mnogo let nazad takogo chuda tekhniki ne bylo dazhe v pomine. Lyudi spokoyno sebe khodili peshkom i yezdili na loshadyakh . No vot v dalekom 1885 godu GotlibDaymler izobrël i zapatentoval dvigatel', a Karl Fridrikh Bents usovershenstvoval yego i zapatentoval avtomobil'. S tekh por zhizn' lyudey izmenilas'. S kazhdym godom razvivalos' avtomobilestroyeniye, izobretalis' vse luchshiye i luchshiye avtomobili, oni usovershenstvovalis' i dopolnyalis'. Segodnya eto uzhe ne prosto mashiny, a nastoyashchiye shedevry, kotoryye imeyut kazhdaya svoyu izyuminku v dizaync, v nachinke i tak daleye. Lyudi pokupayut sebe avto, chtoby idti v nogu so vremenem, chtoby sozdat' sebe komfort, chtoby oshchutit' sebya «belym» chelovekom. Pozvolit' sebe kupit' avtomobil' mozhet kazhdyy chelovek, glavnoye k etomu nuzhno stremit'sya. Ved' ne obyazatel'no pokupat' novuyu, doroguyu marku avto, mozhno kupit' srednyuyu, i ne obyazatel'no novuyu, no, chtoby ona byla na khodu.

Text 9 (colloquial)
Groza doma - fignya, dazhe yesli eto odnoetazhnyy derevenskiy dom. Sovershenno ne strashno, prosto krasivo. A vot groza na prirode… Vot eto real'no strashnovato: vrode kak pod bombèzhkoy, i v printsipe izvestno, chto, yesli dolbanèt v samoye blizhaysheye derevo, mozhet i ubit'. No, po moim nablyudeniyam, vsem absolyutno po figu. Boleye togo, kogda kupayus' na dache, vizhu, chto lyudi pri priblizhenii grozy nikuda osobo ne toropyatsya, i dazhe umudryayutsya perezhdat' sil'nuyu grozu pryamo na lugu. Eto uzhe prosto povedeniye kamikadze, no… lyudi u nas khrabryye… Let desyat' nazad byla u babuli v derevne. Leto, byli na senokose, tut tuchi i grom. Bystren'ko sobralis' i domoy. Po puti dozhd' nas zastal. Koroche, vbezhali domoy, tol'ko po komnatam razbrelis' pereodet'sya, tut babusya orèt: vsem stoyat', ne dvigaytes' i ne dyshite. Ya zamerla telom, a bashkoy kruchu, lyubopytno vsè-taki. Smotryu - sharovaya molniya. Koroche, pokruzhilas' ona po zalu i v pol. V obshchem, opasnoye eto delo…

References

1. Yagunova, E.V.: Kommunicativnaja i smyslovaja structura texta i ego vosprijatije (Communicative and sense structure of the text and text comprehension). Voprosy jasykoznanija, 6, pp. 32–49 (2007) (in Russian)
2. Petrova, T.: Context factor in discourse comprehension: evidence from norm and schizophrenia. In: SGEM2014 Conference on Psychology and Psychiatry, Sociology and Healthcare, Education, SGEM2014 Conference Proceedings, vol. 3, pp. 193–200, 1–9 Sept 2014. www.sgemsocial.org

3. Laane, M.-A.: Readability of research writing and text variables in readability formulas. In: 11th International Symposium Parnu 2012—Topical problems in the field of electrical and power engineering and—Doctoral School of Energy and Geotechnology II (2012)
4. DuBay, W.H.: The Principles of Readability. Impact Information, Costa Mesa, CA (2004)
5. Ashby, J., Rayner, K., Clifton, C.: Eye movements of highly skilled and average readers: differential effects of frequency and predictability. Quart. J. Exp. Psychol. **58A**, 1065–1086 (2005)
6. Rayner, K.: Eye movements and the perceptual span in beginning and skilled readers. J. Exp. Child Psychol. (41), 211–236 (1986)
7. Adler, M.: Kak chitat' knigi (How to read books) Rukovodstvo po chteniju velikih proizvedenij/ Prod.: Mann, Ivanov and Ferber (2011) (in Russian)
8. Just, M., Carpenter, P.: A theory of reading: from eye fixations to comprehension. Psychol. Rev. **87**(4), 329–354 (1980)
9. Uspenskij, B.: Ego Loquens: Jazyk i communicativnoje prostranstvo (Language and communicative sphere), Moscow, 320 p. (2007)

Monitoring Dementia with Automatic Eye Movements Analysis

Yanxia Zhang, Thomas Wilcockson, Kwang In Kim, Trevor Crawford, Hans Gellersen and Pete Sawyer

Abstract Eye movement patterns are found to reveal human cognitive and mental states that can not be easily measured by other biological signals. With the rapid development of eye tracking technologies, there are growing interests in analysing gaze data to infer information about people' cognitive states, tasks and activities performed in naturalistic environments. In this paper, we investigate the link between eye movements and cognitive function. We conducted experiments to record subject's eye movements during video watching. By using computational methods, we identified eye movement features that are correlated to people's cognitive health measures obtained through the standard cognitive tests. Our results show that it is possible to infer people's cognitive function by analysing natural gaze behaviour. This work contributes an initial understanding of monitoring cognitive deterioration and dementia with automatic eye movement analysis.

Keywords Machine learning · Eye movements analysis · Health monitoring · Dementia · Cognitive function

---

Y. Zhang (✉) · T. Wilcockson · K.I. Kim · T. Crawford · H. Gellersen · P. Sawyer
Lancaster University, Bailrigg, UK
e-mail: yazhang@lancaster.ac.uk

T. Wilcockson
e-mail: t.wilcockson@lancaster.ac.uk

K.I. Kim
e-mail: k.kim@lancaster.ac.uk

T. Crawford
e-mail: t.crawford@lancaster.ac.uk

H. Gellersen
e-mail: h.gellersen@lancaster.ac.uk

P. Sawyer
e-mail: p.sawyer@lancaster.ac.uk

I. Czarnowski et al. (eds.), *Intelligent Decision Technologies 2016*,
Smart Innovation, Systems and Technologies 57,
DOI 10.1007/978-3-319-39627-9_26

1 Introduction

Healthy cognitive function is essential to live an independent life. Decline in cognitive health can impact how we perform in daily activities, including walking, making food and interacting with people. A major cause of cognitive decline is dementia, a condition that currently affects around one in six people at the age of 80. Increasing life expectancy means that the number of people who develop dementia will increase. Taking the UK as an example, the number of people living with the condition is predicted to increase from the current figure of 850,000 to over 2 million by 2051 [17].

Although dementia is currently irreversible and ultimately fatal, obtaining an early diagnosis can help maintain quality of life by treating debilitating side effects, such as depression. Moreover, when improved therapies do eventually become available, it is likely that they will have to be administered before the damage to the brain becomes so severe as to render the therapy ineffective. Currently, diagnosis of dementia or of its harbinger, Mild Cognitive Impairment (MCI), is usually performed using paper-based cognitive tests such as the Montreal Cognitive Assessment (MoCA [18]). These are designed to be administered in a clinical setting such as a memory clinic but this can be stressful for the subject and yield poor ecological validity. Worse, many subjects don't refer themselves for a health check until the disease is well advanced. There is therefore a strong interest in developing new techniques for detecting cognitive decline that do not suffer from these disadvantages.

One strand of work seeks to test for deficits in the same cognitive domains (memory, executive function, motor control and so on) that are tested by the paper tests, for example, using everyday computer tasks as proxies for tasks in the tests [15]. However, our work builds upon studies that have shown that eye movements are a bio-marker for dementia [1, 7, 8].

We are engaged on a programme of research in which our goal is to develop ambient eye-tracking systems for the detection of cognitive decline. This work seeks to identify a set of eye movement features that are correlated to variations in cognitive capability among people with cognitive impairment and healthy people. This will deepen our understanding of the link between cognitive health and eye movements and provide insights that we can exploit in the design of our envisioned ambient eye-tracking system.

The primary contribution of this paper is to introduce a computational method that analyses natural gaze behaviour automatically to predict cognitive function. Following a review of eye movements analysis, we describe an experiment that record 15 participants' gaze data while they are watching videos. We then report the computed statistical eye movement features and correlate them with the cognitive assessment scores obtained through paper cognitive tests. The paper concludes with a summary of the findings and discusses the extent to which our results could be used for health monitoring.

2 Related Work

Eye movements have been shown to reflect a combination of top-down cognitive processes (e.g., the observer's task, interest and goals) and bottom-up perceptual processes (e.g., influences of low level image properties) [13]. A growing body of evidence supports the use of eye movements for predicting physical activities being performed [5] and the human performers' internal mental states [4]. Eye movements are also found to provide a sensitive marker of cognitive change or deterioration [7, 8]. Patients with dementia lose the efficient control of attention and have an impairment of inhibitory control and error-correction that exceeds the effects of normal ageing [8]. In anti-saccade tasks, error frequency is correlated with dementia severity, demonstrating a potential for eye tests that provides quantitative measures for dementia diagnosis [7].

2.1 Decoding Mental and Cognitive States

A number of studies have examined specific types of eye movements and their relation to variations in mental and cognitive states. Di Stasi et al. investigated saccade as a diagnostic measure of mental workload [9]. They tested 18 subjects in a virtual driving task with three complexity levels and found that saccadic peak velocity decreased as the mental workload increased. Their results suggest that saccadic peak velocity could be a useful diagnostic index for the assessment of operators' mental workload and attentional state in hazardous environments. Schleicher et al. [19] examined changes in a variety of oculomotoric variables (e.g., blink duration) as a function of increasing sleepiness.

Recently, there has been an increasing amount of work looking into machine learning methods to automatically decode mental and cognitive states [4]. Steichen et al. analysed eye gaze patterns in interactive visualisation tasks using a number of classification methods [20]. Their results showed that using simple machine learning on eye tracking metrics can infer a number of tasks and user characteristics. A number of studies demonstrated the possibility of using eye movement features to classify mental states in scene viewing tasks [12, 14, 16]. Kardan et al. recorded eye movements from 72 participants while performing three tasks: visual search, scene memorization, and aesthetic preference [16]. They used statistical features (mean, standard deviation, and skewness) of fixation durations and saccade amplitudes, as well as the total number of fixations. Their results showed that eye movement distributional properties can classify mental states both within and across individuals. Eye movements are also found to be signatures of implicit navigational and information search intention [14], interaction intents in command issuing [2].

2.2 Health Applications

Eye movement studies have shown that people with neurological conditions exhibit abnormal viewing patterns [1]. With the advancement of computational tools and eye tracking hardware, there is growing interest in using eye movements for health applications.

Benson et al. investigated which eye movement tests (smooth pursuit, fixation stability, and free-viewing tasks) alone and combined can best discriminate Schizophrenia cases from control subjects [3]. Their results showed that a boosted tree model achieved perfect separation of the 88 training cases from 88 control subjects. Its predictive validity on retest assessments and novel cases and control subjects was 87.8 %. However, a probabilistic neural network model was superior and could discriminate all cases from controls with near perfect accuracy at 98.3 % when evaluated on the whole data set of 298 assessments.

Tseng et al. devised a high-throughput, low-cost method to classify clinical populations where participants simply watched television [21]. Based on a computational model of visual attention, they extracted 224 quantitative features from the patients and controls' eye tracking data. Using machine learning in a work flow inspired by microarray analysis, they identified critical features that differentiate patients from control subjects which classified Parkinson's disease versus age-matched controls with 89.6 % accuracy (chance 63.2 %), and attention deficit hyperactivity disorder (ADHD) versus fetal alcohol spectrum disorders (FASD) versus control children with 77.3 % accuracy (chance 40.4 %). Similarly, Crabb et al. tested the hypothesis that age-related neurodegenerative eye disease (e.g., glaucoma) can be detected by examining patterns of eye movement recorded while a person naturally watches a movie [6]. They proposed a novel method to generate saccade density maps from scanpaths of eye movements recording and used kernel principal component analysis (KPCA) for feature extraction. They found that the generated saccadic maps can contain a signature of vision loss which can separate patients from healthy peers at reasonable accuracy. These results demonstrated that by automatic analysing eye movement patterns during visual activities (e.g., scene viewing, video watching) can access individual's cognitive health.

3 Methods

Our goal is to correlate eye movement patterns during video watching with an individual's assessment score from a battery of cognitive tests. We are building on what is known about the eye movement features of cognitively impaired people to see if they can be artificially stimulated and observed by video-watching tasks. In this experiment, we aim to find out which features are important in predicting the cognitive scores.

3.1 Participants

This study collected data from 15 participants (8 female and 7 male) with 9 older control participants (mean 66.11, std 9.57), 3 young controls (mean 26.33, std 4.04) and 3 MCI patients (mean 71.67, std 2.52). MCI is interesting because the sufferer will exhibit cognitive decline greater than that expected for their age, but which is not yet so severe as to significantly inhibit their day-to-day functioning. Someone with MCI is at high risk of developing dementia within 5 years. All participants participated voluntarily. Older adult controls were over 55 years old and recruited from a local church, younger adult controls were recruited from a local university, and MCI participants were recruited from National Health Service (NHS) Memory Services.

3.2 Apparatus

An EyeLink Desktop 1000 eye-tracker (SR Research Ltd., Ontario, Canada) was used at 500 Hz. Participants sat approximately 55 cm away from the monitor (60 Hz) (see Fig. 1). Their dominant eye was determined using the Miles test and tracked accordingly. Experimenter Builder software Version 1.10.1630 was used to control the stimulus events during the eye-tracking task.

3.3 Stimuli and Tasks

Each participant in the experiment underwent a memory test (FCSRT-IR—see the end of this sub-section) and then performed an eye movement task that required them to watch four short videos. Each video lasted 40 s. Three of these videos were viewed on three occasions each. The fourth video was only viewed once and required par-

Fig. 1 The experiment recorded participants' eye movements using a remote eye tracker while watching short video clips

Table 1 Videos viewed during the study

Video	Sessions
1: Coronation of the Queen Elizabeth II	i Free view
	ii How many bald men are in the room?
	iii What are the colours of the clothes within the room?
2: Neil Armstrong landing on the moon	i Free view
	ii How many legs has the moon lander?
	iii Can you see the astronauts' faces?
3: Gordon Brown and family leaving Downing Street after losing the general election in 2010	i Free view
	ii What are each member of Gordon Brown's family wearing?
	iii How many windows are there on the buildings?
4: Hovis: an old advertisement for Hovis bread	i Fixate on the boy with the bicycle in each scene

ticipants to fixate on one object for the duration of the video. Prior to the viewing of each video the participant was given instructions related to the video. In the *free view* session a participant was asked to freely process the video in order to obtain a measure of attention-catching objects that are highly salient. In the second two *instructed* sessions a specific question was asked which was designed to direct the top-down control of eye gaze to non-salient objects that are not necessarily attention-catching. The entire experiment lasted about an hour. The participants watched the videos in a pre-defined order, as we are only interested in eye movement patterns irrespective of the viewing content. In total, we collected 10 video trials per participant. Table 1 describes the instructions on the videos viewed on three occasions.

The participants' answers were recorded. However, we were primarily interested in differences in eye movement patterns toward salient stimuli in the different viewing conditions rather than whether the participant answered the question correctly (in fact, all participants did answer the questions accurately). Across participant groups the eye movement data was investigated in association with performance in the Free and Cued Selective Reminding Test with Immediate Recall (FCSRT-IR [10]). The FCSRT-IR is a measure of memory which is not confounded by normal age-related changes in cognition and has been associated with preclinical and early dementia. Participants are asked to memorise line drawings of easily recognised objects (e.g., grapes) which belong to unique category cues (e.g., fruit). A measure of free recall and a measure of cued recall is obtained by calculating the correct responses (both out of a total of 48).

3.4 Data Analysis

Feature extraction: Eye movements recorded before and after video watching and during blinks were removed. Because we are interested in eye movement patterns rather than prolonged fixations, we filtered out fixations that are longer than 1500 ms and shorter than 10 ms and saccades with amplitude larger than 100°. We extracted the distribution features from the fixation and saccade data per trail. These include the mean, standard deviation, skewness (a measure of symmetry) of the fixation duration, saccade amplitude, average and peak velocity, as well as the number of fixations over each video viewing. 13 features were computed from each video trial. We extract these features because [16] showed that distribution properties of fixations are effective features for classifying mental states.

Prediction model: We used the Pearson correlation coefficients to identify important features that are correlated to the memory function as assessed by the FCSRT-IR measure. We further applied three linear regression models: least-squares regression, ridge regression, and LASSO regression [11] to test our hypothesis that automatic analysing eye movement patterns can predict individual's memory capability. If we denote the input feature vector and the corresponding output memory measure as $\mathbf{x} = [x_1, x_2, ..., x_{13}]^\top$ and y, respectively, all three regression models can be expressed as:

$$\hat{y} := f(\mathbf{x}) = \mathbf{x}^\top \mathbf{w}.$$

Least-squares regression f_S is obtained by minimising the training error: For given n training examples of input feature vectors and the corresponding outputs $\{(\mathbf{x}^1, y^1), \ldots, (\mathbf{x}^n, y^n)\}, f_S$ minimises

$$\mathcal{E}[f] = \sum_{i=1}^{n} (y^i - f(\mathbf{x}^i))^2.$$

In general, simple least-squares regression can overfit to data, i.e., it fits perfectly to the training data but generalizes poorly. Ridge regression overcomes this problem by introducing a *Tikhonov regularization*: The solution f_R minimizes the sum of squared error and a ridge penalty which measures the smoothness of a function: $\mathcal{E}[f] + \lambda_R \|f\|_2^2$ where λ_R is a hyper-parameter. Finally LASSO replaces the L2 penalty in the ridge regression with $L1$ regularizer: LASSO estimator f_L minimises [11]

$$\mathcal{L}[f] = \sum_{i=1}^{n} (y^i - f(\mathbf{x}^i))^2 + \lambda_L \|f\|_1, \tag{1}$$

By trading empirical risk off with L1 penalty, LASSO tends to set some elements of the resulting coefficient vector $\mathbf{w}_L$ zero. This leads to automatic feature selection as the features corresponding to non-zero coefficients in $\mathbf{w}_L$ can be regarded as more *influential* in constructing the prediction.

To evaluate the regression models, training was applied on 70 % of the trials. We use the remaining 30 % of the trials for testing. The model parameters λ in LASSO and Ridge regression were selected with tenfold cross validations. We conducted 100 iterations of randomly sampling the training and testing sets.

4 Results

We first summarise the FCSRT-IR memory scales (out of 48) across subjects. The statistics show that free recall memory scores in the MCI group (mean = 19.67, std = 10.07) are lower than those in the healthy young (mean = 38.00, std = 4.36) and old control groups (mean = 35.33, std = 4.27).

4.1 Correlation with Memory Scale

The first goal of our analysis is to estimate the significance of the memory function differences per statistical feature extracted from the eye movements. Table 2 shows a set of correlations of eye movement features to free memory recall scale. Pearson's r ($-1 \leq r \leq 1$) indicates the strength and direction of the correlation, where 1 is total positive correlation, 0 is no correlation and -1 indicates a perfect negative correlation.

Table 2 Correlations between eye movement features and free recall measures

Feature	Free view		Instructed	
	r	p-value	r	p-value
Mean fixation duration	−0.1248	0.4140	−0.2485	0.0182[a]
Std fixation duration	−0.1228	0.4216	−0.0949	0.3735
Skewness fixation duration	0.3696	0.0125[a]	−0.0248	0.8166
Fixation count	−0.0685	0.6549	0.0772	0.4697
Mean saccade amplitude	0.1728	0.2564	0.0991	0.3527
Std saccade amplitude	−0.2653	0.0782	−0.0056	0.9571
Skewness saccade amplitude	−0.1439	0.3457	−0.0099	0.9263
Mean saccade average velocity	0.4943	0.0006[c]	0.3973	0.0001[c]
Std saccade average velocity	0.0157	0.9182	0.2737	0.0090[b]
Skewness saccade average velocity	−0.0527	0.7305	0.1240	0.2441
Mean saccade peak velocity	0.0679	0.6578	0.2254	0.0326[a]
Std saccade peak velocity	−0.0409	0.7895	0.2281	0.0306[a]
Skewness saccade peak velocity	−0.0073	0.9623	0.1459	0.1700

Notes: [a]$p < 0.05$, [b]$p < 0.01$, [c]$p < 0.001$

There is a significant positive correlation between the skewness of fixation duration and free memory recall scale ($r = 0.3696$, $p = 0.0125$) in the free view condition. The distribution of fixation durations of subjects with high memory scores exhibits a long tail to the right (longer durations). The majority of fixation durations are concentrated to the left (shorter durations). While in the instructed conditions, we find a significant negative correlation between the mean of fixation duration and free memory recall scale ($r = -0.2485, p = 0.0182$). This indicates that the average of fixation time is shorter for subjects with high memory scores.

There is a strong significant positive correlation between the mean of saccade average values of velocity in both free view ($r = 0.4943, p = 0.0006$) and instructed conditions ($r = 0.3973$, $p = 0.0001$). This shows that subjects with low memory scores showed slower saccade motion on average. This result is also reflected in the instructed condition where the mean saccade peak velocity is found to be positively correlated with the memory scale ($r = 0.2254, p = 0.0326$). The standard deviations of the average ($r = 0.2737, p = 0.0090$) and peak ($r = 0.2281, p = 0.0306$) saccade velocity are positively correlated with the memory recall scale, which indicates that the distribution of the saccade speed is more spread in subjects with low memory score.

4.2 Automatic Prediction of Memory Capability

The second goal of our analysis is to construct an optimal model based on the input eye movement to predict an individual's memory capability (The memory score has a total of 48). We fit three linear regression models to predict memory score given a set of input features extracted from eye movement recordings. Figure 2 illustrates the boxplots of the average absolute residuals (difference between the predict score and ground truth) for the three regression models in both free view and instructed conditions. Among all three models, lasso achieves the best accuracy in the free view condition with a mean absolute residual of 5.52 (std = 2.84).

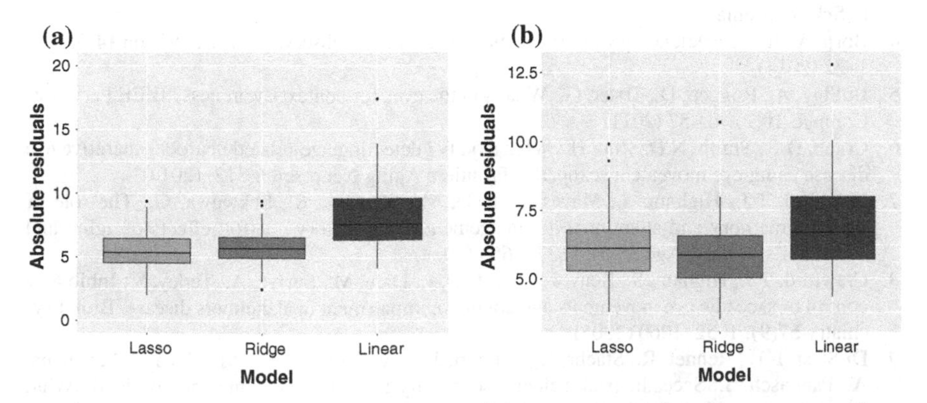

Fig. 2 Boxplots of the mean absolute residuals across three regression models

5 Conclusion

In this paper, we investigated the link between cognitive health and eye movements during visual activities (e.g., video watching). We first identified a set of eye movement features that correlate to people's memory capability, and then demonstrated that automatic eye movement analysis can predict individual's memory function score (from the standard cognitive tests). These findings provide insights into designing visual tests. Designers can focus on specific eye movement features to assess one's memory health. Our proposed predictive model can potentially be used as a new tool for quantifying cognitive health, without the need of undergoing standard tests at clinics and can be executed in people's home environment. However, our analysis is currently limited because our collected data is from a small sample of subjects in a lab environment. In the future, we intend to evaluate our model on bigger data sets, using ambient eye trackers in naturalistic environments. Other than memory capabilities, we will also investigate the relations between eye movements and other cognitive aspects, such as executive function, attention, language skills and more.

Acknowledgments The work described in this paper is funded by EPSRC project EP/M006255/1 Monitoring Of Dementia using Eye Movements (MODEM).

References

1. Anderson, T.J., MacAskill, M.R.: Eye movements in patients with neurodegenerative disorders. Nat. Rev. Neurol. **9**(2), 74–85 (2013)
2. Bednarik, R., Vrzakova, H., Hradis, M.: What do you want to do next: a novel approach for intent prediction in gaze-based interaction. In: Procedings of ETRA 2012, ETRA '12, pp. 83–90. ACM, New York, NY, USA (2012)
3. Benson, P.J., Beedie, S.A., Shephard, E., Giegling, I., Rujescu, D., Clair, D.S.: Simple viewing tests can detect eye movement abnormalities that distinguish schizophrenia cases from controls with exceptional accuracy. Biol. Psychiatry **72**(9), 716–724 (2012). Cortical Inhibition Deficits in Schizophrenia
4. Borji, A., Itti, L.: Defending yarbus: eye movements reveal observers' task. J. Vision **14**(3(29)), 1–22 (2014)
5. Bulling, A., Roggen, D., Trster, G.: What's in the eyes for context-awareness? IEEE Pervasive Comput. **10**(2), 48–57 (2011)
6. Crabb, D.P., Smith, N.D., Zhu, H.: What's on tv? detecting age-related neurodegenerative eye disease using eye movement scanpaths. Frontiers Aging Neurosci. **6**(312) (2014)
7. Crawford, T.J., Higham, S., Mayes, J., Dale, M., Shaunak, S., Lekwuwa, G.: The role of working memory and attentional disengagement on inhibitory control: effects of aging and alzheimer's disease. Age **35**(5), 1637–1650 (2013)
8. Crawford, T.J., Higham, S., Renvoize, T., Patel, J., Dale, M., Suriya, A., Tetley, S.: Inhibitory control of saccadic eye movements and cognitive impairment in alzheimers disease. Biol. Psychiatry **57**(9), 1052–1060 (2005)
9. Di Stasi, L.L., Renner, R., Staehr, P., Helmert, J.R., Velichkovsky, B.M., Cañas, J.J., Catena, A., Pannasch, S.: Saccadic peak velocity sensitivity to variations in mental workload. Aviat. Space Environ. Med. **81**(4), 413–417 (2010)

10. Grober, E., Buschke, H.: Genuine memory deficits in dementia. Dev. neuropsychol. **3**(1), 13–36 (1987)
11. Hastie, T., Tibshirani, R., Friedman, J.: The Elements of Statistical Learning. Springer New York Inc., New York, NY, USA, Springer Series in Statistics (2001)
12. Henderson, J.M., Shinkareva, S.V., Wang, J., Luke, S.G., Olejarczyk, J.: Predicting cognitive state from eye movements. PLoS ONE **8**(5), e64937 (2013)
13. Itti, L., Koch, C.: Computational modelling of visual attention. Nat. Rev. Neurosci. **2**(3), 194–203 (2001)
14. Jang, Y.M., Lee, S., Mallipeddi, R., Kwak, H.W., Lee, M.: Recognition of human's implicit intention based on an eyeball movement pattern analysis. In: Lu, B.L., Zhang, L., Kwok, J. (eds.) Neural Information Processing. Lecture Notes in Computer Science, vol. 7062, pp. 138–145. Springer, Heidelberg (2011)
15. Jimison, H., Jessey, N., McKanna, J., Zitzelberger, T., Kaye, J.: Monitoring computer interactions to detect early cognitive impairment in elders. In: 1st Transdisciplinary Conference on Distributed Diagnosis and Home Healthcare, 2006. D2H2, pp. 75–78. IEEE (2006)
16. Kardan, O., Berman, M.G., Yourganov, G., Schmidt, J., Henderson, J.M.: Classifying mental states from eye movements during scene viewing (2015)
17. Knapp, M., Prince, M., Albanese, E., Banerjee, S., Dhanasiri, S., Fernandez, J., Ferri, C., Snell, T., Stewart, R.: Dementia uk: report to the alzheimer's society. Kings College London and London School of Economics and Political Science (2007)
18. Nasreddine, Z.S., Phillips, N.A., Bdirian, V., Charbonneau, S., Whitehead, V., Collin, I., Cummings, J.L., Chertkow, H.: The montreal cognitive assessment, moca: a brief screening tool for mild cognitive impairment. J. Am. Geriatr. Soc. **53**(4), 695–699 (2005)
19. Schleicher, R., Galley, N., Briest, S., Galley, L.: Blinks and saccades as indicators of fatigue in sleepiness warnings: looking tired? Ergonomics **51**(7), 982–1010 (2008)
20. Steichen, B., Conati, C., Carenini, G.: Inferring visualization task properties, user performance, and user cognitive abilities from eye gaze data. ACM Trans. Interact. Intell. Syst. **4**(2), 11:1–11:29 (2014)
21. Tseng, P.H., Cameron, I., Pari, G., Reynolds, J., Munoz, D., Itti, L.: High-throughput classification of clinical populations from natural viewing eye movements. J. Neurol. **260**(1), 275–284 (2013)

Eye Movement Evidence of Cognitive Strategies in SL Vocabulary Learning

Irina Blinnikova and Anna Izmalkova

Abstract This paper presents the results of an experimental study modeling the situation of non-contextual foreign language vocabulary learning. Subjects (n = 31) memorized words of a foreign language, which were presented in pairs with their Russian translations. Eye movements were recorded during the trial. After the presentation subjects recalled the foreign words, and they also provided a post hoc report about the learning strategies they used. As a result, 3 main techniques of foreign vocabulary learning were distinguished ("graphical", "phonemic" and "semantic" techniques), characterized by emphasis on the corresponding level of processing. Strong patterns of interrelation between the mnemonic techniques, recall score and eye movement characteristics were observed.

Keywords Eye movements · Foreign language vocabulary learning · Paired associates · Levels of processing

1 Introduction

Our study was aimed at defining strategies of paired associate vocabulary learning. Second language vocabulary learning is a part of second language acquisition (SLA). SLA refers both to the study of individuals and groups who are learning a language subsequent to learning their first one (L1) as young children, and to the process of learning that language. The additional language is called a second language (L2), even though it may actually be the third, fourth, or tenth to be acquired [24: 2]. The studies on SLA demonstrated that learners do apply different learning

I. Blinnikova (✉) · A. Izmalkova
Lomonosov Moscow State University, Moscow, Russia
e-mail: blinnikovamslu@hotmail.com

A. Izmalkova
e-mail: mayoran@mail.ru

I. Czarnowski et al. (eds.), *Intelligent Decision Technologies 2016*,
Smart Innovation, Systems and Technologies 57,
DOI 10.1007/978-3-319-39627-9_27

strategies [1, 16]. But learning strategies are not to be understood as personal styles of learning. We understand strategies as a particular configuration of cognitive processes and mechanisms, which enable problem-solving and which change under certain conditions.

1.1 The Studies of Bilingual Lexicon

Cognitive theories seek to explain SLA in terms of mental representations and information processing. In contemporary research the problem of second language (SL) vocabulary learning is considered as bilingual[1] lexicon formation, which is viewed as mental "storage" of word forms and meanings. Research in the field of the bilingual mental lexicon began in the 1980s, when a number of models appeared describing the organization and links of lexical units of different languages. According to the word-association model, for instance, there is a direct connection between L1 and L2 lexical units—words are connected as translation equivalents [11]; other models, such as the concept-mediation model, postulated that these connections are meaning-mediated [20]. Contemporary research on the bilingual mental lexicon is flourishing (for reviews and textbooks see: [17, 19]).

Cognitive psychology and psycholinguistic research address two aspects of the bilingual mental lexicon problem: word recognition and semantic level access [4], and mental lexicon formation and organization [6, 9]. The research on mental lexicon formation mostly deals with defining the role of two groups of factors in the vocabulary acquisition process: external factors, such as presentation format, context and instruction [12]; and the use of learning strategies [1, 9, 16].

Most of methods of the bilingual lexicon investigation are based on the analysis of the lexicon organization via performance indicators (see [19]), such as reaction and execution time, percentage of correct answers. The tasks used here include lexical decision, semantic priming, picture naming, semantic categorization tasks and variations of the Stroop test. Apart from the experimental procedures there is another group of methods, which simulate ecologically valid tasks, such as narrative methods, reading and translation. In recent research these tasks have been accompanied by the recording of objective characteristics of visual material processing as it occurs, such as eye tracking (see [23]). In our study the eye tracking technology will be employed to tackle the problem of bilingual lexicon formation.

[1]In the majority of contemporary works on bilingualism the term "bilingual" is used to refer to a SL speaker of any proficiency level (see, for example, Pavlenko 2009).

1.2 Eye-Tracking Technologies in the Studies of Bilingual Lexicon

Eye-tracking has a number of advantages compared to traditional research methods of the bilingual lexicon: first, registration of eye movements can be used in analysis of natural tasks, such as reading and visual word recognition; secondly, recognition time and other performance indicators of text comprehension can be assessed through their indirect manifestations; finally, eye movement registration makes it possible to obtain objective measures of cognitive resources distribution that can be associated with the subjects' free report [22].

In the majority of eye-tracking research on SL acquisition, a reading technique has been used. The most common eye-movement measures used in a variety of tasks that include SL acquisition and processing are (see [7, 21, 23]).

Eye tracking technique has been broadly used in bilingual studies in tackling such issues as reading SL texts, lexical representation in bilinguals, grammatical and discourse processing. Some of the issues in bilingual lexicon studies that benefit from eye movement recording are:

- Processing difficulties in reading the SL context [21, 22];
- Lexical representation and access in bilinguals: semantic constraint effect in homograph recognition (words with similar forms, but different meanings in L1 and L2) and cognate facilitation (words with overlapping orthographical and semantic representations in L1 and L2) [13, 25, 26];
- Research on grammatical and discourse processing revealed, for instance, that bilinguals perform a complete syntactic parsing of sentences when reading in the second language, and they do so in a manner similar to native speakers [5].

One of the possible applications of bilingual lexicon studies is a project, iDICT, developed by Finnish computer scientists in collaboration with SensoMotoric Instruments [8]. The intellectual interface is designed to detect eye movement characteristics associated with processing difficulties (regressions, fixation duration) and to provide translation when needed.

2 Experiment

In our study, we consider a relatively simple procedure of foreign vocabulary learning: what takes place when new words are presented in pairs with their translation into the subjects' native language.

In this situation, a foreign word must be linked to the meaning given by the word of the native language, in order to be recalled later. Several considerations influenced our focus on the study of this particular learning task. Firstly, it is the main source of vocabulary learning at the initial stages of second language acquisition in adults (in bilingual second language acquisition children learn new words mostly from the context). The basis of the bilingual lexicon in adults is formed by the non-contextual method, which speeds up language acquisition. Secondly, this method allows discovering the basic connection patterns of the new word form with its meaning. Thirdly, paired associate vocabulary learning has been studied much less than contextual vocabulary acquisition.

Primary research on paired associate learning, with the use of eye tracking technique, was carried out by McCormack et al. [14]. However, there have been very few subsequent studies of paired associate learning with eye movements recording since then (especially compared to contextual vocabulary acquisition and recognition). Our work is intended to (a) fill the gap in the eye movement research of vocabulary learning and of paired associate learning in general, and (b) reveal with the help of eye tracking technologies, cognitive strategies used in paired associate vocabulary learning.

We used the paired associate technique introduced by Calkins [2]: the subjects were presented with pairs of stimuli, and then, based on the test stimulus (one of the pair), they had to recall the second stimulus. In our study the stimuli were pseudo-words and their "translation" into the native language of the subjects (Russian). We suggest that the effectiveness of memorizing new words is linked to (a) the mnemonic method used, (b) the order of word presentation, and (c) eye-movement characteristics.

2.1 Participants and Experimental Setup

Subjects. 31 university students, aged 18–26, experienced English learners (B1–C1 level of English), 23 females, 8 males.

Stimuli. 40 slides, each presenting one pair of words: Russian word and pseudoword. All Russian words were concrete nouns with frequency index from 8.5 to 22.8 per million, consisting of 7 letters. Pseudowords were made in Wuggi program [10] on the basis of English words (e.g. "consike", "remwoud", "stalore"—all the words were amendable to English phonetical rules). We checked "association strength" of the stimuli in preliminary research, which lead to exclusion of some words. Namely, if most of the subjects produced the same association to a word, it was excluded; as were excluded the words, where most of the subjects failed to provide associations. From the rest of the words we chose 40 pseudowords for the experiment.

Apparatus. The slides were presented on 19″ computer screen, in Courier New font (monospace), font size = 48. Eye movements were recorded with EyeLink 1000 eye tracker (SR Research), monocular, desktop mount, 500 Hz, camera to eye distance was 55–65 cm.

Processing. The data were processed in SR DataViewer 1.11.1 and SPSS'19 programs.

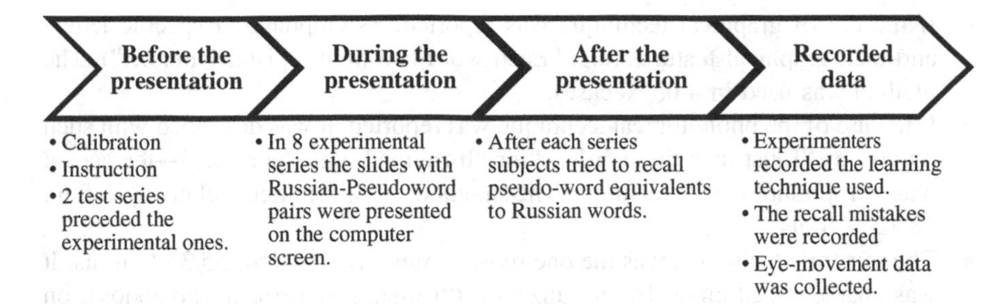

Before the presentation
During the presentation
After the presentation
Recorded data
• Calibration
• Instruction
• 2 test series preceded the experimental ones.
• In 8 experimental series the slides with Russian-Pseudoword pairs were presented on the computer screen.
• After each series subjects tried to recall pseudo-word equivalents to Russian words.
• Experimenters recorded the learning technique used.
• The recall mistakes were recorded
• Eye-movement data was collected.

Fig. 1 The general scheme of the experiment

2.2 *Procedure*

The experiment was applied individually and took approximately 60 min. Before starting the experiment, calibrations were performed. The instructions were given both orally by the experimenter and written on the screen. Subjects were told that during the experiment they would need to learn as many words of a pseudo-foreign language as they could.

The slides were presented on a computer screen, and eye movements were recorded during the presentation. The experiment included 10 series (2 test series and 8 experimental series). Each series comprised 4 slides with word pairs (presented randomly). Presentation time was 5 s per slide. The slides were separated by blank screen with fixation cross in the middle (presented for 1 s). Half of the series had Russian-pseudoword presentation order (a Russian word was presented to the left of a pseudoword), in the other half presentation order was reverse—a pseudoword was presented to the left of a Russian word.

30 s after each series, subjects were given forms on which to provide pseudo-word equivalents to Russian words. Then subjects were given correct answers and they had to provide a free report of the memorizing process (Fig. 1).

2.3 *Measures*

Qualitative recall measures. We rated recall score on a scale from 1 to 14: letter written—1 point (max-7), letter in the right place—1 point (max-7).

Quantitative characteristics of processing and recall

Defining word memorization techniques. We elicited learning techniques on the basis of subjects' post hoc report. In 12.5 % cases (124/992 trials) subjects did not provide the report. All other cases could be attributed to either graphical, or phonological, or semantic techniques, according to the following descriptions:

- If the use of graphical technique was reported, Ss emphasized specific letters and their graphical features (e.g. "Each word had high-register letters…"). The method was used in 4.03 % cases.
- If the use of the phonological technique was reported, it was described with such phrases as "I just read the words…" or "It rhymed with "pore"…"—the accent was on phonological features. The method was reported relatively often: 28.12 % trials.
- The semantic technique was the one most commonly reported: 55.35 % trials. It was characterized either by visualization ("I imagined тетрадь (notebook), on which was written consike (pseudo-word)…") or a chain of associations (for instance, auditory association to the pseudo-word (e.g. consike-конь (kon', a horse) resulted in the report, "I imagined a horse eating a notebook…"). The latter has been described by Atkinson as the "keyword method" [1].

These techniques are described according to the "depth" of processing involved, as defined in the levels of processing theory [3, 27].

Analyzing recall mistakes. The mistakes were also classified according to the levels of processing:

- "graphical" (mixing up letters similar in writing, e.g. "consile" instead of "consike");
- "phonological" (mixing up similarly read letters according to reading rules in the English language, e.g. "konsike" instead of "consike");
- "semantic" (mistakes in associations, e.g. if a keyword was "kon" (a horse), the mistake could be e.g. "ovessike" instead of "consike"[2]);
- "other" (mixed up words, missed letters, etc.).

Eye movement characteristics. In total 992 trials (878 valid) were recorded. Fixation count, dwell time on Russian and pseudoword AOIs, transition count (number of "switches" between the words), regressive eye movements within AOIs were analyzed.

3 Results and Discussion

The effect of conscious use of learning techniques was revealed: $F(4, 992) = 12.81$, $p < 0.01$ (recall score in the trials with no report of the technique was significantly lower than with the use of any learning method). The coefficient of contingency between the techniques used and the mistakes made was 0.316 ($p < 0.01$).

[2]This requires explanation: the word "consike" might be interpreted by Russian speakers as something associated with a horse (the combination of letters "con" looks like the Russian word "*kon*" (horse). This semantic construct is preserved in the memory. However, recalling the word, a subject may reproduce it as "ovessike", which means something associated with oats, and hence with a horse ("*oves*" means "oats" in Russian).

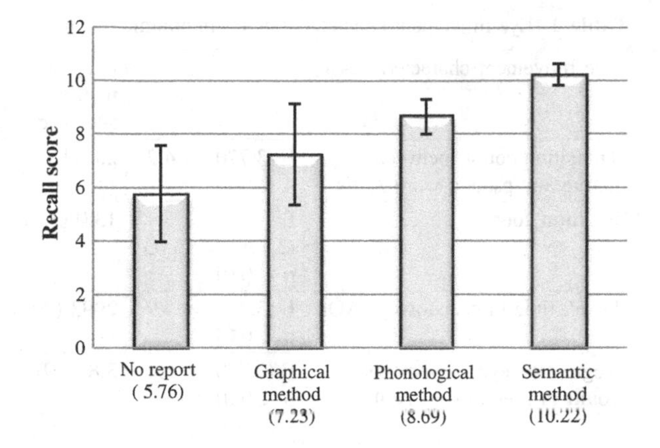

Fig. 2 Mean recall score (on the scale from 1 to 14) in the use of different techniques (mean values are given below the strategies, mean errors are marked as *error bars*)

Significant distinctions were found in the recall score depending on the techniques used: F(4,878) = 17.1 ($p < 0.01$) (see Fig. 2). The recall score increased with the "depth" of processing involved: when graphical technique was reported, the mean recall score was 7.23/14 (std. dev. = 5.72); for the phonological—8.69/14 (std. dev. = 5.47); and for the semantic—10.22/14 (std. dev. = 4.50). The high recall score in methods including a visual component, as compared to "mechanical" methods, is consistent with the results of Paivio [18] and Atkinson [1].

3.1 Eye Movement Characteristics in Different Learning Techniques

We excluded the trials where no technique was reported from data processing (leaving 770 trials). Significant differences in eye movement characteristics were found in mnemonic techniques (see Table 1).

The graphical technique was characterized by a higher fixation count and regressive eye movements on the pseudo-words and a moderate amount of transitions between AOIs (see Fig. 3). Such an eye-movement pattern can be explained by "fractional" perception of the words with emphasis on specific letters.

When subjects reported the use of the phonological method they made more transitions between AOIs, while making a minimal amount of fixations and regressive eye movements (see Fig. 4). This may be due to memorising phonetically.

When using the semantic method of memorization, few transitions between Russian and pseudo-words and a relatively large number of fixations on pseudo-words were observed (see Fig. 5). This may be due to the fact that association-based memorizing is supposed to focus on the semantic level of information processing, which leads to a relatively rapid memorization of Russian words and, consequently, a longer dwell time on pseudo-words. This asymmetrical distribution of attention does not lead to recall mistakes associated with the

Table 1 Eye movement characteristics in mnemonic techniques

Eye movement characteristics		Graphical method Mean (σ)	Phonological method Mean (σ)	Semantic method Mean (σ)
Transition count (between native and pseudo-word AOIs)	F (3,770) = 4.2 p < 0.01	2.8 (1.1)	2.9 (1.3)	2.6 (1.2)
Fixation count	F (3,770) = 13.03 p < 0.01	13.0 (3.2)	11.7 (3.5)	12.3 (3.3)
Dwell time in pseudoword AOI (ms)	F (3,770) = 5.9 p < 0.01	2993 (890)	2737 (863)	2970 (872)
Regressive eye movements count in pseudoword AOI	F (3,770) = 5.4 p < 0.01	3.8 (1.9)	2.9 (1.7)	3.3 (1.6)

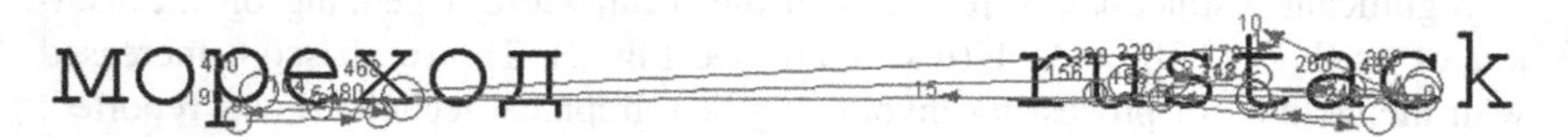

Fig. 3 S's eye movements (graphical technique used): the *circles* represent fixations (duration of the fixations are given above the *circles* and also correspond to the radii), the *arrows* represent saccades (the order numbers of the saccades are given above the *arrows*)

Fig. 4 S's eye movements (phonological technique used): the *circles* represent fixations (duration of the fixations are given above the *circles* and also correspond to the radii), the *arrows* represent saccades (the order numbers of the saccades are given above the *arrows*)

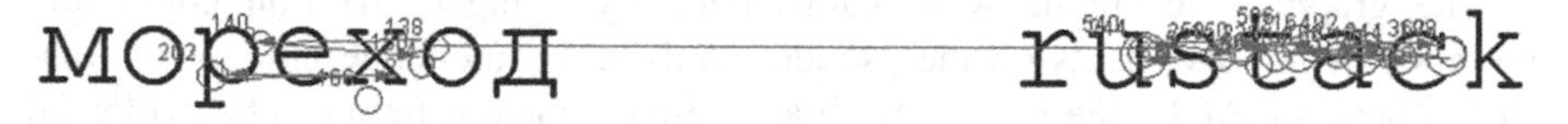

Fig. 5 S's eye movements (semantic technique used): the *circles* represent fixations (duration of the fixations are given above the *circles* and also correspond to the radii), the *arrows* represent saccades (the order numbers of the saccades are given above the *arrows*)

formation of ties with the Russian word: if the subject has completely or partially reproduced the pseudo-word, in the majority of cases (438 of 519) it has been properly related to the Russian word.

Thus, three eye-movement patterns, characteristic of the three techniques defined, were revealed. Our next step was to check whether similar eye-movement patterns can be distinguished irrespective of the techniques. We used cluster

Table 2 Cluster analysis results

	Cluster 1 (354 cases)	Cluster 2 (416 cases)
Transition count	3.14	2.44
Dwell time in pseudoword AOI (ms)	2262.23	3587.68
Dwell time in native word AOI (ms)	1180.96	695.38

Table 3 The distribution of the trials between clusters according to the learning strategies used

	Graphical method (%)	Phonological method	Semantic method
Cluster 1	50	54 % series	43 % series
Cluster 2	50	46 % series	57 % series

analysis to categorize the experimental series on the basis of S's eye movement characteristics, namely, transition and dwell time on native and foreign words. The initial 3-cluster solution which we used did not yield any results that we could match with the techniques. Therefore we opted for a 2-cluster solution, which is given in the details below (see Table 2).

The first cluster, as opposed to the second cluster, included a series with higher transition count, shorter fixation duration on foreign words, and longer fixation duration on native words. The coefficient of contingency between the clusters and the techniques that the subjects used was 0.12 ($p < 0.01$). The source of distinctions was in the phonological and the semantic method (See Table 3).

Therefore the distinguished eye movement patterns are partly verified by cluster analysis, however, further data has to be collected to make predictions of the learning method based on eye movements.

3.2 *Eye Movement Characteristics Depending on the Localization of the First Fixation*

We varied the order of presentation to provoke an even distribution of the first fixations between a Russian word and a pseudo-word. Eye movement characteristics in different learning strategies depended on the localization of the first fixation in a way that differed from the results already reported.

When graphical and phonological techniques were reported, the difference in transition count depending on the localization of the first fixation was not significant. This can be attributed to the lack of emphasis on semantic processing. Since the main focus was on memorizing the visual characteristics of the foreign words (in the graphical technique), and remembering the words by their phonetic link with the Russian word (in the phonological technique), it was not so critical which word was the first to be fixated.

When the semantic technique was used and the first fixation was made on the pseudo-word, the number of "transitions" was significantly higher—2.8 (1.1), than when the first fixation was made on the Russian word—2.3 (1.3); F = 27.9; $p < 0.01$. If we consider the process of memorizing foreign words in terms of inclusion of a new form into an existing logogen (in the terms of Morton [15]), this trend can be explained by the fact that when the learning started with the Russian word, the meaning was already given and it did not require any rehearsal, while in the opposite case the subject saw the new form first and then its meaning, and after that had to go back to the pseudo-word, as it was the word that had to be remembered.

4 Conclusions

We found and described three strategies of memorizing of new SL vocabulary units: the Graphical strategy, the Phonological strategy and the Semantic strategy. These strategies were manifested in mnemonic techniques, reproduction errors and distinctive eye movement patterns of the subjects. This is in accordance with the levels of processing theory [3, 27].

The results obtained show the influence of the conscious use of mnemonic techniques in recall activity—recall score increases with increasing "depth" of processing. Recall was significantly better when the technique included a visual component compared to "mechanical" methods of memorizing, which accords with the results of Paivio [18] and Atkinson [1].

The techniques used for non-contextual foreign vocabulary learning are reflected in eye movements. When the use of graphical technique was reported, a higher fixation count was observed, and more regressive eye movements on the pseudo-words and a moderate amount of transitions between AOIs were made. Such an eye-movement pattern can be explained by "fractural" perception of the words. When the use of the phonological technique was reported, subjects made more transitions between AOIs, while making a minimal amount of fixations and regressive eye movements. This may be due to learning the words by their phonetic link. When using the semantic method of memorization, few transitions between Russian and pseudo-words and a relatively large number of fixations on pseudo-words were observed.

The recall score was connected with the use of a conscious learning method and, in particular, the use of the semantic method, which scores higher on the "depth" of semantic processing. Eye movements in this technique were connected with the localization of the first fixation. The acquired data open up prospects for the evaluation of online learning strategies.

References

1. Atkinson, R.: Mnemotechnics in second-language learning. Am. Psychol. **30**(8), 821 (1975)
2. Calkins, M.: Association. Psychol. Rev. **1**, 476–483 (1894)
3. Craik, F., Lockhart, R.: Levels of processing: a framework for memory research. J. Verbal Learn. Verbal Behav. **11**(6), 671–684 (1972)
4. Dijkstra, T., Van Heuven, W.: Word recognition in the bilingual brain. In: The Handbook of the Neuropsychology of Language, pp. 451–471(2012)
5. Frenck-Mestre, C., Pynte, J.: Syntactic ambiguity resolution while reading in second and native languages. Q. J. Exp. Psychol. **50**(1), 119–148 (1997)
6. Hernandez, A., Li, P.: Age of acquisition: its neural and computational mechanisms. Psychol. Bull. **133**(4), 638 (2007)
7. Holmqvist, K. et al.: Eye tracking: A comprehensive guide to methods and measures. Oxford University Press (2011)
8. Hyrskykari, A.. Utilizing eye movements: overcoming inaccuracy while tracking the focus of attention during reading. Comput. Hum. Behav. **22**(4), 657–671 (2006)
9. Izmalkova, A., Blinnikova, I.: Eye movement correlates of cognitive strategies in foreign language vocabulary learning. In: Perception, vol. 44, pp. 131–131. Sage Publications Ltd. (2015)
10. Keuleers, E., Brysbaert, M.: Wuggy: a multilingual pseudoword generator. Behav. Res. Methods **42**(3), 627–633 (2010)
11. Kirsner, K., Smith, M., Lockhart, R., King, M., Jain, M.: The bilingual lexicon: language-specific units in an integrated network. J. Verbal Learn. Verbal Behav. **23**(4), 519–539 (1984)
12. Kroll, F., De Groot, A.: Handbook of Bilingualism: Psycholinguistic Approaches. Oxford University Press (2005)
13. Libben, M., Titone, D.: Bilingual lexical access in context: evidence from eye movements during reading. J. Exp. Psychol. Learn. Mem. Cogn. **2**(35), 381 (2009)
14. McCormack, P., Haltrecht, E., Hannah, T.: Monitoring eye movements during the learning of successive paired-associate lists. J. Verbal Learn. Verbal Behav. **6**(6), 950–953 (1967)
15. Morton, J.: The logogen model and orthographic structure. Cogn. Process. Spell. 117–133 (1980)
16. Nassaji, H.: L2 vocabulary learning from context: strategies, knowledge sources, and their relationship with success in L2 lexical inferencing. Tesol. Q. **37**(4), 645–670 (2003)
17. Okasha, S.: Conceptual Transfer in the Bilingual Mental Lexicon. Trafford Publishing (2012)
18. Paivio, A.: Dual coding theory: retrospect and current status. Can. J. Psychol./Revue canadienne de psychologie **3**(45), 255 (1991)
19. Pavlenko, A.: The bilingual mental lexicon: interdisciplinary approaches 70. Multilingual Matters (2009)
20. Potter, M., So, K., Von Eckardt, B., Feldman, L.: Lexical and conceptual representation in beginning and proficient bilinguals. J. Verbal Learn. Verbal Behav. **23**(1), 23–38 (1984)
21. Rayner, K.: Eye movements in reading and information processing: 20 years of research. Psychol. Bull. **3**(124), 372 (1998)
22. Rayner, K., Reichle, E.D., Pollatsek, A.: Eye movement control in reading and the EZ reader model. Cogn. Process. Eye Guidance 131–162 (2005)
23. Roberts, L., Siyanova-Chanturia, A.: Using eye-tracking to investigate topics in L2 acquisition and L2 processing. Stud. Second Lang. Acquis. **35**(02), 213–235 (2013)
24. Saville-Troike, M.: Introducing Second Language Acquisition. Cambridge University Press (2006)

25. Schwartz, A., Kroll, J., Diaz, M.: Reading words in Spanish and English: mapping orthography to phonology in two languages. Lang. Cogn. Process. **22**(1), 106–129 (2007)
26. Van Hell, J., Dijkstra, T.: Foreign language knowledge can influence native language performance in exclusively native contexts. Psychon. Bull. Rev. **9**(4), 780–789 (2002)
27. Velichkovsky, B.: Levels of perception: new evidence from eye-tracking and brain-activity analysis. Percept. ECVP Abstr. **35** (2006)

Application of Eye Tracking for Diagnosis and Therapy of Children with Brain Disabilities

Katarzyna Harezlak, Pawel Kasprowski, Michalina Dzierzega and Katarzyna Kruk

Abstract Children affected by brain disabilities require a lot of attention and care to improve their life. The quicker the support will be introduced the better effect can be achieved. One of important impairments resulted from the brain disability is a cerebral visual one. In case of children, communicating with whom is difficult or impossible, assessment of vision quality is very challenging and eye tracking methods may prove very useful. The research discussed in this paper was devoted to development of a workspace, which may support the effort of therapists working on improving the quality of disabled children's life. The important element of this solution is the implicit calibration procedure, making eye movement registration possible. Additionally, there were several stimuli developed and tested with cooperation of therapists from one of associations for children with developmental disabilities. Initial tests confirmed usefulness of the elaborated solution, which facilitates children's vision assessment based on the eye movement signal and may be used for a further children therapy.

Keywords Eye tracking · Brain disabilities · Diagnosis · Therapy

1 Introduction

Nowadays one of the most common type of vision impairments present in children with other disabling conditions is cerebral visual impairments (CVI), which may coexist with other causes of visual impairments such as ocular and ocular motor based [1]. Coupled impairments very often occur already in prenatal life and result

K. Harezlak · P. Kasprowski (✉) · M. Dzierzega · K. Kruk
Silesian University of Technology, Gliwice, Poland
e-mail: pawel.kasprowski@polsl.pl; kasprowski@polsl.pl

K. Harezlak
e-mail: katarzyna.harezlak@polsl.pl

I. Czarnowski et al. (eds.), *Intelligent Decision Technologies 2016*,
Smart Innovation, Systems and Technologies 57,
DOI 10.1007/978-3-319-39627-9_28

from the hypoxic—ischaemic brain injury, chromosomal disorders or adverse influences acting on the child during the fetal period. Complex disability may also be caused by meningitis, infections of the central nervous system epilepsy taking place in postnatal period. The pattern of the cerebral injury depends on the maturity of the brain at time of its damage.

Children with complex disability require systematic brain stimulation. For this reason many visual, auditory and tactile stimuli should engage children's attention. Brain receiving pieces of information from various senses organizes them by recognizing, analyzing and the integration. However, the first important step is to gain the knowledge to what extend children with impairments are able to use vision and for how long they can keep their attention at an environmental object. It may significantly improve work of educational teams and therapists.

Even weak visual abilities may serve as a starting point to a vision enhancement entailing overall intellectual development. Thus children with complex disability including visual impairment should be assessed in terms of functional vision and subsequently subjected to a vision therapy. The aim of the vision therapy is to stimulate vision through providing conditions favorable viewing (i.e. presenting objects possible to be noticed by a child) and provoking the development of basic skills through exercises of eye visual motor system.

Physiologically, vision may be divided into following elements: light, color, contrasts, movement and stereopsis. Assessment of the functional vision relies on determining conditions in which a child is able to see and perceive objects [5]. During tests the size, contrast and an object distance are taken into account to check following visual functions:

- Fixation—ability to keep eyesight on a visual stimulus,
- Eyeballs motility—ability to trace a moving stimulus with eyes,
- Functional visual acuity—a distance from which a child recognizes a character of a given size. Visual acuity is the quantitative measure of the ability to identify black symbols on a white background at a standardized distance, when the size of the symbols varies,
- Contrast sensitivity—the impact of a presented level of a contrast on a child's visual ability,
- Field of vision—an area, within which a child is able to see the presented object. The observation of child's responses to an environmental object may reveal interesting information about impairments.

Such an assessment in case of young non—verbal children with CVI is challenging, while behavior ability, sensitivity to environmental changes and eye movement disorders are taken into account. Thus, determination of CVI, influenced by different cognitive and motor levels may be a difficult task and flexibility during its evaluation is very important.

2 Eye Tracking Support

As a consequence of the previously—presented reasons, two main challenges for treatment of children with complex impairments come to the fore. At first it is very important to determine whether a child is able to see a visual stimulation. Secondly, a therapy requires the usage of stimuli that are interactive—i.e. change in response to child's reactions. In both areas eye tracking utilizing a specialized device seems to be a promising solution.

Because an eye tracker is able to track a gaze point (a place where a child is looking at), gathered data may be used to analyze children's reactions to visual stimuli [4, 7]. Thus, in the diagnosis phase, it is possible to present to children several stimuli differing in colors, contrast and speed and use an eye tracker to check, which of the stimuli is the most interesting for them. During the therapy phase children may work with a computer display and an application that is adapted to their abilities and actively responds to their eye movements. Another important advantage of this method is that all children's sessions may be stored and then analyzed offline by therapists.

It is worth noticing that there are different types of eye trackers. Head-mounted eye tracker allows participants to move their head freely during recordings, yet may not be adjusted to a child's head and may disturb a child. This problem is not present in the case of remote eye trackers, but it must be remembered that turning a head away from an eye tracker causes loss of an eye movement signal. The aforementioned problems are strengthened when children with such impairments like cerebral palsy are taken into consideration, because of the lack of ability to control head movements.

The research discussed in this paper was devoted to development of the workspace, which may support the effort of therapists working on improving the quality of disabled children's life. For this purpose, the group of therapists from an association for children with developmental disabilities was involved to elaborate the appropriate set of stimulations useful in their daily job. Initial experiments showed that this workspace may be helpful in both the diagnosis and therapy phases.

3 Workspace Description

The workspace created for experiments with children is presented in Fig. 1. It consists of one big display ('stimulation screen'), which is used to present a stimulation to children and an additional display—for an operator/therapist but not visible to children—which shows some additional information including: gaze position of a child and eye location in an eye tracker's camera. This display is called the 'control screen'. The operator is able to invoke stimuli and observe children's reactions. She or he can also use the application to trigger special actions such as sounds or an object's movement on the stimulation screen.

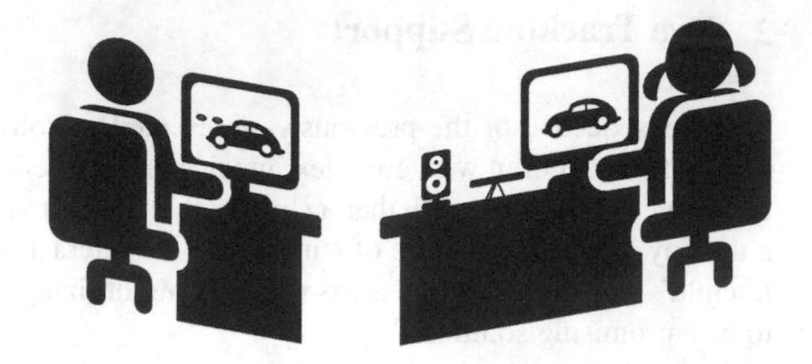

Fig. 1 Workspace used during experiments. It consists of the stimulation screen (*right*) visible to a subject and the control screen (*left*) visible to an operator

3.1 Calibration

Every eye tracking session should start with a calibration, during which a function mapping gaze to screen coordinates is built. In such a process a user is expected to follow with eyes a stimulus appearing in various places of a screen. Because children generally, and especially children with impairments, don't tend to cooperate [10] it is not possible to use the same scenario for calibrating them. There are different possibilities how to deal with this problem. For instance during the studies presented in [6] the calibration was limited to a presentation of only two points. In work described in [3] a small, visually attractive sounding toy at one of the five predefined spatial positions was displayed. However, both solutions are not feasible for children with CVI as they don't cooperate at all. For this reason a new solution had to be invented.

It was observed during our previous experiments that, when an eye tracker is calibrated for one person, it is possible to utilize an output it produces when another person is using the same eye tracker. Obviously, the signal must be recalibrated to show true gaze points of this new person, however, even before this recalibration it is possible to see that eye is moving and determine the direction of this movement [8].

To calibrate the eye tracker properly it is necessary to have some true gaze positions and corresponding eye tracker output. Therefore the idea was to perform an implicit calibration made by an operator [2]. Subsequently, the operator uses the control screen to observe both a stimulation and eye tracker output. When a new object appears on a screen and the operator sees child's reaction (eye tracker output changes) it may be assumed that now the child is looking at this object. The operator needs only to click the object on the control screen and the application registers both the click coordinates and the current eye tracker output.

The gaze calculation module (GCM) has two inputs: uncalibrated gaze coordinates from an eye tracker and coordinates of operator's clicks (Fig. 2). At the beginning the GCM returns gaze coordinates as it receives from the eye tracker. When an operator clicks with mouse any place on a screen, this information is sent to GCM and registered together with the eye tracker output obtained in the same moment as a new *calibration point* (CP). When at least four CPs are available, the recalibration model is calculated. This model maps an eye tracker output to gaze coordinates on the screen. Such a model consists of two functions:

Fig. 2 Simplified Gaze calculation module schema

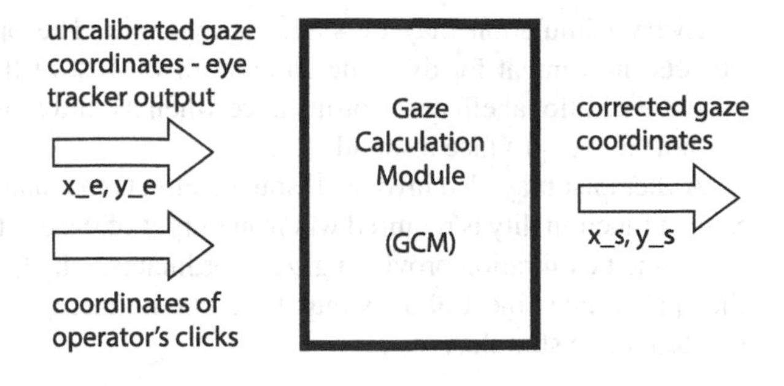

$$x_s = f(x_e, y_e) \tag{1}$$

$$y_s = f(x_e, y_e)$$

where x_e and y_e represent data obtained from an eye tracker and x_s and y_s are an estimated gaze coordinates on a screen.

There are multiple regression functions possible to use [9]. In this study the polynomial quadratic function was used (2).

$$x_s = A_x x_e^2 + B_x y_e^2 + C_x x_e y_e + D_x x_e + E_x y_e + F_x \tag{2}$$

$$y_s = A_y x_e^2 + B_y y_e^2 + C_y x_e y_e + D_y x_e + E_y y_e + F_y$$

The coefficients of the function were calculated based on calibration points (CP) using Levenberg-Marquardt algorithm. Every new point (a new click by an operator) causes recalculation of the coefficients. After collecting at least four such points, the application is able to calculate a calibration function and use it to show child's gaze point more accurately. Of course the accuracy increases when more points are available [8].

3.2 *Working Scenario*

The work starts with a classic calibration made by an operator. Then a child is seated at the stimulation screen and the operator sits at the control screen. The operator may now start different stimulations and observe child's reactions. The control screen displays the same image as the stimulation screen, together with information if and where eyes are visible by the eye tracker and about gaze positions. The gaze positions presented are not accurate, so the operator may use the procedure described in the previous section to calibrate the signal.

Every stimulation may be static or dynamic. The operator can start and stop objects' movement for dynamic stimulations. Most of the stimulations are able to produce additional effects—for instance when a child looks at some object (like e.g. animal) it may produce a sound.

A therapist may also invoke all sounds effects manually by clicking a mouse key. Such a functionality is required when an output of the eye tracker in conjunction with an implicit calibration provided gaze coordinates, which did not match a position of the appropriate object of the simulation, but the therapist is convinced that the child reacted to the stimulus.

4 Initial Experiments

The research group accounted for the 3 children including two visually impaired with cerebral palsy and one child without any impairment. The first two participants involved in the experiment were chosen based on the following criterions—they were children with complex impairments with deep visual perception problem on one hand, and there were the conviction that children were able to see at all, on the other hand. For the purpose of the experiment description, eye movement signals gathered for these children are denoted as follows:

- Sub1—the healthy girl (12 years old)
- Sub2—the girl with deficits in the field of visual attention and with cerebral palsy (5 years old)
- Sub3—the boy with deficits in the field of visual perception and with cerebral palsy (8 years old)

The experiments were conducted with the consent of those children's parents.

In order to achieve the research goal, the five stimuli were developed, which were consulted with therapists working daily with the chosen children. Their experience in field of children's interests and abilities was helpful in preparing appropriate incentives:

- S1—a car that moves from one side of a screen to another. Gaze focused on the car triggers the sound effect representing an engine whirr.
- S2—a bike moving between sides of a screen. Similarly to the previous case, focusing on the bike results in triggering the sound of thrown derailleurs.
- S3—a flower in a pot and a watering can. At the beginning the flower is withered. When a subject looks at the watering can the animation presenting watering the flower is invoked, which results in the flower straighten. If a gaze is focused on the flower, its petals grow up.
- S4—a boat, when focused the sound of a sea is invoked.
- S5—waves with a fish, which floats through a screen and disappears. Observing waves makes the fish appearing once again but on the other side of the screen.

Each experiment consisted of two basic steps: the calibration process realized by a therapist and one of the described stimuli: S1, S2, S3, S4 and S5. The eye tracker had to be calibrated by the therapist because in case of two children it was impossible to explain rules of this process. After the first step, a child was placed in the front of the stimulation display while a therapist took place in the front of the control display ensuring a stimulation preview and its management (see Fig. 1). The operator screen was outside of a child's vision field, not to disturb her/him during a test.

5 Analysis of Results

Due to the paper space limitation, results obtained during the tests will be only discussed in regard to two stimuli—the car (S1) and the flower (S3).

5.1 Car Stimulation (S1)

One healthy child (*Sub*1) and one child with visual impairments and cerebral palsy (*Sub*2) took part in the experiments based on the car stimulation. The map of the example set of registered fixations is shown in Fig. 3—on the left.

Additionally, data from *Sub*1 was presented in Fig. 4—for OX axis on the left and for OY axis on the right, respectively. Coordinates are given in screen pixels and in the function of time expressed in seconds. The green area represents the moving object whereas the black line represents the eyes position in the specific time. It can be observed that for the most of the time of the experiment eyes were visible for the recording device. Furthermore, it is visible, that the child traced with eyes the moving object during the whole experiment.

In the next figure (Fig. 5) recordings from *Sub*2, with usage of the same division, were shown. The red area and black line play the same role as in the previous charts and represent the moving object and eyes position respectively. Analyzing these charts, it may be noticed that, during the first 12 s of the stimulation, the

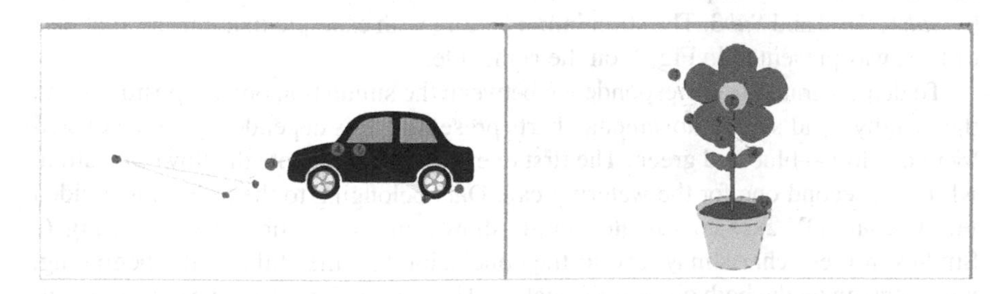

Fig. 3 The fixation maps of example stimuli S1 (*left*) and for S3 (*right*)

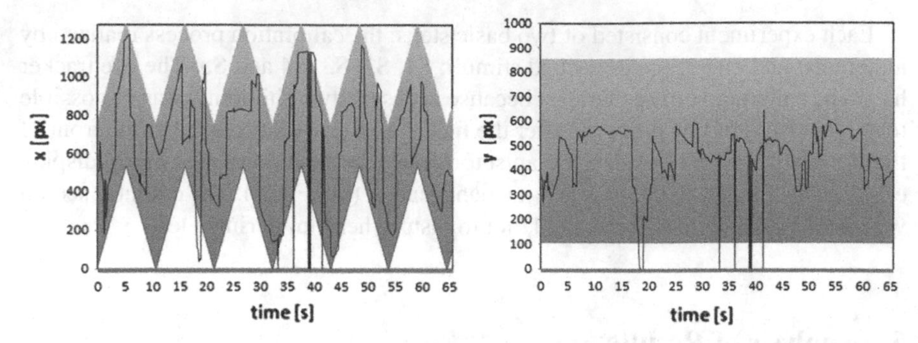

Fig. 4 Eye movement recordings for stimulation S1 for *Sub*1 for OX (*left*) and OY (*right*) axes

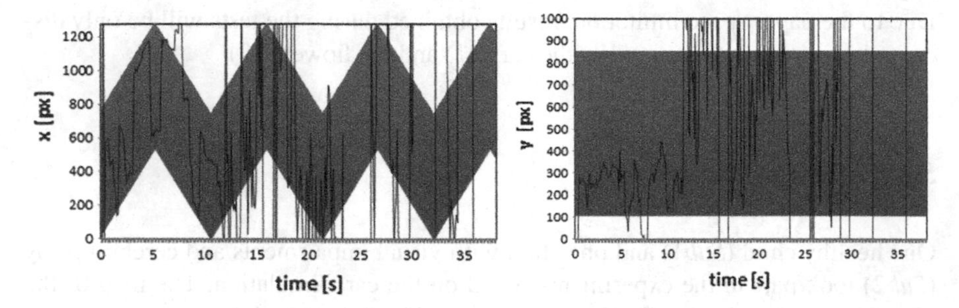

Fig. 5 Eye movement recordings for stimulation S1 for *Sub*2 for OX (*left*) and OY (*right*) axes

eyes movement coincides with the moving object area. After that time the schematic search from one edge of the screen to another is observed. It may indicate that the child became bored or her attention was distracted.

5.2 *Flower Stimulation (S3)*

The second experiment used the stimulation S3—the picture of the flower in the pot. All three children took part in this experiment, with eye movement signals denoted by *Sub*1, *Sub*2 and *Sub*3. The stimulation picture with example fixations for that part of tests was presented in Fig. 3, on the right side.

To demonstrate the correspondence between the simulation objects positions and the healthy child's eyes movement, charts presenting this dependency were covered by two colors—blue and green. The first one was used to denote the flower location, while the second one for the watering can. Data belonging to the *Sub*1 was divided into OX and OY axes coordinates and is shown, in the function of time, in Fig. 6. Studies of these charts may lead to the conclusion that the child, at the beginning, was focusing on the both objects alternately. Then, after about 10 s, the child concentrated gaze on the watering can. Such a behavior triggered the animation representing

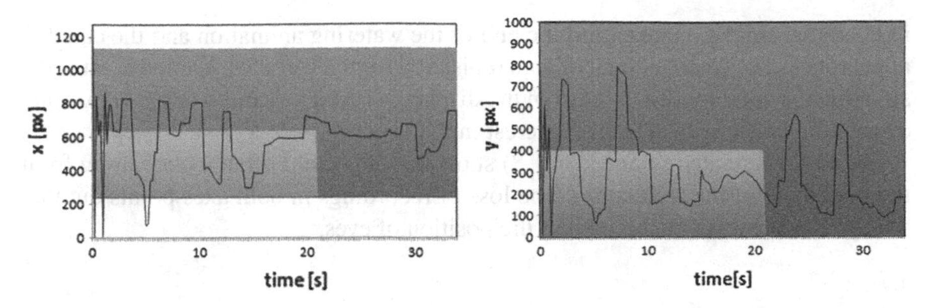

Fig. 6 Eye movement recordings for flower stimulation (S3) and *Sub*1 for OX (*left*) and OY (*right*) axes

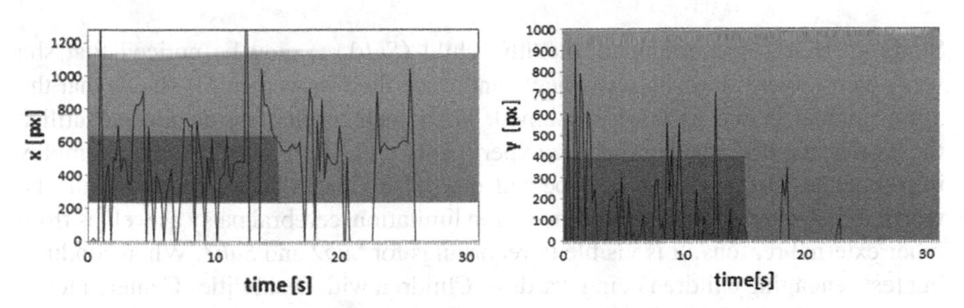

Fig. 7 Eye movement recordings for flower stimulation (S3) and *Sub*2 for OX (*left*) and OY (*right*) axes

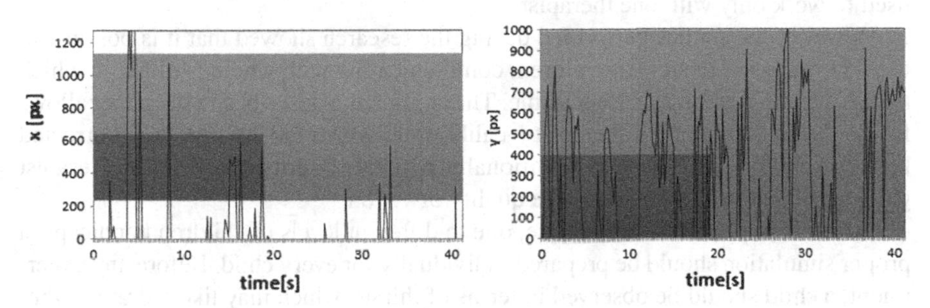

Fig. 8 Eye movement recordings for flower stimulation (S3) and *Sub*3 for OX (*left*) and OY (*right*) axes

watering the flower, after which the can disappeared. In Fig. 6—on the right side—it is visible that the child moved eyes towards the flower observing the effect of the watering (the flower straightening).

Figures 7 and 8 contain results of measurements for *Sub*2 and *Sub*3. To distinguish charts for the healthy child and for those with impairments, there were different colors used for simulation objects placement—orange one was used to mark the space used by the flower and red one for the watering can. Data obtained for *Sub*2 (Fig. 7) indicate the child's interest until the end of the watering animation and the can disappearance. The small amount of data registered for the vertical eye movement is the sign that the child looked away from the display. It could be caused or by an external distracting factor or by a loss of interest in the experiment.

Recordings related to *Sub*3 (Fig. 8) show that this child often looked away from the beginning of the simulation. The loss of recordings in both axes points out that the eye tracker was not able to find the position of eyes.

6 Summary

Studying charts concerning the healthy child (*Sub*1) it may be noticed that she was able to focus on displayed objects and trace their movement. It shows that the application operated as it was assumed. With such results, we decided to utilize the workspace described earlier for experiments with children affected by complex impairments. However, it was expected that in this part of tests results would be worse, because might result from the vision limitation, cerebral palsy as well as from other external reasons. It is visible in recordings for *Sub*2 and *Sub*3. When conducting tests engaging children being wards of Children with Disabilities Center, a lot of tension and distraction was caused by new conditions they have been put into. Other possible problem might be the presence of additional people, because children were used to work only with one therapist.

However, the studies performed during the research showed that it is possible to use an eye tracker device for children, communicating with whom is difficult, which makes a child's calibration impossible. The application described in the paper allows for the initial—done by a therapist—calibration, which during tests was corrected based on child's eyes behavior. Additionally, with the described workspace, therapist gains an objective tool to assess the quality of vision.

Which is important to emphasize, due to different levels of children handicaps, a proper simulation should be prepared individually for every child. Before any experiment, a child should be observed in terms of things, which may likely draw his/her attention. Different image attributes as color, contrast, type of objects should be taken into account. It may help to prepare a simulation that would be able to activate and engage a child and focus his/her attention on simulation tasks. Additionally, before the start of a 'core' experiments, children should be acquainted with an environmental setup to get used to new conditions and devices. The workspace will be extended with new stimuli developed in the cooperation with therapists for other children being under their care.

Acknowledgments We acknowledge the support of Silesian University of Technology grant BK /263/RAu2/2016.

References

1. Bartuccio, M., Maino, D.: Visual Diagnosis and Care of the Patient with Special Needs. Wolters Kluwer Health (2012). https://books.google.pl/books?id=8Pv4qBTaNXEC
2. Brolly, X.L., Mulligan, J.B.: Implicit calibration of a remote gaze tracker. In: Conference on Computer Vision and Pattern Recognition Workshop, 2004. CVPRW'04, pp. 134–134. IEEE (2004)
3. Corbetta, D., Guan, Y., Williams, J.L.: Infant eye-tracking in the context of goal-directed actions. Infancy **17**(1), 102–125 (2012)
4. Duchowski, A.: Eye tracking methodology: Theory and practice, vol. 373. Springer Science & Business Media (2007)
5. Dutton, G., Bax, M.: Visual impairment in children due to damage to the brain: clinics in developmental medicine. Clinics in Developmental Medicine. Wiley (2010). https://books.google.pl/books?id=bM0sk_-O6qAC
6. Frank, M.C., Vul, E., Saxe, R.: Measuring the development of social attention using free-viewing. Infancy **17**(4), 355–375 (2012)
7. Holmqvist, K., Nyström, M., Andersson, R., Dewhurst, R., Jarodzka, H., Van de Weijer, J.: Eye Tracking: a Aomprehensive Guide to Methods and Measures. Oxford University Press (2011)
8. Kasprowski, P., Harezlak, K.: Using non-calibrated eye movement data to enhance human computer interfaces. In: Intelligent Decision Technologies, pp. 347–356. Springer (2015)
9. Kasprowski, P., Harezlak, K., Stasch, M.: Guidelines for the eye tracker calibration using points of regard. In: Information Technologies in Biomedicine, vol. 4, pp. 225–236. Springer (2014)
10. Oakes, L.M.: Advances in eye tracking in infancy research. Infancy **17**(1), 1–8 (2012)

Intelligent Decision Technologies for Water Resources Management

Forecasting Domestic Water Consumption Using Bayesian Model

Wojciech Froelich and Ewa Magiera

Abstract In this paper, we address the problem of forecasting domestic water consumption. A specific feature of the forecasted time series is that water consumption occurs at random time steps. This substantially limits the application of the standard state-of-the art forecasting methods. The other existing forecasting models dedicated to predicting water consumption in households rely on data collected from questionnaires or diaries, requiring additional effort for gathering data. To overcome those limitations, we propose in this paper a Bayesian model to be applied for the forecasting of the domestic water consumption time series. The proposed theoretical approach has been tested using real-world data gathered from an anonymous household.

Keywords Forecasting time series · Domestic water consumption · Bayesian networks

1 Introduction

Forecasting water consumption at the household level is a problem that can be addressed from different perspectives. First, on the basis of the forecasted water consumption, a corresponding system of forecasted bills/charges can be applied. In this case, the resident pays her bills calculated on the basis of forecasted water consumption. This results in more accurate forecasting, less frequent reading of water meters, and the charges paid by the consumer are fair. For this purpose, the monthly time scale is usually applied.

W. Froelich (✉) · E. Magiera
Institute of Computer Science, University of Silesia, Sosnowiec, Poland
e-mail: wojciech.froelich@us.edu.pl

E. Magiera
e-mail: ewa.magiera@us.edu.pl

I. Czarnowski et al. (eds.), *Intelligent Decision Technologies 2016*,
Smart Innovation, Systems and Technologies 57,
DOI 10.1007/978-3-319-39627-9_29

From the perspective of water conservation, the forecasted values of water consumption can be compared with actual values. In the cases when the actual value is lower then the forecasted value, the water conservation process can be recognized; otherwise, the consumer may be warned of excessive water usage. For the purposes of water conservation, the considered time scale in which forecasts are made should be detailed. The households should be equipped with smart water meters that are enabled to perform readings at daily, hourly, or even more detailed temporal scales. Furthermore, to assess water conservation with respect to the type of consumption (e.g., showering, dish-washing, drinking or other activities), water metering should be made separately for each of the corresponding areas of consumption. The role of water efficient technology such as smart meters in reducing water consumption throughout the day is crucial [4].

Forecasting of water consumption is subject to two variables as described in the literature: (a) the analysis of domestic water consumption [5, 15] and (b) the forecasting of time series [7, 8].

A detailed analysis of water consumption at households was made in [16] and more recently in [17]. Water consumption in the kitchen was investigated [20], where water withdrawal was related to different take events, e.g. drinking, cooking, etc. In [4] the patterns of water consumption in the household were analyzed on an hourly scale. A household water saving model based on Bayesian network was proposed [13]. In [9] a general architecture of the decision support system based on forecasting water consumption was described. More recently, the potential of water conservation during dish washing has been investigated in a dissertation [10].

Knowledge about physical properties of households, e.g., number of rooms, type of house/flat, garden, was used for the analysis of residential water consumption [5]. Also in [5] the problem of missing data in water meter readings was investigated. A practical decision support system for water conservation at household has been developed by IBM. On the basis of the available report [12] it can be noted that one of the applied solutions was based on the recognition of trend of water consumption. An overview of the literature on the analysis of domestic water consumption was made [5, 15].

The problem of forecasting domestic water consumption was investigated in several publications. In [11] the daily and hourly patterns of water consumption in households were analyzed. A general review of selected forecasting models used at the urban and household levels was made in [11]. Recently, it has been shown that the bottom-up approach to the forecasting of urban water demand is superior to the top-down approach based on bulk water meters [2].

In particular, Bayesian networks have been used for the forecasting of time series [1, 22]. Recently, an overview of Bayesian forecasting approaches has been reported [3]. Our previous studies [6, 18] applied Bayesian models for the forecasting of water demand at the city level. Due to the different characteristics of the data and the domain of application, those publications are not directly related to the approach presented in this paper.

Most of the existing forecasting models reported in the literature rely on information gathered from households using questionnaires, including information about household occupancy, household income, or even from daily diaries prepared by residents. Significant additional effort is required to gather such data and these data are usually not anonymous. By contrast, our approach utilizes only the near real-time readings of water meters.

For the purpose of forecasting domestic water consumption, we design a structure of the Bayesian network. In addition, we investigate a related problem, the dependence between the equal-length discretization technique and its influence on the accuracy of forecasting.

The paper is organized as follows. Section 2 provides basic notions on Bayesian networks. In Sect. 4, the forecasting accuracy measures selected for this research are reviewed. Section 3 describes our contribution: the adaptation of Bayesian network to the forecasting of domestic water consumption. The results of experiments are presented in Sect. 5. Section 6 concludes the paper.

2 Bayesian Networks—Basic Notions

Bayesian network (BN) is a knowledge representation tool that is proposed in this paper for the application of forecasting of domestic water consumption. The BN is able to relate the symbolic values of explanatory variables to the discretized numerical values of the forecasted time series. BN is a transparent graphical model that explicitly depicts the probabilistic dependencies discovered in data.

A Bayesian network is a triple $BN = (X, DAG, P)$, where X is a set of random variables, $DAG = (V, E)$ is a directed acyclic graph, and P is a set of conditional probability distributions [14]. Each node of the graph $v_i \in V$ is related to the discrete random variable $X_i \in X$. The edges from the set $E \subseteq V \times V$ of DAG correspond to conditional dependence between random variables. $P(X_i|X_{pa(i)})$ is the conditional probability distribution for each $X_i \in X$, where: $pa(i)$ is the set of conditioning variables of the X_i. The posterior probability of unknown X_i given the set of evidence variables $XE \subset X$ is calculated using the BN model.

The learning of the Bayesian network consists of two steps: (1) structure learning and (2) learning of parameters, i.e., the probability distributions P.

To learn the structure of BN we selected already implemented state-of-the-art algorithms: Hill-Climbing—a greedy search on the space of the directed graphs; Tabu Search—a Hill-Climbing with the additional mechanism of escaping from local optima; Max-Min Hill-Climbing—a hybrid algorithm combining the Max-Min algorithm and the Hill-Climbing algorithm; and Restricted Maximization—a generalized implementation of Max-Min Hill–Climbing.

To learn the parameters, Maximum Likelihood Estimation (MLE) has been selected. The details of the applied algorithms are available in the documentation of 'bnlearn' library [21] of the R package [19].

3 Bayesian Approach to the Forecasting of Domestic Water Consumption

Let $\{W(t)\}$ denote the considered water consumption time series, where $W(t) \in \Re$ is a real-valued variable and $t \in [1, 2, \dots, n]$ is a discrete time scale of the length $n \in \aleph$. We define a set of the random variables related to the considered time series. Let $Hour(t) \in [1, 24]$ be a variable indicating the hour of day. Similarly, $Day(t) \in [1, 7]$ is a variable indicating the day of week.

The analysis of water consumption data revealed that the variations in water consumption are related to the part of the day. It is possible to distinguish 5 intervals in which the water consumption can be analyzed: morning, day, afternoon, evening and night. The determination of these particular periods with respect to people's behaviour is a complex problem. For this paper, we decided to make an arbitrary mapping of those intervals to the hours of the day, according to observations of the behavior of the inhabitants in the single household that was considered in our research. The assumed mapping is given in Table 1. Further investigation of the problem of that mapping is left for future research.

The partitioning of the day led to the introduction of the related variable $POD(t) \subset \{morning, day, afternoon, evening, night\}$ assuming symbolic values related to the parts of the day.

To enable the application of the standard Bayesian network for the forecasting of water consumption we discretize the domain of $W(t)$. For the purpose of this paper we decided to use the simplest, equal-length discretization method. The discretized version of $W(t)$ is denoted as $W_d(t) \in \aleph$, where the values of $W_d(t)$ are the positive integers specifying the intervals to which the actual values of $W(t)$ belong. The parameter $k \in \aleph$ determines the number of the discretization intervals applied.

The lagged value $W_d(t-1)$ was assumed as the last of the explanatory variables. Due to a very short forecasting interval of 30 s, the lack of short-term seasonality confirmed by the plot of correlation function (ACF), there was no reason for using higher order lags. The final set of the variables is given in Table 2.

Table 1 Temporal partitioning of day

Period	Start[h]	End[h]
Morning	4.00	9.00
Day	9.00	14.00
Afternoon	14.00	18.00
Evening	18.00	23.00
Night	23.00	4.00

Table 2 The set of variables

Variable	Description
Hour(t)	Hour
Day(t)	Day of week
POD(t)	Part of day
$W_d(t-1)$	Lagged water consumption
$W_d(t)$	Water consumption

4 Forecasting Accuracy Measures

To evaluate forecasting accuracy of the discretized version of time series, i.e., $W_d(t)$ we use two measures already applied in our previous study [18]. To asses the ratio of perfect forecasts PFR (related to all forecasts made) the formula (1) is used:

$$PFR = \frac{\sum_{t=1}^{n} 1|W'_d(t) = W_d(t)}{n}, \tag{1}$$

where $W'_d(t)$, $W_d(t)$ denote the forecasted and actual values of water consumption respectively; $n \in \aleph$ is the final time step of the forecasted time series.

To calculate the accumulated forecasting error the discrete mean absolute error (DMAE) accuracy measure is selected (2):

$$DMAE = \frac{\sum_{t=1}^{n} |W'_d(t) - W_d(t) + 1| \cdot d}{n}, \tag{2}$$

where d denotes the length of the applied discretization interval.

To evaluate the forecasting accuracy, the time series was partitioned into learning and testing periods. We applied the concept of a growing window. Following this assumption, the learning period began at the first time step and finished at time $t-1$ of the historical time series. This way, the length of the growing window increases as time continues. The BN model was retrained at every time step and the prediction was made 1-step ahead, i.e., for the step t. The minimum length of the growing window was set to 20 steps (days).

5 Experiments

For the purpose of this study we have collected data readings from smart water meters installed in the kitchen of the anonymous household. All experiments described in this section were performed using the 'bnlearn' library [21] of the R package [19].

5.1 Source Data

Source data were collected from the initial period of the project lifetime, from 28 November 2014 to 21 February 2015. Due to the short period of gathering data, the collected time series did not cover monthly variations of water consumption. The data were gathered from two smart meters measuring the water intake of the dishwasher and that used by the sink. The readings were made in time intervals of 30 s.

Figure 1a depicts the mean water usage of the dishwasher for every day of the week. As can be noted, the dishwasher was used only on Fridays, Saturdays and Sundays with a very low variation of water usage. For that reason we focused our attention only on forecasting water consumption at the sink. As Fig. 1b shows, the most water consumption occurred just after and before weekends, on Mondays and Fridays, respectively. As Fig. 1c shows, the highest mean water consumption can be recognized in the afternoon with very low consumption at night. When considering short-term water usage, a peak can be observed at about 5 AM, see Fig. 1d.

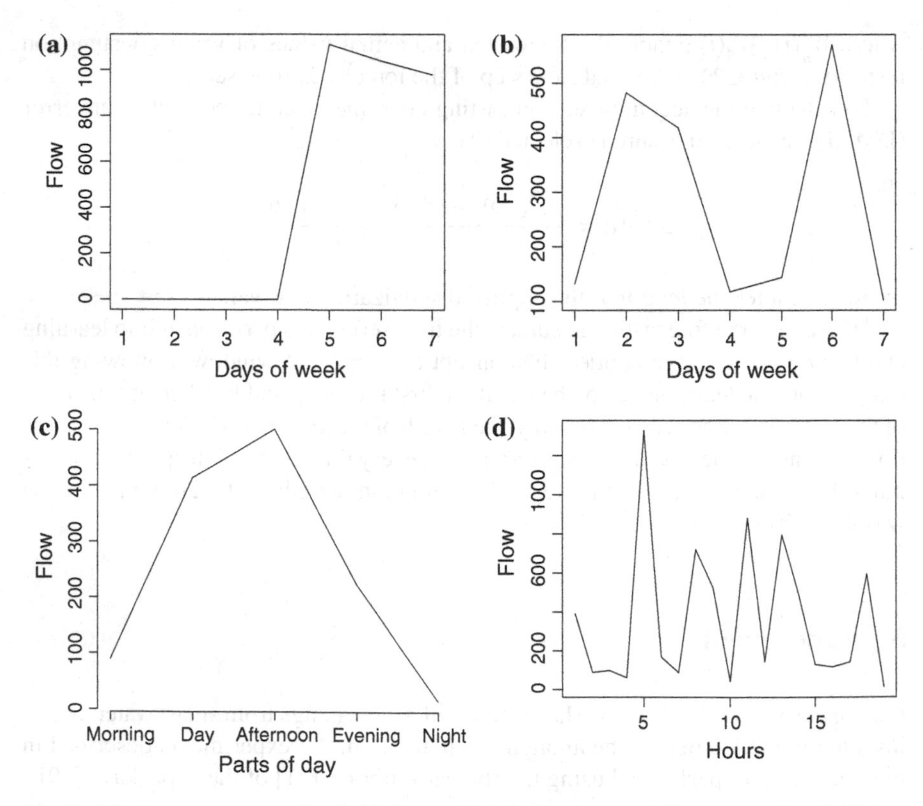

Fig. 1 Mean water consumption in kitchen. **a** Dishwasher—mean daily consumption, **b** Sink—mean daily consumption, **c** Sink—mean daily pattern, **d** Sink—mean hourly consumption

Table 3 Distribution of water consumption during the week

	Sunday	Monday	Tuesday	Wednesday	Thursday	Friday	Saturday
Morning	66.32	97.85	29.40	178.78	0.0	35.625	101.54
Day	172.29	1702.16	696.66	49.0	7.0	7.0	193.16
Afternoon	45.16	430.40	689.12	19.85	277.25	1166.3	53.5
Evening	212.99	273.99	394.92	109.76	15.5	17.5	81.73
Night	0.0	0.0	14.0	0.0	0.0	0.0	0.0

Table 3 presents the distribution of water consumption with respect to the days of the week and the parts of the day. Peaks for water usage were recorded on Monday during the day and on Friday afternoon. The content of Table 3 can be interpreted as an individual, weekly consumption profile of the individual water consumer. Further analysis of these type of profiles for other points of water intake placed in kitchen is planned. After gathering data from the other households, the analysis of data will be performed with respect to other users.

5.2 Experimental Results

First, the structure of the BN was learned automatically using all of the algorithms given in Sect. 2. However, the obtained values of DMAE were high. An attempt was undertaken to design the structure of the BN. Following our previous experiences with the forecasting of time series using Bayesian networks [6, 18], we assumed $W_d(t)$ as a single dependent variable. The rest of the variables that played the role of evidence variables directly influenced $W_d(t)$. This way, only forward inference within the BN was assumed. Table 4 describes the investigated BNs with the related evidence variables.

We started the investigations with a simple BN1 containing only a single evidence variable and the conditional distribution $W_d(t)|W_d(t-1)$ assigned to the child node. In the second step we complemented the network by adding other variables. This

Table 4 Forecasting accuracy for the designed structures of BNs

Bayesian network	Variable
BN1	$W_d(t-1)$
BN2	$W_d(t-1)$, $Day(t)$
BN3	$W_d(t-1)$, $POD(t)$
BN4	$W_d(t-1)$, $Hour(t)$
BN5	$Hour(t)$, $POD(t)$, $Day(t)$
BN6	$W_d(t-1)$, $POD(t)$, $Day(t)$, $Hour(t)$

Table 5 Forecasting accuracy DMAE(PF) for the designed structures of BNs

k	BN1	BN2	BN3	BN4	BN5	BN6
10	227.05(0.85)	233.47(0.83)	210.20(0.88)	265.56(0.77)	292.84(0.70)	309.69(0.67)
20	124.35(0.82)	132.38(0.78)	112.32(0.86)	142.00(0.75)	168.88(0.69)	186.93(0.60)
30	110.98(0.70)	115.26(0.67)	93.33(0.74)	126.76(0.64)	147.35(0.55)	175.70(0.49)
40	91.66(0.63)	88.65(0.61)	75.41(0.68)	107.10(0.55)	117.93(0.51)	137.79(0.42)
50	96.59(0.55)	97.23(0.52)	75.89(0.62)	111.03(0.47)	124.35(0.44)	148.10(0.35)

way BN2, BN3, and BN4 were constructed. Because the addition of *POD*(*t*) was shown to be the most beneficial in terms of DMAE, we complemented BN3 further by adding first the variable *Day*(*t*) and then, *Hour*(*t*). In this way we obtained BN5 and BN6. The obtained results of DMAE with the PFR given in brackets are shown in Table 5. The parameter k denotes the number of equal-length intervals used for the discretization.

Table 5 shows that the BN3 is the best model and that further addition of evidence variables does not help.

As expected, the ratio of perfect forecasts decreases while increasing the number of intervals and thus making those intervals shorter. On the other hand, the value of DMAE increases reaching it's lowest values for $k > 30$. To further investigate this dependency, we repeated the experiment with BN3 for all values of $k \in [5, 50]$. The results are shown in Fig. 2. This way it has been confirmed that the lowest value of DMAE are obtained for $k > 30$. Then the DMAE exhibits random variation. On the other hand, to obtain the PFR as high as possible, for the considered data, we selected $k = 30$ for future research.

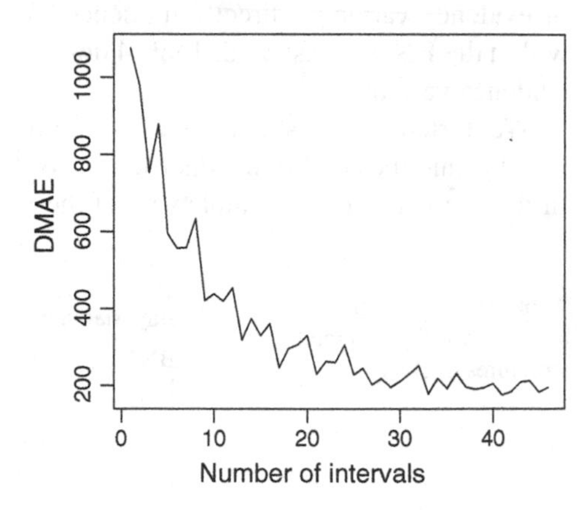

Fig. 2 Discrete mean absolute error

6 Conclusions

In the presented study, we investigated the application of standard Bayesian networks to the forecasting of water consumption. We designed a Bayesian network model that is able to efficiently forecast a discretized time series of domestic water consumption. We have discovered that the application of the evidence variable related to the parts of day augments good forecasting accuracy. Further work is required to compare the obtained results with the other forecasting methods. After collecting more data covering seasonal and monthly variations of water consumption, theoretical and experimental research will be extended. The planned research will also include the analysis of user specific water consumption profiles.

Acknowledgments The work was supported by ISS-EWATUS project which has received funding from the European Union's Seventh Framework Programme for research, technological development and demonstration under grant agreement no. 619228.

References

1. Abramson, B., Brown, J., Edwards, W., Murphy, A., Winkler, R.L.: Hailfinder: a Bayesian system for forecasting severe weather. Int. J. Forecast. **12**(1), 57–71 (1996)
2. Bennett, C., Stewart, R.A., Beal, C.D.: Ann-based residential water end-use demand forecasting model. Expert Syst. Appl. **40**(4), 1014–1023 (2013)
3. Biondi, D., Luca, D.D.: Performance assessment of a bayesian forecasting system (bfs) for real-time flood forecasting. J. Hydrol. **479**, 51–63 (2013)
4. Carragher, B.J., Stewart, R.A., Beal, C.D.: Quantifying the influence of residential water appliance efficiency on average day diurnal demand patterns at an end use level: A precursor to optimised water service infrastructure planning. Resources, Conservation and Recycling, pp. 81–90 (2012)
5. Fox, C., McIntosh, B., Jeffrey, P.: Classifying households for water demand forecasting using physical property characteristics. Land Use Policy **26**(3), 558–568 (2009)
6. Froelich, W.: Forecasting daily urban water demand using dynamic gaussian bayesian network. In: Proceedings of the BDAS 2015, pp. NA–NA. Springer, Ustro, Poland (2015)
7. Froelich, W., Papageorgiou, E.I.: Extended evolutionary learning of fuzzy cognitive maps for the prediction of multivariate time-series. In: Fuzzy Cognitive Maps for Applied Sciences and Engineering—From Fundamentals to Extensions and Learning Algorithms, pp. 121–131. Springer (2014)
8. Froelich, W., Salmeron, J.L.: Evolutionary learning of fuzzy grey cognitive maps for the forecasting of multivariate, interval-valued time series. Int. J. Approx. Reason. **55**(6), 1319–1335 (2014)
9. Froukh, M.: Decision-support system for domestic water demand forecasting and management. Water Res. Manag. **15**(6), 363–382 (2001)
10. Fuss, N.A.: Determination and verification of possible resource savings in manual dishwashing, Ph.D. thesis. Rheinische Friedrich-Wilhelms-Universitt Bonn, 1st edn. (2011)
11. Gato, S.: Forecasting Urban Residential Water Demand, Ph.D. thesis. School of Civil, Environmental and Chemical Engineering Science, RMIT University, 1st edn. (2006)
12. IBM: Smart Water Pilot Study Report. IBM Research, 1st edn. (2011)
13. Inman, D.: The use of Bayesian networks to facilitate implementation of water demand management strategies, Ph.D. thesis. Cranfield University, School of Applied Sciences (2008)

14. Jensen, F.V.: Bayesian Networks and Decision Graphs. Springer (2001)
15. Jorgensen, B., Graymore, M., O'Toole, K.: Household water use behavior: an integrated model. J. Env. Manag. **91**(1), 227–236 (2009)
16. Krants, H.: Matter that Matters.A study of household routines in a process of changing water and sanitation arrangements. Department of Water and Environmental Studiesy, 1st edn. (2005)
17. Linkola, L.: Behaviorally based modeling of domestic water use. Masters thesis. Leiden Unversity, Delft Univesity of Technology, 1st edn. (2011)
18. Magiera, E., Froelich, W.: Application of bayesian networks to the forecasting of daily water demand. In: Proceedings of the KES IDT-15, Sorrento, Italy. pp. NA–NA. Springer, 17–19 July 2015
19. Package, R.: http://www.r-project.org
20. Richter, C., Stamminger, R.: Water consumption in the kitchen a case study in four european countries. Water Res. Manag. **26**(6), 1639–1649 (2012)
21. Scutari, M.: Bayesian network structure learning, parameter learning and inferenceg (2014). http://www.bnlearn.com/
22. Vlachopoulou, M., Chin, G., Fuller, J.C., Lu, S., Kalsi, K.: Model for aggregated water heater load using dynamic bayesian networks. In: Proceedings of the DMIN'12 International Conference on Data Mining, pp. 1–7 (2012)

An Evaluation of the Instruments Aimed at Poland's Water Savings

Krzysztof Berbeka and Malgorzata Palys

Abstract This paper evaluates the effectiveness and cost efficiency of a water abstraction tax fitted to the scarcity of surface water resources. The modelling of the hypothetical consequences of the proposed taxation scheme were conducted using several databases. These databases describe the availability of water resources in Poland, as well as the present schemes, the level of taxation and the economic conditions of Polish municipal water providers. All four scenarios were taken into consideration to meet the criteria of a better fitting of the unit intake tax to the scarcity of water at the local level. However, the progression of the rates and total fiscal effects were different. The proposed instrument was found to have a minor influence on the water operators and a very small influence on the end users; this was due to the low level of taxation at the present time.

Keywords Sustainable urban water management · Abstraction charges · Water tariffs · Municipal water prices · Demand management · Taxation of environmental resources

1 Introduction

Water scarcity is an important concern in Europe. Several terms are used to describe this scarcity, including quantitative and qualitative deficits. These terms are very often confused. Furthermore, local discrepancies are extended to the entire region. Such a procedure overestimates the real impact of the deficits. However, the existing projections of water availability, in the context of global climate changes, are pessimistic.

K. Berbeka (✉) · M. Palys
Institute of Computer Science, University of Silesia, Sosnowiec, Poland
e-mail: krzysztof.berbeka@uj.edu.pl

M. Palys
e-mail: malgorzata.palys@us.edu.pl

I. Czarnowski et al. (eds.), *Intelligent Decision Technologies 2016*,
Smart Innovation, Systems and Technologies 57,
DOI 10.1007/978-3-319-39627-9_30

Due to the global climate change Europe will face a local drops in water resource availability. Furthermore, the irregularity of water availability over the seasons will increase the vulnerability at the short term water stress. In addition, an increase in the frequency of extreme events (e.g., droughts and floods) will create additional incentives for sustainable water management.

This paper aims to evaluate the selected methods of water savings in the municipal sectors. The choice of a municipal water supply investigation was not conducted by chance; the expected reliability of a water provision in this sector is much higher than it is in the agricultural or industrial sectors. The effectiveness and cost efficiency of the proposed solutions will be verified using the data describing Polish municipal water operators. However, the conclusions will be more general, and therefore, applicable to other countries.

2 Theoretical Background

The assessment of several instruments aimed at water savings is a multidimensional task. At the minimum, three basic criteria should be taken into account:

1. Effectiveness (as a proxy of reaching the target),
2. Cost efficiency (as the cost of obtaining defined in the advanced target),
3. Efficiency (as the effect divided by the cost of their obtaining).

In the case of the challenge of a heavy water deficit, the first criteria can be skipped. It is impossible to use more water than is available. Therefore, the target understands that "water savings" is not very important. However, between a heavy imbalance of water and a full level of availability, there are many intermediate conditions. Therefore, a rough checking of whether the discussed instrument is unable to reach the target is necessary. Several factors must be taken into account: that some undertakings have to be planned in advance, that expected targets have to be defined and that cost efficiency is the most important criteria.

The standard comparison of cost efficiency occurs between two basic instruments: administrative instruments and economic instruments. Such comparison indicates an advantage in the economic instruments (more precise: the same target can be obtained at a lower cost) [8]. A short explanation of this advantage is illustrated in Fig. 1a–c.

Let's assume that the "target" means water savings and the curve, MC_i, describes the marginal cost of the activities aimed at water savings. The economy consists, in this case, of only two entities described by two curves: MC_1 and MC_2. The total target, *TT* (amount of water savings), is requested at level $2 * Q_0$, taking into account that the comparability of the costs of the total target have to be the same in both cases. Therefore, $2 * Q_0 = Q_1 + Q_2$. Because the total cost of obtaining the target is equal to the area below the curve of the marginal cost, the total cost (*TC*) in the entire economy is lower in the case of using the taxes. The difference is equal to the polygon *AEFC* (Fig. 1c).

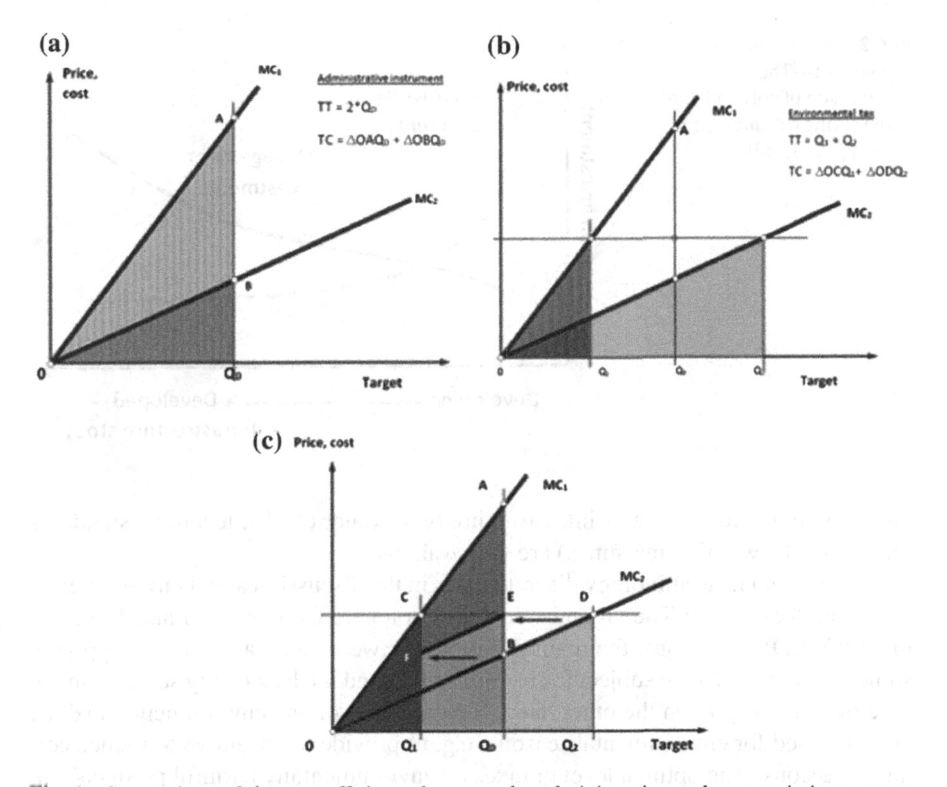

Fig. 1 Comparison of the cost efficiency between the administrative and economic instruments (*Source* Pearce 1990, p. 95)

Such assumptions may not be appropriate in some specific areas, such as water resource management. The reasons for this are: higher transaction costs in the case of environmental taxation (these costs are skipped in the rough analysis presented in Fig. 1a–c) and the lack of a serious marginal cost differentiation between the single entities (meaning that the curve MC_1 is very close to MC_2 and others MC_i). In such a case, the total cost savings presented in Fig. 1c is very low.

It's possible to see some similarities between Fig. 1a–c and the Sadoff and Grey approach (Fig. 2), despite the different starting point. The increasing marginal costs of reaching the target are visible as the decreasing rate of returns from this investment.

The applicability of both instruments (administrative and economic) depends on the level of implementation. For example, in relation to the level of water operators, both instruments are fully applicable. It's possible to introduce permits for water intakes. In addition the taxation of resource consumption is available. At the level of single client (household) taxation, it is fully applicable, but the administrative instruments are quite limited. Daily or monthly limits of water availability for consumption is beyond of the scope (in European circumstances) of this investigation,

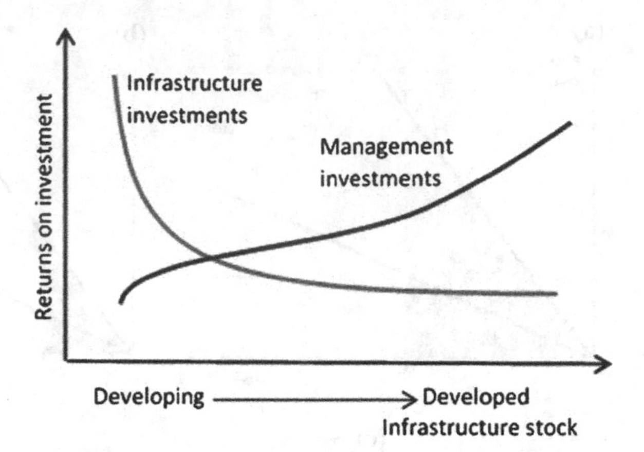

Fig. 2 Returns on investment—The comparison of soft and hard undertakings (*Source* Grey (2007) p. 561, [6])

but some standards for water infrastructure (e.g., water clocks, technical standards for the toilets, washing machines) are still available.

There are some terminology discrepancies in the discussion about environmental taxes and fees. Using The Organisation for Economic Co-operation and Development (OECD) definitions, the main difference between taxes and fees is supplying some services which are subject for fee imposition and the lack of any services in the case of taxation [7]. On the other hand, the definition of an "environmental tax" is: "tax imposed for environmental reasons, e.g. to provide an incentive to reduce certain emissions to an optimal level or taxes on environmentally harmful products". In this context, an interpretation of payments imposed on a water intake is quite controversial. In the case of fees imposed per each unit of withdrawal of water, there is an obvious situation of the provision of the service. Therefore, the term "fee" should be used.

On the other hand, sometimes the payment has a flat rate characteristic and is aimed at decreasing the negative pressure on the environment. Such a situation fits with the definition of an environmental tax. In the case of payments for a water intake, both the variants of charging (flat and volumetric) are common; as such, the terms taxes and fees are roughly equivalent.

3 Analysis of the Methodology and the Modelling Approach

The first hypothesis under investigation was the verification of the effectiveness and efficiency of the taxation imposed at the macro level on water abstraction. The second was similar, but concerned the verification of the pricing system imposed at the end-users level.

The main assumptions of changing the existing system of taxation on water resources was the following:

(a) The unit intake tax should refer to the scarcity of water resources.
(b) The scarcity should be estimated at the local level. The lack of data availability is only a concern to the more detailed disaggregation (such a precise level fits the discrepancies of water availability).
(c) In opposition to the modelling of air pollutant dispersion, the regular grid (independent of the size) is inappropriate. The proper unit is connected with the hydrological disaggregation of surface waters. Therefore, the calculation was conducted at the level of the combined water bodies.
(d) The proxy of the scarcity is calculated as a share of the annual withdrawal, related to the disposable resources (mean annual runoff). The methodology of such an approach was presented in Smakhtin [9].
(e) The modelling is only related to the surface water, because the recovery rate of the underground water resources is more difficult to predict. In addition, the influence of global climate change on groundwater availability is much more smoothing over time and more difficult to predict than surface water.

The present differentiation of the surface water abstraction fee in Poland differs according to the scarcity; the range of differentiation is very limited. In fact, due to the large scale of the areas (*Nomenclature of Territorial Units for Statistics*2 aggregated into 3 groups), the differentiation doesn't correspond to the real scarcity of water. The multiplayer factor ranges from 1, 0 to 1, 2 of the basic unit tax. The new proposals of the tax differentiation is much more aggressive and follows the scarcity index (more than 1000 cells, in the Polish case). Scenarios 1–4, included in the calculation, are presented in Table 1. The scenarios taken into consideration were prepared using the following assumption:

(a) Very small intake (in relation to the available resources) are safe from the environmental point of view. Therefore, the unit charges should be reduced ($S1, S4$).
(b) Second and third scenarios reflect a more fund-raising approach. In the case of a small intake, the present charges are binding. In the case of a serious withdrawal, the aggressive multiplayers are applied.
(c) The coefficients in the last scenario were prepared to gain neutral fiscality at the macro level.

The modelling of the economic and environmental consequences on the side of water operators and the budget will focus on the water operators and the end users. In detail, the following aspects will be reviewed:

(a) Fiscal neutrality of the new approach (changes in the total amount of imposed fees).
(b) Consequences for the water operators defined as changes in water production costs, with special attention paid to the extreme situations (highest increase in the costs).

Table 1 The new proposals of differentiation

Scarcity index in (%)	Scenario 1	Scenario 2	Scenario 3	Scenario 4
	Multiplayer factors of the present unit tax			
Below 1	0, 80	1, 00	1, 00	0, 65
(<1–2)	0, 90	1, 00	1, 00	0, 80
(<2–5)	1, 00	1, 20	1, 20	1, 00
(<5–10)	1, 20	1, 40	1, 50	1, 10
(<10–30)	1, 50	2, 00	2, 50	1, 40
(<30–50)	2, 50	3, 00	4, 00	1, 90
Over 50	3, 00	5, 00	5, 00	3, 00

The basic unit of the surface water intake fees in Poland

(c) Increase in the final price of water provisions for the end-users.
(d) Hypothetical demand reaction at two levels: water operators and end-users.

The relationship between the increase in the unit abstraction tax and the increase in the price for end-users is not obvious. Two aspects have to be included in such modelling. The first is the difference between the amount of withdrawal water and billed water caused by the water losses and consumption for technical purposes (e.g., maintenance of the pipes). The second is the Value Added Tax (VAT) imposed on the service of water provision (7 %). Finally, using data provided by Central Statistical Office of Poland (GUS), it was possible to construct an aggregated multiplier index, including the mentioned relationship [2]. The interpretation of the calculated value (1, 31) is as follows: the increase of the tax imposed on the 1 m^3 abstracted water by one financial unit caused an increase in the final price of the 1 m^3 billed water by 1, 31.

4 Data Collection

The necessary data was compiled from several different sources. The most challenging task was related to the compilation of the data describing the payments for using the environmental services (so called environmental taxes/fees). Such payments are collected in Poland at the regional level (NUTS-2). Some aggregation at the national level was conducted annually, but in this process, the subject of the payments (e.g., water intake, discharge of any kind of pollutants), was lost.

The process of the full compilation of all of the data at national level was conducted only once, during the dedicated project for the figures from 2010 [5]. The results of the mentioned project were used for purposes of the present research. Finally, from 17 000 entities who paid for using the environmental services, 276 payments for the municipal water intake were extracted.

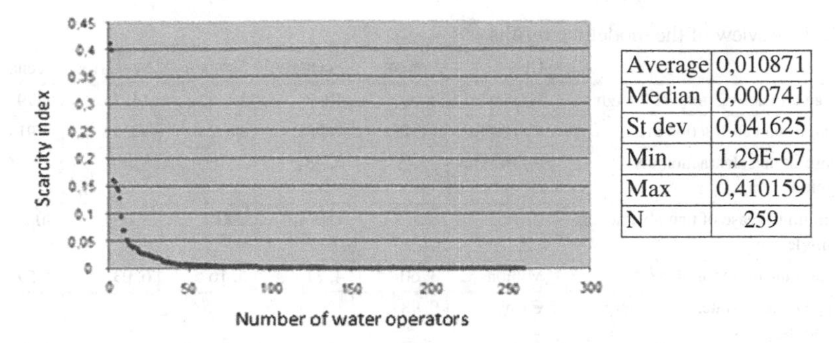

Fig. 3 The scarcity index of the surface water intake for the municipal water supply in Poland. (*Source* Own calculation)

The second database consisted of the administrative permits for water intake associated with the place of withdrawal (in numeric format). This data consisted of 395 permits for a municipal water intake.[1]

The third database was constructed due to the requirements of the Water Framework Directive (every 5 year period) and consisted of the financial data of municipal operators (1000 entities of data from 2010) [4]. The compilation of all of the data was necessary for the purpose of the present project. The possibility of the supplementation of the missing data was quite limited. Only a lack in the information concerning the place of water intake was possible to fulfil due to the ability to use Geographic Information System (GIS) software (ArcInfo 9.0).

The resulting database consisted of 178 municipal water operators. This sample covered 66 % of the surface water intake (in quantitative format) used for municipal purposes in Poland in 2010.

The data describing the present prices of water (household tariffs) were taken from the public server "cena-wody" (water-prices). The sample consisted of 921 entities. The data were from 2015 [3]. The recalculation of all of the prices from the Polish currency to € was made using the exchange rate from 2015: the annual average published by the Central Bank of Poland (NBP), 1 €= 4, 18 PLN.

5 Modelling

The scarcity index (annual withdrawal/ mean annual runoff) is presented in Fig. 3.

The application of the new set of abstraction fees for water operators in Poland is summarized in Table 2. The following conclusions can be formulated:

[1]The number of permits was not correlated with the number of water operators. Sometimes one operator has more than one permit, because he or she uses water from different sources (e.g., rivers, lakes, underground resources).

Table 2 Review of the modelling results

Category	Unit	Present	Scenario 1	Scenario 2	Scenario 3	Scenario 4
Average unit abstraction tax (weighted)	€/1000 m^3	8, 80	10, 31	12, 62	14, 75	8, 79
Median of unit abstraction tax	€/1000 m^3	11, 25	9, 53	11, 78	12, 32	8, 91
Increase of unit abstraction tax (weighted average)	€/1000 m^3		2, 48	3, 78	3, 49	−9, 76
Maximum increase of unit abstraction tax (single entity)	%		150 %	200 %	300 %	90 %
Total amount of imposed tax	Millions €	3, 60	4, 21	5, 16	6, 03	3, 59
Average price of water for households, gross	€/m^3	0, 98				
Median price of water for households, gross	€/m^3	0,89				
Increase in the price caused by changes in the abstraction tax	c€/m^3		0, 25	0, 38	0, 35	−0, 98
	%		0, 25 %	0, 39 %	0, 36 %	−1, 00 %

(*Source* present prices of water [3]; other values—own calculation)

(a) Imposing the new abstraction fees (instead of the existing ones) caused changes (an increase) in the total sum of the imposed surface water abstraction charges. However, fiscal neutrality was also possible (*S*4).
(b) The differentiation of the present and projected charges analyzed on a case by case basis indicated very high discrepancies. The changes differed from a drop of 46 % to an increase of 300 % at the single company level, depending on the variant of the charging scheme.
(c) The implementation of the new tax system will change the water production costs between the operators. This is also the case with fiscal neutrality or a minor change in the total imposed charges. With the new system, the average share is almost the same (2, 3 % of the total costs). The extreme values for the single entities ranged by 10 %.
(d) Taking into account a very limited share of the abstraction charges in the total production costs, the influence of the discussed changes in the final prices of water were found to be marginal (below 1 c €/m^3 in all scenarios).
(e) The demand reaction to the price changes was below the accuracy of the modelling, regardless of the scenario. Therefore, the statement that projected the changes of the abstraction taxes had no influence on the final consumption being justified.

6 Discussion

It is possible to obtain strong fund-raising effects using aggressive taxation (*S*2, *S*3). However, the justification of such a high increase in the unit tax is difficult. Water abstraction in amounts not exceeding 1 % of the available resources places quite mar-

Table 3 Pros and cons of the proposed system of the water intake taxation

Pros	Cons
Incentive system. The vulnerability is fitted to the local conditions. Better implementation of the User Pay's Principle	Influence is not as strong as expected, due to the low share of abstraction taxes in the total production cost/final price
The aggregated fiscal effect of the new system is quite flexible. It is possible to modify the taxation scheme following the scarcity index with fiscal neutrality. Using a more aggressive scheme is also possible to increase the fiscal effect	Serious changes in the financial burdens between the water operators
The analyzed scenarios are generally not affected by the final price of water	The new system is not sensitive to the specific seasonal water intake
	The new system smooths (over time) the dynamics related to climate change. The annual run-off is the average from the multi-year period
	The new system is not sensitive to the specific short time events (droughts) which also influence the balance of the demand and supply

ginal pressure on the environment. Therefore, high taxation has no relationship with the negative effects. Such a solution is very fair from the optimal taxation level in the Pareto sense.

Until now, the assumption that available resources are equal to the mean of the annual run-off have been applied. The proper definition should be corrected by excluding from the total run-off the amount of water which is necessary for environmental purposes. Unfortunately, such data disaggregated into level of combined water bodies are still unavailable. This correction will increase the unit abstraction tax (Table 3).

The analyzed idea of better fitting the abstraction charges to the scarcity of water provides moderate incentives to sustainable water management. The difference between single water operators in the amount of water abstracted and the billed water is approximately 23 % (average for Polish water providers [2]). This suggests some space for conservation. The annual consumption per capita in the household sector in Polish cities is also strongly differentiated; the weighted average is 35, 8 m^3/capita, the minimum is 21, 6 and the maximum is 45, 6 m^3/capita [2]. Such discrepancies also create the possibility of savings. However, the discussed instrument does not provide a strong enough price incentive. A single change in the water price by 0, 4 % is invisible to the consumers and is associated with a quite low demand price elasticity index (−0, 22) [1]; hence, it will not create a serious opportunity for a drop in water consumption.

7 Conclusions

The better fitting of the abstraction taxes to the scarcity of water meets the basic principles of the sustainable consumption of environmental resources. In the long term, such a policy causes a better fitting of the demand for water with the available resources. The present level of water intake taxation in Poland is low, in the context of the total production costs. Therefore, the projected changes have a minor influence on the total abstraction of water. On the level of water operators, some conservation undertakings are still possible, but, in spite of such activities, the expected demand reaction of the end users will be close to the zero-level. The taxation is more effective than the permitting system. However, the serious demand reaction at the final clients level requires a very strong increase in taxation (using factor $* 3 : 5$). Such changes face hard political affordability.

Acknowledgments The work was supported by ISS-EWATUS project which has received funding from the European Union's Seventh Framework Programme for research, technological development and demonstration under grant agreement no. 619228.

References

1. Bartczak, A., Kopanska, A., Raczka, J.: Residential water demand in a transition economy: evidence from Poland. Water Sci. Technol. Water Supply, **9**(5), 509–516 (2009)
2. Central Statistical Office: Environment Protection 2015, Warsaw (2015)
3. Ceny Wody (water prices): http://www.cena-wody.pl/index.php?p=9&brutto=1
4. Cost Recovery for water services in water bodies: Not published paper prepared for President of National Water Management Authority (in Polish) PectoreEco, Gliwice (2013)
5. Economic Analysis of water use in Poland: Not published paper prepared for President of National Water Management Authority (in Polish) MGGP, Kraków (2010)
6. Grey, D., Sadoff, C.: Sink or swim? Water security for growth and development. Water Policy, **9**, 561 (2007)
7. OECD: http://www.oecd.org/ctp/glossaryoftaxterms.htm#T (2016). Accessed Jan 2016
8. Pearce, D.W., Turner, R.K.: Economics of Natural Resources and the Environment. Harvester Wheatsheaf, New York (1990)
9. Smakhtin, V., Revanga, C., Doll, P.: Taking into Account Environmental Water Requirements in Global scale Water Resources Assessments. IWMI The Global Podium (2005)

Soft Computing Approaches for Urban Water Demand Forecasting

Konstantinos Kokkinos, Elpiniki I. Papageorgiou, Katarzyna Poczeta, Lefteris Papadopoulos and Chrysi Laspidou

Abstract This paper presents an integrated framework for water resources management at urban level which consists of a Neuro-Fuzzy and Fuzzy Cognitive Map-based, (FCM) decision support system (DSS) based on multiple objectives and multiple disciplines for planning and forecasting. The proposed DSS has as primary goals to: (a) adaptively control the water pressure of the water distribution system by forecasting the water demand at the urban level and (b) to reduce leakage of the water network by controlling the water pressure. The system follows a model-driven architecture with the inclusion of the FCM-based models and a spatio-temporal model for arranging all data. The validation of the proposed learning algorithms is made for two case studies that comprise different water supply characteristics and correspond to different locations in Europe.

K. Kokkinos · L. Papadopoulos
Information Technologies Institute, CERTH, 6th km Charilaou-Thermi Rd., 57001 Thermi, Greece
e-mail: konst.kokkinos@gmail.com

L. Papadopoulos
e-mail: lefteris1029@gmail.com

E.I. Papageorgiou (✉)
Department of Computer Engineering, Technological Education Institute/University of Applied Sciences of Central Greece, Lamia, Greece
e-mail: epapageorgiou@teiste.gr

E.I. Papageorgiou
Faculty of Business Economics, Hasselt University, Hasselt, Belgium

K. Poczeta
Department of Information Systems, Kielce University of Technology, al. Tysiąclecia Państwa Polskiego 7, 25-314 Kielce, Poland
e-mail: k.piotrowska@tu.kielce.pl

C. Laspidou
Department of Civil Engineering, University of Thessaly, Nea Ionia, Greece
e-mail: laspidou@civ.uth.gr

I. Czarnowski et al. (eds.), *Intelligent Decision Technologies 2016*,
Smart Innovation, Systems and Technologies 57,
DOI 10.1007/978-3-319-39627-9_31

Keywords Fuzzy Cognitive Maps • Neuro-Fuzzy • Water management • Forecasting • Prediction • Decision support

1 Introduction

Many important decisions on sustainable and optimum utilization of water resources depend on the prediction of water demand. Decision-makers are specifically interested in the water demand forecasting since thorough research has shown that, the accurate estimation of water demand is critical for water management as well as water resources planning [1]. On the other hand, DSSs are tools commonly used in many fields such as businesses, industry, engineering, government and so on for improving the consistency and the quality of decision-making. However, when it comes to planning and scheduling tasks, it is rather critical that, decision factors must be predicted accurately so that decision-makers can select the nearest optimum choice from the available alternatives.

Targeting to water resource management, DSSs are technical tools intended to provide valid and sufficient information to decision makers related to the interdependent parameters that affect the optimal function of a water network at urban level. From our point of view, a DSS for water management should include seven distinct components: (a) a data acquisition system, (b) a user-data model interface, (c) a data repository, (d) a set of data analysis tools, (e) a set of interlinked and usually inter-dependent models, (f) the DSS core engine and (g) the interface to produce and illustrate results.

Thus far the modelling and the DSS development have been paramount in order to attack the problem of urban water resources management in a holistic way. Various model-based methodologies are implemented as components of environmental integrated frameworks aiming to carry out the decision-making especially in the case scenario when there is a high interlinking of complex physical processes [2], water demand management [3, 4], climate change effect on water resources and water supply [5, 6]. But as [7] states, model-based methodologies for environmental applications present a certain degree of complexity which limits their use to experts in the modeling and the usage field. Therefore only water company experts have the knowledge to accurately use these systems.

To be more focused to the main idea of this work, our DSS is based on a set of neuro fuzzy and FCM models which to the best of our knowledge have never used before into a water resource management decision support systems. The only close research work to this topic is the work by [8] which is based on an modeling approach using Genetic Algorithms. Also relatively to software implemented the work by [9] illustrated a web tool that is used both for simulated and real scenarios in the use case of Barcelona, Spain water network.

As mentioned before, the main forecasting models used by our DSS are based on neuro fuzzy and FCM models. In recent years, different methods and approaches to urban water demand forecasting have been proposed. Fuzzy and neuro-fuzzy models for short-term water demand forecasting were presented in [10]. In [11], an adaptive neuro-fuzzy inference system (ANFIS) was applied for monthly water consumption modeling based on the socio-economic and climatic factors. The results presented in [12] demonstrate that the ANFIS model is superior to a Mamdani fuzzy inference system (MFIS) for prediction of water consumption time series. Recently, different ANFIS models were investigated for urban water demand prediction in Skiathos Island, Greece [13]. The results seem to be promising for this case study.

A different soft computing approach, named Fuzzy Cognitive Maps was also applied to time series modeling [14–19]. FCMs are dynamic networks with learning capabilities, whereas more and more data is available to model the problem, the system becomes better at adapting itself and reaching a solution [20, 21]. They gained momentum due to their dynamic characteristics and learning capabilities. These capabilities make them essential for modeling and prediction tasks as they improve the performance of these tasks [14–16].

A hybrid algorithm for fuzzy time series prediction based on fuzzy c-means clustering, fuzzy cognitive map and genetic algorithm has been presented in [22, 23]. Evolutionary FCMs for univariate time series prediction have been proposed in [17]. In the case of multivariate time series prediction, the FCMs with genetic-based learning algorithms have been previously proposed in [18, 19]. The case study of the water demand prediction was selected as a representative one to show the capabilities of the proposed FCM algorithm due to the availability of real data measured from the water supply network of Skiathos Island, Greece.

The important purpose of our work is to integrate all the aforementioned models into the proposed DSS in order to forecast water demand, optimize pressure control, reduce leakages in the urban water distribution network and in general accommodate all scenarios, strategies and water sustainability management policies. The remainder of the paper is organized as follows. Section 2 presents the logical architecture of our implemented integrated water management framework at urban level which is an online system that has as a major component the DSS proposed above. In Sect. 3, two soft computing methods are shortly presented with exemplary results from two pilot case studies namely the Skiathos Island, Greece and the city of Sosnowiec, Poland. Finally, Sect. 4 does a comparative analysis of our results and concludes the paper.

2 Field of Study and DSS Architecture

There is a great interest for the establishment of efficient and sustainable water allocation policies by the competent authorities due to the growing scarcity and competition for water across sectors in the same city. These policies must be based on criteria provided by experimental verification of water management methodologies that are applied in Decision Support Systems (DSS) for equitable water allocation without compromising water sustainability. Along these lines, we describe an artificial intelligence model driven integrated DSS for water demand forecasting and water pressure management in urban water distribution networks. The main objective of this approach is to illustrate a prototype that benefits from innovative soft computing forecasting techniques and the use of the EPANET hydraulic model to produce the spatiotemporal interdependency curve between water demand and water pressure at urban level.

From the research topic point of view, the water pressure management at urban level involves the fragmentation of the water network and its reticulation into a set of District Meter Areas, (DMA) which form the spatial characteristics of the area under study. A DMA may be a small section of the water network or the whole network (for small cities/villages). It usually accommodates between 250 and 2500 residential hydrometer connections and uses an appropriate flow-controlled pressure reducing valve (PRV) which is fed by a single water main [1]. In our study, we analyze two pilot cities with diverse characteristics not only relatively to the water demand historical profiles but also relatively to the number of interdependent factors that affect this water demand amounts. In the first case the pilot city of Sosnowiec, Poland is analyzed. More specifically we have chosen a small water sub network forming a DMA of about 250 hydrometers. The area is relatively flat with no special elevation transitions and refers to a field with a vast majority of residential divisions and a small amount of businesses. Studying the historical time series of the aggregated water consumption for the area we concluded that the only factor that affects the total daily water consumption is the type of days we study (type 1 = working days, type 2 = holidays or weekends). The second use case refers to the pilot city of Skiathos Island, Greece. The characteristics of Skiathos differentiating from the first case include: (a) A rather old network of closed pipes with a huge amount of leaks that keep increasing in the last 4 years, (b) A much bigger set of hydrometers (around 3200) out of which a great amount changes status every year (status 1 = active, status 2 = inactive) due to the fact that, most of them refer to summer houses that are inhabited only for 6 months, (c) A direct dependence of the weather/climate and social economics factors to the water consumption of the island (during winter we have an average flow rate of around 70 m^3/h whereas during summer this value becomes around 110 m^3/h). The aforementioned arrangements for both pilot cities ensure that, the water pressure in the DMAs varies according to the water demand experienced however, it is assumed that it never drops under a certain threshold. The hydrometers have been geo-located in a GIS-component allowing multi-criteria processing on hydrometer level. Additionally the functionality of grouping hydrometers and the creation of land zones allows inspection on a macroscopic level.

Most of DSSs rely on two main approaches: the Model-Driven DSS and the Knowledge-Driven DSS [24]. The Knowledge-Driven DSS approach [25] focuses on a creation of a knowledge base for problem reasoning manipulation and on finding a solution through the use of the inference engine. On the contrary, the Model-Driven DSS approach [1] is the one that was chosen here and it is depicted in Fig. 1. More specifically, the overall idea of such a DSS is the analysis of the data stored in the database focusing mostly on the spatiotemporal characteristics of

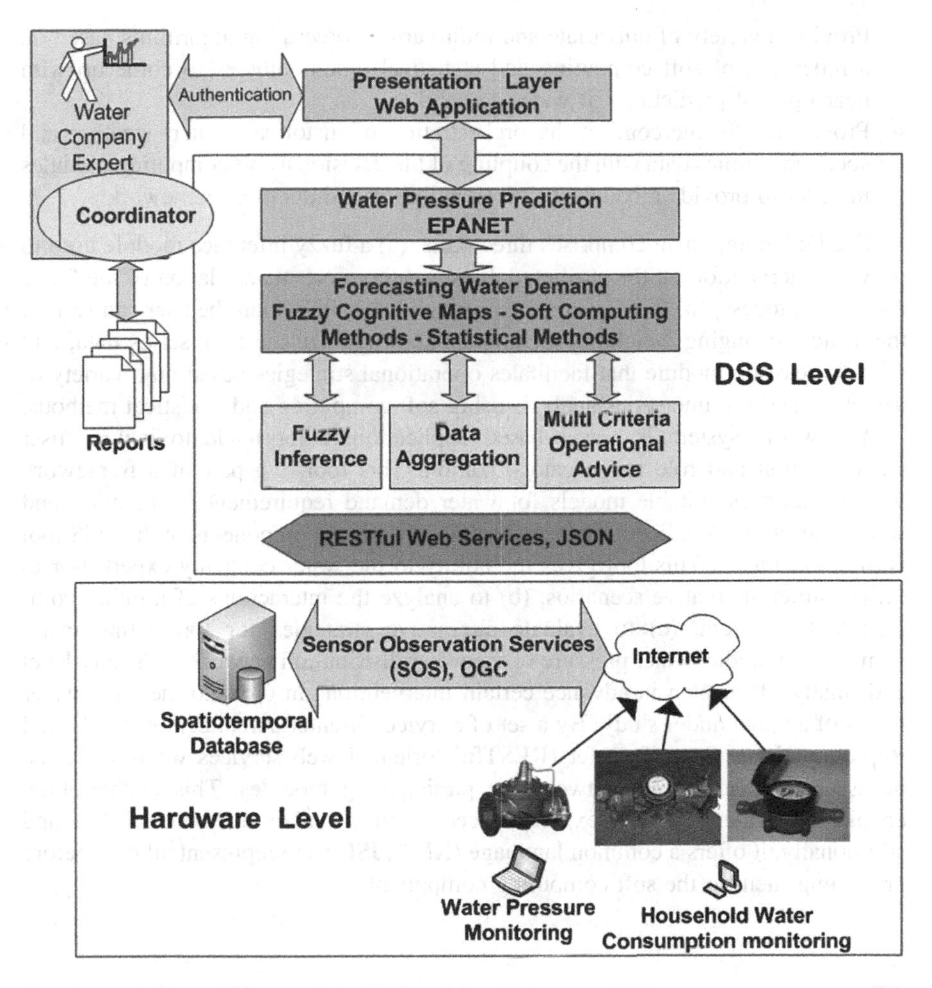

Fig. 1 The basic architecture of the DSS tool for water management

the problem and processing quantitative models that authorizes us to establish patterns. As the schematic in Fig. 1 shows, the whole implementation is based on the integration of various components that are layered up to the presentation top layer that aim to:

- Create a dashboard of functionalities to monitor, process, transfer, alter and store time series of urban water related data either on a city or a household level (flow rate, inlet and outlet water pressure, water consumption for households, weather conditions etc.).
- Manage and optimize decision making processes based on forecasting of water demand on a city level and predicting the water pressure of the whole water distribution system aiming into a reduction to satisfy certain water sustainability criteria.

- Provide a variety of univariate and multivariate forecasting algorithms based on a repertoire of soft computing and statistical methodologies to come up with near optimal predictions of water demand.
- Procedures to interconnect the orchestration of historical, water, weather and socio economics data with the coupling of the decisional soft computing modules in order to provide a collaborative modeling and simulation framework.

The DSS component comprises three parts: (a) a fuzzy inference module used to provide intervention on the attribute tables and therefore manipulation of the fuzzy inference process, (b) the data aggregation module with which the user can restrict the system changing weighting techniques and organize the time series data, and (c) a forecasting module that facilitates operational strategies based on a variety of use cases and on time series analysis using soft computing and statistical methods.

The whole system is a web-based application, tailor made to include user authentication and role usage categorization. This tool is a part of a framework which integrates suitable models for water demand requirements calculation and water allocation. A collection of screenshots of various components of the DSS tool is given in Fig. 2. This tool gives the ability to the water company expert user to (a) construct alternative scenarios, (b) to analyze the interactions of a rather complex water system, (c) to evaluate alternative strategies for forecasting water demand and predict water pressure of the water distribution network at the city level and finally (d) to plan in advance certain interventions in order to meet the water needs of an area under study. By a set of Service Oriented Architecture (SOA) and Representational State Transfer (RESTful) oriented web services we provide the necessary interoperability between the participating modules. This methodology enables the water company expert to access and visualize all historical data and additionally, it offers a common language (XML, JSON) to represent all data before promoting them to the soft computing component.

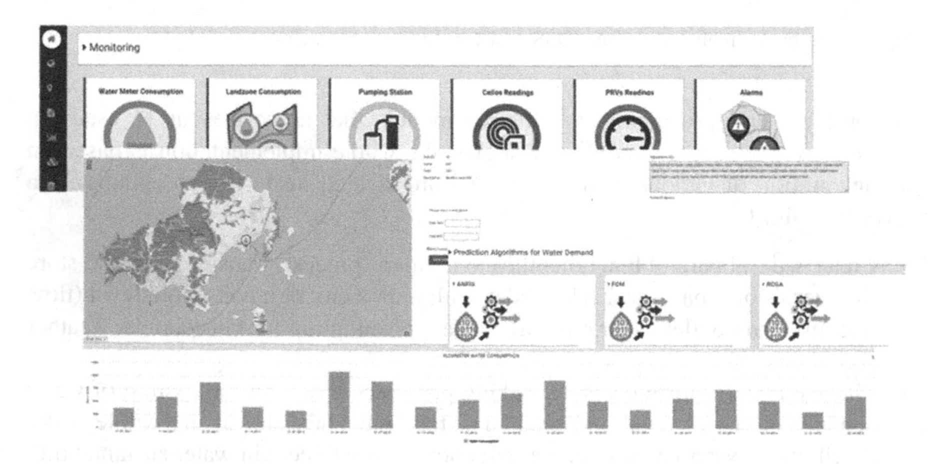

Fig. 2 A puzzle of screenshots of the DSS tool for water management

3 Soft Computing Algorithms for Water Demand Forecasting

Consumer water demand is the major decision factor in water DSSs whose forecast is critical for operating costs optimization for water suppliers [26]. Soft computing algorithms like FCMs and neuro-fuzzy have been previously applied in daily water demand prediction [i.e. 13, 19, 27, 28]. In this study, we show the applicability of two soft computing approaches, the evolutionary-based FCM and the ANFIS method for historical time series prediction in the case of Skiathos (Skiathos is one of the two pilot case cities for the urban management).

A multivariate analysis in a monthly base for the case study of Skiathos [28] using some benchmark forecasting algorithms like ARIMA and ANNs, was accomplished showing that the dominating predictors for water demand are found among meteorological and touristic variables (high daily and mean temperature, rain and wind, monthly increase of population etc.). In this paper, we use these predictors, as input variables (concepts) to forecast the daily prediction of water demand for one-step ahead.

Time series data from January of 2013 to September 2015 of Skiathos, Greece, for short-term urban water demand forecasting, was used for both models' construction, testing and validation. The dataset has been divided to 80 % training set and 20 % testing set. Approximately 6 months were used as the testing period for both soft computing approaches.

3.1 *Evolutionary-Based Fuzzy Cognitive Map Algorithm*

The input data in the multivariate analysis include time series of daily water pumping, daily mean and high temperature, daily precipitation, and daily arrivals in the island [27]. The algorithm of real-coded genetic learning for Fuzzy Cognitive Maps as described in previous work [18, 19, 29] is used to develop the FCM model from historical data. The steps of the real-coded genetic algorithm for FCM construction are described analytically in [29].

The parameters of the genetic algorithm were assigned on the basis of literature and numerous trials [18, 29]. In this case study, the values of genetic algorithm parameters were defined: size of the initial population = 200, maximum number of populations = 500, probability of crossover = 0.5, probability of mutation = 0.05, error = 0.001. Ranking selection, uniform crossover and Mühlenbein's mutation were used. To ensure the survival of the best individual in each population, elite strategy was applied.

The FCM was initialized, learned and tested on the basis of the provided real water demand values from the island of Skiathos. An exemplary FCM model after learning using real coded genetic algorithm is shown in Fig. 2a. Next, using the testing set and following the neural network process for one-step ahead prediction,

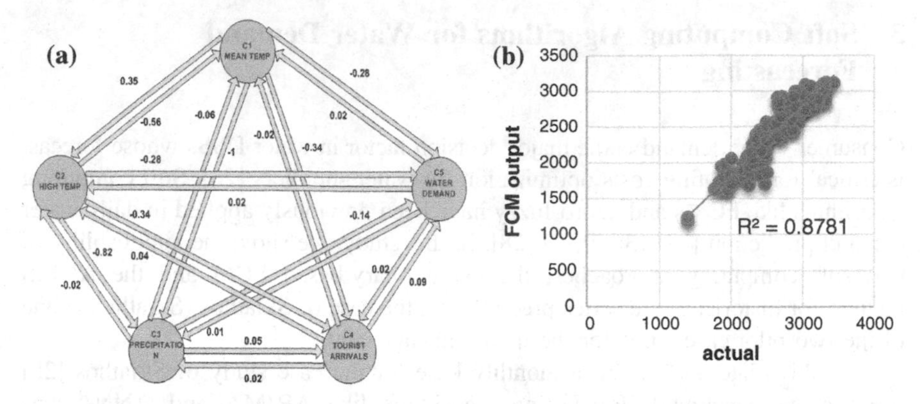

Fig. 3 **a** Fuzzy Cognitive Map model for water demand prediction. **b** R^2 plot

139 daily predicted values were calculated. Figure 2b depicts the R^2 value of the testing period which was calculated up to 0.8781; this is really an acceptable result for this type of prediction. Figure 3 illustrates the comparative plots of actual water demand and FCM forecast versus time for the testing period (April 2014–Sept 2015).

3.2 Neuro-Fuzzy Algorithm for Water Demand Forecasting

The neuro-fuzzy algorithm of ANFIS was used for the same dataset of Skiathos island for water demand prediction. Different ANFIS configuration sets were initially investigated. These settings included the number of membership functions (MFs), types of MFs (triangular, trapezoidal, bell-shaped, Gaussian and sigmoid), types of output MFs (constant or linear), optimization methods (hybrid or back propagation) and the number of epochs [30, 31]. After a large number of experiments, the best configuration of ANFIS was found. The parameters of the best configuration were 2 MFs for gbell type of MFs, constant output, hybrid optimization and 100 epochs.

The analysis was held for the same dataset, which was divided into a training period of 80 % data (2, 5 years) and a testing period of 20 % (6 months) data. The applied methodology gave promising results; indicatively, the derived model gave R^2 values of the testing period up to 0.846, a fact that proves the chosen variables to be good predictors [26].

Figures 4 and 5 show that both methods are adequate and can produce quite accurate water demand predictions. To sum up, we conducted multivariate analysis for the pilot case study of Skiathos island using two different soft computing approaches, the fuzzy cognitive maps and the neuro-fuzzy inference systems. The results of both proposed approaches are really useful for demand forecasting, as they are comparable and easy to be used in practice. Both of these algorithms have

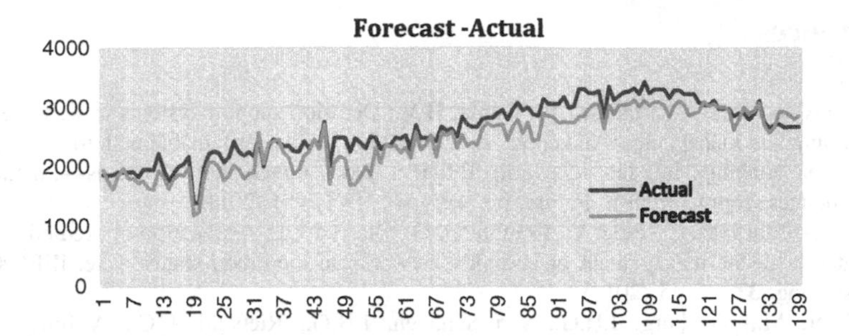

Fig. 4 Comparative plots of actual water demand and FCM forecast

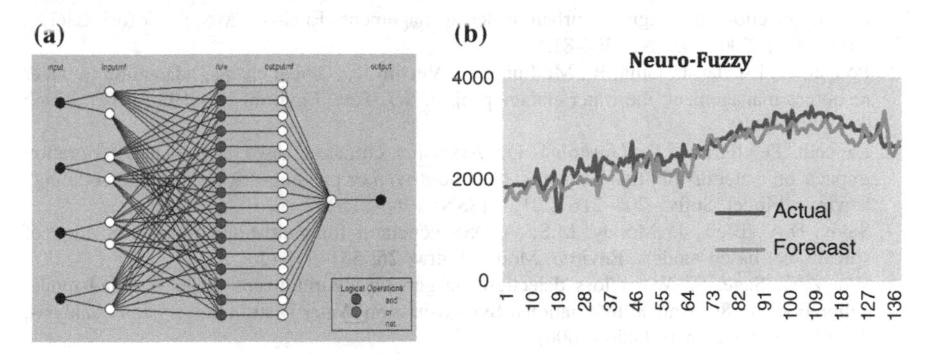

Fig. 5 **a** Neuro Fuzzy model for water demand prediction. **b** Comparative plots of real and predicted daily demand

already implemented in the urban DSS framework. The analysis is expected to benefit in terms of taking into account the importance of touristic activity and meteorological variables related to water demand. Such a tool would be quite useful in the case of a dramatic change in tourism or climate.

4 Conclusions

In this paper, we provided the general framework for an integrated system following a model-driven architecture. The decision support system aims to monitor and manage water resources at urban level with the inclusion of soft computing methods for prediction of water demand and a spatio-temporal model for arranging all data. Two soft computing approaches based on fuzzy cognitive map and neuro-fuzzy models for multivariate daily prediction of water demand in a highly touristic Mediterranean resort were presented. Additional water profiles will also be used to provide an automatic water pressure regulation of the water distribution network and therefore reduce leakages. The work is an undergoing process of the ISS-EWATUS project.

References

1. Serrat-Capdevila, A., Valdes, J.B., Gupta, H.V.: Decision support systems in water resources planning and management: stakeholder participation and the sustainable path to science-based decision making. In: Jao, C. (ed.) Efficient Decision Support Systems—Practice and Challenges from Current to Future, pp. 423–440. INTECH (2011)
2. Ocampo-Martinez, C., Puig, V., Cembrano, G., Quevedo, J.: Application of predictive control strategies to the management of complex networks in the urban water cycle. IEEE Control Syst. Mag. **33**, 15–45 (2013)
3. Bakker, M., Vreeburg, J.H.G., van Schagen, J.H.G., Rietveld, L.C.: A fully adaptive forecasting model for short-term drinking water demand. Environ. Model. Softw. **48**, 141–151 (2013). ISSN: 1364-8152
4. Makropoulos, C.K., Natsis, K., Liu, S., Mittas, K., Butler, D.: Decision support for sustainable option selection in integrated urban water management. Environ. Model. Softw. **23**(12), 1448-1460, (2008). ISSN: 1364-8152
5. Pouget, L., Escaler, I., Guiu, R., Mc Ennis, S., Versini, P.: Global change adaptation in water resources management: the water change project. Sci. Total Environ. 186–193 (2012). ISSN: 0048-9697 440
6. Laucelli, D., Berardi, L., Giustolisi, O.: Assessing climate change and asset deterioration impacts on water distribution networks: demand-driven or pressure-driven network modeling? Environ. Model. Softw. 206–216 (2012). ISSN: 1364-815237
7. Savić, D.A., Bicik, J., Morley, M.S.: A DSS generator for multi objective optimization of spreadsheet-based models. Environ. Model. Softw. **26**, 551–561 (2011)
8. Wu, Z.Y., Sage, P.: Water loss detection via genetic algorithm optimization based model calibration. In: 8th Annual International Symposium on Water Distribution System Analysis, 27–30 Aug, Cincinnati, Ohio (2006)
9. Quevedo, J., Cugueró, M., Pérez, R., Nejjari, F., Puig, V., Mirats, J.: Leakage location in water distribution networks based on correlation measurement of pressure sensors. In: IWA Symposium on Systems Analysis and Integrated Assessment. San Sebastian (2011)
10. Tabesh, M., Dini, M.: Fuzzy and neuro-fuzzy models for short-term water demand forecasting in Tehran. Iran. J. Sci. Technol. Trans. B Eng. **33**(B1), 61–77 (2009)
11. Yurdusev, M.A., Firat, M.: Adaptive neuro fuzzy inference system approach for municipal water consumption modeling: an application to Izmir, Turkey. J. Hydrol. **365**, 225–234 (2009)
12. Firat, M., Turkan, M.E., Yurdusev, M.A.: Comparative analysis of fuzzy inference systems for water consumption time series prediction. J. Hydrol. **374**, 235–241 (2009)
13. Mellios, N., Kofinas, D., Papageorgiou, E., Laspidou, C.: A multivariate analysis of the daily water demand of Skiathos Island, Greece. In: Implementing the Artificial Neuro-Fuzzy Inference Sysytem (ANFIS), IAHR 2015, E-Proceedings of the 36th International Association for Hydro-Environment Engineering and Research World Congress, 28 June–3 July. Hague, Netherlands (2015)
14. Song, H., Miao, C., Roel, W., Shen, Z.: Implementation of fuzzy cognitive maps based on fuzzy neural network and application in prediction of time series. IEEE Trans. Fuzzy Syst. **18** (2), 233–250 (2010)
15. Homenda, W., Jastrzebska, A., and Pedrycz, W.: Modeling time series with fuzzy cognitive maps. In: 2014 IEEE International Conference on Fuzzy Systems (FUZZ-IEEE), Beijing, China, pp. 2055–2062 (2014)
16. Homenda, W., Jastrzebska, A., Pedrycz, W.: Nodes selection criteria for fuzzy cognitive maps designed to model time series. Adv. Intell. Syst. Comput. **323**, 859–870 (2015)
17. Salmeron, J.L., Froelich, W., Papageorgiou, E.I.: Application of fuzzy cognitive maps to the forecasting of daily water demand. Presented at ITISE 2015 (International Work-Conference on Time Series), 1–3 July, Granada, Spain (2015)
18. Papageorgiou, E., Froelich, W.: Multi-step prediction of pulmonary infection with the use of evolutionary fuzzy cognitive maps. Neurocomputing **92**, 28–35 (2012)

19. Papageorgiou, E., Poczeta, K., and Laspidou, C.: Application of fuzzy cognitive maps to water demand prediction. In: 2015 IEEE International Conference on Fuzzy Systems (FUZZIEEE), Istanbul, pp. 1–8 (2015)
20. Kosko, B.: Neural Networks and Fuzzy Systems: A Dynamical Systems Approach to Machine Intelligence. Prentice Hall, New York (1992)
21. Papageorgiou, E.I.: Fuzzy Cognitive Maps for Applied Sciences and Engineering from Fundamentals to Extensions and Learning Algorithms. Springer (2014)
22. Lu, W., Pedrycz, W., Liu, X., Yang, J., Li, P.: The modeling of time series based on fuzzy information granules. Exp. Syst. Appl. **41**, 3799–3808 (2014)
23. Lu, W., Yang, J., Liu, X.: The hybrids algorithm based on fuzzy cognitive map for fuzzy time series prediction. J. Inf. Comput. Sci. **11**(2), 357–366 (2014)
24. Girard, M., Stewart, R.A.: Implementation of pressure and leakage management strategies on the gold coast, Australia: case study. J. Water Resour. Plann. Manage. **133**, 210–217 (2007)
25. Anzaldi, G., Rubion, E., Corchero, A., Sanfeliu, R., Domingo, X., Pijuan, J., Tersa, F.: Towards an enhanced knowledge-based Decision Support System (DSS) for Integrated Water Resource Management (IWRM). Procedia Eng. **89**, 1097–1104 (2014)
26. Billings, B.R., Jones, V.C.: Forecasting Urban Water Demand, vol. 7. American Water Works Associations (2008)
27. Kofinas, D., Mellios, N., Papageorgiou, E., Laspidou, C.: Urban water demand forecasting for the island of Skiathos. Procedia Eng. **89**, 1023–1030 (2014)
28. Laspidou, C.S., Kofinas, D., Mellios, N., Papageorgiou, E., Froelich, W., Magiera, E.: Urban water demand forecasting for the Island of Skiathos using multivariate analysis. In: IWA Water IDEAS Conference Proceedings, Bologna, 22–24 Oct, Italy (2014)
29. Stach, W., Kurgan, L., Pedrycz, W., Reformat, M.: Genetic learning of fuzzy cognitive maps. Fuzzy Sets Syst. **153**(3), 371–401 (2005)
30. Jang, J.S.R., Sun, C.T., Mizutani, E.: Neuro-Fuzzy and Soft Computing, p. 607. Prentice Hall, Englewood Cliffs (1997). ISBN: 0-13-261066-3
31. Pulido-Calvo, I., Gutierrez-Estrada, J.C.: Improved irrigation water demand forecasting using a soft-computing hybrid model. Biosyst. Eng. **102**(2), 202–218 (2009)

Decision Making Theory for Economics

Ranking Alternatives by Pairwise Comparisons Matrix with Fuzzy Elements on Alo-Group

Jaroslav Ramík

Abstract The decision making problem considered in this paper is to rank n alternatives from the best to the worst, using the information given by the decision maker in the form of an n by n pairwise comparisons matrix. Here, we deal with a pairwise comparisons matrix with fuzzy elements (PCF matrix). Fuzzy elements of the pairwise comparisons matrix are applied whenever the decision maker is not sure about the value of his/her evaluation, or, the elements of the PCF matrix are aggregated crisp evaluations in a group decision making problem. We investigate pairwise comparisons matrices with elements from an abelian linearly ordered group (alo-group) over a real interval. We propose a method starting from construction of fuzzy elements of a reciprocal PCF matrix, calculating its consistency and resulting into computation of the priority vector associated to the ranking of the alternatives. Illustrating examples are presented and discussed.

Keywords Ranking alternatives · Pairwise comparisons matrix · Reciprocity · Consistency · Fuzzy elements

1 Introduction

A *decision making problem (DM problem)* which forms an application background in this paper can be formulated as follows:

Let $X = \{x_1, x_2, \ldots, x_n\}$ be a finite set of alternatives ($n > 1$). The decision maker's aim is to rank the alternatives from the best to the worst (or, vice versa), using the information given by the decision maker in the form of an $n \times n$ pairwise comparisons matrix.

Fuzzy sets being the elements of the pairwise comparisons matrix can be applied whenever the decision maker is not sure about the preference degree of his/her

J. Ramík (✉)
Faculty of Business Administration in Karviná, Silesian University in Opava,
Univerzitní nám. 76, 733 40 Karviná, Czech Republic
e-mail: ramik@opf.slu.cz

I. Czarnowski et al. (eds.), *Intelligent Decision Technologies 2016*,
Smart Innovation, Systems and Technologies 57,
DOI 10.1007/978-3-319-39627-9_32

evaluation of the pairs in question. Fuzzy elements may be taken also as the aggregations of crisp pairwise comparisons of a group of decision makers in the group DM problem. Decision makers acknowledge fuzzy pairwise preference data as imprecise knowledge about regular preference information. Fuzzy elements include crisp elements and also interval elements.

Usually, an ordinal *ranking* of alternatives is required to obtain the "best" alternative(s), however, it often occurs that the decision maker is not satisfied with the ordinal ranking among alternatives and a cardinal ranking i.e. *rating* is then required.

The former works that investigated the problem of finding a rank of the given alternatives based on some pairwise comparisons matrices are e.g. [6–9, 15]. In [15] some simple linear programming models for deriving the priority weights from various interval fuzzy preference relations are proposed. Ramik and Korviny in [13] investigated inconsistency of pairwise comparisons matrix with fuzzy elements based on geometric mean. In [11, 12], the author presented a general approach for pairwise comparisons matrices with fuzzy number elements based on alo-groups which unifies the previous approaches.

The recent paper is in some sense a continuation of [11]. In this paper, we propose a three step method starting from construction of a reciprocal PCF matrix, calculating its consistency, and resulting into computation of the priority vector associated to ranking and/or rating of the alternatives. Illustrating examples are presented and discussed.

2 Preliminaries

We say that a fuzzy set A of $\mathbf{R}^* = \mathbf{R} \cup \{-\infty\} \cup \{+\infty\}$ is a *fuzzy interval*, whenever its membership function μ_A satisfies the following condition: there exist $a, b, c, d \in \mathbf{R}^*$, $-\infty \leq a \leq b \leq c \leq d \leq +\infty$, such that

$$\begin{aligned} \mu_A(t) &= 0 \ \text{ if } t < a \text{ or } t > d, \\ \mu_A &\ \text{ is strict increasing and continuous on } [a, b], \\ \mu_A(t) &= 1 \ \text{ if } b \leq t \leq c, \\ \mu_A &\ \text{ is strict decreasing and continuous on } [c, d]. \end{aligned} \tag{1}$$

A fuzzy interval A is *bounded* if $[a, d]$ is a compact interval.

Given $\alpha \in]0, 1]$, let A be a fuzzy set of X. The set $[A]_\alpha = \{x \in X \mid \mu_A(x) \geq \alpha\}$ is called the *α-cut of fuzzy set A*.

For $X = \mathbf{R}^*$, a bounded fuzzy interval A of $\mathbf{R}^*$ and $\alpha = 0$, we define the *zero-cut* of A as $[A]_0 = [a, d]$. A bounded fuzzy interval A is the *fuzzy number* if $b = c$.

An *abelian group* is a set, G, together with an operation $\odot$ (read: operation *odot*) that combines any two elements $a, b \in G$ to form another element in G denoted by $a \odot b$, see [2], or, [4]. The symbol $\odot$ is a general placeholder for a concretely given operation. $(G, \odot)$ satisfies the following requirements known as the *abelian group axioms*, particularly: *commutativity*, *associativity*, there exists an *identity element*

$e \in G$ and for each element $a \in G$ there exists an element $a^{(-1)} \in G$ called the *inverse element to a*.

The *inverse operation* $\div$ to $\odot$ is defined for all $a, b \in G$ as follows

$$a \div b = a \odot b^{(-1)}. \tag{2}$$

An ordered triple $(G, \odot, \leq)$ is said to be *abelian linearly ordered group, alo-group* for short, if $(G, \odot)$ is a group, $\leq$ is a linear order on G, and for all $a, b, c \in G$

$$a \leq b \text{ implies } a \odot c \leq b \odot c. \tag{3}$$

If $\mathcal{G} = (G, \odot, \leq)$ is an alo-group, then G is naturally equipped with the order topology induced by $\leq$ and $G \times G$ is equipped with the related product topology. We say that $\mathcal{G}$ is a *continuous alo-group* if $\odot$ is continuous on $G \times G$.

By definition, an alo-group $\mathcal{G}$ is a lattice ordered group. Hence, there exists $\max\{a, b\}$, for each pair $(a, b) \in G \times G$. Nevertheless, a nontrivial alo-group $\mathcal{G} = (G, \odot, \leq)$ has neither the greatest element nor the least element.

Because of the associative property, the operation $\odot$ can be extended by induction to n-ary operation.

$\mathcal{G} = (G, \odot, \leq)$ is *divisible* if for each positive integer n and each $a \in G$ there exists the (n)th root of a denoted by $a^{(1/n)}$, i.e. $\left(a^{(1/n)}\right)^{(n)} = a$.

Let $\mathcal{G} = (G, \odot, \leq)$ be an alo-group. Then the function $\|.\| : G \to G$ defined for each $a \in G$ by

$$\|a\| = \max\{a, a^{(-1)}\} \tag{4}$$

is called a *$\mathcal{G}$-norm.*

The operation $d : G \times G \to G$ defined by $d(a, b) = \|a \div b\|$ for all $a, b \in G$ is called a *$\mathcal{G}$-distance.* Next, we present the well known examples of alo-groups, see also [4], or, [12].

Example 1 Additive alo-group $\mathcal{R} = (]-\infty, +\infty[, +, \leq)$ is a continuous alo-group with: $e = 0$, $a^{(-1)} = -a$, $a^{(n)} = a + a + \cdots + a = n.a$.

Example 2 Multiplicative alo-group $\mathcal{R}^+ = (]0, +\infty[, \bullet, \leq)$ is a continuous alo-group with: $e = 1$, $a^{(-1)} = a^{-1} = 1/a$, $a^{(n)} = a^n$. Here, by $\bullet$ we denote the usual operation of multiplication.

Example 3 Fuzzy additive alo-group $\mathcal{R}_a = (]-\infty, +\infty[, +_f, \leq)$, see [12], is a continuous alo-group with: $a +_f b = a + b - 0.5$, $e = 0.5, a^{(-1)} = 1 - a, a^{(n)} = n.a - (n-1)/2$.

Example 4 Fuzzy multiplicative alo-group $\mathbf{]0, 1[_m} = (]0, 1[, \bullet_f, \leq)$, see [3], is a continuous alo-group with:

$$a \bullet_f b = \frac{ab}{ab + (1-a)(1-b)}, e = 0.5, a^{(-1)} = 1 - a.$$

3 PCF Matrices

In this paper we shall investigate $n \times n$ pairwise comparisons matrices with elements being bounded fuzzy intervals of the alo-group $\mathcal{G}$ over an interval G of the real line $\mathbf{R}$. We call them shortly *PCF matrices*. Moreover, we assume that all diagonal elements of the PCF matrix are crisp, particularly they are equal to the identity element of $\mathcal{G}$.

Considering this general approach based on alo-group may be useful as various approaches known from the literature are unified. This fact has been demonstrated on the previous 4 examples, where the well known cases of alo-groups are shown. Particularly, all concepts and properties which will be presented bellow can be easily applied to any of four presented alo-groups. In practice, the type of alo-group will depend on the particular DM problem. In some DM problems, it is more appropriate to apply e.g. additive alo-group (see Example 1), in other cases, the multiplicative alo-group (Example 2) is more natural from an interpretation point of view. Sometimes, also fuzzy alo-groups (Examples 3 and 4) are useful.

Notice that elements of PCF matrices may be crisp and/or fuzzy numbers, also crisp and/or fuzzy intervals, fuzzy intervals with bell-shaped membership functions, triangular fuzzy numbers, trapezoidal fuzzy numbers etc. Such fuzzy elements may be either evaluated by individual decision makers, or, they may be made up of crisp pairwise evaluations of decision makers in a group DM problem, see e.g. [1].

3.1 Reciprocity of PCF Matrices

Reciprocity of a PC matrix is a natural property defining the evaluation of couples of alternatives in the reverse order. Here, we define reciprocity for PCF matrices, our approach will cover the classical definitions of reciprocity presented e.g. in [10, 12].

Let $C = \{\tilde{c}_{ij}\}$ be a PCF matrix. As each element of C is a fuzzy interval, we have that each α-cut of C is a closed interval. Let us denote

$$[c^L_{ij}(\alpha), c^R_{ij}(\alpha)] = [\tilde{c}_{ij}]_\alpha.$$

Here, $c^L_{ij}(\alpha)$ is the left end point of this interval, $c^R_{ij}(\alpha)$ is the right end point of the interval.

Let $C = \{\tilde{c}_{ij}\}$ be an $n \times n$ PCF matrix, $\alpha \in [0, 1]$. C is said to be *α-$\odot$-reciprocal*, (sometimes *strong α-$\odot$-reciprocal*), if the following condition holds:

For every $i, j \in \{1, 2, \ldots, n\}$ and for every $c_{ij} \in [\tilde{c}_{ij}]_\alpha$ there exists $c_{ji} \in [\tilde{c}_{ji}]_\alpha$ such that

$$c_{ij} \odot c_{ji} = e. \quad (5)$$

$C = \{\tilde{c}_{ij}\}$ is said to be *$\odot$-reciprocal*, if C is α-$\odot$-reciprocal for all $\alpha \in [0, 1]$.

It can be easily shown, see e.g. [12], that $C = \{\tilde{c}_{ij}\}$ is α-$\odot$-reciprocal if and only if for all $i, j \in \{1, 2, \ldots, n\}$,

$$[c_{ji}^L(\alpha), c_{ji}^R(\alpha)] = [(c_{ij}^R(\alpha))^{(-1)}, (c_{ij}^L(\alpha))^{(-1)}]. \tag{6}$$

When evaluating fuzzy elements of a PCF matrix $C = \{\tilde{c}_{ij}\}$, only one of the membership functions of symmetric elements $\tilde{c}_{ij}$ and $\tilde{c}_{ji}$, $i \neq j$, should be constructed, the other should naturally satisfy condition (6). Then the PCF matrix becomes α-$\odot$-reciprocal.

Example 5 Consider the additive alo-group $\mathcal{R} = (\mathbf{R}, \odot, \leq)$ with $\odot = +$, see Example 1. Let PCF matrices $C = \{\tilde{c}_{ij}\}$ be given as follows:

$$C = \begin{bmatrix} (0;0;0) & (1;2;5) & (4;5;8) \\ (-5;-2;-1) & (0;0;0) & (3;4;5) \\ (-9;-5;-4) & (-5;-4;-3) & (0;0;0) \end{bmatrix}.$$

Here, C is 3×3 PCF matrix, particularly, PCF matrix with triangular fuzzy number elements and the usual "linear" membership functions.

For each $\alpha \in]0, 1[$ condition (6) is not satisfied for fuzzy elements $\tilde{c}_{13} = (4;5;8)$ and $\tilde{c}_{31} = (-9;-5;-4)$ in matrix C, hence, C is not α-+-reciprocal.

3.2 Consistency of PCF Matrices

Rationality and compatibility of a decision making process can be achieved by the consistency property of PCF matrices.

Let $\mathcal{G} = (G, \odot, \leq)$ be as earlier a divisible and continuous alo-group, $C = \{c_{ij}\}$ be a crisp PCF matrix, where $c_{ij} \in G \subset \mathbf{R}$ for all $i, j \in \{1, 2, \ldots, n\}$. The following definition is well known, see e.g. [3].

A crisp PCF matrix $C = \{c_{ij}\}$ is *$\odot$-consistent* if for all $i, j, k \in \{1, 2, \ldots, n\}$

$$c_{ik} = c_{ij} \odot c_{jk}. \tag{7}$$

The following equivalent condition for consistency of crisp PCF matrices is popular e.g. in AHP, see [3].

A crisp PC matrix $C = \{c_{ij}\}$ is $\odot$-consistent if and only if there exists a vector $w = (w_1, w_2, \ldots, w_n)$, $w_i \in G$ such that

$$w_i \div w_j = c_{ij} \text{ for all } i, j \in \{1, 2, \ldots, n\}. \tag{8}$$

Now, we extend the definition to non-crisp PCF matrices as follows, see also [12].

Let $\alpha \in [0, 1]$. A PCF matrix $C = \{\tilde{c}_{ij}\}$ is said to be *α-$\odot$-consistent*, if the following condition holds:

For every $i, j, k \in \{1, 2, \dots, n\}$, there exists $c_{ik} \in [\tilde{c}_{ik}]_\alpha$, $c_{ij} \in [\tilde{c}_{ij}]_\alpha$ and $c_{jk} \in [\tilde{c}_{jk}]_\alpha$ such that

$$c_{ik} = c_{ij} \odot c_{jk}. \tag{9}$$

The matrix C is said to be $\odot$-*consistent*, if C is α-$\odot$-consistent for all $\alpha \in [0, 1]$. If for some $\alpha \in [0, 1]$ the matrix C is not α-$\odot$-consistent, then C is called α-$\odot$-*inconsistent*.

Evidently, if C is crisp, then the previous definitions are equivalent.

Moreover, for $\alpha, \beta \in [0, 1], \alpha \geq \beta$, if $C = \{\tilde{c}_{ij}\}$ is α-$\odot$-consistent, then it is β-$\odot$-consistent.

It can be shown, see e.g. [12], that a PCF matrix $C = \{\tilde{c}_{ij}\}$ is α-$\odot$-consistent if and only if there exists a vector $w = (w_1, w_2, \dots, w_n)$ with $w_i \in G, i \in \{1, 2, \dots, n\}$, such that for all $i, k \in \{1, 2, \dots, n\}$ it holds

$$c_{ik}^L(\alpha) \leq w_i \div w_k \leq c_{ik}^R(\alpha). \tag{10}$$

4 Priority Vectors, Inconsistency of PCF Matrices

The concept of priority vector for ranking the alternatives is based on the optimal solution of the following optimization problem:

(P1)

$$\alpha \longrightarrow \max; \tag{11}$$

subject to

$$c_{ij}^L(\alpha) \leq w_i \div w_j \leq c_{ij}^R(\alpha) \text{ for all } i, j \in \{1, 2, \dots, n\}, \tag{12}$$

$$\bigodot_{k=1}^{n} w_k = e, \tag{13}$$

$$0 \leq \alpha \leq 1, w_k \in G, \text{ for all } k \in \{1, 2, \dots, n\}. \tag{14}$$

If optimization problem (P1) has a feasible solution, i.e. system of constraints (12)–(14) has a solution, then (P1) has also an optimal solution. Let α^* and $w^* = (w_1^*, \dots, w_n^*)$ be an optimal solution of problem (P1). Then α^* is called the $\odot$-*consistency grade of* C, denoted by $g_\odot(C)$, i.e.

$$g_\odot(C) = \alpha^*. \tag{15}$$

Here, $w^* = (w_1^*, \dots, w_n^*)$ is called a *$\odot$-priority vector of C*. This vector is associated with the ranking of alternatives as follows:

$$\text{If } w_i^* > w_j^* \text{ then } x_i \succ x_j,$$

where $\succ$ stands for "is better than".

If optimization problem (P1) has no feasible solution, then we define

$$g_\odot(C) = 0. \tag{16}$$

In that case, the priority vector is not defined at this moment. It will be defined later on.

Generally, problem (P1) is a nonlinear optimization problem that can be efficiently solved e.g. by the dichotomy method, which is a sequence of optimization problems, see e.g. [10]. For instance, given $\alpha \in [0, 1]$, $\odot = +$, problem (P1) can be solved as an LP problem (with variables $w_1, \dots, w_n$).

It was proven in [12], that if all entries $\tilde{c}_{ij}$ of a PCF matrix C are fuzzy numbers and $w^* = (w_1^*, \dots, w_n^*)$ is an optimal solution of (P1), i.e. $\odot$-priority vector of C, then w^* is unique.

The optimal solution α^* and $w^* = (w_1^*, \dots, w_n^*)$ of problem (P1) should be unique as the decision makers usually ask for unique decision, or, a unique ranking of the alternatives. A sufficient condition for uniqueness of the priority vector $w^* = (w_1^*, \dots, w_n^*)$ is that all elements $\tilde{c}_{ij}$ of the PCF matrix C are fuzzy numbers, particularly, that the core of each $\tilde{c}_{ij}$,

$$Core(\tilde{c}_{ij}) = \{t \in G | \mu_{\tilde{c}_{ij}}(t) = 1\},$$

is a singleton. However, this is not the case of PCF matrices where the entries are fuzzy intervals (i.e. trapezoidal fuzzy numbers). Then the uniqueness is not secured and multiple solutions of (P1) could exist. In practical decision making problems such cases usually require reconsidering evaluations of some elements of the PCF matrix.

Let $C = \{\tilde{c}_{ij}\}$ be a PCF matrix, $\alpha \in [0, 1]$. If there exists a triple of elements $i, j, k \in \{1, 2, \dots, n\}$ such that for any $c_{ij} \in [\tilde{c}_{ij}]_\alpha$, any $c_{ik} \in [\tilde{c}_{ik}]_\alpha$, and any $c_{kj} \in [\tilde{c}_{kj}]_\alpha$:

$$c_{ik} \neq c_{ij} \odot c_{jk}, \tag{17}$$

then the PCF matrix C is *α-$\odot$-inconsistent.* If for all $\alpha \in [0, 1]$ the PCF matrix C is α-$\odot$-inconsistent, then we say that C is *$\odot$-inconsistent.* By this definition, for a given PCF matrix C and given $\alpha \in [0, 1]$, C is either α-$\odot$-consistent, or, C is α-$\odot$-inconsistent.

Notice, that for a PCF matrix C problem (P1) has no feasible solution, if and only if C is $\odot$-inconsistent, i.e. C is α-$\odot$-inconsistent for all $\alpha \in [0, 1]$.

It is an important task to measure an intensity of $\odot$-inconsistency of the given PCF matrix. In some cases, a PCF matrix can be "close" to some $\odot$-consistent matrix, in the other cases $\odot$-inconsistency can be strong, meaning that the PCF matrix can be "far" from any $\odot$-consistent matrix.

The $\odot$-inconsistency of C will be measured by the minimum of the distance of the "ratio" matrix $W = \{w_i \div w_j\}$ to the "left" matrix $C^L = \{c^L_{ij}(0)\}$ and "right" matrix $C^R = \{c^R_{ij}(0)\}$, as follows.

Let $w = (w_1, \ldots, w_n)$, $w_i \in G$, $i, j \in \{1, \ldots, n\}$. Denote

$$\begin{aligned} d_{ij}(C,w) &= \quad e \qquad \text{if } c^L_{ij}(0) \leq w_i \div w_j \leq c^R_{ij}(0), \\ &= \min\{\|c^L_{ij}(0) \div (w_i \div w_j)\|, \|c^R_{ij}(0) \div (w_i \div w_j)\|\}, \text{ otherwise.} \end{aligned} \tag{18}$$

Here, by $\|.\|$ we denote the norm defined in Sect. 2.

We define the maximum deviation to the matrix $W = \{w_i \div w_j\}$:

$$I_{\odot}(C,w) = \max\{d_{ij}(C,w) | i,j \in \{1,\ldots,n\}\}. \tag{19}$$

Now, consider the following optimization problem.

(P2)

$$I_{\odot}(C,w) \longrightarrow \min; \tag{20}$$

subject to

$$\bigodot_{k=1}^{n} w_k = e, \tag{21}$$

$$w_k \in G, \text{ for all } k \in \{1,2,\ldots,n\}. \tag{22}$$

The $\odot$-*inconsistency index of PCM matrix* C, $I_{\odot}(C)$, is defined as

$$I_{\odot}(C) = \inf\{I_{\odot}(C,w) | w_k \text{ satisfies (21), (22)}\}. \tag{23}$$

If $w^* = (w^*_1, \ldots, w^*_n)$ is an optimal solution of (P2), then

$$I_{\odot}(C) = I_{\odot}(C, w^*).$$

If there exists a feasible solution of (P1), then $\odot$-inconsistency index of PCM matrix C, $I_{\odot}(C)$, is equal to e, i.e. $I_{\odot}(C) = e$.

It is clear that an optimal solution of (P2) exists, the uniqueness of the optimal solution of (P2) is, however, not saved. Depending on the particular operation $\odot$, problem (P2) may have multiple optimal solutions which is an unfavorable fact from the point of view of the decision maker. In this case, the decision maker should reconsider some (fuzzy) evaluations of pairwise comparisons matrix.

Now, we define a priority vector also in case of $g_{\odot}(C) = 0$, i.e. if no feasible solution of (P1) exists. In contrast to the case of $g_{\odot}(C) > 0$, this priority vector cannot become an α-$\odot$-consistency vector of C for some $\alpha > 0$.

Let C be an $\odot$-inconsistent PCM matrix. The optimal solution $w^* = (w_1^*, \dots, w_n^*)$ of (P2) will be called the *$\odot$-priority vector of C.*

Now, we summarize the obtained results. Let $C = \{\tilde{c}_{ij}\}$ be a PCF matrix. Then exactly one of the following two cases occurs:

- Problem (P1) has a feasible solution. Then consistency grade $g_{\odot}(C) = \alpha$, for some $\alpha, 0 \le \alpha \le 1$, and $I_{\odot}(C) = e$.
- Problem (P1) has no feasible solution. Then consistency grade $g_{\odot}(C) = 0$, C is $\odot$-inconsistent, and $I_{\odot}(C) > e$.

Example 6 Let $X = \{x_1, x_2, x_3\}$ be a set of alternatives, let $C = \{\tilde{c}_{ij}\}$ be a PCF matrix on the fuzzy multiplicative alo-group $]\mathbf{0}, \mathbf{1}[_{\mathbf{m}} = (]0, 1[, \bullet_f, \le)$, with:

$$a \bullet_f b = \frac{ab}{ab + (1-a)(1-b)}, e = 0.5, a^{(-1)} = 1 - a, \tag{24}$$

$$\|a\| = max\{a, 1-a\}.$$

Fuzzy multiplicative alo-group $]\mathbf{0}, \mathbf{1}[_{\mathbf{m}}$ is divisible and continuous. For more details and properties, see Example 4, [5, 12]. Let

$$C = \begin{bmatrix} (0.5;0.5;0.5) & (0.6;0.7;0.8) & (0.75;0.8;0.9) \\ (0.2;0.3;0.4)) & (0.5;0.5;0.5) & (0.7;0.75;0.8) \\ (0.1;0.2;0.25) & (0.2;0.25;0.3) & (0.5;0.5;0.5) \end{bmatrix},$$

Here, C is a 3×3 PCF matrix, particularly, PCF matrix with elements on $]0, 1[$. C is a $\bullet_f$-reciprocal PCF matrix (noncrisp), the elements of C are triangular fuzzy numbers. There is an optimal solution of the corresponding problem (P1) with C, the consistency grade $g_{\bullet_f}(C) = 0.6$. The $\bullet_f$-priority vector w^* of C is $w^* = (0.586, 0.302, 0.112)$, hence $x_1 \succ x_2 \succ x_3$. The inconsistency index $I_{\bullet_f}(C) = 0.5$.

5 Conclusion

This paper deals with the problem of ranking a finite number of alternatives by pairwise comparisons matrices with fuzzy elements. Fuzzy elements of the pairwise comparisons matrix are usually applied whenever the decision maker is not sure about the value of his/her evaluation of the relative importance of elements in question. In comparisons with PC matrices investigated in the literature, here we investigate pairwise comparisons matrices with elements from abelian linearly ordered group (alo-group) over a real interval (PCF matrices). We deal with the reciprocity and consistency of pairwise comparisons matrices with fuzzy intervals which is an extension of the well known concept in crisp case and which is used for ranking the alternatives. Such an approach allows for unifying the additive, multiplicative and also fuzzy approaches known from the literature. The concept of reciprocity and consistency of pairwise comparisons matrices with triangular fuzzy numbers has been already studied in the former author's works as well as in the other literature on this subject, see e.g. [8, 10, 12, 14, 15] and others. Here, we apply some properties of PCF matrices, particularly the property of reciprocity and consistency, to the problem of ranking the alternatives. Moreover, we also solve the problem of calculating special indexes: consistency grade and inconsistency index. Several numerical examples have been presented to illustrate the concepts and derived properties.

Acknowledgments This research has been supported by GACR project No. 14-02424S.

References

1. Bilgic, T., Turksen, I.B.: Measurement of membership functions: theoretical and empirical work. In: Dubois, D., Prade, H. (eds.) Fundamentals of Fuzzy Sets, pp. 195–227. Kluwer Academic Publishers, New York (2000)
2. Bourbaki, N.: Algebra II. Springer, Heidelberg-New York-Berlin (1998)
3. Cavallo, B., D'Apuzzo, L., Squillante, M.: About a consistency index for pairwise comparison matrices over a divisible alo-group. Int. J. Intell. Syst. **27**, 153–175 (2012)
4. Cavallo, B., D'Apuzzo, L.: A general unified framework for pairwise comparison matrices in multicriteria methods. Int. J. Intell. Syst. **24**(4), 377–398 (2009)
5. Cavallo, B., D'Apuzzo, L.: Deriving weights from a pairwise comparison matrix over an alo-group. Soft Comput. **16**, 353–366 (2012)
6. Leung, L.C., Cao, D.: On consistency and ranking of alternatives in fuzzy AHP. Eur. J. Oper. Res. **124**, 102–113 (2000)
7. Mahmoudzadeh, M., Bafandeh, A.R.: A new method for consistency test in fuzzy AHP. J. Intell. Fuzzy Syst. **25**(2), 457–461 (2013)
8. Mikhailov, L.: Deriving priorities from fuzzy pairwise comparison judgments. Fuzzy Sets Syst. **134**, 365–385 (2003)
9. Mikhailov, L.: A fuzzy approach to deriving priorities from interval pairwise comparison judgements. Eur. J. Oper. Res. **159**, 687–704 (2004)
10. Ohnishi, S., Dubois, D., et al.: A fuzzy constraint based approach to the AHP. In: Uncertainty and Intelligent Information Systems, pp. 217–228. World Sci., Singapore, 2008 (2008)
11. Ramik, J.: Isomorphisms between fuzzy pairwise comparison matrices. Fuzzy Optim. Decis. Making **14**, 199–209 (2015)
12. Ramik, J.: Pairwise comparison matrix with fuzzy elements on alo-group. Inf. Sci. **297**, 236–253 (2015)
13. Ramik, J., Korviny, P.: Inconsistency of pairwise comparison matrix with fuzzy elements based on geometric mean. Fuzzy Sets Syst. **161**, 1604–1613 (2010)
14. Salo, A.A.: On fuzzy ratio comparison in hierarchical decision models. Fuzzy Sets Syst. **84**, 21–32 (1996)
15. Xu, Z.S., Chen, J.: Some models for deriving the priority weights from interval fuzzy preference relations. Eur. J. Oper. Res. **184**, 266–280 (2008)

An Analysis on Three Inflection Points on Four Economic Phases

Takafumi Mizuno and Eizo Kinoshita

Abstract Economic cycles can be represented in a sequence of four economic phases. These four phases are Thetical economy, Bubble economy, Bubble collapse economy, and Antithetical economy. This is a description of an economic cycle in Kinoshita's economic scheme: Thetical economics and Antithetical economics. In this paper, we describe mechanisms of three inflection points of these four phases. We focus on the consumption propensity, and we construct a model that a change of the consumption propensity causes transitions of these phases. Through the modeling and its analyses of inflection points, we clear a nature of economic phases. A market enlarges its economic scale in phases before a bubble collapse, and the market optimizes its economic activities in phases after the collapse.

1 Introduction

Kinoshita provided a macroeconomic frame; Thetical economics and Antithetical economics [1]. Each economics is a set of theories of economy. This frame arises from his analysis on bubble economies [2]. Kinoshita claimed that we are ruled by theories of Thetical economics before a bubble collapse, while we are ruled by theories of Antithetical economics after the bubble collapse.

Kinoshita describes an economic cycle with four economic phases [3]. They are Thetical economy, Bubble economy, Bubble collapse economy, and Antithetical economy. The economic cycle of a market starts which the Thetical economy. At a certain time, enthusiasm of the markets leads us into the Bubble economy. Suddenly, excess consumptions break the Bubble economy. We are in the Bubble

T. Mizuno (✉) · E. Kinoshita
School of Urban Science, Meijo University, Gifu, Japan
e-mail: tmizuno@meijo-u.ac.jp

E. Kinoshita
e-mail: kinoshit@meijo-u.ac.jp

I. Czarnowski et al. (eds.), *Intelligent Decision Technologies 2016*,
Smart Innovation, Systems and Technologies 57,
DOI 10.1007/978-3-319-39627-9_33

collapse economy until ends of the market's confusion. Then, Antithetical economy starts. He states that a large economic expansion needs to restart new Thetical economy after overcoming the Antithetical economy.

Kinoshita and Mizuno displayed a mechanism a collapse of bubble economy [4]. That is a transition from Bubble economy to Bubble collapse economy. In the mechanism, a variable plays an important role of the collapse. The variable is the consumption propensity in macroeconomics. Mizuno and Kinoshita show that if consumption propensity coefficient is larger than 1, then increasing investments decreases productions. The research, however, treats consumption propensity as exogenous variable.

In this research, we describe a mechanism of three inflection points: a start of Bubble economy, a change from Bubble economy to Bubble collapse economy (bubble collapse), and the end of Bubble collapse economy.

1.1 A Model of Rationalities of Economic Agents

In economics, researchers are modeling economic agents: nation people, corporations, or governments. Researchers model these agents as objects that have well disciplined, act rationally, and they can obtain all information of markets. The model is quite unrealistic, but macroeconomics stands on the model. We cannot understand economic statistics or its observations without the model.

Kinoshita modeled a rationality of economic agents using representations of OR's linear programing [5]. He refers to his model as "managements." In this paper, we refer to the model as micro model.

In Thetical economy, a rationality of a corporation is represented as

$$\max \sum_{j \in J} c_j x_j, \tag{1}$$

$$s.t.$$

$$\forall i \in I, \quad \sum_{j \in J} a_{ij} x_j \leq b_i. \tag{2}$$

J is a set of products which are made by the corporation, and I is a set of items which are materials of products or accounting subjects when the corporation produces its product. In other words, the corporation consumes items of I, and produces products of J. x_j is an amount of a product j of the corporation, c_j is profit of one unit of the product j. a_{ij} is an amount of the item i, the corporation needs the amount of i when the corporation produces the product j. b_i is limit of debt for an accounting subject i. A cost of producing a product j is

$$cost(j) = x_j \sum_{i \in I} a_{ij}. \tag{3}$$

The sum of the term $\sum_{j\in J} c_j x_j$ in the Eq. (1) for all corporations is the supply Y in macroeconomics. While the term $\sum_{j\in J} a_{ij} x_j$ in the Eq. (2) is classified into the consumption C or into the investment I. The classification depends on kinds of the corporation and kinds of the accounting subjects of i.

In Antithetical economy, a rationality of a corporation is represented as

$$\min \sum_{i\in I} u_i b_i, \tag{4}$$

s.t.

$$\forall j \in J, \quad \sum_{i\in I} u_i a_{ij} \geq c_j. \tag{5}$$

u_i is a mortgage rate or an interest rate of an accounting subject i.

2 Our Assumptions

We describe three inflection points of transitions of economics phases in this paper. The consumption propensity coefficient a plays an important roles in our descriptions.

In studies of macroeconomics, the variable a is often treated as an exogenous variable. In this paper, we assume the variable a is an endogenous variable, and we describe inflection points with varying the variable.

Structures of the variable have been analyzed by researchers who study consumption functions or preference functions. There are many concrete models of consumption functions, and there are many procedures to detect preference functions of individuals.

We put a simplest assumption that there is a special level in a market. And the consumption propensity depends on the rate of the supply Y to the level. Closing the supply to the level increases the consumption propensity.

3 The First Inflection Point: From Thetical Economy to Bubble Economy

The supply Y creates demands in Thetical economy.

$$Y \equiv C + I, \tag{6}$$
$$C = aY \tag{7}$$

where a is consumption propensity, C is national demands, and I is investments. This is a simple macroeconomic model.

In the start of Bubble economy, a level of supply Y_g appears. The level Y_g is higher than the supply Y. The consumption propensity a depends on the level;

$$a = \frac{Y}{Y_g}. \tag{8}$$

Increasing Y increases the consumption propensity.

From a view of with the micro model, growing the consumption propensity beyond 1 needs a condition. There is a special product k which holds the condition:

$$\exists k \in I \cap J, \quad \text{and} \quad b_k \to \infty. \tag{9}$$

and a term $\sum_{j \in J} a_{kj} x_j$ of all corporations are added into the consumption C in macro-economics.

4 The Second Inflection Point: From Bubble Economy to Bubble Collapse Economy

When the supply Y exceeds the level Y_g, the consumption the propensity coefficient $a = Y/Y_g$ is larger than 1. Then the bubble collapses. The level Y_g acts as an upper limit of the supply Y in the Bubble economy.

After the collapse, the economy will be into Bubble collapse economy.

From the view with micro model, a changing value

$$b_k : \infty \to \text{a finite value}. \tag{10}$$

is trigger of the collapse.

5 The Third Inflection Point: From Bubble Collapse Economy to Antithetical Economy

Under the Bubble collapse economy, investment efficiency $\partial Y / \partial I$ is less than zero, and the consumption propensity is greater than 1. They are extraordinary situations. They are canceled throughout the economic phase.

A supply level Y_o appears in the economy. The consumption propensity is

$$a = \frac{Y_o}{Y} \tag{11}$$

The Bubble collapse economy ends with positive investment I. Or $Y = Y_o$. Then Antithetical economy starts.

Under the Antithetical economy, the micro model is the Eq. (4) with the Eq. (5). Corporations do not try to enlarge their profits. They want to clear their debts. These settlements of their debts are swaps debts of corporations for credits of banks. With a macro view, the exchange does not increase the supply Y.

Under Antithetical economy, we can represent the supply Y of macroeconomics as

$$Y = \min\{C + I, \quad Y_o\} \tag{12}$$

Even if Y is sufficiently large,

$$C + I = Y_o \quad \text{(constant)}. \tag{13}$$

The micro model is modified as

$$\min \sum_{i \in I} u_i b_i, \qquad \text{((4) reprint)}$$

$$s.t.$$

$$\forall j \in J, \quad \sum_{i \in I} u_i a_{ij} \geq c_j, \qquad \text{((5) reprint)}$$

$$\sum_{j \in J} c_j x_j \leq \text{constant}. \tag{14}$$

In other words, corporations make effort to optimize their production activities in Antithetical economy.

6 Discussions

We can interpret mechanisms of inflections of four phases in this paper as activities of a market. In this section, the market means a general term of a set of economic agents.

In the Thetical economy, the market enlarges its market scale. The scale is represented as the supply Y of macroeconomics. This expansion changes in the Bubble economy. The market seems to enlarge the market scale to reach a goal. The goal is the level of Y_g in the Bubble economy. And the accomplishing the goal destroys the bubble. In the Bubble collapse economy, the market calms down its economic activities. The market leads its state toward a moderate state. The moderate state is represented as Y_o in the Bubble collapse economy. After the Bubble collapse economy, because of shocks of the collapse with many debts, the market wants to clean its debts without expanding the market scale. In other words, the market optimizes its structures in Antithetical economy. The market will complete the optimization in time. The micro model in the Antithetical economy and the micro model in the Thetical economy have a duality. They have a same solution when the solution exists. After the optimization, the market will retain its state notwithstanding the market's efforts to enlarge its profits.

From a macro view, Antithetical economy will end when the upper limit Y_o will be removed, or

$$Y \leftarrow \min\{C + I, \quad \infty\} \tag{15}$$

From a view of the micro model, because the term $\left[\max \sum_{j \in J} c_j x_j\right]$ of a corporation in the Eq. (1) reaches an optimal value, the corporation cannot obtain profits over the optimal value. Increasing profits beyond the optimum value needs changes structures of the corporation. We can represent one of new structures simply

$$\max \left[\sum_{j \in J} c_j x_j + \sum_{j \in K} c_j x_j\right], \tag{16}$$

$s.t.$

$$\forall i \in I, \quad \sum_{j \in J \cup K} a_{ij} x_j \leq b_i. \tag{17}$$

K is a set of new items which does not exists ex-economic phases. It implies that the corporation must develop unprecedented goods or services.

We treat each limit of debt, which is represented in the variable b_i in the micro model, as each corporation's constraint. Change of the limit triggers the Bubble economy occurs and its collapse in our model. The limit comes from the outside of corporations. In the actual world, banks decide the limit. Therefore, in our models, decision-makings of banks are very important for births and collapses of the Bubble economy.

7 Conclusions

We describe mechanisms of inflection points on four economic phases in this paper. This reinforces Kinoshita's economic scheme: Thetical economics and Antithetical economics. We focus on changes of the consumption propensity as a cause of transitions of economic phases. In the mechanism, we assume that the consumption propensity depends on the supply of macroeconomics. Through the description of these mechanisms, we obtain an insight of Kinoshita's scheme. The Thetical economics is a set of theories of expanding markets' scale, while the Antithetical economics is a set of theories of optimizing markets' activities.

We now have an economic model that consists of economic phases with mechanisms of their transitions. Next step of our study is statistical verifications of validity of the model.

References

1. Kinoshita, E.: A proposal of primal and dual problems in macro-economics. China-USA Bus. Rev. **10**(2), 115–124 (2011)
2. Kinoshita, E.: Why bubble economy occurs and crashes?-repeated history of economic growth and collapse. Chin. Bus. Rev. **10**(2), 102–111 (2011)
3. Kinoshita, E.: A proposal of thetical economy and antithetical economy mechanism of occurrence and collapse of bubble economy. J. Bus. Econ. **3**(2), 117–130 (2012)
4. Kinoshita, E., Mizuno, T.: Analysing mechanism of an economic phase. China-USA Bus. Rev. **12**(11), 1025–1032 (2013)
5. Kinoshita, E.: Thetical and antithetical business management. J. Bus. Econ. **6**(6), 1086–1096 (2015)

Economic Rationalities and Governmental Actions on the Thetical Economics and the Antithetical Economics

Eizo Kinoshita and Takafumi Mizuno

Abstract Kinoshita has been proposed a macroeconomic paradigm. It consists of the Thetical economics and the Antithetical economics. In this paper, we describe behavioral principles of corporations and the government on the paradigm. The description is a model based on linear programming of Operations Research. And we discuss governmental activities using the model. In conclusion, we state that the government must adopt monetary policy in an economic phase which is dominated by the Thetical economics, and the government must adopt fiscal policy in another economic phase which is dominated by the Antithetical economics.

1 Introduction

Kinoshita has been proposed a new macroeconomic paradigm. He asserts that there are two economics: the Thetical economics and the Antithetical economics [1, 5, 6]. Each economics is a set of theories that dominate economic activities. His insight divides economic cycles roughly into four phases. In the division, a bubble collapse plays a central role [2, 3]. If Say's law is valid in an economic phase, then the Thetical economics dominates the phase. He refers to the phase as the Thetical economy. While if the Keynes's effective demand is effective in an economic phase, then the Antithetical economics dominates the phase. He refers to the phase as the Antithetical economy. Enthusiasm of economic agents leads an economic phase to the Bubble economy. And its collapse makes the phase the Bubble collapse economy. An economic cycle is the sequence of the Thetical economy, the Bubble economy, the Bubble collapse economy, and the Antithetical economy.

E. Kinoshita (✉) · T. Mizuno
School of Urban Science, Meijo University, Gifu, Japan
e-mail: kinoshit@meijo-u.ac.jp

T. Mizuno
e-mail: tmizuno@meijo-u.ac.jp

I. Czarnowski et al. (eds.), *Intelligent Decision Technologies 2016*,
Smart Innovation, Systems and Technologies 57,
DOI 10.1007/978-3-319-39627-9_34

We use symbols as follows. A symbol Y expresses the national supply, a symbol C expresses the national consumption, and I expresses the national investment. The national supply is a total of profit of all corporations in the nation. The national consumption is a total of consumption of all economic agents in the nation. And the national investment is a total of investment of all economic agents in the nation. The national demand is a total of demands of all economic agents in the nation; the national demand is represented in $C+I$.

The Thetical economics is valid when Say's law is effective. Say's law implies that the national supply is always equivalent to the national demand;

$$Y \equiv C+I. \tag{1}$$

Because the total of demands is bigger than the total of supply, if a corporation supplies a product then consumers buy the product immediately.

The Antithetical economics is a set of theories when Say's law is not possible. In the situations, the national supply often does not equal the national demand. The national demand is smaller than the national supply. It means that there are unsold products in markets. A purpose of the macroeconomics is searching equilibrium states of the national supply and the national demand by changing the national demand. If a equilibrium state exists, then

$$Y = C+I. \tag{2}$$

Controlling demand side $C+I$ decides the equilibrium state of the national supply. So Keynes referred to $C+I$ as effective demand. The Antithetical economics is valid when the effective demand is effective.

We can say roughly that the Say's phase is prosperity and the Keynes's phase is a recession.

Kinoshita's macroeconomic frame is a one-way cycle of economic phases. Our economic state changes from the Say's phase to the Keynes's phase via a bubble birth and its collapse. If we succeed resurrection of markets, then the economic state restarts from the Say's phase.

Kinoshita constructs microeconomic foundations of his macroeconomic model. In usual studies of the macroeconomics, economic agents are modeled simply. All economic agents expand their profits, they are well-disciplined, they can acquire all information of markets, and their behavior is rational. Kinoshita gives a concrete mathematical model of the rationality. He represents the rationality with linear programing techniques of OR (Operations Research) [4].

The rough division of macroeconomic cycles and the representation of microeconomic rationalities can provide concise descriptions and smart analyses for studies of the macroeconomics.

In this paper, we described a model of business managements' rationality and a model of government administrations' obligation. And we argue a decision making of governmental actions; Which policy must we choose, monetary policy or fiscal policy?

2 Representations of Economic Rationalities

Economic agents have different strategies which are different for each economic phase. There are dualities between strategies for Thetical economy and strategies for Antithetical economy (Fig. 1). We describe a model of their behavioral principles with regard to economic rationalities. The model must represent their dualities. The model is represented in linear programming statements, dualities of rationalities are represented in primary problem and dual problem.

In this paper, we treat corporations and the government as economic agents. Corporations have same rationalities in each economic phase. In this section, we describe a model of rationality of corporations for Thetical economy and Antithetical economy.

In the Thetical phase, corporations maximize their business profits. The maximization is represented in a primal problem of linear programming. The primal problem of a corporation is formulated in

$$\max \sum_{j=1}^{n} c_j x_j, \tag{3}$$

s.t.

$$\sum_{j=1}^{n} a_{ij} x_j \leq b_i, \quad i = 1, 2, \ldots, m, \tag{4}$$

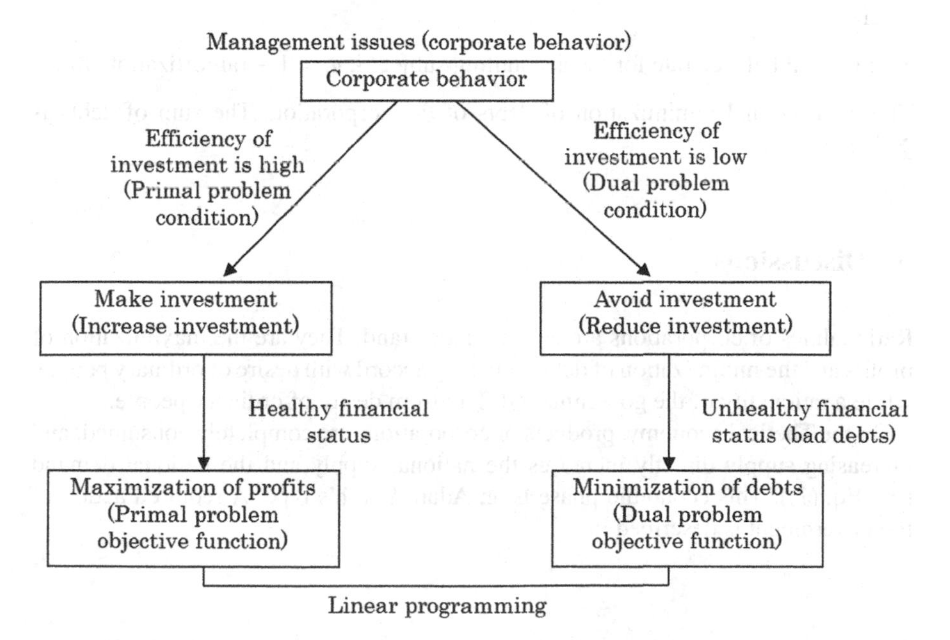

Fig. 1 Corporations' behavior [1]

$$x_j \geq 0, \quad j = 1, 2, \ldots, n. \tag{5}$$

Following list is correspondence of variables and its meanings.

x_j : The number of units of a product j made by the corporation.
c_j : The amount of profits of one unit of a product j.
a_{ij} : Costs in an account subject i to produce the product j for one unit.
b_i : The amount debts of an account subject i.

This is a rational maximization of profits of the corporation. The amount of profits is $\sum_{j=1}^{n} c_j x_j$.

The Antithetical economy arises after the Bubble collapse economy with shocks of the collapse and huge debts. In the phase, corporations minimize their debts. The minimization is represented in a dual problem of aforementioned primal problem. The dual problem of the corporation is represented in

$$\min \sum_{i=1}^{m} u_i b_i, \tag{6}$$

s.t.

$$\sum_{i=1}^{m} u_i a_{ij} \geq c_j, \quad j = 1, 2, \ldots, n, \tag{7}$$

$$u_i \geq 0, \quad i = 1, 2, \ldots, m, \tag{8}$$

where

u_i : Unpaid balance rate for the accounting subject i; $u_i = 1 - (\text{amortization\_rate})$.

This is a rational minimization of debts of the corporation. The sum of debts is $\sum_{i=1}^{m} u_i b_i$.

3 Discussions

Rationalities of corporations are easy to understand. They are the maximization of profits and the minimization of debts, and they accord with desire of ordinary people, while a rationality of the government differs from desire of ordinary people.

In the Thetical economy, products of corporations are completely consumed, and increasing supply directly increases the national supply and the national demand (the Eq. (1)). This economic phase is an Adam Smith's type. A required action of the government is described in

$$\min \sum_{j=1}^{\nu} \gamma_j \chi_j, \tag{9}$$

$$\sum_{j=1}^{\nu} \alpha_{ij} \chi_j \geq \beta_i, \quad i = 1, 2, \ldots, \mu, \tag{10}$$

$$\chi_i \geq 0, \quad j = 1, 2, \ldots, \nu. \tag{11}$$

Following list is correspondence of variables and its meanings.

χ_j :A rate of the remainder of national loans for an administrative service j. Increasing the rate increases expenses of the service j.

γ_j :Demand for funds as national loans for an administrative service j.

α_{ij} :Satisfaction of a resident i when the government gives the resident one unit of costs of a service j.

β_j :A desiring level of total services of the government for a resident i.

This is a minimization of national loans. The formulation means that the government is required its finance reform.

Let us look again the rationalities of corporations. To enlarge profit of corporations, the government can control indirectly variables: c_j and b_i in Eqs. (3) and (4). The variable c_j is profit of some product and the variable b_i is the amount of debts of corporations. They are controlled by interest rate of debts and the amount of money in banks. We can understand instinctively that corporations need money to enlarge their productions, and the government give corporations the money. These activities are generally referred to as monetary policy.

In the Antithetical economy, the government has to control demand side to balance the supply and demand (the Eq. (2)). This economic phase is a Keynesian type. A required action of the government is described in

$$\max \sum_{i=1}^{\mu} y_i \beta_i, \tag{12}$$

$$\sum_{i=1}^{\mu} y_i \alpha_{ij} \leq \gamma_j, \quad j = 1, 2, \ldots, \nu, \tag{13}$$

$$y_i \geq 0, \quad i = 1, 2, \ldots, \mu, \tag{14}$$

where

y_i : The amount of public money to increase satisfaction by one unit for a resident i.

This is a maximization of residents' satisfaction. The formulation means that the government is required fiscal stimulus. It is generally referred to as fiscal policy (or financial policy).

Table 1 Summary of duality in two economic phases

	The Thetical economy	The Antithetical economy
Say's law	Supply creates demand	Not effective
Effective demand	Not effective	Demands create supply
Corporations' rationality	Maximization of profit	Minimization of debts
Governmental norm	Monetary policy	Fiscal policy

Unlike rationalities of corporations, activities of the government are unnatural. The government derives its activities based on relations of the macroeconomics (Eqs. (1) and (2)) and rationalities of corporations (Eqs. (3)–(8)). In other words, corporations will act rationally, while the government must act normatively.

We summarize duality relationship between the Thetical economy and the Antithetical economy in Table 1.

4 Conclusions

In this paper, we describe rationalities of economics agents in the framework of the Thetical economics and the Antithetical economics. The framework consists of theories of the macroeconomics with microeconomic models. Conventional macroeconomic models, such as consumption functions, are often derived from empirical laws without microeconomic models for rationalities of economic agents. We provide microeconomic models of corporations and the government that model is derived from their rationalities. Their rationalities are directly represented in linear programming statements of OR. Especially for the government, that is its norm.

And we discuss the norm. In the Thetical economy, the government must adopt monetary policy. While in the Antithetical economy, the government must adopt fiscal policy.

We keep abstract arguing in this paper. We must discuss fact fitting of our models with actual statistics in future works.

References

1. Kinoshita, E.: A proposal of primal and dual problems in macro-economics. China-USA Bus. Rev. **10**(2), 115–124 (2011). ISSN: 1537-1514
2. Kinoshita, E.: Why bubble economy occurs and crashes?–repeated history of economic growth and collapse. Chin. Bus. Rev. **10**(2), 102–111 (2011). ISSN: 1537-1506
3. Kinoshita, E.: A proposal of thetical economy and antithetical economy-mechanism of occurrence and collapse of bubble economy. J. Bus. Econ. **3**(2), 117–130 (2012). ISSN: 2155-7950
4. Kinoshita, E.: Thetical and antithetical business management. J. Bus. Econ. **6**(6), 1086–1096 (2015)

5. Kinoshita, E.: Trap of economics the world has fallen in—a survey of Kinoshita theory in macroeconomics. Eur. Sci. J. 75–80 (2015). ISSN: 1857–7881 (Print), ISSN: 1857–7431
6. Kinoshita, E.: A proposal of thetical economy and antithetical economy by using operations research techniques. Eur. Sci. J. **11**(19), 29–48 (2015). ISSN: 1857-7881(Print), e ISSN: 1857-7431

Improvement of the Weights Due to Inconsistent Pairwise Comparisons in the AHP

Kazutomo Nishizawa

Abstract One of the most important problems in the Analytic Hierarchy Process (AHP) is consistency of pairwise comparisons by the decision maker. This study focuses on the comparison methods to be used when the weights of the alternatives and criteria in AHP are inconsistent. In general, the weights in AHP use the principal eigenvector of the pairwise comparison matrix. However, for example, due to the decision maker's misunderstandings, inconsistencies in pairwise comparisons sometimes arise. The consistency of the pairwise comparison matrix is usually determined using Consistency Index (CI) values. In the traditional AHP, when judged inconsistent, repeating the pairwise comparison is usually recommended. However, if the repeated comparison is arbitrarily performed, the results will not be optimal. In fact, to obtain the overall evaluation of alternatives, we often use inconsistent weights, even given the inconsistencies in the latter. Another method for judging the consistency of the pairwise comparison is to use a directed graph. Cycles in a directed graph represent comparison inconsistencies. Therefore in this paper, based on the principal eigenvalue and cycles in the directed graph of the pairwise comparison matrix, a method of correcting the principal eigenvector taking into consideration consistency is proposed.

Keywords AHP · Pairwise comparison · Consistency Index · Directed graph · Cycles

1 Introduction

In the Analytic Hierarchy Process (AHP) [9] and the Analytic Network Process (ANP) [10], some problems, for example rank reversal and weight normalization, were pointed out and were improved [1–4, 6–8, 11].

K. Nishizawa (✉)
Nihon University, 1-2-1 Izumicho, Narashino, Chiba 275-8575, Japan
e-mail: nishizawa.kazutomo@nihon-u.ac.jp

I. Czarnowski et al. (eds.), *Intelligent Decision Technologies 2016*,
Smart Innovation, Systems and Technologies 57,
DOI 10.1007/978-3-319-39627-9_35

In the AHP, the pairwise comparison matrix $\boldsymbol{A}$, which consists of n alternatives, is constructed by the decision maker. In this paper, we assume that $\boldsymbol{A}$ consists of complete comparisons. Through AHP, we calculate the principal eigenvalue and corresponding eigenvector of $\boldsymbol{A}$ using the power method. The weights of alternatives in the AHP usually use the principal eigenvector.

In this study, the weights of alternatives taking into account the consistency of comparisons are considered.

In the AHP, we usually evaluate comparisons as "consistent" or "inconsistent" based on Consistency Index (CI) values. CI is calculated using Eq. (1) based on the principal eigenvalue.

$$CI = (\lambda_{max} - n)/(n - 1), \tag{1}$$

where λ_{max} is the principal eigenvalue of $\boldsymbol{A}$. In general, if $CI < 0.1$ then we consider $\boldsymbol{A}$ to be consistent and if $CI > 0.1$ then it is considered inconsistent.

Another method of determining consistency involves using the directed graph of $\boldsymbol{A}$ [5]. In the pairwise comparison, if alternative "i" is better than alternative "j", then we indicate " i → j". If alternatives "i" and "j" are equally important, then we indicate " i — j". In directed graph of $\boldsymbol{A}$, there is either no cycle or there are some cycles of various length. If directed graph of $\boldsymbol{A}$ has no cycle, and thus is compliant with the transitive law, then $\boldsymbol{A}$ is considered to be consistent. If some cycles of length three are observed in the directed graph of $\boldsymbol{A}$, thus being compliant with the circulation law, then we consider it as inconsistent. If there is a cycle of length m ($m > 3$) in the directed graph of the complete comparisons, such cycle always includes some cycles of length three. Therefore, it should be considered only the cycles of length three.

For example, two kinds of pairwise comparison are illustrated. In these cases, the simplest pairwise comparisons are carried out. These are called binary comparisons. Using parameter θ, we can construct comparison matrix $\boldsymbol{A}$. If alternative "i" is better than alternative "j", then the element of $\boldsymbol{A}$, that is $\boldsymbol{a}_{ij} = \theta$, and $\boldsymbol{a}_{ji} = 1/\theta$. In these examples, assume that three alternatives a1, a2 and a3 are being compared.

First, the consistent comparison matrix is shown in Eq. (2).

$$\boldsymbol{A} = \begin{bmatrix} 1 & \theta & \theta \\ 1/\theta & 1 & \theta \\ 1/\theta & 1/\theta & 1 \end{bmatrix} \tag{2}$$

Calculating as $\theta = 2$, we get the principal eigenvalue $\lambda_{max} = 3.0536$ and we get corresponding eigenvector in Eq. (3). In this study, $\boldsymbol{w}$ is not normalized.

$$\boldsymbol{w} = \begin{bmatrix} 1.000000 \\ 0.629961 \\ 0.396850 \end{bmatrix} \tag{3}$$

From Eq. (1), we obtain $CI = 0.0268$. Thus $\boldsymbol{A}$ is considered consistent.

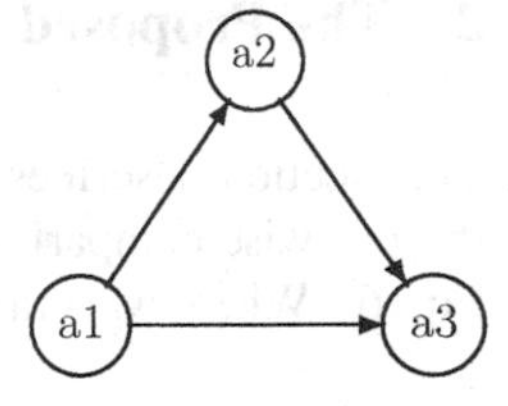

Fig. 1 Directed graph of no cycle

Another method to determine consistency is through use of the directed graph of comparison matrix. The directed graph of $\boldsymbol{A}$ is drawn in Fig. 1 based on Eq. (2).

From Fig. 1, $\boldsymbol{A}$ has no cycle, and thus complies with the transitive law. Thus $\boldsymbol{A}$ is considered consistent.

The next example is of inconsistency. The comparison matrix is shown in Eq. (4).

$$\boldsymbol{A} = \begin{bmatrix} 1 & \theta & 1/\theta \\ 1/\theta & 1 & \theta \\ \theta & 1/\theta & 1 \end{bmatrix} \tag{4}$$

Calculating as $\theta = 2$, we get the principal eigenvalue $\lambda_{max} = 3.5000$ and we get the corresponding eigenvector in Eq. (5).

$$\boldsymbol{w} = \begin{bmatrix} 1.000000 \\ 1.000000 \\ 1.000000 \end{bmatrix} \tag{5}$$

From Eq. (1), we obtain $CI = 0.2500$. Thus $\boldsymbol{A}$ is considered inconsistent.

Figure 2 shows the directed graph of $\boldsymbol{A}$ based on Eq. (4).

In Fig. 2, there is one cycle of length three. Thus the directed graph of $\boldsymbol{A}$ complies with the circulation law, $\boldsymbol{A}$ is considered inconsistent.

In the traditional AHP, for comparisons judged inconsistent, repeat of pairwise comparisons is usually recommended. However, a repeat of pairwise comparisons usually results in arbitrary evaluations, so it is not a good method. Therefore in this paper, for the case of inconsistent comparisons, an improvement method is proposed.

This paper consists of following the sections. Section 2 describes the proposed method. Examples of the proposed method are illustrated in Sect. 3. And finally, in Sect. 4, the results obtained through the proposed method are discussed and the study concluded.

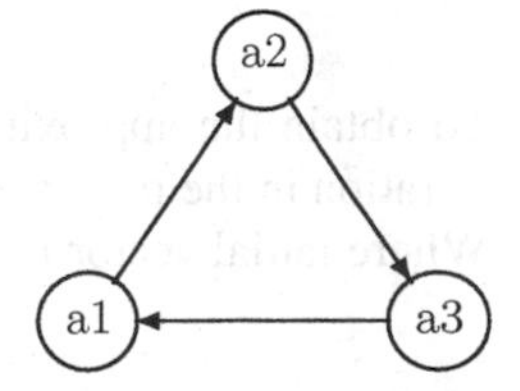

Fig. 2 Directed graph with one cycle

2 The Proposed Method

This section describes the proposed method. In the case of perfect consistency of the pairwise comparison, the element of comparison matrix $\boldsymbol{A}$, that is a_{ij} supports Eq. (6). Where w_i is the weight of alternative ai.

$$a_{ij} = w_i/w_j \tag{6}$$

In the perfectly consistent $\boldsymbol{A}$, as shown, we get $\lambda_{max} = n$. On the other hand, in the inconsistent case, we get $\lambda_{max} > n$.

To correct the weights of alternatives for consistency, in this study, the corrective parameter α is defined in Eq. (7). α means rough consistency.

$$\alpha = n/\lambda_{max} \tag{7}$$

Based on α, the corrected weight w'_i is calculated using Eq. (8). Where k_i is the number of cycles which are related to alternative ai on the directed graph of $\boldsymbol{A}$.

$$w'_i = \alpha^{k_i+1} w_i \tag{8}$$

In the case of perfect consistency, that is $\lambda_{max} = n$, we get $\alpha = 1$ from Eq. (7). In this case $\boldsymbol{w}' = \boldsymbol{w}$. If there is no cycle in the directed graph, that is, $k_i = 0$ but $\lambda_{max} > n$, we get $\boldsymbol{w}' = \alpha \boldsymbol{w}$ from Eq. (8).

The procedure of the proposed method is as follows.

P1: Calculate the principal eigenvalue and corresponding eigenvector of the pairwise comparison matrix $\boldsymbol{A}$ using the power method.
P2: Calculate the corrective parameter α using Eq. (7).
P3: Find cycles of length three the directed graph of $\boldsymbol{A}$.
P4: Count the number of cycles k_i which are related to the alternative ai.
P5: Correct the weights using Eq. (8).

If the weights of alternatives are calculated from $\boldsymbol{A}$ using the geometric mean, unfortunately we do not have λ_{max}. Therefore we need to calculate the approximate eigenvalue $\bar{\lambda}$ using the following well known procedure.

Through the geometric mean, the weights of alternatives w_i is obtained using Eq. (9).

$$w_i = \sqrt[n]{\prod_{j=1}^{n} a_{ij}} \tag{9}$$

To obtain the approximate eigenvalue, we perform calculations only once for the iteration in the power method. From Eq. (10), the approximate vector $\boldsymbol{x}$ is obtained. Where initial vector $\boldsymbol{w}$ is calculated using Eq. (9).

$$\boldsymbol{A}\boldsymbol{w} = \boldsymbol{x} \tag{10}$$

If the power method is carried completion, we get $\boldsymbol{A}\boldsymbol{w} = \lambda_{max}\boldsymbol{w}$. However when once iteration, $\boldsymbol{x}$ in Eq. (10) is the approximate vector, then different eigenvalues λ_i are obtained using Eq. (11).

$$\lambda_i = x_i / w_i \tag{11}$$

The approximate eigenvalue $\bar{\lambda}$ is obtained by calculating the average of λ_i from Eq. (12).

$$\bar{\lambda} = \frac{1}{n}\sum_{i=1}^{n}\lambda_i \tag{12}$$

Using $\bar{\lambda}$ instead of λ_{max} in Eq. (7), α is determined.

3 Examples

In this section, using the proposed method, three examples are illustrated. The directed graph in Example 1 shows no cycles, Example 2 has one cycle and Example 3 has four cycles. Applying the proposed method to these examples, each principal eigenvector is corrected.

3.1 Example 1

The first example consists of five alternatives, a1 to a5. Comparison matrix $\boldsymbol{A}_1$ is shown in Eq. (13).

$$\boldsymbol{A}_1 = \begin{bmatrix} 1 & 3 & 2 & 1/2 & 1 \\ 1/3 & 1 & 5 & 1/4 & 1/2 \\ 1/2 & 1/5 & 1 & 1/4 & 1/3 \\ 2 & 4 & 4 & 1 & 2 \\ 1 & 2 & 3 & 1/2 & 1 \end{bmatrix} \tag{13}$$

Through the proposed procedure “P1”, we get $\lambda_{max} = 5.3813$ and we get corresponding eigenvector in Eq. (14). However $\boldsymbol{w}_1$ is not normalized.

$$\boldsymbol{w}_1 = \begin{bmatrix} 0.567933 \\ 0.365566 \\ 0.179137 \\ 1.000000 \\ 0.533289 \end{bmatrix} \tag{14}$$

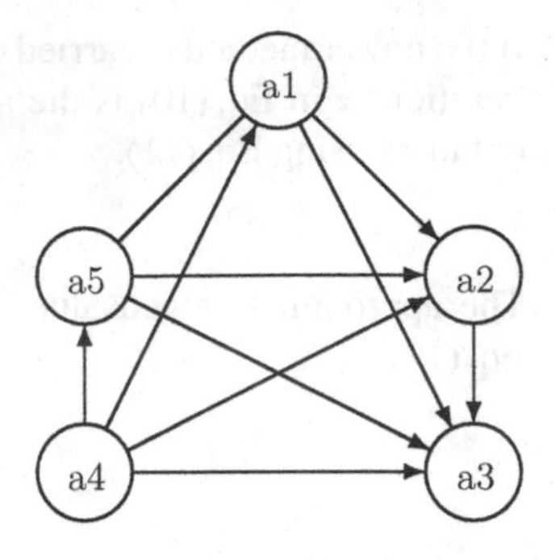

Fig. 3 Directed graph of $\boldsymbol{A}_1$

In this example, we get $CI = 0.0953$ from Eq. (1), however we dither over consistency because $CI \fallingdotseq 0.1$.

Next, Fig. 3 shows the directed graph of $\boldsymbol{A}_1$.

Since no cycle is observed, it appears to be consistent.

Through the proposed procedure "P2", we get $\alpha = 0.929151$ using Eq. (7). Because no cycle is observed in Fig. 3, we get $k_i = 0$ through the proposed procedures "P3" and "P4". Through the proposed procedure "P5", we get corrected vector $\boldsymbol{w}_1{}'$ in Eq. (15) using Eq.(8).

$$\boldsymbol{w}_1{}' = \alpha \boldsymbol{w}_1 = \begin{bmatrix} 0.527696 \\ 0.339666 \\ 0.166445 \\ 0.929151 \\ 0.495506 \end{bmatrix} \tag{15}$$

Through the proposed method, in this example, there is no change in the ordering of alternatives.

Based on the weights obtained through geometric mean using Eq. (9), we get $\bar{\lambda} = 5.368394$ using Eq. (12) and we get $\alpha = 0.931377$.

3.2 *Example 2*

The next example consists of five alternatives, a1 to a5. Comparison matrix $\boldsymbol{A}_2$ is shown in Eq. (16).

$$\boldsymbol{A}_2 = \begin{bmatrix} 1 & 1/4 & 1/6 & 1/8 & 1/7 \\ 4 & 1 & 1/2 & 1/3 & 1/3 \\ 6 & 2 & 1 & 1/2 & 2 \\ 8 & 3 & 2 & 1 & 1/2 \\ 7 & 3 & 1/2 & 2 & 1 \end{bmatrix} \tag{16}$$

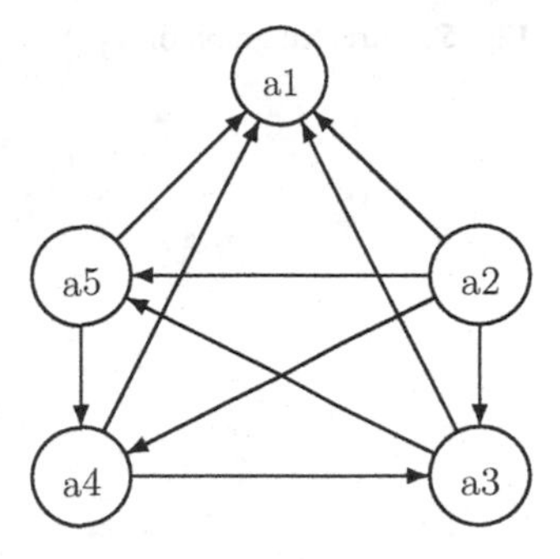

Fig. 4 Directed graph of $\boldsymbol{A_2}$

Table 1 Cycles—alternatives in $\boldsymbol{A_2}$

Cycles\Alternatives	a1	a2	a3	a4	a5
(a3-a5-a4)	0	0	1	1	1
The number of related cycles (k_i)	0	0	1	1	1

Using P1, we get $\lambda_{max} = 5.3518$ and we get corresponding eigenvector $\boldsymbol{w_2}$ in Eq. (17).

$$\boldsymbol{w_2} = \begin{bmatrix} 0.116882 \\ 0.363940 \\ 0.902340 \\ 0.995339 \\ 1.000000 \end{bmatrix} \tag{17}$$

From Eq. (1), we get $CI = 0.0880$ in this example. As in Example 1, we dither over consistency because $CI \fallingdotseq 0.1$. Using P2, we get $\alpha = 0.934258$.

Next, Fig. 4 shows the directed graph of $\boldsymbol{A_2}$.

In Fig. 4, there is one cycle of length three, (a3-a5-a4). So it is seems to be inconsistent. Table 1 is obtained from Fig. 4.

Through P3 and P4, we get k_i from Table 1 and through P5, we get corrected vector $\boldsymbol{w_2}'$ in Eq. (18) using Eq. (8).

$$\boldsymbol{w_2}' = \begin{bmatrix} \alpha^1 \times 0.116882 \\ \alpha^1 \times 0.363940 \\ \alpha^2 \times 0.902340 \\ \alpha^2 \times 0.995339 \\ \alpha^2 \times 1.000000 \end{bmatrix} = \begin{bmatrix} 0.109198 \\ 0.340014 \\ 0.787597 \\ 0.868770 \\ 0.872839 \end{bmatrix} \tag{18}$$

The results using the proposed method, in this example, show that there is no change in the ordering of alternatives.

We get $\bar{\lambda} = 5.341311$ and $\alpha = 0.936100$.

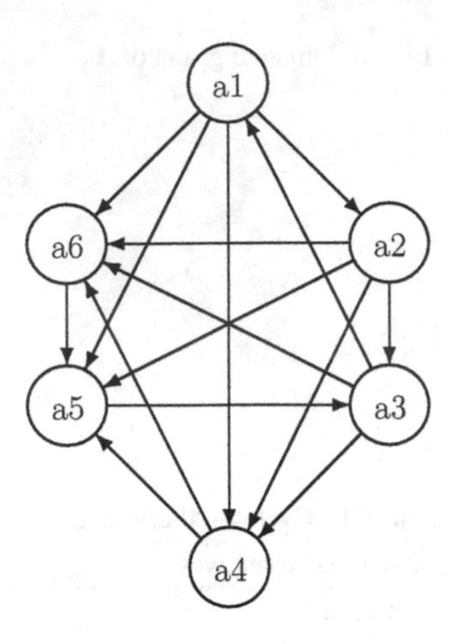

Fig. 5 Directed graph of A_3

3.3 Example 3

In the next example, the binary comparisons consists of six alternatives, a1 to a6. The comparison matrix is shown in Eq. (19).

$$A_3 = \begin{bmatrix} 1 & \theta & 1/\theta & \theta & \theta & \theta \\ 1/\theta & 1 & \theta & \theta & \theta & \theta \\ \theta & 1/\theta & 1 & \theta & 1/\theta & \theta \\ 1/\theta & 1/\theta & 1/\theta & 1 & \theta & \theta \\ 1/\theta & 1/\theta & \theta & 1/\theta & 1 & 1/\theta \\ 1/\theta & 1/\theta & 1/\theta & 1/\theta & \theta & 1 \end{bmatrix} \tag{19}$$

Using P1, calculating as $\theta = 2$, we get $\lambda_{max} = 6.7587$ and we get corresponding eigenvector $\boldsymbol{w_3}$ in Eq. (20). Then $CI = 0.1517$ is obtained and we consider $\boldsymbol{A_3}$ as inconsistent.

$$\boldsymbol{w_3} = \begin{bmatrix} 1.000000 \\ 0.974743 \\ 0.869358 \\ 0.618222 \\ 0.570360 \\ 0.498701 \end{bmatrix} \tag{20}$$

Using P2, we get $\alpha = 0.887741$.

Next, Fig. 5 shows the directed graph of $\boldsymbol{A_3}$.

In Fig. 5, there are four cycles of length three, (a1-a2-a3), (a1-a5-a3), (a3-a4-a5) and (a3-a6-a5), so as a result $\boldsymbol{A_3}$ is considered inconsistent.

Table 2 is obtained based on Fig. 5.

Using P3 and P4, we get k_i from Table 2, and through P5, we get corrected vector $\boldsymbol{w_3}'$ in Eq. (21) using Eq. (8).

Table 2 Cycles—alternatives in A_3

Cycles\Alternatives	a1	a2	a3	a4	a5	a6
(a1-a2-a3)	1	1	1	0	0	0
(a1-a5-a3)	1	0	1	0	1	0
(a3-a4-a5)	0	0	1	1	1	0
(a3-a6-a5)	0	0	1	0	1	1
The number of related cycles (k_i)	2	1	4	1	3	1

$$\boldsymbol{w_3}' = \begin{bmatrix} \alpha^3 \times 1.000000 \\ \alpha^2 \times 0.974743 \\ \alpha^5 \times 0.869358 \\ \alpha^2 \times 0.618222 \\ \alpha^4 \times 0.570360 \\ \alpha^2 \times 0.498701 \end{bmatrix} = \begin{bmatrix} 0.699614 \\ 0.768179 \\ 0.479325 \\ 0.487211 \\ 0.354237 \\ 0.393018 \end{bmatrix} \tag{21}$$

The results obtained through the proposed method demonstrate, in this example, that the order of alternatives of $\boldsymbol{w_3}'$ are different from $\boldsymbol{w_3}$. The order in $\boldsymbol{w_3}$ is a1 > a2> a3> a4> a5> a6, however, in $\boldsymbol{w_3}'$, it is a2> a1 >a4 >a3> a6 >a5.

We get $\bar{\lambda} = 6.739464$ and $\alpha = 0.890278$.

4 Conclusion

In this study, a method for correcting inconsistent pairwise comparisons in AHP was proposed. The proposed method is as follows.

1. Using the principal eigenvalue of pairwise comparison matrix $\boldsymbol{A}$, the corrective parameter α was defined.
2. Using α and the number of cycles of length three in the directed graph of $\boldsymbol{A}$, a corrective procedure was proposed.

Applying the proposed method to the three examples, the following results could be obtained.

1. In the case of inconsistencies, i.e. those represented by cycles of length three in directed graph of $\boldsymbol{A}$, the order of alternatives in terms of priority were changed.
2. The proposed method offers promising results for determining the overall evaluation of alternatives demonstrating inconsistency.

Topics for further study are as follows.

1. Appropriateness of the corrective parameter α.
2. The evaluation of the proposed method that includes application to other examples demonstrating inconsistency is required.

References

1. Belton, V., Gear, T.: On a short-coming of saaty's method of analytic hierarchies. Omega **11**, 228–230 (1983)
2. Belton, V., Gear, T.: The legitimacy of rank reversal—a comment. Omega **13**, 143–145 (1985)
3. Kinoshita, E., Nakanishi, M.: Proposal of new AHP model in light of dominant relationship among alternatives. J. Oper. Res. Soc. Jpn. **42**, 180–197 (1999)
4. Kinoshita, E., Sugiura, S.: A comparison study of dominant AHP and similar dominant models. J. Res. Inst. Meijo Univ. **7**, 115–116 (2008)
5. Nishizawa, K.: A Consistency Improving Method in Binary AHP. J. Oper. Res. Soc. Jpn. **38**, 21–33 (1995)
6. Nishizawa, K.: Normalization method based on dummy alternative with perfect evaluation score in AHP and ANP. Intell. Decis. Technol. **1**(SIST 15), 253–262 (2012)
7. Nishizawa, K.: Improving of the weight normalization method on alternatives in AHP and ANP. Smart Digital Futures 2014, pp. 155–163. IOS Press (2014)
8. Nishizawa, K.: The Improvement of Pairwise Comparison Method of the Alternatives in the AHP. Intell. Decis. Technol. **1**(SIST 39), 483–491 (2015)
9. Saaty, T.L.: The Analytic Hierarchy Process. McGraw-Hill, New York (1980)
10. Saaty, T.L.: The Analytic Network Process. RWS Publications, Pittsburgh (1996)
11. Schoner, B., Wedley, W.C., Choo, E.U.: A unified approach to AHP with linking pins. Eur. J. Oper. Res. **13**, 384–392 (1993)

Super Pairwise Comparison Matrix in the Dominant AHP

Takao Ohya and Eizo Kinoshita

Abstract We have proposed an SPCM (Super Pairwise Comparison Matrix) to express all pairwise comparisons in the evaluation process of D-AHP (the dominant analytic hierarchy process) or the multiple dominant AHP as a single pairwise comparison matrix. This paper shows that the evaluation value resulting from the application of LLSM (the logarithmic least-squares method) to an SPCM matches the evaluation value determined by the application of D-AHP to the evaluation values obtained from each pairwise comparison matrix by using the geometric mean.

Keywords Super pairwise comparison matrix · The dominant AHP · Logarithmic least-squares method

1 Introduction

AHP (the Analytic Hierarchy Process) proposed by Saaty [1] enables objective decision making by top-down evaluation based on an overall aim.

In actual decision making, a decision maker often has a specific alternative (regulating alternative) in mind and makes an evaluation on the basis of the alternative. This was modeled in D-AHP(the dominant AHP), proposed by Kinoshita and Nakanishi [2].

If there are more than one regulating alternatives and the importance of each criterion is inconsistent, the overall evaluation value may differ for each regulating alternative. As a method of integrating the importance in such cases, CCM (the

T. Ohya (✉)
School of Science and Engineering, Kokushikan University, Tokyo, Japan
e-mail: takaohya@kokushikan.ac.jp

E. Kinoshita
Faculty of Urban Science, Meijo University, Gifu, Japan
e-mail: kinoshit@urban.meijo-u.ac.jp

I. Czarnowski et al. (eds.), *Intelligent Decision Technologies 2016*,
Smart Innovation, Systems and Technologies 57,
DOI 10.1007/978-3-319-39627-9_36

concurrent convergence method) was proposed. Kinoshita and Sekitani [3] showed the convergence of CCM.

Ohya and Kinoshita [4] proposed an SPCM (Super Pairwise Comparison Matrix) to express all pairwise comparisons in the evaluation process of the dominant analytic hierarchy process (AHP) or the multiple dominant AHP (MDAHP) as a single pairwise comparison matrix.

Ohya and Kinoshita [5] showed, by means of a numerical counterexample, that in MDAHP an evaluation value resulting from the application of the logarithmic least-squares method (LLSM) to an SPCM does not necessarily coincide with that of the evaluation value resulting from the application of the geometric mean multiple dominant AHP (GMMDAHP) to the evaluation value obtained from each pairwise comparison matrix by using the geometric mean method.

Ohya and Kinoshita [6] showed, using the error models, that in D-AHP an evaluation value resulting from the application of the logarithmic least squares method (LLSM) to an SPCM necessarily coincide with that of the evaluation value resulting obtained by using the geometric mean method to each pairwise comparison matrix.

Ohya and Kinoshita [7] showed the treatment of hierarchical criteria in D-AHP with super pairwise comparison matrix.

Ohya and Kinoshita [8] showed the example of using SPCM with the application of LLSM for calculation of MDAHP.

This paper shows that the evaluation value resulting from the application of LLSM to an SPCM agrees with the evaluation value determined by the application of D-AHP to the evaluation value obtained from each pairwise comparison matrix by using the geometric mean.

2 D-AHP and SPCM

This section explains D-AHP and an SPCM to express the pairwise comparisons appearing in the evaluation processes of D-AHP and MDAHP as a single pairwise comparison matrix. Section 2.1 outlines D-AHP procedure and explicitly states pairwise comparisons, and Sect. 2.2 explains an SPCM that expresses these pairwise comparisons as a single pairwise comparison matrix.

2.1 Evaluation in D-AHP

The true absolute importance of alternative $a(a=1, \ldots, A)$ at criterion $c(c=1, \ldots, C)$ is v_{ca}. The final purpose of the AHP is to obtain the relative value (between alternatives) of the overall evaluation value $v_a = \sum_{c=1}^{C} v_{ca}$ of alternative a. The procedure of D-AHP for obtaining an overall evaluation value is as follows:

D-AHP

Step 1: The relative importance $u_{ca} = \alpha_c v_{ca}$ (where α_c is a constant) of alternative a at criterion c is obtained by some kind of methods. In this paper, u_{ca} is obtained by applying the pairwise comparison method to alternatives at criterion c.

Step2: Alternative d is the regulating alternative. The importance u_{ca} of alternative a at criterion c is normalized by the importance u_{cd} of the regulating alternative d, and u_{ca}^d $(= u_{ca}/u_{cd})$ is calculated.

Step3: With the regulating alternative d as a representative alternative, the importance w_c^d of criterion c is obtained by applying the pairwise comparison method to criteria, where, w_c^d is normalized by $\sum_{c=1}^{C} w_c^d = 1$.

Step4: From u_{ca}^d, w_c^d obtained at Steps 2 and 3, the overall evaluation value $t_a = \sum_{c=1}^{C} w_c^d u_{ca}^d$ of alternative a is obtained. By normalization at Steps 2 and 3, $u_d = 1$. Therefore, the overall evaluation value of regulating alternative d is normalized to 1.

2.2 SPCM

The relative comparison values $r_{c'a'}^{ca}$ of importance v_{ca} of alternative a at criteria c as compared with the importance $v_{c'a'}$ of alternative a' in criterion c', are arranged in a (CA × CA) or (AC × AC) matrix. This is proposed as an SPCM $R = (r_{c'a'}^{ca})$ *or* $(r_{a'c'}^{ac})$.

In a (CA × CA) matrix, index of alternative changes first. In a (CA × CA) matrix, SPCM's $(A(c-1)+a, A(c'-1)+a')$ th element is $r_{c'a'}^{ca}$.

In a (AC × AC) matrix, index of criteria changes first. In a (AC × AC) matrix, SPCM's $(C(a-1)+c, C(a'-1)+c')$ th element is $r_{a'c'}^{ac}$.

In an SPCM, symmetric components have a reciprocal relationship as in pairwise comparison matrices. Diagonal elements are 1 and the following relationships are true:

If $r_{c'a'}^{ca}$ exists, then $r_{ca}^{c'a'}$ exists and

$$r_{ca}^{c'a'} = 1/r_{c'a'}^{ca}, \tag{1}$$

$$r_{ca}^{ca} = 1. \tag{2}$$

Pairwise comparison at Step 1 of D-AHP consists of the relative comparison value $r_{ca'}^{ca}$ of importance v_{ca} of alternative a, compared with the importance $v_{ca'}$ of alternative a' at criterion c.

Pairwise comparison at Step 3 of D-AHP consists of the relative comparison value $r_{c'd}^{cd}$ of importance v_{cd} of alternative d at criterion c, compared with the importance $v_{c'd}$ of alternative d at criterion c', where the regulating alternative is d.

SPCM of D-AHP or MDAHP is an incomplete pairwise comparison matrix. Therefore, the LLSM based on an error model or an eigenvalue method such as the Harker method [9] or two-stage method is applicable to the calculation of evaluation values from an SPCM.

3 SPCM + LLSM in the Dominant AHP

This section shows that an evaluation value resulting from the application of the LLSM to an SPCM agrees with the overall evaluation value resulting from the application of D-AHP to the evaluation value obtained by application of the geometric mean method to each pair-wise comparison matrix.

In Sect. 3.1, an overall evaluation value is obtained by applying D-AHP to evaluation values that are obtained by applying the geometric mean method to each pairwise comparison matrix. In Sect. 3.2, an overall evaluation value is obtained by applying the LLSM to an SPCM to show that it agrees with the overall evaluation value obtained in Sect. 3.1.

Hereinafter, the regulating alternative in D-AHP is assumed to be alternative 1. This assumption can generally be satisfied by renumbering alternatives.

3.1 D-AHP and Geometric Mean Method

Pairwise comparison at Step 1 of D-AHP consists of the relative comparison value $r^{ca}_{ca'}$ of importance v_{ca} of alternative a as compared with the importance $v_{ca'}$ of alternative a' from the view point of criterion c. R^A_c is the pairwise comparison matrix between alternatives from the view point of criterion c, whose (a, a') th element is $r^{ca}_{ca'}$. Therefore, the relative importance u_{ca} of alternative a at criterion c resulting from the application of the geometric mean method to the pairwise comparison matrix R^A_c becomes the geometric mean of the values in row a of R^A_c. In other words, u_{ca} is calculated with following equation.

$$u_{ca} = \left(\prod_{a'=1}^{A} r^{ca}_{ca'} \right)^{1/A}, \quad c = 1, \ldots, C, \ a = 1, \ldots, A. \tag{3}$$

At Step 2, this value is normalized by u_{c1} and we obtain the following equation:

$$u^1_{ca} = u_{ca}/u_{c1} = \left(\prod_{a'=1}^{A} r^{ca}_{ca'} \right)^{1/A} / \left(\prod_{a'=1}^{A} r^{c1}_{ca'} \right)^{1/A}, \quad c = 1, \ldots, C, a = 1, \ldots, A. \tag{4}$$

Pairwise comparison at Step 3 of D-AHP consists of the relative comparison value $r_{c'1}^{c1}$ of importance v_{c1} of alternative 1 at criterion c, compared with the importance v_{c1} of alternative 1 at criterion c', R_1^C is the pairwise comparison matrix between criteria of the dominant alternative 1 whose (c, c') th element is $r_{c'1}^{c1}$. Therefore, the relative importance w_{c1} of alternative 1 at criterion c resulting from the application of the geometric mean method to the pairwise comparison matrix R_1^C becomes the geometric mean of the values in row c of R_1^C, where w_{c1} is normalized to $\sum_{c=1}^{C} w_c^1 = 1$ as shown in the following equation:

$$w_c^1 = \left(\prod_{c'=1}^{C} r_{c'1}^{c1}\right)^{1/C} / \sum_{c''=1}^{C} \left(\prod_{c'=1}^{C} r_{c'1}^{c''1}\right)^{1/C}, \quad c = 1, \ldots, C. \tag{5}$$

At step 4, with u_{ca}^1, w_{c1} shown in Eqs. (4) and (5), the overall evaluation value $u_a = \sum_{c=1}^{C} w_c^1 u_{ca}^1$ of alternative a is

$$v_{ca} = w_c^1 u_{ca}^1 = \frac{\left(\prod_{c'=1}^{C} r_{c'1}^{c1}\right)^{1/C} \left(\prod_{a'=1}^{A} r_{ca'}^{ca}\right)^{1/A}}{\sum_{c=1}^{C} \left(\prod_{c'=1}^{C} r_{c'1}^{c1}\right)^{1/C} \cdot \left(\prod_{a'=1}^{A} r_{ca'}^{c1}\right)^{1/A}} \tag{6}$$

3.2 *LLSM Application to SPCM in the Dominant AHP*

For existing pairwise comparison values $r_{c'a'}^{ca}$ $(c < c', a < a')$ in an SPCM, an error model is assumed as follows:

$$r_{c'a'}^{ca} = \varepsilon_{c'a'}^{ca} \frac{v_{ca}}{v_{c'a'}}. \tag{7}$$

where $\varepsilon_{c'a'}^{ca}$ is error term, v_{ca} is non-random but unobservable parameters. Taking the logarithms (base e) of both sides gives

$$\ln r_{c'a'}^{ca} = \ln v_{ca} - \ln v_{c'a'} + \ln \varepsilon_{c'a'}^{ca} \tag{8}$$

To simplify the equation, logarithms will be represented by over dots as $\dot{r}_{c'a'}^{ca} = \log r_{c'a'}^{ca}$, $\dot{v}_{ca} = \log v_{ca}$, $\dot{\varepsilon}_{c'a'}^{ca} = \log \varepsilon_{c'a'}^{ca}$. Using this notation, Eq. (8) becomes

$$\dot{r}_{c'a'}^{ca} = \dot{v}_{ca} - \dot{v}_{c'a'} + \dot{\varepsilon}_{c'a'}^{ca}, c, c' = 1, \ldots, C, \ a, a' = 1, \ldots, A \tag{9}$$

From Eqs. (1) and (2), we have followings.

If $\dot{r}^{ca}_{c'a'}$ exists, then $\dot{r}^{c'a'}_{ca}$ exists and

$$\dot{r}^{c'a'}_{ca} = -\dot{r}^{ca}_{c'a'} \tag{10}$$

$$\dot{r}^{ca}_{ca} = 0. \tag{11}$$

There are two types of pairwise comparison in the dominant AHP: $r^{ca}_{ca'}$ at Step 1 and $r^{c1}_{c'1}$ at Step 3. In the least-squares method, therefore, $\hat{v}_{ca}$ is obtained from the pairwise comparison $\dot{r}^{ca}_{ca'}(a=1,\ldots,A-1,a'=a+1,\ldots,A,c=1,\ldots,C)$ and $\dot{r}^{c1}_{c'1}(c=1,\ldots,C-1,c'=c+1,\ldots,C)$ to minimize

$$S=\sum_{c=1}^{C}\sum_{a=1}^{A-1}\sum_{a'=a+1}^{A}(\dot{r}^{ca}_{ca'}-\hat{v}_{ca}+\hat{v}_{ca'})^2+\sum_{c=1}^{C-1}\sum_{c'=c+1}^{C}(\dot{r}^{c1}_{c'1}-\hat{v}_{c1}+\hat{v}_{c'1})^2. \tag{12}$$

In Eq. (12), the first term $\sum_{c=1}^{C}\sum_{a=1}^{A-1}\sum_{a'=a+1}^{A}(\dot{r}^{ca}_{ca'}-\hat{v}_{ca}+\hat{v}_{ca'})^2$ associates the relative comparison value $r^{ca}_{ca'}$ of importance v_{ca} of alternative a, compared with the importance $v_{ca'}$ of alternative a' from the view point of criterion c, and the second term $\sum_{c=1}^{C-1}\sum_{c'=c+1}^{C}(\dot{r}^{c1}_{c'1}-\hat{v}_{c1}+\hat{v}_{c'1})^2$ associates the relative comparison value $r^{c1}_{c'1}$ of importance v_{c1} of alternative 1 at criterion c, compared with the importance $v_{c'1}$ of alternative 1 at criterion c'.

As Eq. (7) shows, only the ratio is important with regard to $\hat{v}_{ca}$ and the constant multiple is arbitrary, becoming an arbitrary additive constant in the logarithm $\hat{v}_{ca}$ form.

From $\frac{\partial S}{\partial \hat{v}_{ca}}=0$, we have

$$\begin{aligned}\frac{1}{2}\frac{\partial S}{\partial \hat{v}_{c1}} &= \sum_{a=1}^{A}(\hat{v}_{c1}-\hat{v}_{ca}-\dot{r}^{c1}_{ca})+\sum_{c'=1}^{C}(\hat{v}_{c1}-\hat{v}_{c'1}-\dot{r}^{c1}_{c'1})\\ &= (A\hat{v}_{c1}-\sum_{a=1}^{A}\hat{v}_{ca}-\sum_{a=1}^{A}\dot{r}^{c1}_{ca})+(C\hat{v}_{c1}-\sum_{c'=1}^{C}\hat{v}_{c'1}-\sum_{c'=1}^{C}\dot{r}^{c1}_{c'1})=0,\\ &\qquad c=2,\ldots,C\end{aligned} \tag{13}$$

$$\begin{aligned}\frac{1}{2}\frac{\partial S}{\partial \hat{v}_{ca}} = \sum_{a'=1}^{A}(\hat{v}_{ca}-\hat{v}_{ca'}-\dot{r}^{ca}_{ca'}) &= (A\hat{v}_{ca}-\sum_{a'=1}^{A}\hat{v}_{ca'}-\sum_{a'=1}^{A}\dot{r}^{ca}_{ca'}),\\ &\quad a=2,\ldots,A,\ c=2,\ldots,C\end{aligned} \tag{14}$$

Second term $\sum_{c'=1}^{C}(\hat{v}_{c1}-\dot{r}^{c1}_{c'1}-\hat{v}_{c'1})$ of Eqs. (13) is obtained from $\frac{\partial}{\partial \hat{v}_{c1}}\sum_{c=1}^{C-1}\sum_{c'=c+1}^{C}(\dot{r}^{c1}_{c'1}-(\hat{v}_{c1}-\hat{v}_{c'1}))^2$.

The fact that

$$\hat{v}_{ca} = \frac{1}{C}\sum_{c'=1}^{C} \dot{r}_{c'1}^{c1} + \frac{1}{A}\sum_{a'=1}^{A} \dot{r}_{ca'}^{ca} - \frac{1}{A}\sum_{a'=1}^{A} \dot{r}_{ca'}^{c1} + Const., \; c=1,\ldots,C, \; a=1,\ldots,A \tag{15}$$

satisfies Eqs. (13) and (14) is easy to confirm using Eqs. (10) and (11). In Eq. (15), Const. is arbitrary constant.

From Eq. (15),

$$\hat{v}_{ca} = Const. \left(\prod_{c'=1}^{C} r_{c'1}^{c1}\right)^{1/C} \frac{\left(\prod_{a'=1}^{A} r_{ca'}^{ca}\right)^{1/A}}{\left(\prod_{a'=1}^{A} r_{ca'}^{c1}\right)^{1/A}}, \; c=1,\ldots,C, \; a=1,\ldots,A \tag{16}$$

In accordance with the normalization at Step 3 of the dominant AHP, the normalized equation $\hat{v}_{ca}(a=1,\ldots,A, c=1,\ldots,C)$ is such that the overall evaluation value of the regulating alternative (alternative 1) will be 1. In other words, $\sum_{c=1}^{C} \hat{v}_{c1} = 1$. Therefore,

$$\hat{v}_{ca} = \frac{\left(\prod_{c'=1}^{C} r_{c'1}^{c1}\right)^{1/C}}{\sum_{c''=1}^{C}\left(\prod_{c'=1}^{C} r_{c'1}^{c''1}\right)^{1/C}} \frac{\left(\prod_{a'=1}^{A} r_{ca'}^{ca}\right)^{1/A}}{\left(\prod_{a'=1}^{A} r_{ca'}^{c1}\right)^{1/A}}, c=1,\ldots,C, \; a=1,\ldots,A \tag{17}$$

From Eq. (17), we see that the overall evaluation value $\hat{v}_a = \sum_{c=1}^{C} \hat{v}_{ca}$ of alternative a agrees with Eq. (6).

As shown above, an evaluation value resulting from the application of the LLSM to an SPCM agrees with the overall evaluation value resulting from the application of the dominant AHP to evaluation values that are obtained by applying the geometric mean method to each pairwise comparison matrix.

4 Conclusion

It is well known that in complete pairwise comparison matrix, the evaluation values applying the geometric mean method agree with the evaluation values resulting from the application of the LLSM. This paper shows that the evaluation values resulting from the application of the LLSM to an SPCM agree with the evaluation values resulting from the application of the dominant AHP to evaluation values that are obtained by applying the geometric mean method to each pairwise comparison matrix.

References

1. Saaty, T.L.: The Analytic Hierarchy Process. McGraw-Hill, New York, NY (1980)
2. Kinoshita, E., Nakanishi, M.: Proposal of new AHP model in light of dominative relationship among alternatives. J. Oper. Res. Soc. Jpn. **42**, 180–198 (1999)
3. Kinoshita, E., Sekitani, K., Shi, J.: Mathematical Properties of Dominant AHP and Concurrent Convergence Method. J. Oper. Res. Soc. Jpn. **45**, 198–213 (2002)
4. Ohya, T., Kinoshita, E.: Proposal of Super Pairwise Comparison Matrix. In: Watada, J., et al. (eds.) Intelligent Decision Technologies, 247–254. Springer, Berlin (2011)
5. Ohya, T., Kinoshita, E.: Super Pairwise Comparison Matrix in the Multiple Dominant AHP. In: Watada, J., et al. (eds.) Intelligent Decision Technologies. Smart Innovation, Systems and Technologies 15, vol. 1, pp. 319–327. Springer, Berlin (2012)
6. Ohya, T., Kinoshita, E.: Super pairwise comparison matrix with the logarithmic least-squares method. In: Neves-Silva, R., et al. (eds.) Intelligent Decision Technologies, Frontiers in Artificial Intelligence and Applications, vol. 255, pp. 390–398. IOS press (2013)
7. Ohya, T., Kinoshita, E.: The treatment of hierarchical criteria in dominant AHP with super pairwise comparison matrix. In: Neves-Silva, R., et al. (eds.) Smart Digital Futures 2014, pp. 142–148. IOS press (2014)
8. Ohya, T., Kinoshita, E.: Using super pairwise comparison matrix for calculation of the multiple dominant AHP. In: Neves-Silva, R., et al. (eds.) Intelligent Decision Technologies, Smart Innovation, Systems and Technologies 39, pp. 493–499. Springer (2015)
9. Harker, P.T.: Incomplete pairwise comparisons in the Analytic Hierarchy Process. Math. Model. **9**, 837–848 (1987)

Interdisciplinary Approaches in Business Intelligence Research and Practice

Heterogeneous NoSQL Databases Abstraction Approach Based on Full Text Search Indexes

Hassen Fadoua and Grissa Touzi Amel

Abstract The exponential growth of unstructured data in the mobile applications, the social networks and the web technologies led to NoSQL database emergence. While this specific class of DBMS provided a better scalability for databases, the lack of a standard DML that unifies and simplifies querying NoSQL data stores is still a hard deal especially in heterogeneous environments. A simple SQL query can turn into a complex map-reduce function in the NoSQL world in order to obtain the same result in the standard SQL DDBMS. With no common convention between the large variety of NoSQL implementations and families, each product implemented its vision of the NoSQL concept. Each implementation covered distinct functional scopes, depending on the target domain and the creation purposes. Meanwhile, many successful NoSQL databases integrated a powerful full text component to enhance their search capabilities. To remedy this variety limitation, we propose a new incremental approach that allows (1) the standardization of NOSQL search queries among heterogeneous NoSQL data stores and (2) NoSQL search queries optimizing. This approach is based on (1) the definition of a new universal engine for full text indexing, (2) incremental synchronization of data and indexes between the stretched sites.

Keywords Nosql • Heterogeneous databases • Standardization • Common language • Fulltext indexes

H. Fadoua (✉) · G.T. Amel
LIPAH, FST, University of Tunis El Manar, Tunis, Tunisia
e-mail: hassen.fadoua@gmail.com

G.T. Amel
e-mail: amel.touzi@enit.rnu.tn

I. Czarnowski et al. (eds.), *Intelligent Decision Technologies 2016*,
Smart Innovation, Systems and Technologies 57,
DOI 10.1007/978-3-319-39627-9_37

1 Introduction

The workload diversity generated by every single system in the modern world has pushed the relational database family into its edge. Along with the exponential growth of unstructured data in mobile applications, the social networks and the web technologies, a compelling need to rescale classical dimensions of backend systems emerged. This specific class of DBMS, easy to deploy, provided a better scalability for databases. The RDBMS family implements ACID properties and guarantee that a series of operations will succeed inside a functional unit called a transaction. This is the main difference between relational databases and the NoSQL family members, even though some NoSQL implementations support another non-traditional transaction definition such as RavenDB [1]. It uses a weak form of isolation called "snapshot isolation". It provides global transactions via an external coordinator, but use is discouraged in its official documentation because the index updates are not atomic. The scalability of the RDBMS is harder to implement due to this concept. To keep data integrity and support transactions, a multi-server RDBMS would need to have a fast backend communication channel to synchronize all possible transactions and writes, while preventing or handling deadlocks. NoSQL solutions usually offer record-level atomicity, but cannot guarantee the transactional criteria of a query. It offers a considerable horizontal scaling abilities and ease sharding. This natively distributed database has great advantages on the scalability tradeoff. However, the lack of a standard DML that unifies and simplifies querying NoSQL databases is one of the reasons behind old system's abhorrence to NoSQL. A simple SQL query can turn into a complex program in the NoSQL. With no common convention between the large variety of NoSQL implementations and families, each product implemented its vision of the NoSQL concept. Each implementation covered distinct functional scopes, depending on target domain and creation purposes. For example, a key value database type is mainly intended for storing session information, user profiles or shopping card data. A NoSQL graph database suits better space problems where the system has interconnected data, such as social networks and recommendation engines. These targeted implementations prefer some functional features and invest more effort on their implementation. Design preferences may discourage combining two features such the case of the transactional spirit versus scalability aims. This need driven implementation method resulted into numerous NoSQL engines released concurrently with great aims but no standard specification behind. Each community or vendor invested in its implementation to satisfy a specific need, but not to cover all the classic database specifications. This variety made management and utilization quite impossible in a heterogeneous environment [8]. Although they all use the Map-Reduce mechanism for data retrieval, they do not offer the same features or at least the same meaning. The write operations have variable syntaxes depending on the NoSQL database family and implementation. Along with this emergence, the traditional search methods such as keyword search and link directories become impractical. One of the most successful features added recently to the leading NoSQL implementations is the full text indexing. This particular search technique enables many forms of fuzzy search such as phonetic algorithms, distance algorithms, stemming and anywhere truncation methods. In the opposite of querying databases, the full text search queries are standards and easy to use.

In this work, we propose a new incremental approach that allows (1) the standardization of NOSQL search queries among heterogeneous NoSQL data stores and (2) NoSQL search queries optimizing. This approach is based on (1) the definition of a new universal engine for the full text indexing, (2) an incremental synchronization process of data and indexes between stretched sites.

2 Related Works

Already aware about NoSQL diversity problems, many works focused in establishing a generic language that may unify all the NoSQL implementations. JSONiq is a query language and treatment designed for the popular model of JSON data. The main inspiration behind JSONiq is XQuery, which has proven so far a successful and productive query language for semi-structured data (especially XML) [4]. UnQL (Unstructured Query Language) is supposed to be an open source query language for NoSQL bases type value documents. However, UnQL does not cover the data definition language (DDL) SQL statements like CREATE TABLE or CREATE INDEX [2]. AQL provides an SQL-like command line interface for Aerospike database, UDF and index management. Aerospike provides AQL to provide an interface similar to tools that utilize SQL. Unfortunately, many basic features are still missing such as limiting the result set which is not yet available in AQL. There are many other attempts described in [3]. However, the full solution is still missing and no implementation covered the diversity problem. The authors in [1] tried to create a strict mapping between different NoSQL families' syntaxes and SQL but it concludes that future works are still needed to include graph database family in BQL (Bridge Query Language). The examined works aiming to answer the "Standard interface" constraint seems to fail because of the increasing number different features for each DBMS and the underlying data models of each NoSQL DBMS. This is the also one of the main reasons that made this work focus only on the search component. In the full text search techniques, many academic and professional works dealt with textual indexes sharding in order to fill the velocity of distributed systems. Solr and Elastic search are two search engines built on Lucene to improve its scalability. However, there are many constraints behind adopting these widely implemented solutions because of the transactional nature of many NoSQL DMBS [7]. Moreover, these search engines act separately from databases which will complicate design of a distributed heterogeneous database management system. Academically speaking, the author of [6] claims to build the efficient way of distributing inverted indexes over Hadoop. The implemented tool (DTPS) has shown acceptable experimental results compared to Lucene (Single Node) and Katta (Open source project). We believe that this result is not accurate because a Lucene indexing time may vary depending on its configuration (store, no-store, tokenized, …). We could not access the source code of DTPS to run our proper tests.

3 Proposed Approach

3.1 Approach Overview

The flowchart of our approach is presented in Fig. 1 and below is the various steps described in details.

Data Stores. This layer is responsible for NoSQL database access. It has a universal driver that tracks any insert, update or delete operation in each database. It can access key-value, column, object or graph database (Fig. 1). This is mainly used to retrieve raw data to be indexed. The next main use case is retrieving the original record once matched on the search indexes.

Full-text Index Management Layer. This component takes any tracked action from the data stores layer and creates the matching document. Building a document from various raw formats is described in details in Sect. 3.2. The generated indexes are distributed context aware. It contains two main components: The "Full Text Indexing Engine" (FTI) and The "Full Text Querying Interface" (FTQ). The FTI writes indexed objects into the "Global Textual Search Indexes" (Fig. 1). The FTQ is responsible of searching among the generated indexes.

The Global Synchronizer (GS). This GS is a fully distributed solution based on a message passing protocol in order to synchronize index repositories and indexed databases in a distributed architecture. For readability sake, we did not mention this

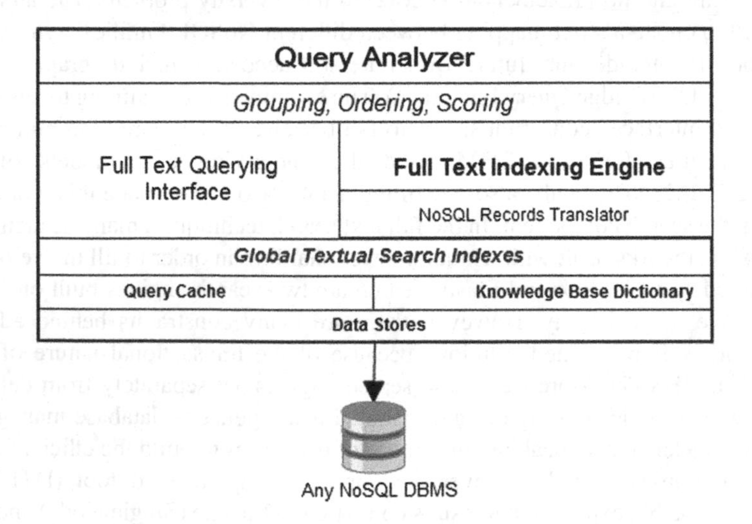

Fig. 1 Architectural overview of the proposed approach

component on Fig. 1. Actually, the GS is a standalone pluggable component. Our work in GS [5] is intended to establish a standard exchange protocol between different databases and synchronize the full text indexes in parallel with databases even in a heterogonous context.

The Query Management Layer. This layer offers a productive interface for users to use basic Lucene-like search operation and to create their own operators. The built queries are shared between the users in order to build a knowledge base of facilities.

3.2 Full Text Indexes Building

The full text indexes are an optimized model of raw data stored inside any NoSQL database. It is optimized because it enables advanced search capabilities and drastically accelerates search queries. Each index item references the raw record using an index identifier (idx_id). This identifier allows the direct access to the raw values once it matches the search criteria. Moreover, these indexes must keep the raw record location to enable the raw object accessor to fetch it from the appropriate location. The textual indexes described in Fig. 2a have a distributed aware structure. It holds a global calculated dictionary that references all sites dictionary. Each site dictionary holds the references for its related records. Because of the different families of records, index types may vary from a store to another. The site dictionary holds homogeneous record types because it reflects a DBMS in one site. The indexes representation of each record must be a faithful transcription of each record. For example, a key-value type record must be represented by field name and its indexed value. In a graph database model, a record describes its content and its relation with other records. This relation may be labeled to enable navigation inside a graph. This relation is treated by the "NoSQL Records Translator" (Fig. 1) as an embedded record. The related nodes are fetched and indexed as collections of nodes inside the main record. The local dictionary is calculated in each CRUD operation

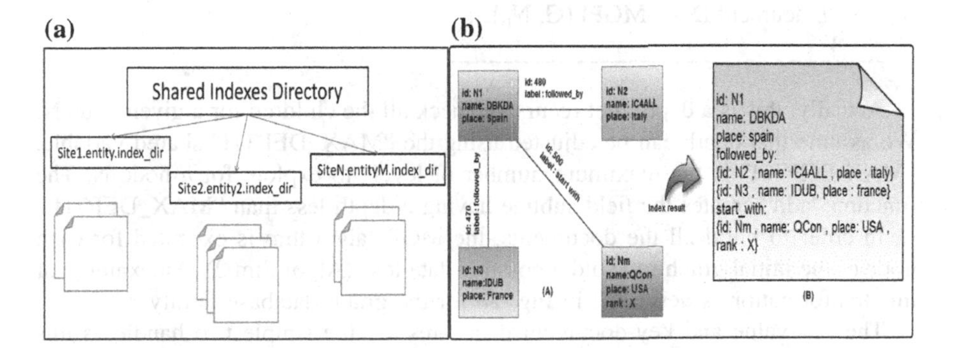

Fig. 2 Tree illustration of a shared index dictionary

on this record. Meanwhile, transforming the raw objects stored inside NoSQL databases into a plain free-text index is a real challenge. This plain index is commonly referred as "document" in the literature. The transformation must handle this variety, maintain the richness of the different data models and avoid performance impacts. The hardest structure to handle is the graph databases. The NoSQL graph databases are the programmatic description of a many to many relationships. Moreover, in a graph database relationship may have different labels. Figure 2b illustrates a single transformation of a graph object (A) to a plain index structure, also called document (B).

Algorithm 1 : Morphing a node to a plain index recursively (MGPI)

```
Input :
Graph G, Node N, Depth L
Output:
if(L < MAX_DEPTH){
Add(N, label);
for (Node Ni in N.getChildren())
{
  L ← L+1;
  MGPI (G, Ni,L);
}
}
```

Algorithm 2 : Morphing a whole graph MWG

```
Input :
Graph G, Root Node N, Depth L=0
Output:
for (Node Ni in N.getChildren())
{
  Document Di ← MGPI (G, Ni,L);
}
```

Actually, this is a depth first recursive check all the children for a given node N. We assume that depth can be adjusted using the "MAX_DEPTH" shared variable. "MAX DEPTH" is the maximum number of levels to explore for a node N. The function "add" creates the field subtree having a depth less than "MAX_DEPTH".

In order to build all the documents, the MGPI algorithm is executed for each node of the initial graph to avoid a possible data loss (Algorithm 2). An example of this tranformation is described in Fig. 2b for the graph database family.

The key-value and key-document datastores are the simplest to handle as this structure is the same in the database and in the flat index repository. The flattening process consists of reading the raw object attributes and then transforming them to field labels. The associative arrays behind the key-value stores would generate as much documents as stored keys. The key-document however would require some more attention. Actually, in the key-value database, "values" are simple and primitive types. However, in key-document, a value is a complex type wich may be trees, collections, dictionaries or any combination of accepted data types. The good news for the transformation process is that there are no integrity constraints even if a related document can be mentioned by its key.

Algorithm 3 : Morphing key document store MKDS

```
Input :
Database Key-Document kd
Index Document d
Output:
Document d
for (Attributes at_i in kd.getAttributes())
{
 if(at_i.getType() is simple_types){
   addAttribute(d,at_i.getFieldName(),at_i.getFieldValue)
 }else{
   MKDS(d,at_i)
 }
}
```

The column family databases have been built for a major performance gain. The principle of this family is dividing items by column, and not by item. This enables compression and duplicate removal from each column. The gain behind this method becomes considerable with huge amount of "almost" repeated data. Actually, this strategy will avoid reading the whole item in case if query asks only for a few fields. This is particularly interesting when dealing with a large number of columns in a classic relational database. In order to profit from this powerful concept, the morph process of a columnar NoSQL database uses the same logic behind this family. Every fragment (column holding an ID and a value) will be stored as an independent document in the resulting indexes. Indeed, this is the same morph process as a key-value data store.

3.3 Search Indexes Synchronization

In the distributed environment, the global index layer must be synchronized in order to offer the same view of the distributed database to every client. In our case, the sites are not just identical copies of a master site, but each site holds a distinct

DBMS with distinct data. This multi-master writer architecture must be handled with care to avoid the data consistency loss with NoSQL stores. The "Global Synchronizer" (GS) is designed for this purpose. The GS propagates any change on every site of the architecture to the rest of the sites. Each site is declared inside every GS, once added to the distributed architecture. The previously created data are handled by the local GS before this site joins the network and starts synchronization. The resulting local dictionary is distributed over the topology and kept inside each global dictionary with a reference to its site. Starting from this event, every change on local indexes is tracked inside a local queue. The GS emits periodically a signal toward all the nodes in the distributed architecture to query for updates executed on other sites. Each GS that receives this query answers with the queue items that were not yet delivered to requesting site. To avoid network flooding, the number of delivered items is limited. The querying interval is another important in throughput management.

4 Experimental Results

The evaluation of the proposed approach is examined in two main axes: update and search performances.

4.1 *Update Performances Evaluation*

The proposed approach must ensure at least a similar performance as the original database. Moreover, the update operation must be executed on indexes in a separate thread so as it does not affect the original update/insert time. Our benchmark is a contact database describing each contact by its phone number, a first name, a last name, the address (village) and the age. The entry dataset holds more than 800.000 differents items. Using a normal client to perform this mass insertion is not the most realistic test strategie for a few reasons: the naive approach of sending commands serially is slow because the connected clients have to pay for the round-trip delay for every query. It is possible to use pipelining, but for mass insertion of many records we had to write new commands while reading replies at the same time to make sure that insertion is as fast as possible. Moreover, only a small percentage of clients support non-blocking I/O. However, to evaluate the real NoSQL datastores performance, we have to treat the ready-to-use massive data as its generation date is on the insert time. Indeed, we implemented the "naive" non-blocking strategy to pipeline the insert operations with 100 parallel lightweight processes (Java threads).

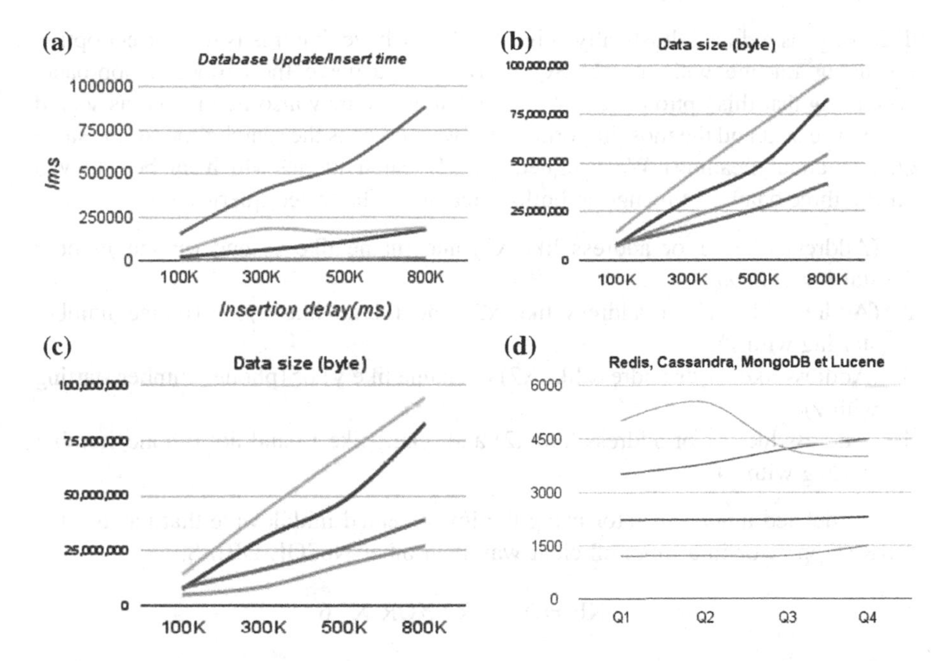

Fig. 3 Experimental results

The initialization of the data store shows that the implemented middleware based on Lucene is the fastest for updates (Fig. 3a). Apache Cassandra was the slowest in our benchmark. The write and the update operations were our first evaluation criteria because the proposed approach suggests adding an extra layer to each type of NoSQL database. This new component must not alter database write performances, and we estimate that the added overhead by the Lucene indexing operation is negligible.

This result allows us to safely update indexes in the same transaction as the database update transaction without breaking down the original performance. The second studied criterion is the disk utilization. Apache Cassandra is the winner in the minimum used storage deal (Fig. 3b). Actually, this is the main idea behind "Columnar" NoSQL databases. Exploding a database to columns allows this family to compress each column and reduce the repeated rows disk consumption. We have to remember here that the used configuration for all the tested databases is the basic scenario without any tuning (Lucene compound files are not enabled). The fields inside the Lucene index repository are stored. The index repository may act as an independent database. This is not really what we need when adding the middleware because the storing feature is already there with the existing NoSQL database. So if we turn off storing the fields content (STORE = NO) in indexes, the full text indexes disk usage is reduced drastically (Fig. 3c). We believe that this is the correct option for using Lucene with any NoSQL DBMS to achieve the proposed approach. Assuming that this option is enabled, the disk usage may also be marked as solved issue. The next and the most important axis we studied is the search time for the same query (natural meaning). We designed four (4) search queries which can be achieved on the three databases using the built-in features. The tested queries are:

1. (Address like x1 or address like x2) and (name like y) and (phone_number starting with z)
2. (Address like x1 or address like x2) and (name like y) or (phone_number starting with z)
3. (Address like x1 or address like x2) or (name like y) or (phone_number starting with z)
4. (Address like x1 or address like x2) and (name like y) and not (phone_number starting with z)

We defined a new operator using the implemented middleware that handles the "LIKE" operator in a more efficient way than other NoSQL DBMS.

$$\mathrm{LIKE} = X * \mathrm{OR} * \mathrm{XOR}\ X\tilde{}6$$

The "LIKE" operator definition for the proposed approach suugests that either the field contains a string that starts or ends with the given value or the distance between a string in the search field and the given value is less than 60 % (the "~" operator denotes the percentage of similarity to X). The search response time gives our implementation of the aparoach has the fastest response (Fig. 3d). Moreover, the search result is wider using the new defined functions. The returned dataset for the same search query is richer using the search engine middleware (Fig. 4).

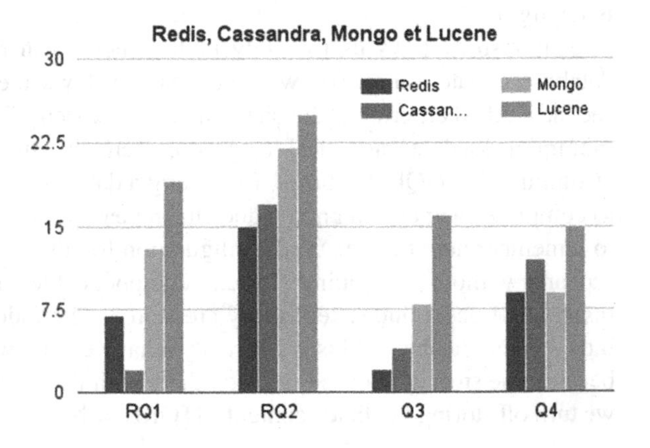

Fig. 4 Result count for the same search query

5 Conclusion and Future Work

In this work we dissected the problem of the syntactic diversity in heterogeneous NoSQL distributed systems and presented NoSQL standardization attempts. Through related works and following a new trend, we introduced the full text indexes layer in order to give end users a standard interface to search NoSQL databases behind the distributed architecture. Moreover, adding full text search indexes enables powerful search techniques. To build a suitable index format, we explained how to deal with different NoSQL record's structures to prepare the search interface for a standard querying mechanism. To handle the distributed data changes, the introduced approach holds a synchronization mechanism based on persisted queues though the global synchronizer (GS). We explained by the end of the article how to query these indexes, despite the different NoSQL data stores behind. During the prototype implementation we discovered some Lucene powerful capabilities that may extend this work into a complete data management language (DML) for heterogonous NoSQL databases.

References

1. Bach, M., Werner, A.: Standardization of NoSQL Database Languages. Beyond Databases, Archit. Struct. **11**, 1037–1054 (2014)
2. Buneman, P., et al.: A query language and optimization techniques for unstructured data. ACM SIGMOD Rec. **25**(2), 505–516 (1996)
3. Curé, O., et al.: Data integration over nosql stores using access path based mappings. Database Exp. Syst. (2011)
4. Florescu, D., Fourny, G.: JSONiq: The history of a query language. IEEE Internet Comput. (2013)
5. Hassen, F., Touzi Grissa, A.: Near real-time synchronization approach for heterogeneous distributed databases. In: DBKDA 2015, the Seventh International Conference on Advances in Databases, Knowledge, and Data Applications, pp. 107–113 (2015)
6. Lee, T., et al.: The efficient implementation of distributed indexing with hadoop for digital investigations on Big Data. Comput. Sci. Inf. Syst. **11**(3), 1037–1054 (2014)
7. Liu, Z., et al.: MUSYOP: towards a query optimization for heterogeneous distributed database system in energy data management. In: International (2014)
8. Sellami, R., et al.: ODBAPI: a unified REST API for relational and NoSQL data stores. Big Data (BigData Congr.) (2014)

Potentials of Image Mining for Business Process Management

Rainer Schmidt, Michael Möhring, Alfred Zimmermann, Ralf-Christian Härting and Barbara Keller

Abstract An enormous amount of data in the context of business processes is stored as images. They contain valuable information for business process management. Up to now this data had to be integrated manually into the business process. By advances of capturing it is possible to extract information from an increasing number of images. Therefore, we systematically investigate the potentials of Image Mining for business process management by a literature research and an in-depth analysis of the business process lifecycle. As a first step to evaluate our research, we developed a prototype for recovering process model information from drawings using Rapidminer.

Keywords Image Mining · BPM · Business Process Management · Object recognition · Picture · Process analysis

1 Introduction

Business Process Management is an important means for adapting enterprises to changing requirements [1]. Most information used for business process management is either represented as formal process models using approaches such as ARIS [2] or BPMN [3] or laid down in written documentation. Although this approach is conceptually sound, the use of formal methods excludes many stakeholders at least partially [4], because they are not familiar with these tools. In general, the need to

R. Schmidt (✉) · M. Möhring
Munich University of Applied Sciences, Munich, Germany
e-mail: Rainer.Schmidt@hm.edu

R.-C. Härting · B. Keller
Aalen University of Applied Sciences, Aalen, Germany

A. Zimmermann
Reutlingen University, Reutlingen, Germany
e-mail: alfred.zimmermann@reutlingen-university.de

I. Czarnowski et al. (eds.), *Intelligent Decision Technologies 2016*,
Smart Innovation, Systems and Technologies 57,
DOI 10.1007/978-3-319-39627-9_38

integrate more information from heterogeneous information sources has been identified as a precondition for improving process quality [5].

At the same, the amount of digitized information has increased in many enterprises significantly [6]. Not only by the advance of smartphones, but also through scanners, cameras etc. images created in huge numbers. Capturing and storing images has become an everyday activity in many enterprises [7]. Images are taken to capture handwritten notes, whiteboards, comments on printed documentation etc. Apps such as OfficeLens [8] allow to capture images very easily and use them within office software. Software for extracting textual information from images has become a standard. Detection of geometric figures in hand-written drawings is available [9], the detection of semantically deep drawings is in close reach. Often information relevant to business process management is contained in the images.

However, up to now this affluence of images is not used as an input for business process management. Thus, the potentials using these images for business process management are not exploited. Therefore, this paper address the following research question: *What are the potentials of Image Mining for Business Process Management?* To explore this research question, we made a systematic literature review and described the basics of Image Mining and Business Process Management. Furthermore, a prototype was designed to show the potentials of one scenario of Image Mining for Business Process Management. The paper is structured as follows: after this introduction, we define basics of Image Mining as well as Business Process Management. In Sect. 3, we define potentials of Image Mining for Business Process Management. A prototype as an example of the potentials is designed and tested in Sect. 4. In Sect. 5, related work is described and the paper concludes with a discussion of the results.

2 Background

2.1 Image Mining

Image Mining extract implicit knowledge, relationships through image data and other implicit patterns from images or image databases according to Zhang et al. [10]. Therefore, Image Mining integrates different research streams and results from Data Mining, Machine Learning, Database Management etc. [10, 11]. Zhang et al. [10] argues that Image Mining is not just an extension to the traditional Data Mining. Image Mining can be interpreted as a unique research field and uses as well as integrates different methods from different research fields [10].

The typical Image Mining process can be divided into different steps [12]. The first step is preprocessing of image data, like loading the image and special segmentations [12]. The next step is feature extraction and transformation [12], where common image attributes (e.g. color, edge, shape, texture) are extracted from the images. The third step are Image Mining techniques [12]. There are different Image

Mining techniques according to Zhang et al. [10] (can may be used in the field of BPM):

- Object Recognition
- Image Retrieval
- Image Indexing
- Image Clustering and Classification
- Association Rule Mining
- Neural network

Object Recognition tries to find known as well as similar objects in different images [10]. Through Image Retrieval users as well as information systems can easily find images e.g. based on different patterns [10, 13]. Further, image retrieval is a process of processing limited information to support users retrieval goals at a short time [12]. Furthermore, approaches to indexing images for implementing a information system to retrieve images as well as image data are necessary [10]. To find out similarities of different images as well as cluster different images according to their individualities an image classification and image clustering is needed [10]. Through Association Rule Mining [10] interesting trends, patterns and (pattern) rules of different images can be extracted, Rule Mining can be applied based on a large database of images or e.g. a combined collection of images [10, 14]. Artificial neuronal networks can be used to mine a large amount of image data for feature extraction [10, 15, 16].

Finally, based on the results of these techniques an evaluation and knowledge creation are the last steps of the typical Image Mining process [12]. Therefore, decisions related to BPM are possible (e.g. analyze of graphical process documentation). General aspects of BPM are described in the next section.

2.2 Business Process Management

Business process management [17] is the method-based application of methods, techniques, and tools to business processes during their lifecycle [1]. There are different definition of the business process lifecycle [1, 17]. The most frequently used phases are design, deployment, operation and optimization. We will use the definition developed in [18].

Process Identification

Starting from a business perspective, the processes are identified that contribute to achieving a business goal [18]. They are also delimited and related to each other. Often, the processes found are integrated into an enterprise architecture. The process identification phase consists of two sub-phases. In the designation phase [18], an understanding of the processes in an organization and their interrelationship shall be achieved. Depending on the abstraction level, different numbers of processes may emerge. In the following evaluation phase, the processes found in the designation phase are prioritized according to their need for modeling, redesign etc.

Process Discovery/Design

The goal of process discovery [18] is to collect information about an existing process and create a documentation of the present state of the process. Using modeling approaches such as BPMN [3] or ARIS [2] the current or strived for state of the process is depicted in one or several models. Process discovery has to cope with three challenges [18]. First, the knowledge about the process is fragmented. Frequently, no single domain expert has a complete picture of the process but only parts of it. Second, the knowledge of the domain experts is often organized from a case-oriented way, but not a process-oriented view. Third, the business domain experts are not familiar with process modeling methods. Instead, they are using ad hoc defined approaches for depicting their knowledge. Therefore, it is important to assure that these informal descriptions are in sync with the formal ones. The prototype presented in section four supports this use case.

Process Analysis

The processes are analyzed using both qualitative and quantitative means [18]. Value-Added Analysis [19], Root Cause Analysis [20] and Issue Documentation are important steps of the qualitative process analysis.

Goal of the value-added analysis is to identify unnecessary process steps and to remove unnecessary steps. To do so, the process steps are classified into value-adding, business value-adding and non-value adding tasks. Then, the non-value adding tasks are eliminated either completely or automated as far as possible.

In the root-cause analysis [20], the relationship between adverse effects on one hand and causal and contributing effects on the other side shall be identified. Cause-effect diagrams and why-why diagrams are frequently used by means of root-cause analysis.

As a result, a register of issues is created. Whenever possible the issues are not only described qualitatively, but also their impact is quantified. Furthermore, the list of issues should be prioritized, pareto charts are important means of doing this.

Quantitative Process Analysis [21] start with capturing data covering the process performance dimensions' time, cost, quality and flexibility. In flow analysis, important performance indicators such as cycle time are determined. Other analysis objects are queues and queue lengths.

Process Redesign/Improvement

Using the register of issues during process analysis changes shall be identified to resolve these issues [22]. If there are multiple ways to resolve an issue, they should be compared. As a result a set of changes is proposed that addresses seven elements [18]: internal and external customers, business operations, business behavior, organization, information, technology and the external environment. However, the possible goals time, cost, quality and flexibility cannot be achieved to the same extent at the same time.

Process Implementation

This phase transforms the as-is process into the to-be process [18]. Process implementation embraces both organization and information systems. Often it is started by selecting a process automation platform such a dedicated business process management system [23], workflow management systems or enterprise resource planning software. In this way manual tasks can be replaced by automated one. These information systems furthermore provide execution transparency and the enforcement of regulatory rules and laws.

Process Monitoring and Controlling

After deployment, the business process the operational phase of the business process starts and process instances are created [18, 24]. They represent the execution of the business process such as business transactions. During the operation phase, data representing is collected for later analysis. This data is used in the optimization phase, in order to find possible improvements. Themes for performance analysis are time, cost, quality and flexibility.

3 Potentials of Image Mining for Business Process Management

Based on a systematic literature review according to Kitchenham [16] in databases like SpringerLink, IeeeXplore, AISel, Sciencedirect ACM digital library with keywords like "Image Mining" AND "BPM" or "Business Process Management" for the last decade, we cannot find research papers which address an overview or broad insights of the use of Image Mining for Business Process Management. Therefore, we define in the following some core aspects and potentials of Image Mining for Business Process Management according to the fundamentals of BPM according to Sect. 2.2 of the paper.

Images can be differentiated into documents, drawings and pictures. Documents contain textual information, that can be recovered using optical character recognition. Drawings contain graph-based information. Approaches for recovering graphical information are just starting, such as graph detection in OneNote Drawings. Pictorial data contains not directly recoverable information, but often allows to detect metadata, such as the types of products depicted etc. During the business process lifecycle, image data are created on many occasions, such as workshops, meetings, documentation etc. There are a number of sources of image data. In many enterprises, paper documents do not travel within the organization, but are scanned on arrival. Other significant sources are mobile phones and tablets with cameras.

Process Identification

A lot of Image data is created in the designation phase, especially during workshops and meetings. Drawings depicting the anticipated process architecture are created. The image data originate from scans and cameras. Object recognition and retrieval can be used to identify and find processes.

Process Discovery/Design

Although powerful process modeling tools are available, plenty of image-based data is created during process discovery and design, because the domain experts are not familiar with process modeling methods and the tools supporting them. Often ad hoc defined approaches are used for depicting processes. A particular challenge is the fragmentation of process knowledge leading to multiple separate images that cover the same process. Furthermore, the case-oriented perspective of the domain experts has to be transformed to a process-oriented view. In addition, checking if the modeled processes are correct (e.g. with a comparison of textual descriptions) can be made through object recognition.

Process Analysis

Image data containing drawings are a major source for value-added analysis in order to identify unnecessary process steps and to remove unnecessary steps. In root-cause analysis, cause-effect diagrams and why-why diagrams are frequently used means. Pareto charts are important means for prioritizing issues found during process analysis. Image-based data supports quantitative process analysis e.g. to capture customer queues and queue lengths. Object as well as image recognition can be used to detect differences in business process models. Furthermore, image clustering and classification can help to e.g. better understand similarities in processes.

Process Redesign/Improvement

During process redesign and improvement, image-based data is created as a result of workshops, meetings etc. They contain suggestions to redesign and improve areas such as internal and external customers, business operations, business behavior, organization, information, technology and the external environment. In this step, e.g. object recognition can be used to check if the improved business process is modeled correct (e.g. in comparison to the specification).

Process Implementation

Image-based data in this phase often contains specifications as well as external information such as of regulatory rules and laws. Image mining can support the process implementation phase by using image mining techniques.

Process Monitoring and Controlling

During process operation, a lot of image data is created in many enterprises. Paper documents do not travel within the organization, but are scanned on arrival.

Another important source are mobile phones and tablets with their built cameras. In production environments, images are used for documenting productions quality. All this data are collected for later analysis and to find possible improvements. Image indexing and retrieval are an important in the means to do this analysis.

4 Prototype: Object Recognition in Business Process Models Through Image Mining

To show the potentials of one area of Image Mining for Business Process Management, we implement a Prototype for object recognition of Business Process Models. The goal of this prototype is to detect business process modeling elements like gateways, activities etc. from images. The detected modeling elements and their order can be further used to interpret the model and to pre-check with other models (e.g. textual descriptions [25]). The prototype was designed based on general prototyping principles according to [26] and implemented through the software Rapid Miner [27] and the Image Mining package [28, 29] (BurgSys). As a modeling notation, we used the EPC notation [2]. The EPC is well known and used in practice and has e.g. not so much modeling elements like BPMN [2, 3].

In the following, we describe the implementation of our prototype for the modeling element "XOR" (exclusive disjunction) of the EPC. For other modeling elements (like activities, AND, OR) the implementations are similar. First two types of images were load into the mining software. One type are images with different representations of XOR, like shown below (Fig. 1).

The other types are images without XOR representations. The next step is the core detector algorithm, which uses these two types of images to learn the differences and to build a model for correct detection of the business process modeling element XOR. As a detector algorithm, we used the fast haar detector, which is very common for detecting objects [29, 30]. The generated model was used to detect the modeling element XOR in the modeled business processes. Therefore, the modeled business processes were load and then discovered based on the model. Finally, the detected XOR were extracted.

To check the possibilities of our prototype, we used the modeled business process of [25] of the business case "important credit application processing" for evaluation (Fig. 2).

The results of the prototype for detecting "XOR" in this sample case is shown in the following figure (Fig. 4). Furthermore, other modeling notation elements like AND, OR, activities etc. can be detected similarly to the described case. The

etc.

Fig. 1 Two examples of XOR representations in images

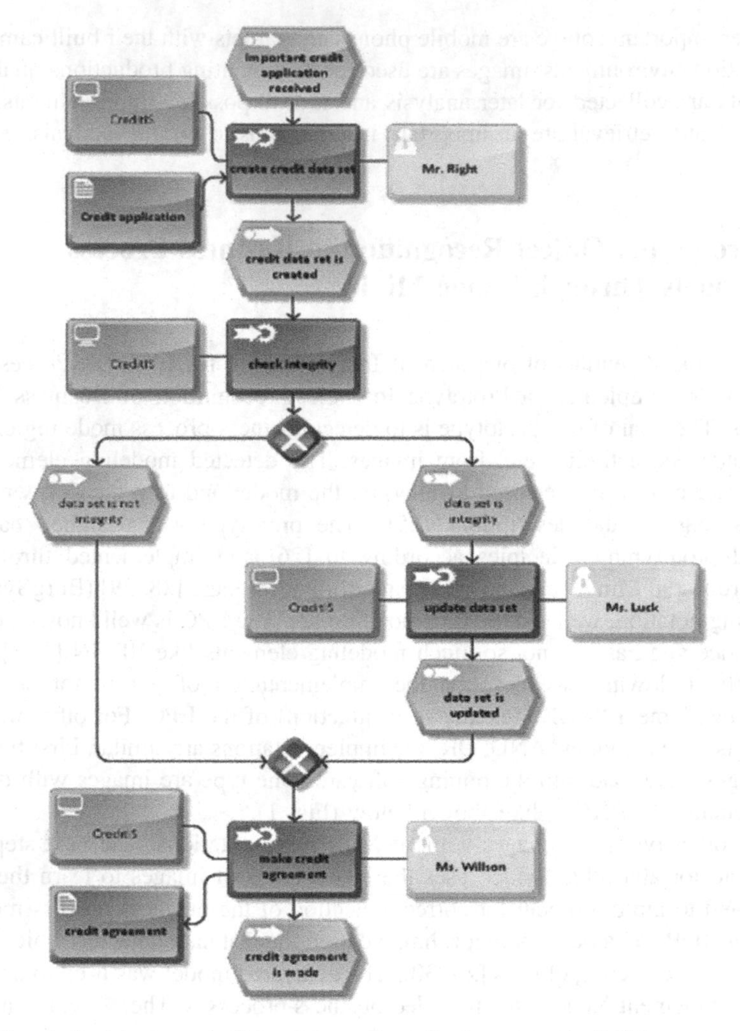

Fig. 2 Sample business process "important credit application processing" [25]

following figure shows an short excerpt of the implementations through Rapid Miner 5.4 [27, 28] (Fig. 3).

As seen in Fig. 4 the two "XOR" are detected correctly for this sample process. Further evaluation experiments show similar results. The results of the prototype can be used e.g. for checking the modeled business process with textual descriptions. Therefore, a comparison of the occurrences of notations elements and their order can be a quality check if the business process was modeled correctly. Furthermore, textual analysis of the process model can be made through Text Mining according to [31, 25] and compared with the results of the prototype.

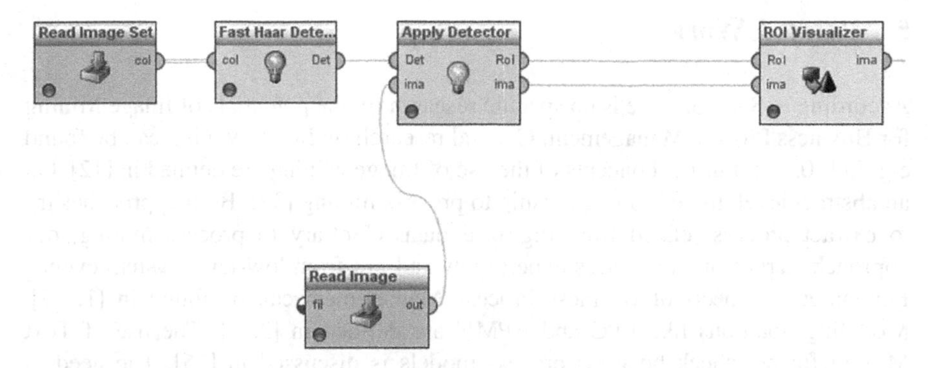

Fig. 3 Excerpt of implementation in Rapid Miner 5

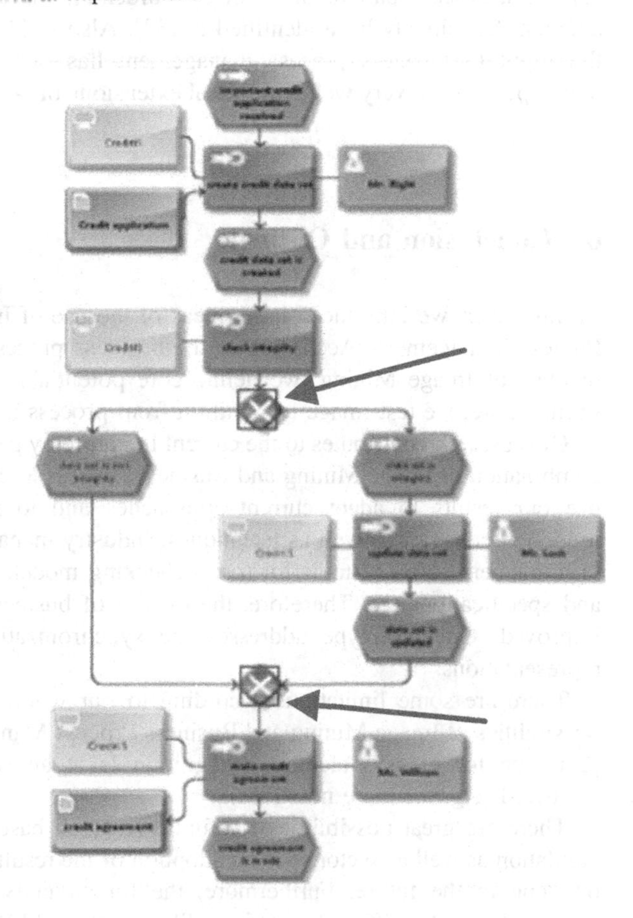

Fig. 4 Correct detected XOR of the sample case (*Sample case* [25])

5 Related Work

According to Sect. 3, there is no specific research on the potentials of Image Mining for Business Process Management. General research on Image Mining can be found e.g. in [10, 13]. Further concepts of the use of Image Mining are defined in [12]. On an abstract level, there is a relationship to process mining [32]. Both approaches try to extract process related from digitized data. Contrary to process mining, our approach starts from a business expert view and not from low-level system events. Fundamental aspects of Business Process Management can be found in [1, 17]. Modeling notations like EPC and BPMN are defined in [2, 3]. The use of Text Mining for pre-check business process models is discussed in [25]. The need to capture a broader part of the reality in order to improve business process management has already been identified in [33]. Also in [4, 5] the need for extending the input to business process management has been described. The concepts developed here fit very well with social extensions of business process management [5].

6 Conclusion and Outlook

In this paper, we introduced basic ideas of the use of Image Mining for Business Process Management. According to the business process lifecycle and the fundamentals of Image Mining, we define core potentials to combine both concepts. Furthermore, we test image recognition from process models.

Our research contributes to the current literature by generating a new view of the combination of Image Mining and Business Process Management. Researchers can use our results to adapt current approaches and to improve business process modeling behavior as well as techniques. Industry managers can use our approach to implement software tools for (pre-) checking modeled business process models and specifications etc. Therefore, the quality of business process projects can be improved. Our prototype addresses the synchronization of image and textual representations.

There are some limitations according to our work. We cannot address to all possibilities of Image Mining and Business Process Management. Furthermore, our prototype for process element recognition is at an early stage and should be improved (e.g. for more notations).

There are great possibilities for future research based on our work. Empirical validation as well as sector specific adoption of the results and the prototype should be done in the future. Furthermore, the functionality of the prototype can be increased e.g. to different notations like UML, BPMN, etc. as well as further specific adoptions.

References

1. Weske, M.: Business Process Management: Concepts, Languages, Architectures. Springer, Berlin, Heidelberg (2007)
2. Scheer, A.-W., Nüttgens, M.: ARIS architecture and reference models for business process management. In: van der Aalst, W., Desel, J., Oberweis, A. (eds.) Business Process Management, pp. 376–389. Springer, Berlin, Heidelberg (2000)
3. White, S.A.: Introduction to BPMN. IBM Coop. 2008–029 (2004)
4. Bruno, G., Dengler, F., Jennings, B., Khalaf, R., Nurcan, S., Prilla, M., Sarini, M., Schmidt, R., Silva, R.: Key challenges for enabling agile BPM with social software. J. Softw. Maint. Evol. Res. Pract. **23**, 297–326 (2011)
5. Schmidt, R., Nurcan, S.: BPM and social software. In: Ardagna, D., Mecella, M., Yang, J., Aalst, W., Mylopoulos, J., Rosemann, M., Shaw, M.J., Szyperski, C. (eds.) Business Process Management Workshops, pp. 649–658. Springer, Berlin, Heidelberg (2009)
6. Brynjolfsson, E.: Understanding the Digital Economy: Data, Tools, and Research: Data, Tools and Research. The MIT Press (2000)
7. Vincent, A., Varghese, G.: Towards A Robust and Stand-Alone System for Binarization and OCR of Document Images (2015)
8. Office Lens—Windows-Apps im Microsoft Store. https://www.microsoft.com/de-de/store/apps/office-lens/9wzdncrfj3t8
9. Team, O.: OneNote—what's new in January 2016. https://blogs.office.com/2016/01/29/onenote-whats-new-in-january-2016/ (2016)
10. Zhang, J., Hsu, W., Lee, M.L.: Image mining: Issues, frameworks and techniques. In: Proceedings of the 2nd ACM SIGKDD International Workshop on Multimedia Data Mining (MDM/KDD'01). University of Alberta (2001)
11. Burl, M.C., Fowlkes, C., Roden, J.: Mining for image content. Syst. Cybern. Inf. Inf. Syst. Anal. Synth. (1999)
12. Mishra, N., Silakari, D.S.: Image mining in the context of content based image retrieval: a perspective. IJCSI Int. J. Comput. Sci. Issues **9**, 98–107 (2012)
13. Stanchev, P., Flint, M.: Using image mining for image retrieval. In: IASTED Conference "Computer Science and Technology," Cancun, Mexico, pp. 214–218 (2003)
14. Ordonez, C., Omiecinski, E.: Discovering association rules based on image content. In: IEEE Forum on Research and Technology Advances in Digital Libraries, 1999. Proceedings, pp. 38–49. IEEE (1999)
15. Carpenter, G., Grossberg, S., Markuzon, N., Reynolds, J.H., Rosen, A.B., et al.: Fuzzy ARTMAP: A neural network architecture for incremental supervised learning of analog multidimensional maps. IEEE Trans. Neural Netw. **3**, 698–713 (1992)
16. Kitchenham, B.: Procedures for performing systematic reviews. Keele UK Keele Univ. **33**, 1–26 (2004)
17. Van der Aalst, W., ter Hofstede, A., Weske, M.: Business process management: a survey. Bus. Process. Manage. 1019–1019 (2003)
18. Dumas, M., La Rosa, M., Mendling, J., Reijers, H.A.: Fundamentals of Business Process Management. Springer, Berlin, Heidelberg (2013)
19. Conger, S.: Six sigma and business process management. In: Handbook on Business Process Management, vol. 1, pp. 127–146. Springer (2015)
20. Wilson, P.F.: Root Cause Analysis: A Tool for Total Quality Management. ASQ Quality Press (1993)
21. Zhu, J.: Quantitative Models for Performance Evaluation and Benchmarking: Data Envelopment Analysis with Spreadsheets. Springer (2014)
22. Bach, V., Brecht, L., Hess, T., Österle, H.: Enabling Systematic Business Change: Integrated Methods and Software Tools for Business Process Redesign. Springer (2013)
23. Hajo, A.: Reijers: Implementing BPM systems: the role of process orientation. Bus. Process Manag. J. **12**, 389–409 (2006)

24. Metzger, A., Leitner, P., Ivanovic, D., Schmieders, E., Franklin, R., Carro, M., Dustdar, S., Pohl, K.: Comparing and combining predictive business process monitoring techniques. IEEE Trans. Syst. Man Cybern. Syst. 45, 276–290 (2015)
25. Schmidt, R., Möhring, M., Härting, R.-C., Zimmermann, A., Heitmann, J., Blum, F.: Leveraging textual information for improving decision-making in the business process lifecycle. In: Neves-Silva, R., Jain, L.C., and Howlett, R.J. (eds.) Intelligent Decision Technologies. Sorrent (2015)
26. Naumann, J.D., Jenkins, A.M.: Prototyping: the new paradigm for systems development. Mis. Q. 29–44 (1982)
27. Akthar, F., Hahne, C.: RapidMiner 5 Operator Reference. Rapid-GmbH (2012)
28. Burget, R., Karasek, J., Smékal, Z., Uher, V., Dostal, O.: Rapidminer image processing extension: a platform for collaborative research. In: Proceedings of the 33rd International Conference on Telecommunication and Signal Processing, pp. 114–118 (2010)
29. Masek, J., Burget, R., Karasek, J., Uher, V., Guney, S.: Evolutionary improved object detector for ultrasound images. In: 2013 36th International Conference on Telecommunications and Signal Processing (TSP), pp. 586–590. IEEE (2013)
30. Viola, P., Jones, M.: Rapid object detection using a boosted cascade of simple features. In: Proceedings of the 2001 IEEE Computer Society Conference on Computer Vision and Pattern Recognition, 2001. CVPR 2001, pp. I–511. IEEE (2001)
31. Tan, A.-H., et al.: Text mining: The state of the art and the challenges. In: Proceedings of the PAKDD 1999 Workshop on Knowledge Discovery from Advanced Databases, p. 65 (1999)
32. Van Der Aalst, W.: Process mining. Commun. ACM **55**, 76–83 (2012)
33. Wang, M., Wang, H.: From process logic to business logic—a cognitive approach to business process management. Inf. Manage. **43**, 179–193 (2006)

Decision Trees as Readable Models for Early Childhood Caries

Vladimir Ivančević, Nemanja Igić, Branko Terzić, Marko Knežević and Ivan Luković

Abstract Assessing risk for early childhood caries (ECC) is a relevant task in public health care and an important activity in fulfilling this task is increasing the knowledge about ECC. Discovering important information from data and sharing it in an understandable format with both experts and the general population could be beneficial for advancing and spreading the knowledge about this disease. After having experimented with association rule mining, we investigate the possibility of using decision trees as readable models in risk assessment. We build various decision trees using different algorithms and splitting criteria, favouring compact decision trees with good predictive performance. These decision trees are compared to the previous ECC models for the same analyzed population, namely a logistic regression model and an associative classifier, as well as to decision trees for caries from other studies. The results indicate flexibility and usefulness of decision trees in this context.

Keywords Early childhood caries · Risk assessment · Decision tree

V. Ivančević (✉) · N. Igić · B. Terzić · M. Knežević · I. Luković
Faculty of Technical Sciences, University of Novi Sad, Novi Sad, Serbia
e-mail: dragoman@uns.ac.rs

N. Igić
e-mail: nemanjaigic@uns.ac.rs

B. Terzić
e-mail: branko.terzic@uns.ac.rs

M. Knežević
e-mail: marko.knezevic@uns.ac.rs

I. Luković
e-mail: ivan@uns.ac.rs

I. Czarnowski et al. (eds.), *Intelligent Decision Technologies 2016*,
Smart Innovation, Systems and Technologies 57,
DOI 10.1007/978-3-319-39627-9_39

1 Introduction

The importance of understanding and treating early childhood caries (ECC) is being slowly acknowledged across the world. This disease represents a significant threat to the sensitive population it targets, namely children under the age of six, as well as to the whole society. Even in many developed countries, which make considerable investments in prevention and treatment, it is difficult to eradicate ECC, as there are various subgroups at high caries risk. This pattern has been observed across different continents and it is present in countries such as USA [1], Brazil [2], and Serbia [3].

Despite numerous medical advances, many open questions about ECC remain. One such issue, which has been recognized in the conclusion paper of the 2014 ECC Conference, is the assessment of caries risk [4]. As there is a large body of investigated risk factors for which there is no reliable confirmatory evidence, additional studies concerning such factors are needed in order to make progress in battling ECC. This problem is further aggravated by the diversity and interrelatedness of considered factors, which include microbiological, dietary, hygienic, and even social risks [4, 5].

In our previous studies on ECC in the South Bačka area (Autonomous Province of Vojvodina, Republic of Serbia), we employed data mining, namely association rule mining, for the following two purposes: (i) to identify potential risk factors for ECC by analyzing associations rules about ECC and candidate risk factors [6]; and (ii) to create an understandable predictive model for ECC by combining the association rules into a rule-based classifier [7]. Our primary motivation was to create readable models from real-world data that could be understood by general population or utilized to facilitate communication between data mining experts, who are looking for valuable patterns, and domain experts, who know how to interpret such information. Association rules appeared especially convenient, as they present information in a straightforward manner, allow for grouping into readable classifiers, have comprehensive character, and lack structural restrictions of decision trees.

In the present study, we are turning attention to decision trees, as they are also readable models that may be used both for explanation and prediction. According to [8], decision trees possess the following three advantages: (i) easy to represent and interpret; (ii) without conflicting knowledge; and (iii) identify and reveal important knowledge features. As they might be simpler to read and apply than typical rule-based classifiers, such as associative classifiers, we considered it worthwhile to investigate readable decision trees that could match the accuracy of previous models.

Our main goal is to construct a decision tree that could be of similar or better predictive performance when compared to the previous models for ECC in the South Bačka area: a logistic regression model [9] and an associative classifier [7]. To this end, we employed traditional tree generation by recursive partitioning and summarization of association rules into a decision tree. We evaluate the resulting trees with respect to their predictive power and structural properties, as well as compare them to other tree-based caries models from similar studies.

Besides Introduction and Conclusion, the paper is organized into four additional sections. In Sect. 2, we describe related results from previous studies on ECC modelling and discuss the usage of decision trees in ECC research. In Sect. 3, we elaborate on the employed data set and the applied methods. In Sect. 4, we present the resulting decision trees, as well as caries models from other studies. In Sect. 5, we review these decision trees and compare them to the models from other studies.

2 Related Work

In this section, we first comment on models from the related studies and then discuss the suitability of decision trees for ECC modelling.

2.1 Previous ECC Models

The initial model of ECC in South Bačka was a logistic regression model composed of five child-related variables (city, gender, birth order, birth weight and use of medical syrups), which was constructed by medical experts and based on the results of a statistical analysis of the ECC data [9]. It was followed by rule-based classifiers composed of association rules about ECC (associative classifiers) [7]. We demonstrated that 6-rule or 8-rule associative classifiers may outperform the initial logistic regression model. Moreover, these models incorporated many factors not directly related to the child, e.g., socioeconomic status of the family and factors concerning parents.

However, this improvement was coupled with a cost. Although understandable, associative classifiers generally do not provide a clear overview of all the included factors as important rules may combine numerous factors and a classifier may incorporate numerous diverse rules. This was controlled to some extent by clustering factors and favouring classifiers with dissimilar factors. Nonetheless, we decided to investigate decision trees as readable models with the hope that their structure could be more readable and their performance comparable to that of the previous models.

2.2 Decision Trees for Caries

Over the past 25 years, there have been at least six studies that utilized decision trees as caries models [10–15]. These studies mostly rely on classification and regression trees (CART) [16], the only exceptions being the use of C 5.0 [17] in [12] and CHAID [18] in [14]. An overview of those models is given in Sects. 4 and 5.

On the other hand, there are other types of decision tree formation. Decision trees that use previously mined association rules are often neglected in practice despite having some promising advantages. For instance, associative classification trees (ACTs) seem to be smaller and slightly more accurate than traditional decision trees [8], which should present sufficient motivation to evaluate them in our search for a compact and accurate model of ECC. Moreover, as we already possess a large set of association rules about ECC, we decided to form ACTs for ECC.

At present, there are various suggestions on how to build such decision trees [8, 19–22]. However, a practical problem with ACTs is that there are few implementations readily available in popular data analysis and mining tools. Given the detailed explanation of ACT creation and the positive evaluation results reported in [8], we chose to implement that particular ACT variant (see Sect. 3).

3 Materials and Methods

In this section, we describe an ECC data set and outline the creation of decision trees.

3.1 Data

The ECC data was recorded by Tušek [9] for a 10 % sample of children from the South Bačka area, which is part of the Autonomous Province of Vojvodina, Republic of Serbia. The data set features 341 records, one for each examined child, across a selection of 36 categorical variables. The output variable indicates ECC presence (30.5 %) or absence (69.5 %), while input variables correspond to potential risk factors such as socioeconomic status, dietary habits, health awareness, and behaviour of the child and the parents. More information about the data set may be found in [6].

3.2 Decision Tree Formation

The decision trees that we created could be divided into the following two groups: traditional decision trees and associative classification trees.

Formation of Traditional Decision Trees. We used RapidMiner Studio [23], a software tool for data mining, to create five traditional decision trees. Their formation is based on recursive partitioning, which, in each step, requires selection of a splitting attribute according to one of the predefined criteria. We used the RapidMiner implementation of the algorithm for decision tree generation (RDT) to create four decision trees, one for each of the four splitting criteria supported in the tool (accuracy, gain ratio, information gain, and the Gini index). We post-pruned each tree and applied the evolutionary computation approach to optimize the confidence parameter. The Gaussian mutation was used to maintain generic diversity across the population of five individuals, while the tournament selection approach with the tournament fraction set to 0.25 was used to perform selection in the population. The fifth decision tree was created using the CHAID implementation in RapidMiner, which utilizes a χ^2-based splitting criterion. In order to obtain more nodes in the tree, we set the minimum node size for split to 18 and the minimum leaf size to 9. For all five trees, the city variable was excluded, as it led to less general models. The maximum tree depth was set to seven, based on the average human capacity to process information (Miller's Law [24]).

Associative Classification Trees. The ACT formation process is described in detail in [8]. The root node contains a starting set of association rules, which is recursively split among the child nodes until one of the predefined termination criteria is fulfilled (homogenous rule set achieved or a specific threshold met). In addition to specifying the set of association rules and the splitting criterion needed to select the attribute for node splitting (confidence gain or entropy gain), ACT formation requires threshold values δ_{SUPPORT}, $\delta_{\text{CONFIDENCE}}$, and $\delta_{\text{TREE HEIGHT}}$, which determine boundary values for the node support, confidence, and height within the tree, respectively. The predictions given by ACTs generally depend on the rule sets in the leaves. We used the Java programming language to implement formation and application of ACTs.

Formation of Associative Classification Trees. The association rules for ACT formation were mined from the ECC data set according to the specification in [6], but the resulting rule set was not pruned. The complete rule set encompassed 18 rules about ECC presence and 88142 rules about ECC absence. We believe that there may be three potential reasons for the disproportionate shares of the two rule subsets: (i) in the examined population, ECC presence is less common than ECC absence (30.5 vs. 69.5 %); (ii) the size of the ECC data set (341 cases) might be too small to reliably detect less frequent patterns about ECC presence; and (iii) although there are numerous potential risk factors for ECC [25], it is possible that many of them have only a minor role in the examined population, which could manifest as a multitude of less frequent (and difficult to detect) patterns. We created three ACTs, one for each implemented splitting criterion: confidence gain, entropy gain, and lift gain. The first two criteria initially yielded poor predictive performance, so we added the lift gain criterion by modifying the confidence gain criterion to take into account the change in the lift of a group of rule sets instead of the confidence change. The threshold values for support, confidence, and height were set to 0.01, 0.8, and 7, respectively.

3.3 Decision Tree Evaluation

We aimed for decision trees of compact structure (readability) and good predictive performance. The readability was evaluated by node count (ND) and tree height (H). The predictive performance was evaluated on the whole data set by accuracy (ACC) and true skill statistic (TSS) [26], which is a measure somewhat similar to kappa [27]. Both TSS and kappa yield values between −1 and 1, with positive values denoting performance better than random guessing, but TSS is simpler to calculate and does not appear to be dependent on prevalence [26]. TSS was used so that we could directly compare the decision trees to the previous ECC models: the logistic regression model (TSS = 0.325) and the associative classifier (TSS = 0.41) [7]. For each tree group, we measured the overall predictive performance associated with each supported splitting criterion as the average accuracy in a stratified tenfold cross-validation (CV ACC). We used the dot program [28] to visualize the selected decision trees.

We also inspected decision trees for caries from other studies [10–15]. In this overview, we considered data set characteristics (population and predictor variables), tree formation procedure (TFP), variable count in the tree (V), tree height (H), and predictive performance as measured by sensitivity (SN) and specificity (SP).

4 Results

The eight resulting decision trees are presented in Table 1. An overview of various decision trees from the other studies and three representative decision trees from the present study, which are denoted by "~", is given in Table 2. With respect to both prediction and structure, we selected for discussion a single tree from each group: the ACT for splitting by lift gain (see Fig. 1) and the CHAID decision tree (see Fig. 2).

Table 1 The resulting decision trees

Tree group	Variable selection	ND	H	ACC (%)	TSS	CV ACC (%)
Traditional	Accuracy	19	5	76	0.23	69
Traditional	Gain ratio	19	6	74	0.15	68
Traditional	Information gain	55	6	85	0.59	66
Traditional	The Gini index	77	6	87	0.68	65
Traditional	CHAID	22	4	75	0.32	69
ACT	Confidence gain	6	2	70	0.00	69
ACT	Entropy gain	4	1	67	0.11	67
ACT	Lift gain	6	2	72	0.17	69

Table 2 Decision trees from caries studies

Study	Data set	TFP	V	H	SN (%)	SP (%)	Remark
[10]	N = 1024 (age 5–9) Portland, ME, USA 32 predictors	CART (Gini)	2	2	62	77	/
	N = 914 (age 5–8) Aiken, SC, USA 32 predictors		9	9	64	86	/
[11]	N = 466 (elementary school age) Rochester & Finger Lakes areas, NY, USA 5 predictors	CART (Gini)	3	3	/	55	/
[12]	N = 500 (age 5–8) Ena & Nakatsugawa areas, Gifu, Japan 12 predictors	C 5.0	5	4	73	77	/
[13]	N = 442 (age 20–64) Osaka, Japan 5 predictors	CART	2	2	34	85	Primary caries
			3	2	72	73	Secondary caries
[14]	N = 1681 (age 1–4) Dundee, Scotland, UK 56 predictors	CHAID	3	3	65	69	High caries-risk (n = 784)
[15]	N = 1322 (age 0–5) KY, USA 7 predictors	CART	3	4	43	78	/
~	N = 341 (age 1–5) South Bačka area, Vojvodina, Serbia 35 predictors	RDT (acc.)	7	5	25	98	/
		CHAID	7	4	41	90	/
		ACT (lift)	2	2	22	95	/

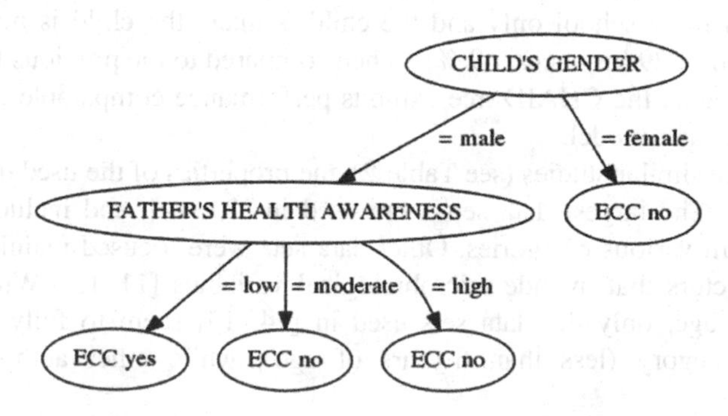

Fig. 1 An ACT obtained when splitting by lift gain

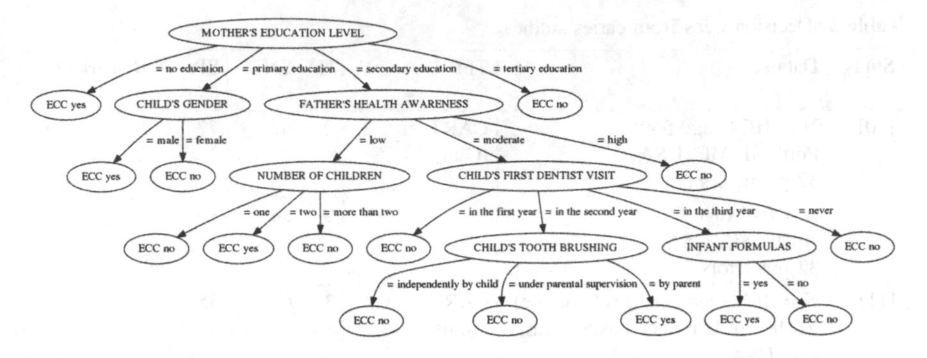

Fig. 2 A traditional decision tree obtained when applying CHAID

5 Discussion

The obtained ACTs are almost trivial and, with the exception of the ACT obtained when splitting by lift gain, have little value (see Table 1). The selected ACT (see Fig. 1) has better performance as it encapsulates an important pattern from a previous study [6]. It seems that the strong disproportion between the rules about ECC presence and absence (18 vs. 88142), prevents formation of a sound ACT.

The traditional decision trees outperform the ACTs in prediction, but are less readable (see Table 1). Splitting by the Gini index or information gain provides strong accuracy. However, the resulting trees are complex and there is most probably overfitting. Consequently, the most convenient traditional decision trees were those formed when splitting by accuracy or applying CHAID. Both decision trees are relatively compact trees of solid accuracy and both include variables about mother's education level and child's gender. Our preference for the CHAID tree was based on its higher TSS value. Some of the relevant variables that were identified in previous studies [6, 7, 9] may be observed in the CHAID tree, namely child's gender and father's health awareness. This decision tree also contains potential patterns as it includes variables concerning oral hygiene and related behaviour. Moreover, as shown in Fig. 2, if the mother has no education, or has completed primary school only and the child is male, the child is more likely to have ECC (20 of 29 instances, 69 %). When compared to the previous ECC models for the same area, the CHAID tree exhibits performance comparable to that of the logistic regression model.

Among the similar studies (see Table 2), the properties of the used data sets vary considerably. The largest data sets were used in [10, 14], and include numerous variables from various categories. Other data sets were focused mainly on narrow sets of predictors that include microbiological variables [11–13]. With respect to population's age, only the data sets used in [14, 15] seem to fully fit the early childhood category (less than 6 years of age), while other authors generally examined data about children aged 5 or more [10–12] and adults [13]. When compared to other data samples, the present data set has the fewest cases, but a relatively large set of potential predictors, and focuses completely on early childhood.

All of the reviewed trees (see Table 2) are generally readable as they contain from two to nine variables. The listed predictive performances of these models should not be compared directly because there are large differences in goals, evaluation methods and data sets across the studies. However, it appears that our selected decision trees have somewhat lower sensitivity, which could be attributed to the absence of microbiological predictors in the available data set. The key strengths of our selected decision trees are their high specificity and the possibility to apply them in practice more easily as they do not include variables that require oral examination of a child.

6 Conclusion

The decision trees appear to be very flexible models since their construction may be finely tuned to provide compact trees or strong predictive performance, depending on the actual needs. They may be well-suited to ECC research and ECC risk assessment, as they provide a readable but powerful common ground between domain experts and data analysts. Although decision trees that are summarized from association rules seem to have great potential, they did not provide satisfactory results. The particularities of the analyzed population and the data set, including data imbalance and low number of association rules for the positive class, might have been the reasons behind the poor performance. Our future research may include creating new algorithms for building associative decision trees that would be more resistant to these problems.

Acknowledgments The research presented in this paper was supported by the Ministry of Education, Science, and Technological Development of the Republic of Serbia under Grant III-44010. The authors are most grateful to Ivan Tušek and Jasmina Tušek for the provided data set and valuable support throughout the study.

References

1. Ghazal, T., Levy, S.M., Childers, N.K., Broffitt, B., Cutter, G.R., Wiener, H.W., Kempf, M.C., Warren, J., Cavanaugh, J.E.: Factors associated with early childhood caries incidence among high caries-risk children. Commun. Dent. Oral Epidemiol. **43**(4), 366–374 (2015)
2. Corrêa-Faria, P., Martins-Júnior, P.A., Vieira-Andrade, R.G., Marques, L.S., Ramos-Jorge, M. L.: Factors associated with the development of early childhood caries among Brazilian preschoolers. Braz. Oral Res. **27**(4), 356–362 (2013)

3. Tušek, I., Carević, M., Tušek, J.: Prevalence of early childhood caries among members of different ethnic groups in the South Bačka area (in Serbian). Vojnosanit. Pregl. **69**(12), 1046–1051 (2012)
4. Garcia, R., Borrelli, B., Dhar, V., Douglass, J., Ramos Gomez, F., Hieftje, K., Horowitz, A., Li, Y., Ng, M.W., Twetman, S., Tinanoff, N.: Progress in early childhood caries and opportunities in research, policy, and clinical management. Pediatr. Dent. **37**(3), 294–299 (2015)
5. Berkowitz, R.J.: Causes, treatment and prevention of early childhood caries: a microbiologic perspective. J. Can. Dent. Assoc. **69**(5), 304–307b (2013)
6. Ivančević, V., Tušek, I., Tušek, J., Knežević, M., Elheshk, S., Luković, I.: Using association rule mining to identify risk factors for early childhood caries. Comput. Methods Programs Biomed. **122**, 175–181 (2015)
7. Ivančević, V., Knežević, M., Tušek, I., Tušek, J., Luković, I.: Human friendly associative classifiers for early childhood caries. In: 7th KES International Conference on Intelligent Decision Technologies (KES-IDT 2015), pp. 243–253. Springer (2015)
8. Chen, Y.-L., Hung, L.T.-H.: Using decision trees to summarize associative classification rules. Expert Syst. Appl. **36**, 2338–2351 (2009)
9. Tušek, I.: The Influence of social environment and ethnicity on caries prevalence in the early childhood (in Serbian). Ph.D. thesis, University of Belgrade (2009)
10. Stewart, P.W., Stamm, J.W.: Classification tree prediction models for dental caries from clinical, microbiological, and interview data. J. Dent. Res. **70**(9), 1239–1251 (1991)
11. Gansky, S.A.: Dental data mining: potential pitfalls and practical issues. Adv. Dent. Res. **17**, 109–114 (2003)
12. Tamaki, Y., Nomura, Y., Katsumura, S., Okada, A., Yamada, H., Tsuge, S., Kadoma, Y., Hanada, N.: Construction of a dental caries prediction model by data mining. J. Oral Sci. **51**, 61–68 (2009)
13. Ito, A., Hayashi, M., Hamasaki, T., Ebisu, S.: Risk assessment of dental caries by using classification and regression trees. J. Dent. **39**, 457–463 (2011)
14. MacRitchie, H.M.B., Longbottom, C., Robertson, M., Nugent, Z., Chan, K., Radford, J.R., Pitts, N.B.: Development of the Dundee Caries Risk Assessment Model (DCRAM)—risk model development using a novel application of CHAID analysis. Commun. Dent. Oral Epidemiol. **40**, 37–45 (2012)
15. Li, H.F.: Data mining and pattern discovery using exploratory and visualization methods for large multidimensional datasets. Ph.D. thesis, University of Kentucky (2013)
16. Breiman, L., Friedman, J.H., Olshen, R.A., Stone, C.J.: Classification and Regression Trees. Wadsworth, Belmont CA (1984)
17. Kuhn, M., Johnson, K.: Applied Predictive Modeling. Springer (2013)
18. Kass, G.V.: An exploratory technique for investigating large quantities of categorical data. Appl. Stat. **29**(2), 119–127 (1980)
19. Säuberlich, F., Gaul, W.: Decision tree construction by association rules. In: 23rd Annual Conference of the Gesellschaft für Klassifikation, pp. 245–253. Springer (2000)
20. Wang, K., Zhou, S., He, Y.: Growing decision trees on support-less association rules. In: 6th ACM SIGKDD International Conference on Knowledge Discovery and Data Mining (KDD'00), pp. 265–269. ACM (2000)
21. Abdekhalim, A., Traore, I., Sayed, B.: RBDT-1: a new rule-based decision tree generation technique. In: International Symposium on Rule Interchange and Applications (RuleML 2009), pp. 108–121. Springer (2009)
22. Peng, Y., Ye, Y., Yin, J.: Decision tree construction algorithm based on association rules. In: 2nd International Conference on Computer Application and System Modeling (ICCASM 2012), pp. 754–756. Atlantis Press (2012)
23. RapidMiner Studio—RapidMiner. https://rapidminer.com/products/studio/
24. Miller, G.A.: The magical number seven plus or minus two: some limits on our capacity for processing information. Psychol. Rev. **63**(2), 81–97 (1956)

25. Fontana, M.: The clinical, environmental, and behavioral factors that foster early childhood caries: evidence for caries risk assessment. Pediatr. Dent. **37**(3), 217–225 (2015)
26. Allouche, O., Tsoar, A., Kadmon, R.: Assessing the accuracy of species distribution models: prevalence, kappa and the True Skill Statistic (TSS). J. Appl. Ecol. **43**, 1223–1232 (2006)
27. Cohen, J.: A coefficient of agreement for nominal scales. Educ. Psychol. Measur. **20**(1), 37–46 (1960)
28. Graphviz—Graph Visualization Software. http://graphviz.org/

Pattern Recognition in Audio and Speech Processing

Music Genre Classification Using a Gradient-Based Local Texture Descriptor

Faisal Ahmed, Padma Polash Paul and Marina Gavrilova

Abstract With the increasing popularity and availability of online music databases that store vast collections of music, automated classification of music genre has attracted significant attention for the management of such large-scale databases. This paper presents a new music genre classification method that utilizes gradient-based texture analysis of the spectrograms constructed from the audio signals. We propose to use gradient directional pattern (GDP)—a robust local texture descriptor that exploits the gradient directional information to encode the local texture properties of an image. The proposed method first computes spectrograms from the audio signals and then applies the GDP operator to construct the feature descriptors that represent micro-level texture details of the spectrograms. We use a support vector machine (SVM) for the classification task. The effectiveness of the proposed method is evaluated using the GTZAN genre collection music database. Our experiments show promising results for the proposed GDP-based spectrogram texture analysis, as compared against some other existing music genre classification methods.

Keywords Music genre classification · Local texture analysis · Spectrogram · Gradient directional pattern (GDP)

1 Introduction

Music genre is one of the most widely-used descriptors for organizing, tagging, and managing large-scale music databases. In this context, automated music genre classification is an interesting research problem in the domain of audio signal processing, which has potential applicability in effective management of both online and offline collections of music [6]. According to Lidy et al. [13], among the different approaches to represent the contents of a music or audio clip, extracting distinctive features from the audio signal is the most commonly used one. However, the

F. Ahmed (✉) · P.P. Paul · M. Gavrilova
Department of Computer Science, University of Calgary, Calgary, Alberta, Canada
e-mail: faahmed@ucalgary.ca

I. Czarnowski et al. (eds.), *Intelligent Decision Technologies 2016*,
Smart Innovation, Systems and Technologies 57,
DOI 10.1007/978-3-319-39627-9_40

subjectivity and ambiguity related to the concept of music genre and the wide varieties of music styles have made robust feature extraction a challenging task. In the study conducted by McKay and Fujinaga [14], even human participants were not able to correctly classify the music genre of 24 % of the total number of samples. Hence, the most crucial aspect of designing a successful music genre classification system is to construct a robust audio feature representation that can effectively minimize the intra-class distance while maximizing the inter-class variations [6].

This paper presents a new music genre classification method that utilizes the gradient directional pattern (GDP)—a robust local texture operator for spectrogram feature description. The motivation is to represent the contents of an audio signal through a spectrogram, from where texture features are extracted to represent the overall appearance. The method is similar to extracting appearance-based texture features from an image, where the texture operator is applied on the spatial domain. In our previous works, the GDP operator has successfully been applied to facial expression recognition [1] and gender classification from facial images [3]. In this work, we evaluate the effectiveness of the GDP-based texture description in capturing micro-level texture details from the audio spectrogram, which can be utilized as the feature descriptor for music genre classification. We use a publicly available music genre database, namely the GTZAN genre collection to evaluate the performance of the proposed GDP-based spectrogram content description. In our experiments, the proposed method achieves promising results, as compared against some existing music genre classification methods.

2 Related Work

Earlier works on music content representation mostly utilized rhythm-specific music features extracted from the audio signal. These features were trained using different machine learning techniques to perform the classification ask. One of the early works on feature-based audio signal representation presented by Dannenberg et al. [7] uses neural network and Naive Bayes classifier to recognize musical styles. Their work focused on characterizing four different types of music improvisation. Later, Tzanetakis and Cook [19] utilized timbral texture, beat, and pitch-specific feature components as music content descriptors. These features were extracted based on short-time Fourier transform (STFT), mel frequency cepstral coefficients (MFCCs), and wavelet transform (WT). Features used in this study were also incorporated into the MARSYAS framework [17], a widely-used system for music genre classification. Some of the other similar approaches involve statistics-based spectrum analysis, histogram and rhythmic pattern extraction from audio signals, as presented in [12, 17]. Li et al. [11] conducted a comparative study on the performances of the traditional timbral texture, rhythmic and pitch-specific audio features against wavelet coefficients-based histogram, where support vector machine (SVM) and linear discriminant analysis (LDA) were utilized to perform the classification task. In their study, Daubechies wavelet coefficient histograms (DWCH) obtained the highest recognition performance. Ezzaidi and Rouat [8] presented another study where Gaussian mixture models (GMM) were trained with mel-frequency cepstral coefficients (MFCCs) to achieve high recognition performance.

In recent years, spectrogram appearance-based music content description has attracted much attention due to its simplicity and computational efficiency. The motivation is to treat the spectrograms as gray images and use appearance-based feature extractors to highlight the texture properties of such images. This in turn, can effectively be used as a content descriptor of the original audio file. Based on this argument, Neammalai et al. [15] applied Fourier transform and image processing techniques on audio spectrograms for classification of music and speech signals. Another approach presented by Costa et al. [5] used gray level co-occurrence matrix (GLCM) descriptor to represent the spectrogram texture. More recently, local binary pattern (LBP) texture operator has been used on spectrogram images for effective music genre classification [6]. Originally introduced by Ojala et al. [16] for texture classification, the LBP operator encodes the local neighborhood of an image by comparing the neighbor gray values with respect to the center, which can effectively capture micro-level texture details, such as edges, corners, spots, etc. However, the performance of the LBP operator deteriorates under the presence of noise and non-monotonic gray scale variations [1, 2]. Hence, Wu and Zhang [20] combined Gabor wavelets-based texture features with LBP to further boost the recognition performance. A similar approach was presented in [4], where Gabor filters were used with local phase quantization (LPQ) texture descriptor to represent the spectrogram texture. Some other similar local texture operators address the limitation of LBP by incorporating extra levels of information in the encoding process (e.g. local ternary pattern (LTP) [18]), applying extra pre-processing steps to enhance the texture information (e.g. Sobel-LBP [21]), or encoding less sensitive gradient information instead of gray scale values (e.g. local directional pattern (LDP) [10], directional ternary pattern (DTP) [2]).

This paper presents an improvement over the LBP-based spectrogram texture analysis for music genre classification. While the LBP texture operator works on gray-scale values which are susceptible to random noise and non-monotonic variations, the proposed GDP method encodes the more stable gradient information instead of gray levels and thus can capture robust spectrogram texture both in smooth and high-textured local regions.

3 Proposed Method

The proposed music genre classification system comprises several components. Since our method is based on the texture analysis of the visual representation of audio signals, the first step is to construct spectrogram images from the original audio samples. Next, gradient directional pattern (GDP) operator is applied on the spectrograms to extract micro-level texture details. Spatial histograms computed from different regions of the GDP encoded spectrogram images are concatenated to form the

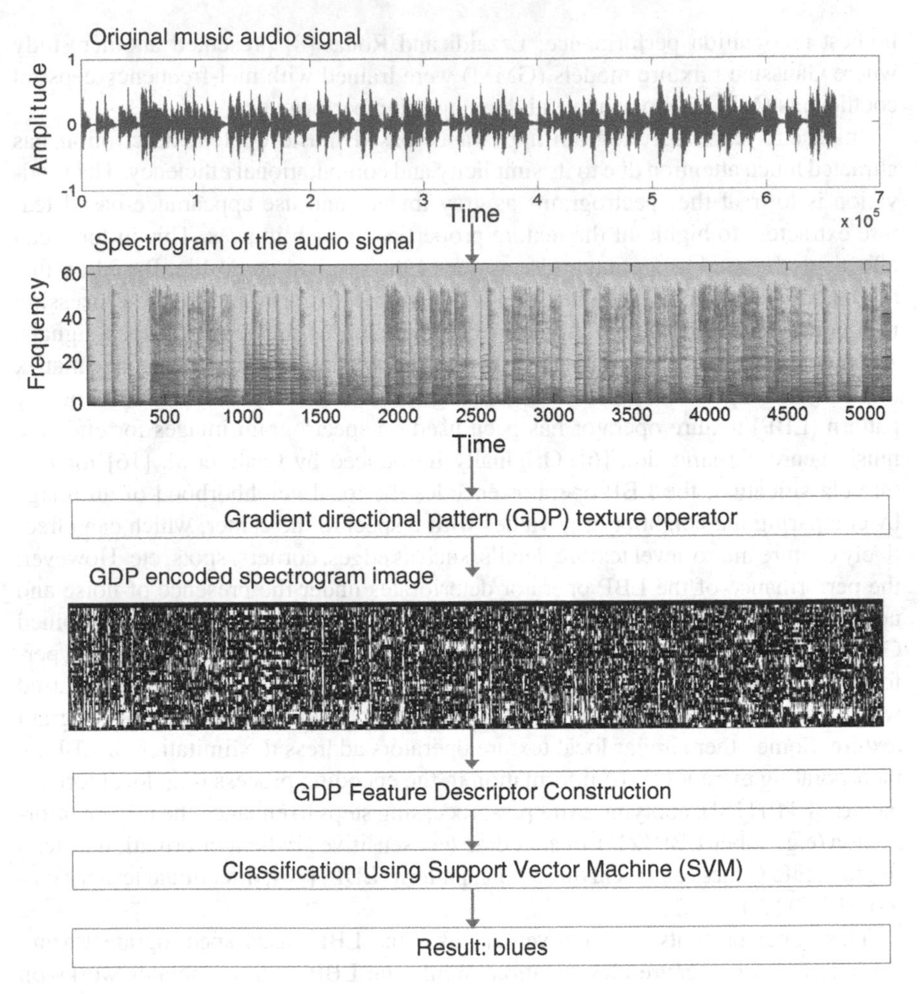

Fig. 1 Components of the proposed music genre classification system

final feature descriptor. Lastly, a support vector machine (SVM) classifier is trained with the GDP feature descriptors to perform the classification task. Figure 1 shows the components of the proposed method.

3.1 Spectrogram Texture Encoding Using GDP

Since image gradients are more robust than gray values, any texture operator that exploits the gradient values tend to be more robust in encoding the local texture information. In this paper, we propose to use gradient directional pattern (GDP) to

Fig. 2 Sobel masks

Horizontal Mask

-1	-2	-1
0	0	0
1	2	1

Vertical Mask

-1	0	1
-2	0	2
-1	0	1

encode the spectrogram texture. The GDP operator utilizes the gradient angle values of each pixel in a local neighborhood in order to isolate important micro-level texture information, such as edges, spots, corners, smooth regions, etc. The GDP operator follows the same encoding approach as LBP, except that the GDP operator thresholds the local region based on the gradient angle values. First, the direction of the gradient vector for each pixel is computed based on the following formula:

$$\alpha(i,j) = \tan^{-1}(G_i/G_j) \tag{1}$$

Here, $\alpha(i, j)$ is the gradient direction angle of the pixel (i, j), and G_i and G_j are the two elements of the gradient vector that can be obtained by applying the Sobel operator on the source image. The Sobel operator comprises a horizontal and a vertical mask which are convoluted with the image in order to obtain the values of G_i and G_j, respectively. The Sobel masks are shown in Fig. 2.

After computing the gradient direction values, the GDP operator encodes a local 3×3 neighborhood by thresholding the neighbor gradient directions with respect to the center gradient α_c and a threshold value t. Any neighbor gradient angle α_i with a value $\alpha_c - t \leq \alpha_i \leq \alpha_c + t$ is encoded as 1 and the rest are encoded as 0. In practice, the GDP operator encodes the neighbors having a similar gradient direction with respect to the center as 1 and the rest of the neighbors as 0. Introducing the threshold t ensures consistent encoding in both high textured and smooth regions. The resultant bit values for each neighbor are then concatenated to form an 8-bit binary pattern and the corresponding decimal value is assigned to the center. This process is continued for every pixel in the spectrogram image as a window mechanism and thus, a GDP encoded image representation is obtained. Formally, the GDP operator can be defined as:

$$GDP(x_c, y_c) = \sum_{p=0}^{P-1} s(GD_p, GD_c)2^p \tag{2}$$

$$s(GD_p, GD_c) = \begin{cases} 1, & GD_c - t \leq GD_p \leq GD_c + t \\ 0, & otherwise \end{cases} \tag{3}$$

Here, GD_c is the gradient direction angle of the center pixel (x_c, y_c), GD_p is the angles of its neighbors, and t is the threshold. Figure 3 illustrates the basic GDP encoding method.

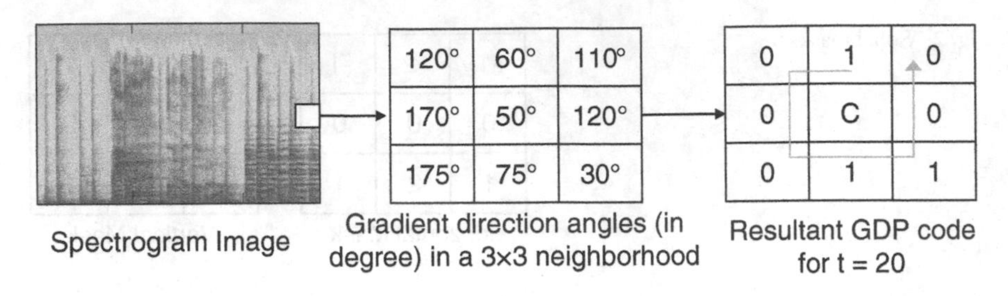

Fig. 3 Illustration of the basic GDP encoding method. Here, the GDP code for the center C is 10001100

3.2 GDP Feature Descriptor Construction

Applying the GDP-based texture encoding on a spectrogram results in an encoded image representation. A global histogram computed from this encoded image can be treated as a feature vector, which can be defined as:

$$H_{GDP}(i) = \sum_{x=1}^{M}\sum_{y=1}^{N} f(GDP(x,y), i), where f(a,i) = \begin{cases} 1, a = i \\ 0, otherwise \end{cases} \tag{4}$$

Here, H_{GDP} is the GDP histogram of an $M \times N$ encoded image and i is the GDP code value. However, such a feature descriptor only contains the occurrence frequency of the GDP micro-patterns and fails to represent any locality information. Many researchers have argued that, inclusion of location information in the feature

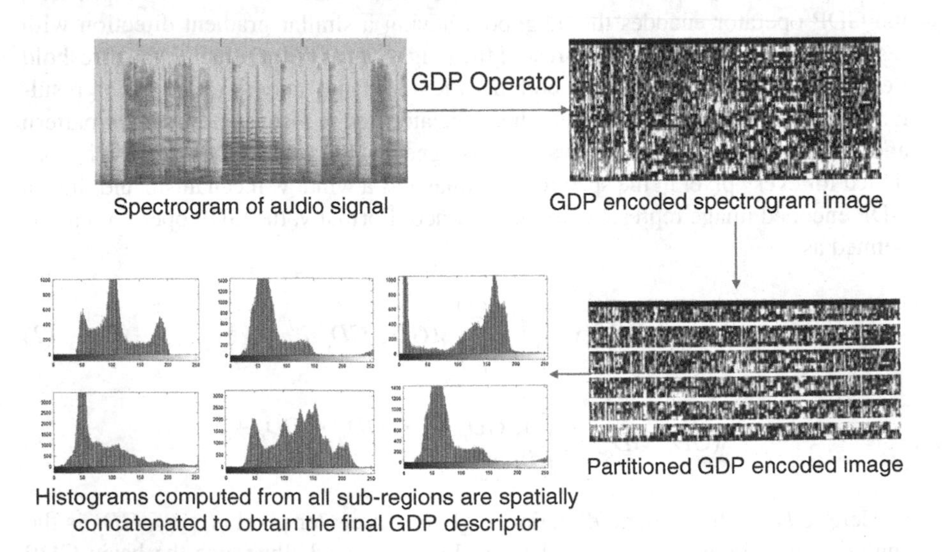

Fig. 4 Construction of the GDP feature descriptor from a spectrogram

descriptor makes it more robust and informative, thus increasing the recognition performance [6]. Hence, in order to incorporate some notion of location information with the GDP-based face feature descriptor, the whole face region is divided into some equal sub-regions and individual local GDP histograms computed for all the sub-regions are spatially concatenated to obtain a final feature descriptor. The GDP feature descriptor construction is illustrated in Fig. 4.

4 Experiments and Results

The performance of the proposed method is evaluated using a publicly available music genre database, namely the GTZAN genre collection database [19]. The database contains 1000 audio tracks, each of which is a 30 s clip. It includes a total of ten different genre, namely blues, classical, country, disco, hiphop, jazz, metal, pop, reggae, and rock. There are 100 samples for each of these categories. All the audio files have the same sampling rate of 22050 Hz Mono 16-bit and are saved in .wav format. For our experiment, we selected 50 audio clips from each of the 10 genre.

A support vector machine (SVM) with a radial-basis function (RBF) kernel is used for the classification task. The performance of the proposed method can be influenced by adjusting the threshold value t and the number of partitions in the spectrogram. Therefore, optimal parameter selection is an important task to obtain the best recognition performance. In order to find the optimal parameter values, the performance of the proposed method is evaluated for different t values: 10, 20, 30, and 40 while the spectrogram is partitioned into 3, 6, and 9 equal-sized regions. A tenfold cross validation is used to evaluate the recognition performance. The tenfold cross-validation is an iterative process where the dataset is divided into 10 equal subsets. At each iteration, the classifier is trained using 9 subsets and the remaining one subset is used for testing. This process is repeated for 10 times and each of the 10 subsets is used as a testing set once. The final cross-validation rate is the average percentage of samples which are correctly classified. Figure 5 shows the recognition performance of the proposed method for different threshold values and number of partitions. From the figure, it can be observed that, the highest recognition performance for the proposed method is found for threshold value of 40 while the spectrogram is partitioned into 8 local regions. Further partitioning the spectrogram or increasing the threshold value results in a decrease in the performance.

The proposed method is also compared against some of the existing music genre classification methods, namely local binary pattern (LBP) [6], Gabor wavelets-based spectrogram features [4], and perceptual linear prediction (PLP) [9]. In all cases, a SVM classifier with the RBF kernel is used for the tenfold cross-validation task. The optimal parameter settings found from Fig. 5 is used for the proposed method. Figure 6 shows the comparison of the cross-validation rates of the proposed method against these existing methods. It can be observed that the proposed method achieves the highest recognition performance of 84.5 %. The effectiveness of the proposed method is due to the utilization of the gradient direction values for encoding the local texture that facilitates a more robust description of the image micro-properties.

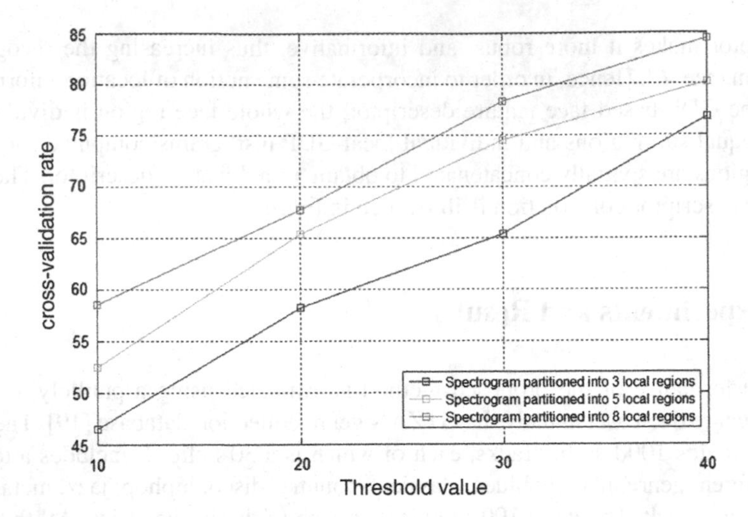

Fig. 5 Recognition performance of the proposed method for different parameter settings

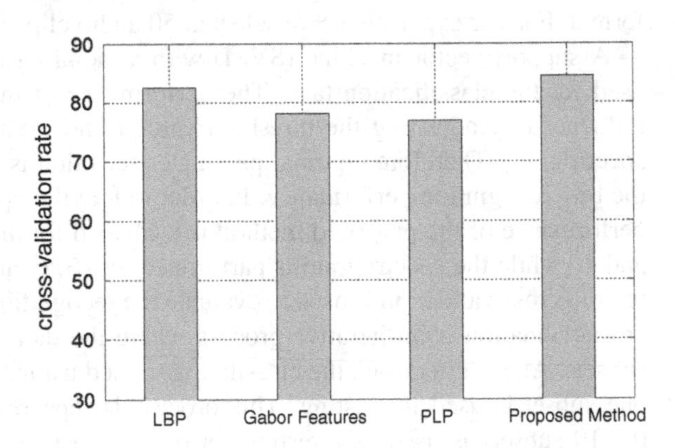

Fig. 6 Recognition performance obtained for different methods

5 Conclusion

In this paper, we propose to use gradient directional pattern (GDP) to represent spectrogram texture for music genre classification. The GDP operator encodes the local texture of a neighborhood by quantizing the direction values of the gradient vector of each neighbor with respect to the center. Utilizing gradient direction values facilitates more informative and robust texture encoding, producing stable GDP patterns in both smooth and high-textured regions. Experimental analysis with a music genre database reveals that, the proposed method can effectively represent and recognize spectrogram texture for music genre classification, as compared against some other existing methods.

Acknowledgments The authors would like to thank NSERC Discovery Grant Project 1028463, NSERC Engage, AITF, and MITACS Accelerate for partial support of this project.

References

1. Ahmed, F.: Gradient directional pattern: a robust feature descriptor for facial expression recognition. IET Electron. Lett. **48**(19), 1203–1204 (2012)
2. Ahmed, F., Kabir, M.H.: Directional ternary pattern (dtp) for facial expression recognition. In: IEEE International Conference on Consumer Electronics, pp. 265–266 (2012)
3. Ahmed, F., Paul, P., Wang, P., Gavrilova, M.: Gender classification from face images based on gradient directional pattern (gdp). In: International Conference on Computational Science and Its Applications, vol. LNCS 9156, pp. 233–243 (2015)
4. Costa, Y., Oliveira, L., Koerich, A., Gouyon, F.: Music genre recognition using gabor filters and lpq texture descriptors. In: Iberoamerican Congress on Pattern Recognition, vol. LNCS 8259, pp. 67–74 (2013)
5. Costa, Y., Oliveira, L., Koerich, A., Gouyon, F.: Music genre recognition using spectrograms. In: International Conference on Systems, Signals and Image Processing, pp. 151–154 (2011)
6. Costa, Y., Oliveira, L., Koerich, A., Gouyon, F., Martins, J.: Music genre classification using lbp textural features. Sig. Process. **92**, 2723–2737 (2012)
7. Dannenberg, R., Thom, B., Watson, D.: A machine learning approach to musical style recognition. In: International Computer Music Conference (1997)
8. Ezzaidi, H., Rouat, J.: Automatic musical genre classification using divergence and average information measures. In: Research report of the world academy of science, engineering and technology (2006)
9. Hermansky, H.: Perceptual linear predictive (plp) analysis of speech. J. Acoust. Soc. Amer. **87**(4), 1738–1752 (1990)
10. Jabid, T., Kabir, M.H., Chae, O.: Robust facial expression recognition based on local directional pattern. ETRI J. **32**(5), 784–794 (2010)
11. Li, T., M, M.O., Li, Q.: A comparative study on content-based music genre classification. In: international ACM SI-GIR conference on research and development in information retrieval, pp. 282–289 (2003)
12. Lidy, T., Rauber, A.: Evaluation of feature extractors and psychoacoustic transformations for music genre classification. In: International Conference on Music Information Retrieval, pp. 71–80 (2005)
13. Lidy, T., Silla, C., Cornelis, O., Gouyon, F., Rauber, A., Kaestner, C., Koerich, A.: On the suitability of state-of-the-art music information retrieval methods for analyzing, categorizing and accessing non-western and ethnic music collections. Signal **90**, 1032–1048 (2010)
14. McKay, C., Fujinaga, I.: Musical genre classification: is it worth pursuing and how can it be improved? In: International Conference on Music Information Retrieval, pp. 101–106 (2006)
15. Neammalai, P., Phimoltares, S., Lursinsap, C.: Speech and music classification using hybrid form of spectrogram and fourier transformation. In: Asia-Pacific Signal and Information Processing Association Annual Summit and Conference, pp. 1–6 (2014)
16. Ojala, T., Pietikainen, M., Maenpaa, T.: Multiresolution gray-scale and rotation invariant texture classification with local binary patterns. IEEE Trans. Pattern Anal. Mach. Intell. **24**(7), 971–987 (2002)
17. Silla, C.N., Koerich, A.L., Kaestner, C.: Feature selection approach for automatic music genre classification. Int. J. Semant. Comput. **3**(2), 183–208 (2009)
18. Tan, X., Triggs, B.: Enhanced local texture feature sets for face recognition under difficult lighting conditions. In: IEEE International Workshop on Analysis and Modeling of Faces and Gestures, LNCS vol. 4778, pp. 168–182 (2007)

19. Tzanetakis, G., Cook, P.: Musical genre classification of audio signals. IEEE Trans. Speech Audio Process. **10**(5), 293–302 (2002)
20. Wu, H., Zhang, M.: Gabor-lbp features and combined classifiers for music genre classification. In: International Conference on Computer and Information Application, pp. 419–422 (2012)
21. Zhao, S., Gao, Y., Zhang, B.: Sobel-lbp. In: IEEE International Conference on Image Processing, pp. 2144–2147 (2008)

Robust Speaker Identification in a Meeting with Short Audio Segments

Giorgio Biagetti, Paolo Crippa, Laura Falaschetti, Simone Orcioni and Claudio Turchetti

Abstract The paper proposes a speaker identification scheme for a meeting scenario, that is able to answer the question "is somebody currently talking?", if yes, "who is it?". The suggested system has been designed to identify during a meeting conversation the current speaker from a set of pre-trained speaker models. Experimental results on two databases show the robustness of the approach to the overlapping phenomena and the ability of the algorithm to correctly identify a speaker with short audio segments.

Keywords Speaker identification · Meeting conversation · Speaker diarization · Overlapping speech

1 Introduction

Speaker identification aims at detecting which speaker a given pool the unknown speech is derived from, and can be considered as a particular case of the more general problem of speaker recognition which is addressed to recognize, indentify or verify individuals using speech [13]. The main tasks involved in a speaker identification system are feature extraction, speaker modeling, and speaker classification.

G. Biagetti · P. Crippa · L. Falaschetti (✉) · S. Orcioni · C. Turchetti
DII – Dipartimento di Ingegneria dell'Informazione, Università Politecnica delle Marche, Via Brecce Bianche 12, 60131 Ancona, Italy
e-mail: l.falaschetti@univpm.it

G. Biagetti
e-mail: g.biagetti@univpm.it

P. Crippa
e-mail: p.crippa@univpm.it

S. Orcioni
e-mail: s.orcioni@univpm.it

C. Turchetti
e-mail: c.turchetti@univpm.it

I. Czarnowski et al. (eds.), *Intelligent Decision Technologies 2016*,
Smart Innovation, Systems and Technologies 57,
DOI 10.1007/978-3-319-39627-9_41

The usually adopted features in speaker identification are the same adopted in speech recognition, namely Mel frequency cepstral coefficients (MFCCs), perceptual linear prediction coefficients, etc. [10]. Among these, MFCCs have shown the best performance due to their particular robustness to the environment and flexibility.

As far as speaker modeling is concerned, assuming an utterance from a speaker is a random sequence of frames, the Gaussian mixture model (GMM) is widely used in speaker identification [11]. Since this model falls into the family of statistical model, it requires training data sampled from the class of speakers to be identified.

For the task of speaker classification, the optimal Bayesian classifier guarantees the minimum classification error by identifying the speaker model which exhibits the maximum GMM a posteriori probability [13].

In the classic speaker indentification scenario it is required that the identification system be able to identify a person when one speaker alone is speaking for a time interval. In the different scenario of a meeting [3, 8, 14] speech from one speaker can abruptly change to, or can be overlapped with, speech from another speaker. In particular overlapping speech can greatly degrade the performance of speaker identification. These problems are common in Speaker Diarization whose main goal is to segment audio into speaker-homogeneous regions with the goal of answering the question “who spoke when?”.

However in diarization system the output is limited to labeling speaker region with number or letters, without detecting the speaker’s indentity. This goal is performed without prior training of specific models, as many of such systems work completely unsupervised. The main operational tasks to be carried out in a speaker diarization system are: *speech activity detection* (to separate speech from non-speech), *segmentation* (to detect speaker changes to segment the audio data), *clustering* (to group the segmented regions together into spoken-homogeneous clusters).

The aim of this paper is to derive a robust speaker identification scheme for a meeting scenario that is able to answer the question “is somebody currently talking?”, if yes, “who is it?”. Thus this task is performed using results from both the field of speaker identification and speaker diarization.

The suggested system has been designed to identify during a meeting conversation the current speaker from a set of pre-trained speaker models. In particular a robust speaker identification algorithm has been adopted in order to mitigate the problem of the overlapping speech.

The paper is organized as follows. Section 2 provides a brief overview of the speaker identification algorithm. Section 3 presents the experimental results carried out on a data base properly designed to simulate a true meeting including overlapping phenomenon and on the AMI Meeting Corpus. Section 4 summarizes the conclusions of the present work.

2 Speaker Classification Algorithm

The speaker classification algorithm used in this work is based on the approach [1] used in a classic speaker identification scenario, and it is summarized in the following.

2.1 Single Frame Classification

We denote with $y[n]$, $n = 0, \ldots, N-1$, a frame representing the power spectrum of the speech signal, extracted from the time domain waveform of the utterance under consideration, through a pre-processing algorithm including pre-emphasis, framing and log-spectrum. Typical duration values for frames range from 20 to 30 ms (usually 25 ms) and a frame is generated every 10 ms (thus consecutive 25 ms frames generated every 10 ms will overlap by 15 ms).

In a Bayesian speaker identification scheme, a group of S speakers is represented by the probability density functions (pdfs)

$$p_s(\mathrm{y}) = p(\mathrm{y} \mid \theta_s), \qquad s = 1, 2, \cdots, S \tag{1}$$

where θ_s are the parameters to be estimated using the training set $\mathcal{W}$.

The objective of classification is to find the speaker model θ_s which has the maximum a posteriori probability for a given frame y belonging to the testing set $\mathcal{Z}$. Using Bayes' theorem and assuming that $p(\theta_s)$ and $p(\mathrm{y})$ are independent of S, it results:

$$\hat{s}(\mathrm{y}) = \underset{1 \leq s \leq S}{\operatorname{argmax}} \left\{ p(\theta_s \mid \mathrm{y}) \right\} = \underset{1 \leq s \leq S}{\operatorname{argmax}} \left\{ p_s(\mathrm{y}) \right\} . \tag{2}$$

The main issue in Bayesian classification is to accurately estimate the pdf $p_s(\mathrm{y})$. To this end the most generic statistical speaker modeling one can adopt for the single speaker is the GMM [11], is given by the equation

$$p(\mathrm{y} \mid \theta_s) = \sum_{i=1}^{F} \alpha_i \, \mathcal{N}(\mathrm{y} \mid \mu_i, \mathrm{C}_i) \tag{3}$$

where α_i, $i = 1, \ldots, F$ are the mixing weights, and $\mathcal{N}(\mathrm{y}|\mu_i, \mathrm{C}_i)$ represents a Gaussian distribution density with mean μ_i and covariance matrix C_i.

$\theta = \{\alpha_1, \mu_1, \mathrm{C}_1, \ldots, \alpha_F, \mu_F, \mathrm{C}_F\}$, (the index s is omitted for the sake of notation simplicity) is the set of unknown parameters to be estimated that specify the Gaussian mixture.

An estimate of θ, with training data $\mathcal{W}$ can be obtained by the *maximum likelihood* (ML)

$$\hat{\theta}_{\mathrm{ML}} = \underset{\theta}{\operatorname{argmax}} \{ \log p(\mathcal{W} \mid \theta) \} \tag{4}$$

however as (4) is difficult to be solved analytically since (4) contains the log of a sum, the usual choice for solving ML estimate of the mixture parameters is the expectation maximization (EM) algorithm.

The EM algorithm, which has been adopted in this work, is based on the interpretation of $\mathcal{W}$ as incomplete data and the set $\mathcal{H}$ as the missing part of the complete data $\mathcal{X} = \{\mathcal{W}, \mathcal{H}\}$. The complete data log-likelihood, i.e. the log-likelihood of $\mathcal{X}$ as though $\mathcal{H}$ was observed, is

$$\log\left[p\left(\mathcal{W}, \mathcal{H}|\theta\right)\right] = \sum_{\ell=1}^{L}\sum_{i=1}^{F} \mathrm{h}_i^{(\ell)} \log\left[\alpha_i \mathcal{N}(\mathrm{y}^{(\ell)}|\mu_i, \mathrm{C}_i)\right] . \quad (5)$$

In general the EM algorithm computes a sequence of parameter estimates $\{\hat{\theta}(p), p = 0, 1, \ldots\}$ by iteratively performing two steps:

- *Expectation step:* compute the expected value of the complete log-likelihood, given the training set $\mathcal{W}$ and the current parameter estimate $\hat{\theta}(p)$. The result is the so-called *auxiliary function*

$$Q\left(\theta|\hat{\theta}(p)\right) = E\left\{\log\left[p\left(\mathcal{W}, \mathcal{H}|\theta\right)\right] |\mathcal{W}, \hat{\theta}(p)\right\} . \quad (6)$$

- *Maximization step:* update the parameter estimate

$$\hat{\theta}(p+1) = \underset{\theta}{\operatorname{argmax}}\left\{Q\left(\theta|\hat{\theta}(p)\right)\right\} \quad (7)$$

  by maximizing the Q-function.

Usually for 8 kHz (16 kHz) bandwidth speech, the vector y has a dimension $N =$ 128 (256). So that a too large amount of training data would be necessary to estimate the pdf $p(\mathrm{y}|\theta)$ and, in any case, with such a dimension the estimation problem is impractical.

The usual choice is to solve this problem is to reduce the vector y to a vector k_M of lower dimension by a linear transform H such that

$$\mathrm{k}_M = \mathrm{H}\,\mathrm{y} , \quad (8)$$

where y is a $N \times 1$ vector, k_M an $M \times 1$ vector, H an $M \times N$ matrix, and $M \ll N$. The vector k_M represents the so-called feature-vector belonging to an appropriate M-dimension subspace [6, 12].

Principal component analysis (PCA) [7] has proven to be an excellent technique for dimensionality reduction in many application areas including data compression, image analysis, visualization to mention just a few. The main property of PCA [5], is that for a set of observed N-dimensional data vectors $y[n]$, $n = 0, \ldots, N-1$, M *principal axes* ϕ_j, $j = 1, \ldots, M$, can be derived such that they are orthonormal axes onto which the retained *variance* under projection is maximal.

The PCA of y is derived from the Karhunen-Love transform (KLT), defined by the couple of equations

$$\mathrm{y} = \boldsymbol{\Phi}\,\mathrm{k}\,, \tag{9}$$

$$\mathrm{k} = \boldsymbol{\Phi}^T \mathrm{y}\,, \tag{10}$$

where $\boldsymbol{\Phi} = [\phi_1, \ldots, \phi_N]$ is an $N \times M$ matrix and $\mathrm{k} = [k_1, \ldots, k_N]^T$ is the transformed random vector.

The M principal axes are identified as those corresponding to the M maximal eigenvalues $\lambda_j, j = 1, \ldots M$ of $\mathrm{R_{yy}}\phi_j = \lambda_j \phi_j\,, j = 1, \ldots, N$, where $\mathrm{R_{yy}}$ is the autocorrelation function. Thus $\boldsymbol{\Phi}$ decomposes as $\boldsymbol{\Phi} = [\boldsymbol{\Phi}_M\,,\ \boldsymbol{\Phi}_\eta]$, and (9) can be rewritten as:

$$\mathrm{y} = \boldsymbol{\Phi}\,\mathrm{k} = \boldsymbol{\Phi}_M\,\mathrm{k}_M + \boldsymbol{\Phi}_\eta\,\mathrm{k}_\eta = \mathrm{x}_M + \eta_\mathrm{y}\,, \tag{11}$$

being $\Phi_M = [\phi_1, \ldots, \phi_M]$ an $N \times M$ matrix, k_M an $M \times 1$ vector. In a similar way (10) becomes:

$$\begin{bmatrix} \mathrm{k}_M \\ \mathrm{k}_\eta \end{bmatrix} = \begin{bmatrix} \boldsymbol{\Phi}_M^T \\ \boldsymbol{\Phi}_\eta^T \end{bmatrix} \mathrm{y}\,. \tag{12}$$

In (11) the term

$$\mathrm{x}_M = \boldsymbol{\Phi}_M\,\mathrm{k}_M\,, \tag{13}$$

represents the truncated expansion, and it is equivalent to the approximations

$$\mathrm{y} \approx \mathrm{x}_M, \qquad \mathrm{k} \approx \mathrm{k}_T = \begin{pmatrix} \mathrm{k}_M \\ 0 \end{pmatrix}, \tag{14}$$

Thus, as k_M is given by $\mathrm{k}_M = \boldsymbol{\Phi}_M^T\,\mathrm{y}$, comparing with (8) yields $\mathrm{H} = \boldsymbol{\Phi}_M^T$.

On the basis of previous results a Bayesian classification scheme which is consistent with PCA can be derived.

Given a group of S speakers, let us define the pdfs $p_s(\mathrm{k}_T) = p(\mathrm{k}_T \mid \theta_s)$, $s = 1, 2, \ldots, S$, where k_T is the truncation of k. Consequently the pdf $p_s(\mathrm{k}_T) = p_s(\mathrm{k}_M)\,\delta(\mathrm{k}_\eta)$, represents an approximation of the pdf in (1). Thus (2) becomes:

$$\hat{s}(\mathrm{y}) = \underset{1 \le s \le S}{\operatorname{argmax}} \left\{ p_s(\mathrm{k}_M)\,\delta(\mathrm{k}_\eta) \right\} = \underset{1 \le s \le S}{\operatorname{argmax}} \left\{ p_s(\mathrm{k}_M) \right\}\,. \tag{15}$$

As you can see comparing (15) with (2), the dimensionality of classification problem is reduced from N to M, with $M < N$.

2.2 Multi Frame Classification

The accuracy of speaker identification can be considerably improved using a sequence of frames instead of a single frame alone. To this end let us refer to a sequence of V frames defined as $Y = \{\mathrm{y}^{(1)}, \ldots, \mathrm{y}^{(V)}\}$, where $\mathrm{y}^{(v)}$ represents the vth frame. Using (15) we can determine the class each frame $\mathrm{y}^{(v)}$ belongs to. Thus the S sets $\mathcal{Z}_s = \left\{\mathrm{y}^{(v)} \mid \mathrm{y}^{(v)} \text{ belongs to class } s\right\}, s = 1, \ldots, S$, are univocally determined.

Given Y, we define the score for each class s as

$$r_s(Y) = \sum_{\mathrm{y}^{(v)} \in \mathcal{Z}_s} p(\mathrm{y}^{(v)}) \tag{16}$$

where $p(\mathrm{y}^{(v)})$ represents the probability achieved by the frame $\mathrm{y}^{(v)}$.

Finally the multi-frame speaker identification is based on:

$$\hat{s}(Y) = \underset{1 \leq s \leq S}{\operatorname{argmax}} \left\{r_s(Y)\right\}. \tag{17}$$

3 Experimental Results

Experiments are conducted using two different corpora, *(i)* a data base called DBT that was specifically designed to subject the algorithm to a severe test, where a large percentage of overlapping speech and different consistency of framing material is considered, *(ii)* the well known AMI Meeting Corpus, as it represents a widely accepted test for the evaluation of speaker diarization system.

3.1 Features Extractor

Figure 1 shows the block diagram of the proposed front-end employed for feature extraction. At the input of the processing chain a voice activity detection block drops all non speech segments from the input audio records, exploiting the energy

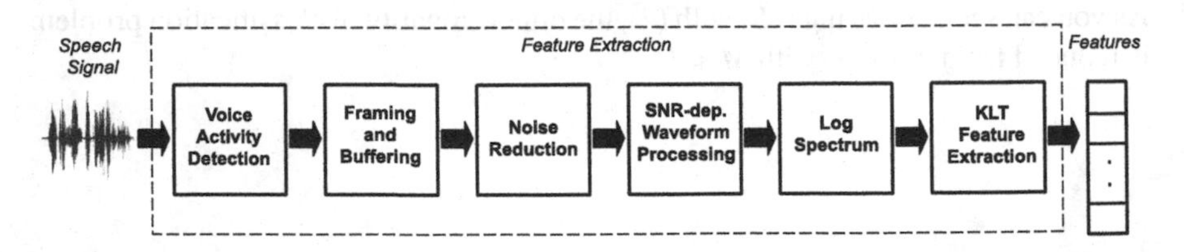

Fig. 1 The proposed front-end for feature extraction

acceleration associated with voice onset. The signal is then divided into overlapping frames of 25 ms (200 samples), with a frame shift of 10 ms (80 samples). Hence buffering is required for storing overlapping regions among frames. Besides, before computing the DKLT features, each frame is cleaned up by a noise reduction block based on the Wiener filter. Further enhancements are then performed by a SNR-dependent waveform processing phase, that weights the input noise-reduced frame according to the positions of its smoothed instant energy contour maxima. It is worth noting that noise reduction introduces an overall latency of 30 ms (3 frames) due to its algorithm requiring internal buffering.

3.2 Experiments on Data Base DBT

A first set of experiments was carried out on a large database, called DBT, which was formed by collecting several audio recordings of five different speakers, two females (A, B) and three males (C, D, E) as reported in Table 1. All recordings extracted are mono, 8 kilosamples per second, 16 bit. The consistency of DBT database in terms of number of frames used for each speaker is reported in Table 1. In order to test several different models, the databases DB1, DB2, and DB3, with different percentage consistency of training subsets, have been derived.

A meeting scenario has beeen simulated by interleaving 45 audio segments extracted from database *liber liber* (http://www.liberliber.it/) to achieve a 20 min audio track. In the conversation the 5 speakers alternate each other with short turn durations. More specifically two audio tracks have been derived: in the former the audio segments follow one another without overlap, in the latter an overlapping of 20 % is taken into account to test the robustness of the algorithm to the overlapping phenomenon.

The widely adopted metric for diarization performance measurement is the Diarization Error Rate (DER). It has been introduced by the NIST in 2000 within the Speaker Recognition evaluation [9] for their speaker segmentation task [4]. The

Table 1 Recordings used for the creation of the identification corpus

Database			DBT	80 % DBT	50 % DBT	20 % DBT
Speaker	Gender	Duration (s)	Model 1	Model 2	Model 3	Model 4
A	F	761	58903	47122	29451	11780
B	F	2593	195591	156472	97795	39118
C	M	251	18867	1509	9431	3773
D	M	838	63713	50970	31856	12742
E	M	1162	91253	73002	45626	18250
Total		5605	428327	342659	214161	85663

Source liber liber (http://www.liberliber.it/). The material was used for training purposes. The consistency of the databases used for modeling is shown in terms of number of frames

DER is defined as the ratio of incorrectly detected speaker time to total speaker time. The metric is computed by mapping the system output speaker segment sets to reference speaker segment sets so as to minimize the total error. By defining the following errors:

- *Speaker assignment errors* (E_{spkr}): percentage of scored time that a speaker ID is assigned to the wrong speaker.
- *Missed detections* (E_{miss}): percentage of scored time that a hypothesized non-speech segment corresponds to a reference speaker segment.
- *False alarm detections* (E_{fa}): percentage of scored time that a hypothesized speaker is labelled as a non-speech in the reference.

the final DER is given by

$$DER = E_{spkr} + E_{miss} + E_{fa} \tag{18}$$

Figures 2 and 3 report for speech without and with overlapping respectively, the value of the DER as a function of the number of frames used in the identification algorithm and for the four models previously defined. As you can see the DER drastically decreses as the sequence length of frame increases, while only minor differences are due to the various models. It is worth to notice that due to robustness of the algorithm, the performance are not appreciably degraded in a meeting conversation with overlapping.

Tables 2 and 3 report the values of the three parameters (E_{spkr}, E_{miss}, E_{fa}) that contribute to the *DER*.

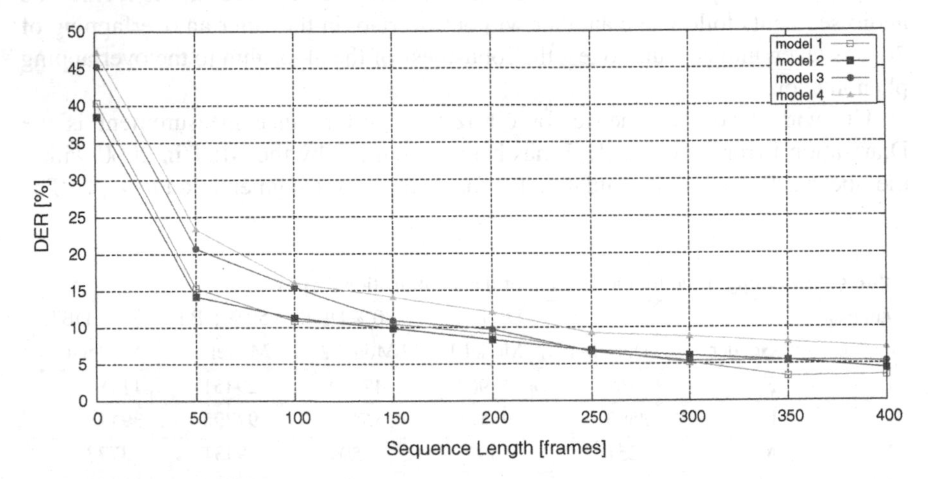

Fig. 2 DER in a meeting conversation without overlapping speech as a function of the sequence length, for different training models

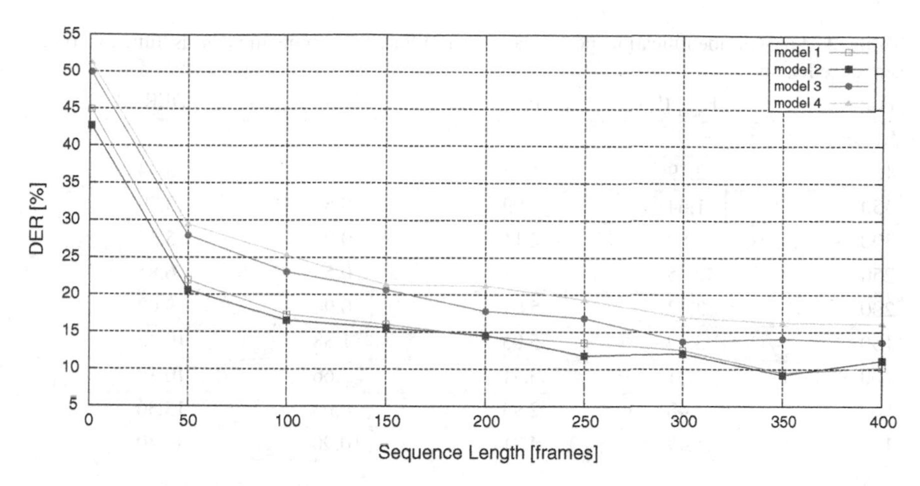

Fig. 3 DER in a meeting conversation with 20 % of overlapping speech as a function of the sequence length, for different training models

3.3 Experiments on AMI Meeting Corpus

A second set of experiments evaluating the speaker identification system were performed using meeting audio data from the *AMI Meeting Corpus* (http://www.idiap.ch/dataset/ami/). AMI is a large, multi-site and multi-disciplinary project with the aim of developing meeting browsing technologies that improve work group effectiveness. As part of the development process, the project is collecting a corpus of 100 h of meetings using instrumentation that yields high quality, synchronized multi-modal recording, with, for technical reasons, a focus on groups of four people [2].

Experiments are conducted with a subset of 20 meetings of the AMI Corpus, belonging to the IDIAP subset ('IS' meetings) of the corpus. This subset comprises 38 meetings, each involving four participants engaged in a scenario-based meeting ranging in duration from 13 to 40 min. The meetings contain approximately 18 % overlapping speech. The AMI meetings are a convenient choice since the 20 meetings are split into five different sessions, each one containing four meetings with the same four participants. One meetings is randomly chosen to train the four speaker models. The classification with a window length variable from 1 to 4 s is used on the rest of the data to perform diarization. The amount of speech used (per speaker) to train the models are shown in Table 4. We also train an additional 60 s room-specific non-speech model. Surprisingly, with only 60 s of speech per speaker the system is able to obtain good performance.

Table 2 Speaker identification performance for different training models as function of the sequence length

Frames	E_{spkr} (%)	E_{miss} (%)	E_{fa} (%)	DER (%)
Model 1				
400	2.19	1.25	0.00	3.44
350	1.64	1.09	0.54	3.29
300	2.82	2.11	0.23	5.17
250	2.15	4.11	0.58	6.85
200	3.13	5.01	0.94	9.09
150	3.52	5.05	1.88	10.46
100	4.31	3.91	2.66	10.89
50	7.48	2.93	4.93	15.36
1	25.18	4.31	10.90	40.40
Model 2				
400	2.19	1.88	0.31	4.38
350	2.19	2.74	0.54	5.48
300	2.35	2.82	0.94	6.11
250	1.95	4.31	0.58	6.85
200	2.35	4.85	1.09	8.30
150	2.82	4.58	2.46	9.87
100	4.15	3.36	3.76	11.28
50	6.26	2.82	5.13	14.22
1	23.15	4.10	11.25	38.51
Model 3				
400	3.13	1.56	0.62	5.32
350	3.01	1.92	0.54	5.48
300	3.05	1.64	0.70	5.40
250	2.54	3.52	0.58	6.66
200	4.23	3.60	1.88	9.71
150	4.34	3.40	3.17	10.93
100	6.11	3.44	5.87	15.43
50	10.30	2.50	7.9	20.72
1	27.76	3.89	13.63	45.30
Model 4				
400	1.88	4.70	0.62	7.21
350	3.29	4.38	0.27	7.95
300	3.52	3.99	1.17	8.69
250	3.91	3.91	1.37	9.20
200	5.01	4.85	2.19	12.06
150	4.58	5.64	3.87	14.10
100	6.58	4.31	5.17	16.06
50	11.48	3.60	8.26	23.35
1	28.74	4.67	13.76	47.18

Table 3 Speaker identification performance for different training models as function of the sequence length

Frames	E_{spkr} (%)	E_{miss} (%)	E_{fa} (%)	DER (%)
Model 1				
400	8.66	1.23	0.30	10.21
350	8.39	1.08	0.00	9.47
300	9.98	2.08	0.46	12.53
250	9.86	3.09	0.58	13.54
200	8.97	4.02	1.39	14.39
150	10.56	3.36	2.08	16.01
100	10.60	3.17	3.55	17.33
50	13.69	2.39	5.84	21.93
1	29.18	4.06	11.64	44.89
Model 2				
400	8.66	2.16	0.30	11.14
350	7.85	1.35	0.00	9.20
300	7.89	3.71	0.46	12.07
250	7.93	3.28	0.58	11.80
200	8.82	4.79	0.92	14.54
150	9.05	4.52	1.97	15.55
100	10.44	2.94	3.17	16.55
50	12.76	2.66	5.14	20.58
1	27.26	4.03	11.38	42.68
Model 3				
400	12.38	0.61	0.61	13.61
350	12.18	0.81	1.08	14.08
300	10.67	2.08	0.92	13.69
250	13.54	2.12	1.16	16.83
200	12.69	2.32	2.78	17.79
150	14.39	2.20	4.06	20.66
100	14.54	1.70	6.80	23.05
50	16.67	1.66	9.59	27.93
1	31.57	3.37	14.96	49.91
Model 4				
400	11.45	3.71	0.92	16.09
350	11.91	3.25	1.08	16.25
300	12.07	4.17	0.69	16.94
250	14.12	3.86	1.35	19.34
200	13.92	4.95	2.32	21.20
150	14.16	3.59	3.59	21.35
100	14.77	4.02	6.50	25.30
50	17.79	2.47	9.16	29.44
1	32.29	4.51	14.37	51.18

The meeting data contain overlapping speech

Table 4 DER for IDIAP AMI Corpus using small training model and short sequences of speech frame

Frames	Model		
	30 (s)	60 (s)	90 (s)
	DER (%)		
400	12.26	11.90	12.98
300	14.87	12.71	14.33
200	18.93	15.87	16.59
100	28.76	25.96	24.61

4 Conclusion

The paper describes a speaker identification scheme that is able to identify a speaker in a meeting, that is when speech from one speaker can abruptly change to, or can be overlapped with, speech from another speaker. Although these problems are common in speaker diarization, in such a case the output is limited to labeling speaker region with number or letters, without detecting the speaker's identity.

Experiments conducted on two distinct database have shown the robustness of the approach in a meeting scenario.

References

1. Biagetti, G., Crippa, P., Curzi, A., Orcioni, S., Turchetti, C.: Speaker identification with short sequences of speech frames. In: Proceedings of the International Conference on Pattern Recognition Applications and Methods, pp. 178–185 (2015)
2. Carletta, J., Ashby, S., Bourban, S., Flynn, M., Guillemot, M., Hain, T., Kadlec, J., Karaiskos, V., Kraaij, W., Kronenthal, M., et al.: The AMI meeting corpus: a pre-announcement. Springer, Berlin (2005)
3. Friedland, G., Vinyals, O.: Live speaker identification in conversations. In: Proceedings of the 16th ACM International Conference on Multimedia, pp. 1017–1018. ACM (2008)
4. Galibert, O.: Methodologies for the evaluation of speaker diarization and automatic speech recognition in the presence of overlapping speech. In: Proceedings of INTERSPEECH, pp. 1131–1134 (2013)
5. Hotelling, H.: Analysis of a complex of statistical variables into principal components. J. Educ. Psychol. **24**(6), 417–441 (1933)
6. Jiang, X.: Linear subspace learning-based dimensionality reduction. IEEE Sig. Process. Mag. **28**(2), 16–26 (2011)
7. Jolliffe, I.T.: Principal Component Analysis. Springer Series in Statistics. Springer, New York (1986)
8. Luque, J., Hernando, J.: Robust speaker identification for meetings: UPC CLEAR'07 meeting room evaluation system. In: Multimodal Technologies for Perception of Humans, pp. 266–275. Springer (2008)
9. NIST: 2000 speaker recognition evaluation—evaluation plan. (2000). http://www.itl.nist.gov/iad/mig/tests/spk/2000/spk-2000-plan-v1.0.htm

10. Reynolds, D.A.: Experimental evaluation of features for robust speaker identification. IEEE Trans. Speech Audio Process. **2**(4), 639–643 (1994)
11. Reynolds, D.A., Rose, R.C.: Robust text-independent speaker identification using Gaussian mixture speaker models. IEEE Trans. Speech Audio Process. **3**(1), 72–83 (1995)
12. Singh-Miller, N., Collins, M., Hazen, T.J.: Dimensionality reduction for speech recognition using neighborhood components analysis. In: Proceedings of INTERSPEECH, pp. 1158–1161 (2007)
13. Togneri, R., Pullella, D.: An overview of speaker identification: accuracy and robustness issues. IEEE Circ. Syst. Mag. **11**(2), 23–61 (2011)
14. Yella, S.H., Bourlard, H.: Overlapping speech detection using long-term conversational features for speaker diarization in meeting room conversations. IEEE/ACM Trans. Audio Speech Lang. Process. **22**(12), 1688–1700 (2014)

Author Index

I. Czarnowski et al. (eds.), *Intelligent Decision Technologies 2016*,
Smart Innovation, Systems and Technologies 57,
DOI 10.1007/978-3-319-39627-9

Printed in the United States
By Bookmasters